White Biotechnology

White Biotechnology

Edited by **Suzy Hill**

SYRAWOOD
PUBLISHING HOUSE

New York

Published by Syrawood Publishing House,
750 Third Avenue, 9th Floor,
New York, NY 10017, USA
www.syrawoodpublishinghouse.com

White Biotechnology
Edited by Suzy Hill

International Standard Book Number: 978-1-68286-060-1 (Hardback)

Printed in the United States of America.

Contents

Permissions

List of Contributors

Preface

White biotechnology is concerned with the practical application of biotechnology in industries. Biofuel production, catabolism, biodegradation of hazardous chemicals, etc. are some of the crucial industrial applications of white biotechnology. This book discusses the fundamentals as well as modern approaches of white biotechnology and also unfolds the innovative aspects of the field while giving the reader a detailed insight into the processing and manufacturing of materials, energy, paper, textiles, etc. This text is indispensable for the students of biotechnology. It will also prove to be a useful tool for scholars and researchers in this field.

This book is the end result of constructive efforts and intensive research done by experts in this field. The aim of this book is to enlighten the readers with recent information in this area of research. The information provided in this profound book would serve as a valuable reference to students and researchers in this field.

At the end, I would like to thank all the authors for devoting their precious time and providing their valuable contributions to this book. I would also like to express my gratitude to my fellow colleagues who encouraged me throughout the process.

Editor

Erythritol production on wheat straw using *Trichoderma reesei*

Birgit Jovanović, Robert L Mach and Astrid R Mach-Aigner[*]

Abstract

We overexpressed the *err1* gene in the *Trichoderma reesei* wild-type and in the cellulase hyperproducing, carbon catabolite derepressed strain Rut-C30 in order to investigate the possibility of producing erythritol with *T. reesei*. Two different promoters were used for *err1* overexpression in both strains, a constitutive (the native pyruvat kinase (*pki*) promoter) and an inducible one (the native *β*-xylosidase (*bxl1*) promoter). The derived recombinant strains were precharacterized by analysis of *err1* transcript formation on D-xylose and xylan. Based on this, one strain of each type was chosen for further investigation for erythritol production in shake flasks and in bioreactor experiments. For the latter, we used wheat straw pretreated by an alkaline organosolve process as lignocellulosic substrate. Shake flask experiments on D-xylose showed increased erythritol formation for both, the wild-type and the Rut-C30 overexpression strain compared to their respective parental strain. Bioreactor cultivations on wheat straw did not increase erythritol formation in the wild-type overexpression strain. However, *err1* overexpression in Rut-C30 led to a clearly higher erythritol formation on wheat straw.

Keywords: Erythritol; Erythrose reductase; *Trichoderma reesei*; Wheat straw; Lignocellulose

Introduction

Erythritol is a four-carbon sugar alcohol, which is applied as flavor enhancer, formulation aid, humectant, stabilizer, thickener, and as low-calorie sweetener, of which the latter is the main utilization. Compared with other polyols yielding about 2 cal/g, erythritol yields only up to 0.2 cal/g, which is due to the fact that erythritol does not undergo systemic metabolism in the human body but is excreted unchanged in the urine (Moon et al. 2010). Additionally, as a small molecule, it is easily absorbed already in the upper intestine and therefore, causes less digestive distress than other sweeteners (Livesey 2001). Since erythritol is not assimilated by *Streptococcus mutans* it is non-cariogenic. Furthermore it has some favorable physical and chemical properties: it is thermally stable (no decomposition or colorization at 200°C for 1 h), better crystallizeable than sucrose, and less hygroscopic (Kasumi 1995). The negative enthalpy of solution leads to a cooling effect when dissolved. The sweetness of erythritol is plain with very weak after-taste. In a 10% (w/v) solution it has 60-80% the sweetness of sucrose. It has a natural occurrence in several foods including beer, sake, wine, soy sauce, water melon, pear, and grape. The tolerance of erythritol by animals and humans was intensively studied (Munro et al. 1998). No adverse toxicological effects were observed. Also no carcinogenic, mutagenic or teratogenic potential or effects on fertility could be detected. Therefore, erythritol is a sugar substitute with a growing market and optimization of its production remains an issue.

Current biotechnological production of erythritol use osmophilic yeasts like *Aureobasidium* sp., *Trichosporonoides* sp., *Torula* sp., and *Candida magnoliae*. As substrate a highly concentrated glucose (typically 40% (w/v)) solution is applied, which is gained from chemically and enzymatically hydrolyzed wheat- and cornstarch. The hydrolyzed starch serves as carbon source and causes a high osmotic pressure that pushes the yeast to produce the osmolyte erythritol (Moon et al. 2010). Although these processes reach 40% (w/w) yields of D-glucose to erythritol conversion, they depend on D-glucose as starting material. With regard to (socio)economical issues, starch-derived D-glucose is not a preferable substrate. Therefore, it would be an interesting alternative to use

*Correspondence: astrid.mach-aigner@tuwien.ac.at
Department for Biotechnology and Microbiology, Institute of Chemical Engineering, Vienna University of Technology, Gumpendorfer Str. 1a, A-1060 Wien, Austria

organisms that can utilize non-food, lignocellulosic biomass for the production of erythritol.

In a previous work (Jovanovic et al. 2013) we characterized the erythrose reductases (Err1) from the filamentous ascomycota *Trichoderma reesei* (telemorph *Hypocrea jecorina* (Kuhls et al. 1996)), *Aspergillus niger*, and *Fusarium graminearium* (telemorph *Gibberella zeae*), which are all very potent degraders of biomass. It turned out that the Err1 of *T. reesei* and *A. niger* showed comparable activities, whereas the Err1 from *F. graminearium* had a considerably lower activity (Jovanovic et al. 2013). In the present study we focused on the potential of producing erythritol in *T. reesei* from lignocellulosic biomass. The native lignocellulose-degrading enzymes of the fungus have already broad application in industry, i.e. in pulp and paper (Buchert et al. 1998; Noé et al. 1986; Welt and Dinus 1995), food and feed (Galante et al. 1993; Lanzarini and Pifferi 1989; Walsh et al. 1993), and textile industries (Koo et al. 1994, Kumar et al. 1994, Pedersen et al. 1992) as well as in biofuel production (Hahn-Hägerdal et al. 2006; Himmel et al. 2007; Ragauskas et al. 2006). As such a strong producer of cellulases and hemicellulases (a genome-wide search using the JGI Genome Portal (http://genome.jgi-psf.org/Trire2/Trire2.home.html) revealed for *T. reesei* 10 celluloytic and 16 xylanolytic enzyme-encoding genes (Martinez et al. 2008)) it is likely that *T. reesei* is able to grow on cheap biowaste material like wheat straw as the sole carbon source. This is supported by former reports on *T. reesei* capable of growing on lignocellulosic material (Acebal et al. 1986; Dashtban et al. 2013).

In this study we used wheat straw that was pretreated by an alkaline organosolve process (Fackler et al. 2012) to remove the lignin up to a residual concentration of about 1% (w/w), which makes the cellulose and hemicellulose more easily accessible for the fungus. We investigated a *T. reesei* wild-type strain and the strain Rut-C30. Rut-30 is a cellulase hyperproducing, carbon catabolite derepressed mutant (Montenecourt and Eveleigh 1979), which is the parental strain of most industrially used *T. reesei* strains (Peterson and Nevalainen 2012; Derntl et al. 2013). In both strains the *err1* gene was overexpressed using either the native, constitutive promoter from the pyruvate kinase encoding gene (*pki*) or the native, inducible promoter from the β-xylosidase 1 encoding gene (*bxl1*). The overexpression strains were screened for enhanced *err1* transcript formation and the best ones where then cultivated on D-xylose and wheat straw for investigating their erythritol production capacity.

Materials and methods
Strains and cultivation conditions
The *T. reesei* strains QM6aΔ*tmus53* (Steiger et al. 2011) and Rut-C30 (ATCC 56765), which was derived from the wild-type strain QM6a by one UV-light and two N-methyl-N'-nitro-N-nitrosoguanidine mutation steps (Montenecourt and Eveleigh 1979), were maintained on 3% malt extract (MEX) agar. The recombinant *T. reesei* strains QPEC1, QBEC2, RPEC1, and RBEC2 generated during this study, were maintained on MEX agar containing 250 μl/l hygromycin B (Merck, Darmstadt, Germany).

Purification of transformed strains by streak out of spores was done on MEX agar containing 250 μl/l hygromycin B and 500 μl/l IGEPAL® CA-630 (Sigma-Aldrich, St. Louis, MO, USA).

Cultivation in shake flasks was performed in 250-ml-Erlenmeyer flasks containing 50 ml Mandels-Andreotti (MA) medium (Mandels 1985) supplemented with 1% (w/v) D-xylose or 1% (w/v) birch-wood xylan. For inoculation 10^9 conidia per liter were used. Growth conditions were pH 5, 30°C, and 160 rpm shaking rate. Mycelia and supernatant were seperated by filtration. For short-term storage, harvested mycelia were shock-frozen and kept in liquid nitrogen, supernatants were kept at -20°C.

Plasmid construction
The *err1* gene and the promoter region of *bxl1* (1.5 kbp upstream *bxl1*, p*bxl1*) from *T. reesei* were amplified from cDNA, which was generated as described below in the according section. Primers were used to introduce restriction sites adjacent to the gene. Primer sequences are given in Table 1. The PCR product was subcloned into pJET-1.2 (Thermo Scientific, Waltham, MA, USA), using chemically competent *Escherichia coli* TOP 10 (Invitrogen, Life Technologies Ltd, Paisley, UK) for plasmid replication.

For the construction of pBJ-PEC1 the vector pRLM$_{ex30}$ (Mach et al. 1994), which contains the *hph* gene flanked by the *pki* promoter (p*pki*) and the *cbh2* terminator, was used. The *hph* gene was removed by *Nsi*I/*Xba*I digestion and subsequently, *err1* that was excised from pJET-1.2 also by *Nsi*I/*Xba*I digestion, was inserted.

For the construction of pBJ-BEC2 p*pki* was excised from pBJ-PEC1 with *Xho*I/*Xba*I digestion and replaced by p*bxl1*, excised from pJET-1.2 with *Sal*I/*Xba*I digestion.

Protoplast transformation
For QM6aΔ*tmus53* protoplast transformation was performed as described in (Gruber et al. 1990). 5 g of either pBJ-PEC1 or pBJ-BEC2 and 1 μg pAN7, which confers hygromycin B resistance (Punt et al. 1987), were co-transformed into the fungal genome.

Biolistic transformation
Rut-C30 was transformed with the Biolistic® PDS-1000/He Particle Delivery System (Bio-Rad Laboratories, Hercules, CA, USA) according to a modified protocol originally described in (Te'o et al. 2002). 5 μg of either pBJ-PEC1 or pBJ-BEC2 and 1 μg pAN7, which confers hygromycin B

Table 1 Oligonucleotides used during the study

Name	Sequence (5' – 3')	Usage
pbxl1_SalI__EcoRI_f	ATATAGTCGACGAATTCAGCTTGTCTGCCTTGATTACCATCC	Vector construction
pbxl1_XbaI_r	ATATATCTAGATGCGTCCGGCTGTCCTTC	Vector construction
err1_XbaI_f	ATATATCTAGAATGTCTTCCGGAAGGACC	Vector construction
err1_Nsi_r	TATATATGCATTTACAGCTTGATGACAGCAGTG	Vector construction
ppki_f	GCACGCATCGCCTTATCGTC	PCR test
qerr1_f	CTTTACCATTGAGCACCTCGACG	RT-qPCR
qerr1_r	GGTCTTGCCCTGCTTCTTGG	RT-qPCR
qact1_f	TGAGAGCGGTGGTATCCACG	RT-qPCR
qact1_r	GGTACCACCAGACATGACAATGTTG	RT-qPCR
qsar1_f	TGGATCGTCAACTGGTTCTACGA	RT-qPCR
qsar1_r	GCATGTGTAGCAACGTGGTCTTT	RT-qPCR

resistance (Punt et al. (Punt et al. 1987)), were co-transformed into the fungal genome.

DNA isolation

Fungal genomic DNA was isolated by phenol-chloroform extraction, using a FastPrep®-24 (MP Biomedicals, Santa Ana, CA, USA) for cell disruption. About 100 mg of mycelia was transferred to 400 μl DNA extraction buffer (0.1 M Tris-HCl pH 8.0, 1.2 M NaCl, 5 mM EDTA) and grounded with glass beads (0.37 g 0.01 0.1 mm, 0.25 g 1 mm, 1 piece 3 mm) using the FastPrep. Afterwards, the mixture was immediately put on 65°C, supplemented with 9 μM RNase A, and incubated for 30 min. Then 200 μl of phenol (pH 7.9) and 200 μl of a chloroform-isoamyl alcohol-mixture (25:1) were added, with vigorous mixing following each addition. Phases were separated by centrifugation (12000 g, 10 min, 4°C) and the aqueous phase was transferred into a new vial. DNA was precipitated by addition of the 0.7-fold volume of isopropanol to the aqueous phase. After 20 min incubation at room temperature (RT) the DNA was separated by centrifugation (20000 g, 20 min, 4°C) and washed with 500 μl ethanol (70%). The air-dried DNA pellet was solubilised in 50 μl Tris-HCl (10 mM, pH 7.5) at 60°C.

RNA isolation and cDNA synthesis

RNA extraction from fungal mycelia was performed with peqGOLD TriFast™ (peqlab, Erlangen, Germany) according to the manufacturer's procedure, using a FastPrep®-24 (MP Biomedicals) for cell disruption. RNA quantity and quality were determined with a NanoDrop 1000 (Thermo Scientific). A 260 nm/280 nm ratio of at least 1.8 was stipulated for further sample processing. cDNA synthesis was performed with RevertAid™H Minus First Strand cDNA Synthesis Kit (Thermo Scientific) according to the manufacturer's procedure using 0.5 μg of RNA.

Transcript analysis

RT-qPCR analysis was performed in a Rotor-Gene Q cycler (Qiagen, Hilden, Germany). The qPCR amplification mixture had a total volume of 15 μl, containing 7.5 μl 2x IQ SYBR Green Supermix (Bio-Rad Laboratories), 100 nM forward and reverse primer, and 2 μl cDNA (diluted 1:100). Primer sequences are given in Table 1. As reference genes *act1* and *sar1* were used (Steiger et al. 2010). All reactions were performed in triplicates. For each gene a no-template control and a no-amplification control (0.01% SDS added to the reaction mixture) was included in each run. The cycling conditions for *act1* and *err1* comprised 3 min initial denaturation and polymerase activation at 95°C, followed by 40 cycles of 15 s at 95°C, 15 s at 59°C and 15 s at 72 s. For *sar1* different cycling conditions were applied: 3 min initial denaturation and polymerase activation at 95°C, followed by 40 cycles of 15 s at 95°C, and 120 s at 64 s. PCR efficiency was calculated from the Rotor-Gene Q software. Relative expression levels were calculated using the equation

$$relative\ transcript\ ratio = E_r^{C_r} \cdot E_t^{-C_t} \cdot E_{r0}^{-C_{r0}} \cdot E_{t0}^{C_{t0}}, \quad (1)$$

where E is cycling efficiency, C is the threshold cycling number, r is the reference gene, t the target gene and a 0 marks the sample which is used as the reference (Pfaffl 2001).

Probe preparation for Southern blot analysis

For the probe preparation 500 ng of *err1* cDNA, 5 μl 10x Klenow buffer (Thermo Scientific), and 6.5 μl 100 μM random hexamer primer (Thermo Scientific) were filled up with double distilled water (ddH$_2$O) to a final volume of 39 μl and incubated at 95°C for 5-10 min. The reaction mixture was put on ice and 5 μl Biotin PCR Labeling Mix (Jena Bioscience, Jena, Germany) and 1 μl Klenow fragment exo- (Thermo Scientific) were added. The mixture was filled up with ddH$_2$O to a final volume of 50 μl and incubated at 37°C for 24 h. For DNA precipitation

10 μl LiCl (4 M) and 200 μl ethanol (96%) were added. After incubation at RT for 15 min, at ice for 15 min, and at -20°C for 1 h, the DNA was separated by centrifugation with 20000 g at 4°C for 30 min. The pellet was washed with 500 μl ethanol (70%), followed by centrifugation with 20000 g at 4°C for 10 min. After drying the pellet at 50°C for about 10 min, it was dissolved in 100 μl ddH$_2$O. The quality of the probe was tested by agarose gel electrophoresis, and the concentration was determined with the NanoDrop 1000 (Thermo Scientific).

Southern blot analysis

For the Southern blot 15 μg chromosomal DNA of each strain used in this study was digested in a triple digestion with 5 μl of each NdeI, SalI, and BglII (each 10 U/μl, Thermo Scientific), using 10x Buffer O (Thermo Scientific). The reaction mixtures were filled up with ddH$_2$O to a final volume of 100 μl and then split into 20 μl aliquots for digestion at 37°C over night (o/n). After digestion, samples were incubated at 70°C for 20 min and the completion of digestion controlled by agarose gel electrophoresis. Digestion aliquots of each sample were pooled and concentrated to a final volume of 10 - 20 μl and applied to a 1% agarose gel. As length standard 5 μl Gene Ruler 1 kb DNA Ladder (Thermo Scientific) was used. Using the Mini-Sub® Cell system (Bio-Rad Laboratories) the gel was run at 80 V for 1 h in TAE buffer, and afterwards incubated in 0.4 M NaOH and 0.6 M NaCl, and then in 0.5 M Tris (pH 7.5) and 1.5 M NaCl for 30 min each. The DNA was transferred to a Biodyne B membrane (Pall Corporation, Port Washington, NY, USA) by a capillary blot with 10x saline-sodium citrate (SSC) buffer (1.5 M NaCl, 0.15 M sodium citrate, pH 7.2) o/n. After blotting, the membrane was incubated in 0.4 M NaOH, and then in 0.2 M Tris (pH 7.5) for 1 min each. Cross-linking was performed with a GS Gene Linker UV chamber (Bio-Rad Laboratories) using program C3 and 150 mJoule on the wet membrane. For pre-hybridization, the membrane was incubated at 65°C for 3 h in 20 ml Southern blot hybridization buffer (25% (v/v) 20x SSC, 10% (v/v) 50x Denhardt's solution, 0.2% (v/v) EDTA (0.5 M, pH 8.0), 0.05 M NaH$_2$PO$_4$, 0.1% (w/v) SDS, 0.5% (w/v) BSA), supplemented with 100 μg/ml single stranded salmon sperm DNA and freshly denatured (10 min at 95°C) probe. For hybridization, the membrane was incubated at 65°C o/n in 10 ml Southern blot hybridization buffer. The membrane was washed twice at RT for 5 min in 50 ml 2x SSC supplemented with 0.1% (w/v) SDS, followed by washing twice at 65°C for 15 min with 50 ml 0.1x SSC, supplemented with 0.1% (w/v) SDS. After incubation at RT for 10 min in Southern blot blocking solution (125 mM NaCl, 17 mM Na$_2$HPO$_4$, 8 mM NaH$_2$PO$_4$, 0.5% (w/v) SDS, pH 7.2), the membrane was incubated light-protected at RT for 30 min in Southern blot blocking solution supplemented with 1 μg/ml Dylight 650-labeled Streptavidin (Thermo Scientific). The membrane was washed 4 times light-protected at RT for 10 min in 50 ml 1:10 diluted Southern blocking solution and then scanned with Typhoon FLA 9500 (GE Healthcare Life Sciences, Buckinghamshire, England) set for Alexa Fluor 647 at 1000 V.

Cultivation in bioreactors

Cultivation was performed in 2-l-bench top bioreactors (Bioengineering AG, Wald, Swiss), containing 1.3 l fermentation medium ((NH$_4$)$_2$SO$_4$ 3.50 g/l, KH$_2$PO$_4$ 5.00 g/l, MgSO$_4$.7H$_2$O 1.25 g/l, NaCl 0.625 g/l, peptone from Casein 1.25 g/l, Tween® 80 0.625 g/l), supplemented with 1.5 ml/l trace element solution (FeSO$_4$.7H$_2$O 0.90 mM, MnSO$_4$.H$_2$O 0.50 mM, ZnSO$_4$.7H$_2$O 0.24 mM, CaCl$_2$.7H$_2$O 0.68 mM), 1.7% (w/v) wheat straw (pretreated by an alkaline organosolv process for lignin removal (Fackler et al. 2012) (Annikki, Graz, Austria)), and Antifoam Y-30 Emulsion (1 ml/bioreactor). For inoculation 10^9 conidia per liter were used. Agitation rate was 500 rpm, temperature was 28°C, and aeration rate was 0.5 vvm.

GC analysis

Mycelia from shake flask cultures were ground under liquid nitrogen. The powder was suspended in 3 ml distilled water and sonicated using a Sonifier® 250 Cell Disruptor (Branson, Danbury, CT, USA) (power 70%, duty cycle 40%, power for 3 min, on ice). Insoluble compounds were separated by centrifugation (20000 g, 10 min, 4°C), the clear supernatant was used for further processing.

Supernatants from shake flask cultures were used directly for further processing.

For samples from cultivation in bioreactors 30 ml of the whole cultivation broth were first mechanically disrupted with a potter for 1 min, then sonicated, and afterwards centrifuged as described above for mycelia from shake flask cultures.

Sample preparation for GC was done in triplicates as follows: 300 μl of the clear supernatant (prepared as described above), supplemented with 10 ng myo-inositol as internal standard, was gently mixed with 1.2 ml ethanol (96%) and incubated for 30 min at RT for protein precipitation. The precipitate was separated by centrifugation (20000 g, 10 min, 4°C). Samples were dried under vacuum and thereafter silylated (50 μl pyridine, 250 μl hexamethyldisilazane, 120 μl trimethylsilyl chloride). For quantitative erythritol determination a GC equipment (Agilent Technologies, Santa Clara, CA, USA) with a HP-5-column (30 m, inner diameter 0.32 mm, film 0.26 μm) (Agilent) was used. The mobile phase consisted of helium with a flow of 1.4 l/min, the column temperature was as follows: 150°C for 1 min, ramping 150 220°C (ΔT 4°C/min), ramping 220 - 320°C (ΔT 20°C/min), 320°C for 6.5 min. Detection was performed with FID at 300°C. The retention times were determined using pure standard substances.

Sodium hydroxide soluble protein (SSP)

2 ml cultivation broth were centrifuged at 20000 g for 10 min at 4°C. The supernatant was discarded and the pellet resuspended in 3 ml 0.1 M NaOH before sonication with a Sonifier® 250 Cell Disruptor (Branson) (power 70%, duty cycle 40%, power 20 s, pause 40 s, 10 cycles, on ice). The sonicated samples were incubated for 3 h at RT. After centrifugation (20000 g, 10 min, 4°C) the supernatant was used to determine protein concentration with a Bradford assay. Therefore, 20 μl diluted sample (1:10 - 1:100) were added to 1 ml 1:5 diluted Bradford Reagens (Bio-Rad Laboratories) and incubated for exactly 10 min at RT before measuring the absorption on a V-630 UV-Vis spectrophotometer (Jasco, Tokio, Japan) at 595 nm. As standard bovine serum albumin in concentrations from 10 - 100 μg/ml was used.

Results

Characterization of err1 overexpression strains

Protoplast transformation of the wild-type strain with the plasmid pBJ-PEC1, introducing err1 under the constitutive pki promoter of T. reesei, yielded 8 recombinant strains (named QPEC1-#). With the plasmid pBJ-BEC2, introducing err1 under the inducible bxl1 promoter of T. reesei, 3 recombinant strains (named QBEC2-#) were received. Biolistic transformation of Rut-C30 with the plasmid pBJ-PEC1 yielded 12 recombinant strains (named RPEC1 #), and with the plasmid pBJ-BEC2 20 recombinant strains (named RBEC2-#) were obtained. Stable insertion of the plasmid into the fungal genome was confirmed by isolation of genomic DNA and a following PCR amplifying a fragment including the introduced promoter and err1. After two rounds of spore streak outs, 3-7 recombinant strains of all four types were chosen for further characterization according to their growth. The selected recombinant strains were cultivated in shake flasks on D-xylose as well as birch-wood xylan followed by transcript analysis of err1. From each type, the strain with the highest transcript rate was chosen for further characterization (Figure 1). From now on strains were termed QPEC1, QBEC2, RPEC1, and RBEC2, respectively. A determination of the copy number of the newly introduced err1 in the four finally selected recombinant strains was performed by Southern blot analysis (Figure 2). Ectopic in tandem integration, which is the most common in T. reesei (Mach and Zeilinger 1998), was observed in all four strains. For QPEC1 and RPEC1 more than 5 additional copies were estimated, for QBEC2 and RBEC2 1-2 additional copies were estimated.

Increased production of erythritol on D-xylose

In order to get first insight in the native erythritol formation in the parental strains and the effect of the err1 overexpression, the strains QPEC1 and RPEC1, as well as their respective parental strains, were cultivated in shake flasks. For this first experiment D-xylose was used as carbon source as all strains grow well on this carbon source, and on the other hand as the monomer of the xylan-backbone it is a main component of lignocellulose, which is aimed to be used finally. Samples were taken after biomass formation was observed and analyzed by gas chromatography (GC) for erythritol production. Separate analysis of the supernatant and the mycelia revealed that no erythritol could be found in the supernatant. The erythritol concentrations detected in the mycelia are presented in Figure 3. For the wild-type and QPEC1 we could demonstrate, that the overexpression strain contained clearly more erythritol than the parental strain, with an increase of 1.6-fold (24 h) and 3.2-fold (30 h) (Figure 3a). For Rut-C30 and RPEC1 the increase of intracellular erythritol concentration in the err1 overexpression strain are not that explicit compared to the wild-type and to QPEC1 (Figure 3b). After 30 h and 36 h the increase in the recombinant strain is 1.2-fold and 1.4-fold, respectively, compared to the parental strain. Compared with the wild-type, both Rut-C30 and RPEC1 contained slightly less erythritol. However, this observation was considered as a preliminary result because the advantages of using Rut-C30 are not necessarily that pronounced on D-xylose than on a lignocellulosic substrate, which finally should be used according to the aim of this study.

Erythritol formation by the wild-type and its err1 overexpression strains on pretreated wheat straw

Experiments to investigate the growth ability on pretreated wheat straw and corresponding erythritol production on this substrate were performed by cultivation in a bioreactor starting with the wild-type strain and its respective err1 overexpression strains, QPEC1 and QBEC2. All three strains were able to grow on wheat straw as sole carbon source, even if inoculated directly with conidia and not with pregrown fungal mycelium. Microscopic analysis of samples taken 8 h after inoculation already showed a high germination rate. Further microscopic samples taken during the fermentation process showed good mycelial growth, strongly branched hyphae, and disappearance of the straw, which is due to enzymatic degradation by the fungus. Samples for investigation of erythritol production were taken 48 h and 72 h after inoculation. Since the cultivation broth contained aside from the mycelia also wheat straw as insoluble compound, it was not possible to separate the mycelia for analysis. Therefore, the whole samples were analyzed for erythritol content. Sodium hydroxide soluble protein

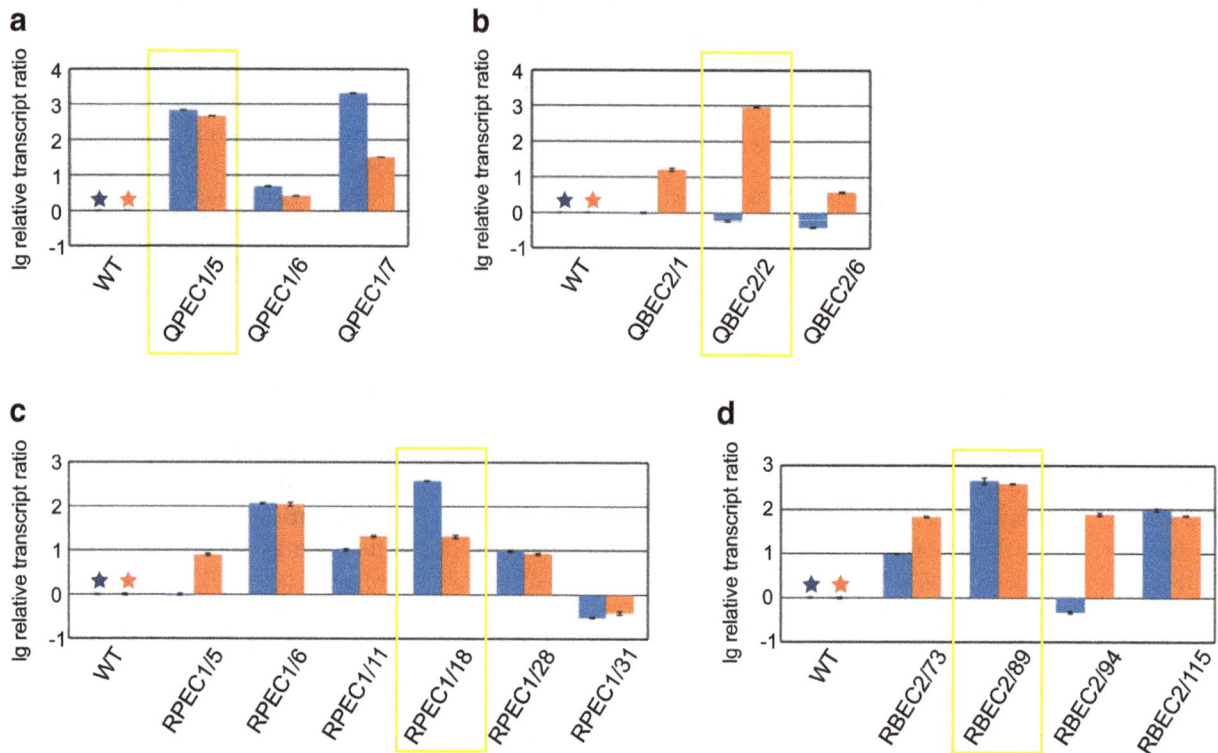

Figure 1 Transcript analysis of *err1* in parental and recombinant *T. reesei* strains. The *T. reesei* wild-type strain (WT) and preselected recombinant strains derived from transformation of the wild-type **(a, b)** or of Rut-C30 **(c, d)**, which are expressing *err1* either under the constitutive *pki* promoter **(a, c)** or under the inducible *bxl1* promoter **(b, d)**, respectively, were cultivated in shake flasks on D-xylose (blue bars) for 30 h (wild-type) or 72 h (Rut-C30) and on birch-wood xylan (red bars) for 48 h (wild-type) or 72 h (Rut-C30). Strains chosen for further experiments are framed in yellow. The transcript analysis was performed by qPCR using *sar1* and *act1* as genes for data normalization and levels always refer to the wild-type strain on the respective carbon source (indicated by a blue and red asterisk). Results are given as relative transcript ratios in logarithmic scale (lg). The values are means from three measurements. Error bars indicate standard deviations. Biological experiments (cultivations) were performed in duplicates.

(SSP) was determined and was used as an indicator for the biomass concentration. From the SSP one can conclude that the strains have a similar growth behavior (Figure 4a). The xylanase activity was similar in the wild-type and in QBEC2, but clearly increased in QPEC1 after 72 h (Figure 4b). In contrast to the results from the shake flask experiments on D-xylose, no increase in production of erythritol in the recombinant strains could be found by GC analysis (Figure 4c), even though transcript analysis of *err1* showed a slight increase in the recombinant strains after 48 h, and an even more pronounced one after 72 h (Figure 4d). Summarizing, these strains can grow on wheat straw and metabolize derived monosaccharides to erythritol. However, overexpression of *err1* did not enhance erythritol formation on wheat straw.

Erythritol formation by Rut-C30 and its *err1* overexpression strains on pretreated wheat straw

As a cellulase hyperproducing strain, Rut-C30 can be expected to better utilize lignocellulosic substrates compared to the wild-type strain. Indeed, an analog

experiment to the one described above, using Rut-C30 and RPEC1 showed more promising results as increased erythritol production in the overexpression strain was observed (Additional file 1). Consequently, a more extensive study drawing samples every 12 h, starting 18 h after inoculation, was conducted with these strains again cultivated in a bioreactor on pretreated wheat straw. The SSP indicated an similar growth behavior for all strains, whereupon RPEC1 after 42 h slightly dropped behind the others (Figure 5a). The same pattern could be observed even more clearly for the xylanase activities (Figure 5b). The course of erythritol concentration is depicted in Figure 5c. One can observe that the parental strain Rut-C30 started slightly faster with erythritol formation. All strains reached their maximum erythritol production after 42 h, whereupon the *err1* overexpression strains showed increased formation compared to their parental strain. Even though RPEC1 and RBEC2 shared nearly the same maximum erythritol concentration, they differed in their time course of production. Erythritol formation by RBEC2 rose faster in the beginning, but also

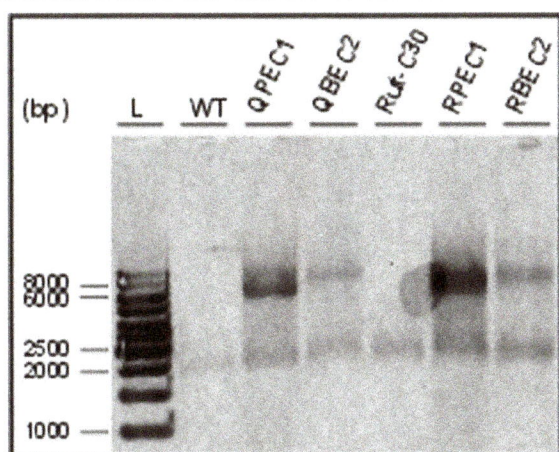

Figure 2 Southern blot analysis of parental and *err1* overexpression *T. reesei* strains. On an agarose gel *NdeI/SalI/BglII*-digested DNA from the wild-type (WT) strain bearing the native *err1* (2340 bp), the thereof derived *err1* overexpression strains QPEC1 (containing the native *err1* (2340 bp) and n+1 inserted fragments (5410 bp)) and QBEC2 (containing the native *err1* (2340 bp) and n+1 inserted fragments (6626 bp)), Rut-C30 bearing the native *err1* (2340 bp), and the thereof derived *err1* overexpression strains RPEC1 (containing the native *err1* (2340 bp) and n+1 inserted fragments (5410 bp)) and RBEC2 (containing the native *err1* (2340 bp) and n+1 inserted fragments (6626 bp)) was separated. n means the band intensity in relation to the native *err1*-containing band. A 1 kb DNA ladder (L) was used for estimation of DNA fragment size; indicated sizes are given in bp. As probe a biotin-labeled fragment containing the structural *err1* gene was used. For visualization Dylight 650-labeled streptavidin was applied and the membrane was scanned with a Typhoon FLA 9500.

of this storage compound when conditions (e.g. carbon source availability) become less favorable. It should also be noticed that the amount of erythritol produced by the recombinant strains was about 10-fold higher compared to the wild-type at the peak of production. The transcript analysis showed constant expression of *err1* for RPEC1 and an increasing expression for RBEC2, which is in good accordance with the type of promoters used. The expression of *err1* in the parental strain first decreased until it reaches a minimum at 42 h after inoculation. Afterwards, it slightly reincreased, but always remained lower than in the overexpression strains (Figure 5d).

Discussion

The by an alkaline organosolve process pretreated wheat straw (Fackler et al. 2012) used in our experiments, turned out to be a very well utilizeable substrate for *T. reesei* cultivation. In contrary to other pretreatment processes, this method does not require any chemicals or catalysts that subsequently inhibit fungal growth. The alcohol, which is used in the process as organic solvent, can be sufficiently removed by washing. The achieved removal of lignin (up to a residual share of 1%) (Fackler et al. 2012) makes the utilizeable cellulose and hemicellulose enough accessible for the fungus so that even direct inoculation with conidia was possible with this substrate as sole carbon source.

The comparison of the recombinant strains with their respective parentals showed that the overexpression of *err1* was successful an led to an increase in erythritol formation. In case of the wild-type and its recombinant strains this effect was more pronounced in shake flask cultivations on D-xylose, whereas Rut-C30 and its recombinant strains yielded better results in the bioreactor cultivation on pretreated wheat straw. Not only the relative increase of erythritol concentration in the recombinant strains compared to the parental strain was higher, but also the total amount of erythritol produced was about

dropped faster after having reached the maximum. After 66 h the erythritol concentration dropped for all three strains to a nearly equal level, so it seems that the overexpression of *err1* does not only boost the formation of erythritol but might also trigger the erythritol consumption

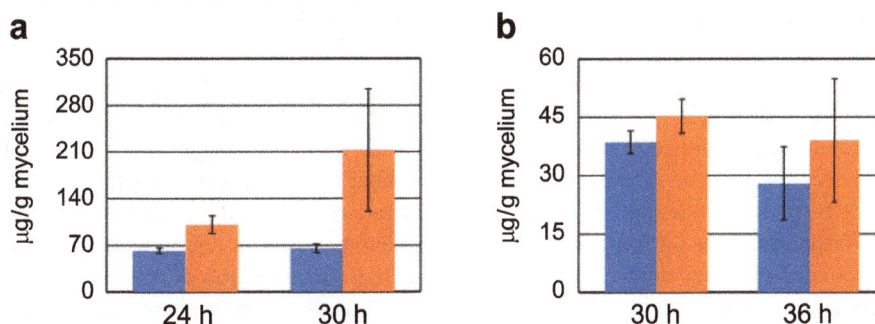

Figure 3 Erythritol production on D-xylose. The *T. reesei* **(a)** wild-type strain (blue bars) and the thereof derived *err1* overexpression strain QPEC1 (red bars) as well as **(b)** Rut-C30 (blue bars) and the thereof derived *err1* overexpression strain RPEC1 (red bars) were cultivated in shake flasks on D-xylose. Samples were taken after the indicated time and erythritol concentration was determined by GC-analysis from cell free extracts. Biological experiments (cultivations) were performed in duplicates. Standard deviations were obtained from two biological duplicates and measurements in triplicates each.

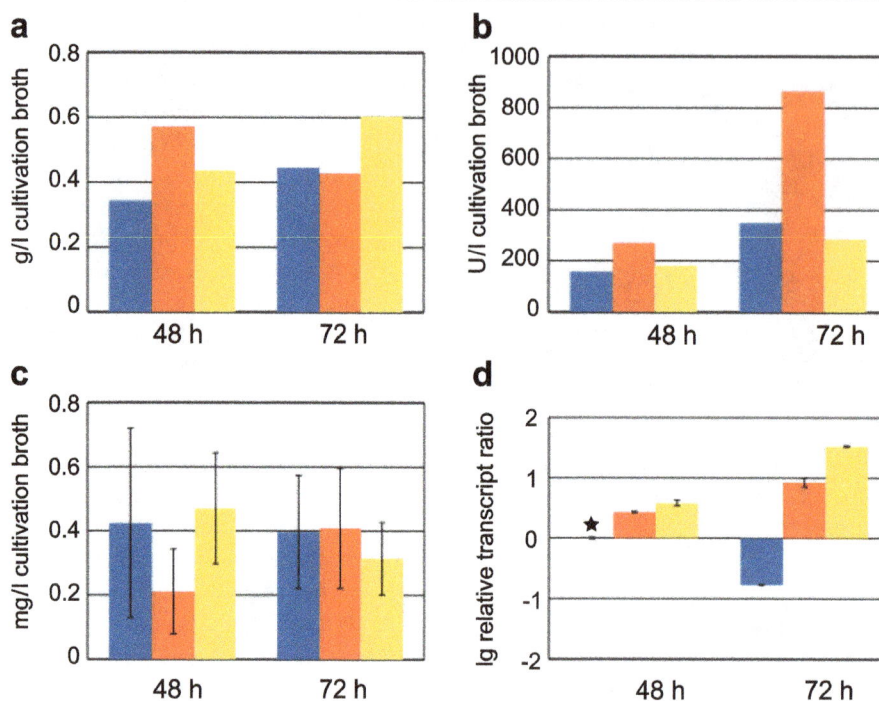

Figure 4 Cultivation of the wild-type and *err1* overexpression strains on wheat straw. The *T. reesei* wild-type strain (blue bars), and the *err1* overexpression strains QPEC1 (red bars) and QBEC2 (yellow bars) were cultivated in bench-top bioreactors on pretreated wheat straw. Samples were taken after 48 and 72 h. **(a)** Sodium soluble protein concentration (given in g/l cultivation broth) was measured in triplicates in cultivation broth samples after cell disruption to indicate biomass formation. Standard deviations were below 5%. **(b)** Xylanase activity (given in U/l cultivation broth) was measured in triplicates in the cultivation supernatants. Standard deviations were below 5%. **(c)** Erythritol concentration (given in mg/l cultivation broth) was measured in triplicates by GC in cultivation broth samples after cell disruption. Error bars indicate standard deviations. **(d)** Transcript analysis of *err1* (given as relative transcript ratio in logarithmic scale (lg)) was performed by qPCR in triplicates using *sar1* and *act1* as genes for data normalization and levels always refer to the wild-type strain cultivated for 48 h (as indicated by an asterisk). Error bars indicate standard deviations. Biological experiments (cultivations) were performed in duplicates.

10-fold increased compared to the wild-type and its recombinant strains. This observation can be explained by the fact that Rut-C30 is a cellulase hyperproducing, carbon catabolite derepressed strain (Montenecourt and Eveleigh 1979), which makes it very likely that it better utilizes a complex substrate like wheat straw. This assumption is supported by the observed increased biomass formation and enhanced xylanase activity produced. Concerning the promoters used, the constitutive *pki* promoter seems to be favorable, since the erythritol production peak was slightly higher and this high level remained for a longer period (54 h). It should be mentioned that an even higher maximum might occur between the samples taken. However, cultivation time turned out to be an important factor for the erythritol formation, since after the peak of production, the erythritol concentration drops about as fast as it rises in the beginning. Accordingly, the elimination of the back reaction can be considered as one of the main targets of further strain improvement. Since we found that in *T. reesei* erythritol is not exported to the media, but accumulated in the cell, presumably, the most efficient way to prevent the back reaction would be to

force the fungus to secrete the erythritol. This strategy would also be favorable in consideration of the osmotic balance of the cell. Taking into account that in case of erythritol production methods using yeasts, erythritol can be found in the supernatant (see e.g. Ryu et al. 2000; Rymowicz et al. 2009; Sawada et al. 2009), in yeasts must exist a transport system for erythritol that probably can be introduced into *T. reesei*. Another strategy to improve erythritol formation could be to reduce the accumulation of other polyols. This would on the one hand provide additional starting material for the erythritol production, and at the same time it would prevent an additional rise of the intracellular osmotic pressure by these substances.

GC analysis of the cultivation broth of Rut-C30 and its recombinant strains grown on wheat straw revealed especially a high accumulation of arabinitol, but also considerable amounts of xylitol (highest concentrations measured were 696 mg arabinitol and 63 mg xylitol per liter fermentation broth). Both substances are metabolites in the interconversion of the pentoses derived from lignocellulose degradation (i.e. L-arabinose, D-xylose) (Figure 6). Overexpression of the L-arabinitol dehydrogenase and

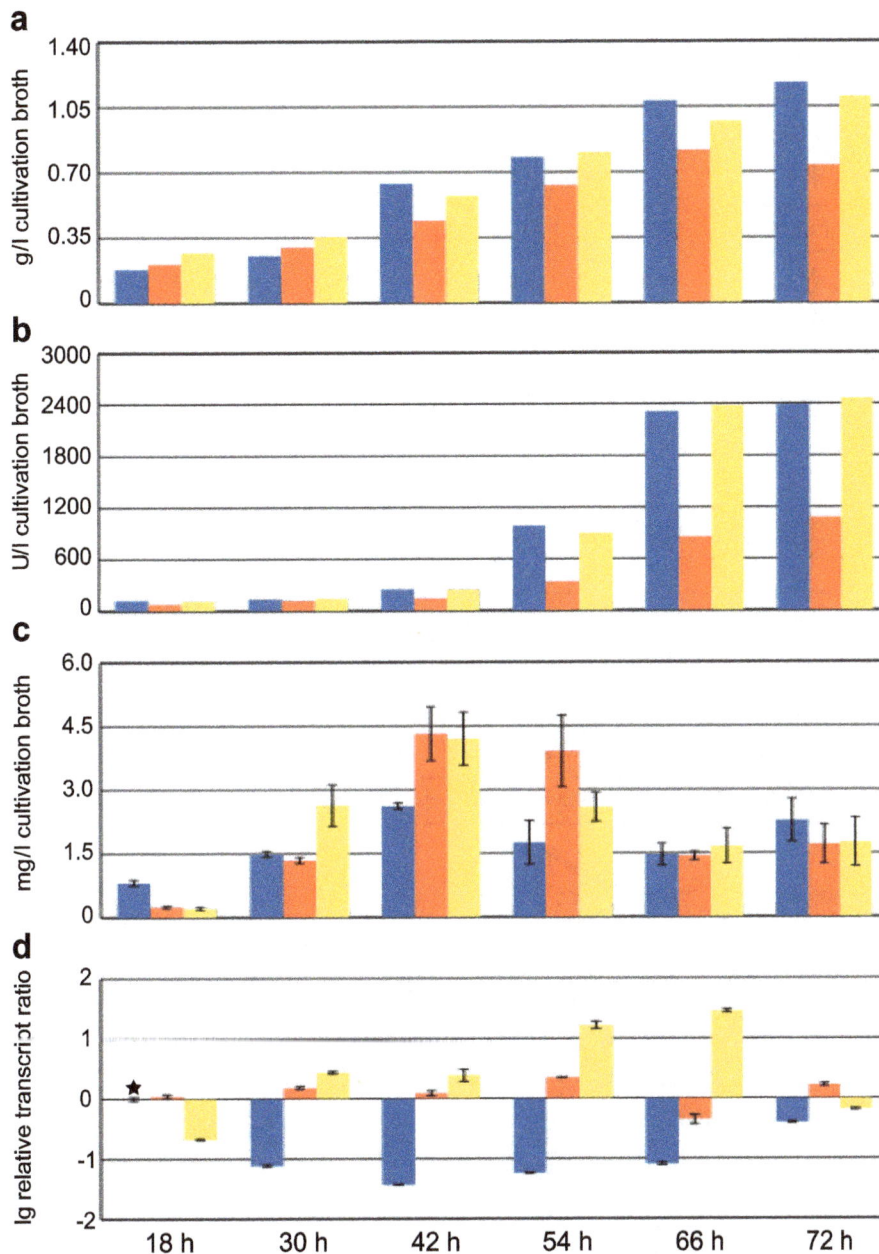

Figure 5 Cultivation of Rut-C30 and *err1* overexpression strains on wheat straw. Rut-C30 (blue bars), and the *err1* overexpression strains RPEC1 (red bars) and RBEC2 (yellow bars) were cultivated in bench-top bioreactors on pretreated wheat straw. Samples were taken after 18, 30, 42, 54, 66, and 72 h. **(a)** Sodium soluble protein concentration (given in g/l cultivation broth) was measured in triplicates in cultivation broth samples after cell disruption to indicate biomass formation. Standard deviations were below 5%. **(b)** Xylanase activity (given in U/l cultivation broth) was measured in triplicates in the cultivation supernatants. Standard deviations were below 5%. **(c)** Erythritol concentration (given in mg/l cultivation broth) was measured in triplicates by GC in cultivation broth samples after cell disruption. Error bars indicate standard deviations. **(d)** Transcript analysis of *err1* (given as relative transcript ratio in logarithmic scale (lg)) was performed by qPCR in triplicates using *sar1* and *act1* as genes for data normalization and levels always refer to Rut-C30 cultivated for 18 h (as indicated by an asterisk). Error bars indicate standard deviations. Biological experiments (cultivations) were performed in duplicates.

the D-xylulose reductase in *T. reesei* might help here to enforce the flux of these two major substrates into the pentose phosphate pathway (PPP) and thus enhance erythritol formation, which is a side product of the PPP. Even if the amounts of erythritol produced by now in *T. reesei* (approx. 5 mg/l) do not reach the current production standards with yeasts, it must be taken into consideration, that these yeast strains are highly mutagenized, and subsequently selected for high erythritol production for many years, and production conditions

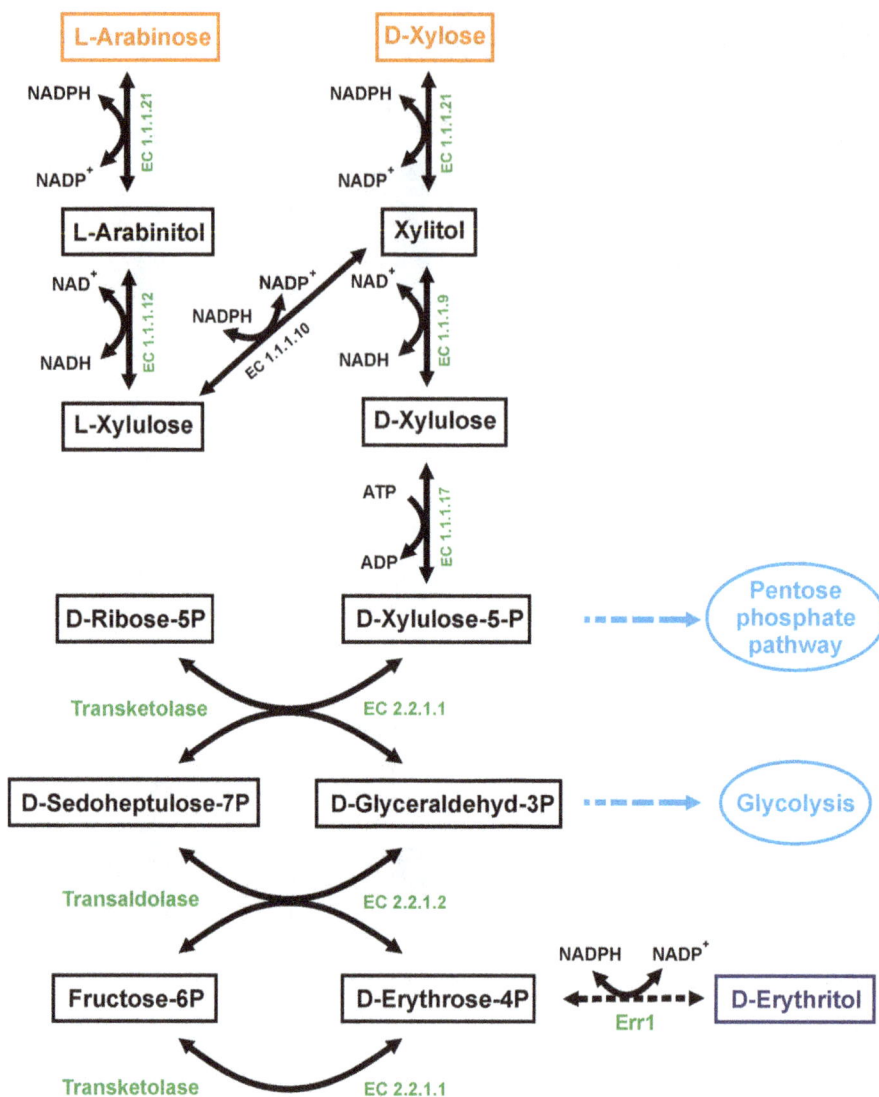

Figure 6 Schematic drawing of metabolic pathways of pentoses and erythritol in *T. reesei*. Metabolites are given in boxes. Monomeric sugars derived from hydrolytic lignocellulose degradation by *T. reesei* are given in orange. The target substance, erythritol, is given in purple. Enzyme names and EC numbers are given in green. Adjacent pathways are indicated in blue. Dashed arrows indicate (possible) involvement of more than one enzyme.

were optimized for decades now. For example, in 1999 for *C. magnolia* 6.9 g/l erythritol formation was reported for the wild-type, while with mutant strains concentrations of up to 25 g/l were obtained (Yang et al. 1999). Years later further process optimization led to a maximum final erythritol concentration of 200 g/l if fed with 70% (w/v) glucose supplemented with yeast extract (Koh et al. 2003). As D-glucose became the almost exclusively used substrate for erythritol production, usually the D-glucose to erythritol conversion (in % (w/w)) is reported. Recently, for the *C. magnolia* NCIM 3470 mutant R23 a yield of 31.1% was reported (Savergave et al. 2011), while with *Trichosporonoides megachiliensis* SN-G42 a conversion of 47% can be achieved (Sawada et al. 2009). However,

the conversion of D-glucose to erythritol can not easily directly be compared to erythritol production from wheat straw due to the very different nature of these substrates. Anyway, the aim of this study was to provide proof-of-concept for the synthesis of erythritol from wheat straw and additional metabolic engineering as described above. Subsequent strain screening might lead to competitive production levels in biomass-degrading fungi like *T. reesei*, with the advantage of using cheap and sustainable substrates.

Concluding this, in the present study we demonstrated that the production of erythritol on the renewable, non-food substrate wheat straw, using *T. reesei* is possible. The alkaline organosolve pretreatment process used for the wheat straw is compatible for subsequent fungal growth and provides an easily utilizeable substrate. Moreover, strain modification by overexpression of *err1* led to increased erythritol formation on this substrate.

Additional file

Additional file 1: Figure S1. Cultivation of Rut-C30 and an *err1* overexpression strain on wheat straw. The Rut-C30 (blue bars) and the *err1* overexpression strain RPEC1 (red bars) were cultivated in bench-top bioreactors on pre-treated wheat straw. Samples were taken after 48 and 72 hours. (a) Sodium soluble protein concentration (given in g/l cultivation broth) was measured in triplicates in cultivation broth samples after cell disruption to indicate biomass formation. Standard deviations were below 5%. (b) Xylanase activity (given in U/l cultivation broth) was measured in triplicates in the cultivation supernatants. Standard deviations were below 5%. (c) Erythritol concentration (given in mg/l cultivation broth) was measured by GC in cultivation broth samples after cell disruption. Standard deviations were obtained from measurements in triplicates. (d) Transcript analysis of *err1* (given as relative transcript ratio in logarithmic scale (lg)) was performed by qPCR using sar1 and act as genes for data normalization and levels always refer to Rut-C30 cultivated for 48 h (as indicated by an asterisk). Standard deviations were obtained from measurements in triplicates. Biological experiments (cultivations) were performed in duplicates.

Competing interests

A European patent entitled 'Method for the production of erythritol' (no. EP20100183799, 5.4.2012) (Mach and Mach-Aigner 2012) was issued.

Acknowledgments

This study was supported by Annikki GmbH, by two grants from the Austrian Science Fund (FWF): [P20192, P24851] given to RLM and ARMA, respectively, by an Innovative Project of Vienna University of Technology (Demo-Tech), and by a doctoral program of Vienna University of Technology (AB-Tec).

References

Acebal C, Castillon M, Estrada P, Mata I, Costa E, Aguado J, Romero D, Jimenez F (1986) Enhanced cellulase production from *Trichoderma reesei* QM 9414 on physically treated wheat straw. Appl Microbiol Biotechnol 24(3):218–223. doi:10.1007/BF00261540

Buchert J, Oksanen T, Pere J, Siika-aho M, Suurnkki A, Viikari L (1998) Applications of *Trichoderma reesei* enzymes in the pulp and paper industry. In: Harman G, Kubicek C (eds) Trichoderma & Gliocladium, vol 2. Taylor & Francis Ltd, London, UK, pp 343–357

Dashtban M, Kepka G, Seiboth B, Qin W (2013) Xylitol production by genetically engineered *Trichoderma reesei* strains using barley straw as feedstock. Appl Biochem Biotechnol 169(2):554–569. doi:10.1007/s12010-012-0008-y

Derntl C, Gudynaite-Savitch L, Calixte S, White T, Mach R, Mach-Aigner A (2013) Mutation of the xylanase regulator 1 causes a glucose blind hydrolase expressing phenotype in industrially used *Trichoderma* strains. Biotechnol Biofuels 6(1):62. doi:10.1186/1754-6834-6-62

Fackler K, Ters T, Ertl O, Messner K (2012) Method for lignin recovery. Patent WO/2012/027767

Galante YM, Monteverdi R, Inama S, Caldini C, De Conti A, Lavelli V, Bonomi F (1993) New applications of enzymes in wine making and olive oil production. Italian Biochem Soc Trans 4:34

Gruber F, Visser J, Kubicek CP, de Graaff LH (1990) The development of a heterologous transformation system for the cellulolytic fungus *Trichoderma reesei* based on a *pyrG*-negative mutant strain. Curr Genet 18(1):71–76

Hahn-Hägerdal B, Galbe M, Gorwa-Grauslund M, Lidén G, Zacchi G (2006) Bio-ethanol - the fuel of tomorrow from the residues of today. Trends Biotechnol 24(12):549–556

Himmel ME, Ding SY, Johnson DK, Adney WS, Nimlos MR, Brady JW, Foust TD (2007) Biomass recalcitrance: engineering plants and enzymes for biofuels production. Science 315(5813):804–807

Jovanovic B, Mach R, Mach-Aigner A (2013) Characterization of erythrose reductases from filamentous fungi. AMB Express 3(1):43. doi:10.1186/2191-0855-3-43

Kasumi T (1995) Fermentative production of polyols and utilization for food and other products in Japan. JARQ 29(1):49–55

Koh ES, Lee TH, Lee DY, Kim HJ, Ryu YW, Seo JH (2003) Scale-up of erythritol production by an osmophilic mutant of *Candida magnoliae*. Biotechnol Lett 25(24):2103–2105

Koo H, Ueda M, Wakida T, Yoshimura Y, Igarashi T (1994) Cellulase treatment of cotton fabrics. Text Res J 64(2):70–74. doi:10.1177/004051759406400202

Kuhls K, Lieckfeldt E, Samuels GJ, Kovacs W, Meyer W, Petrini O, Gams W, Börner T, Kubicek CP (1996) Molecular evidence that the asexual industrial fungus *Trichoderma reesei* is a clonal derivative of the ascomycete *Hypocrea jecorina*. PNAS 93(15):7755–7760

Kumar A, Lepola M, Purtell C (1994) Enyme finishing of man-made cellulosic fabrics. Text Chem Color 26(10):25–28

Lanzarini G, Pifferi P (1989) Enzymes in the fruit juice industry. In: Cantarelli C, Lanzarini G (eds) Biotechnology applications in beverage production. Elsevier Applied Food Science Series, Springer Netherlands, pp 189–222. doi:10.1007/978-94-009-1113-0_13

Livesey G (2001) Tolerance of low-digestible carbohydrates: a general view. Br J Nutr 85:7–16

Mach R, Mach-Aigner A (2012) Method for the production of erythritol. Patent EP20100183799

Mach R, Zeilinger S (1998) Genetic transformation of *Trichoderma* and *Gliocladium*. In: Kubicek CP, Harman GE (eds) Trichoderma & Gliocladium. Taylor & Francis Ltd, London, UK, pp 225–224

Mach R, Schindler M, Kubicek CP (1994) Transformation of *Trichoderma reesei* based on hygromycin B resistance using homologous expression signals. Curr Genet 25(6):567–570

Mandels M (1985) Applications of cellulases. Biochem Soc Trans 13(2):414–416

Martinez D, Berka RM, Henrissat B, Saloheimo M, Arvas M, Baker SE, Chapman J, Chertkov O, Coutinho PM, Cullen D, Danchin EGJ, Grigoriev IV, Harris P, Jackson M, Kubicek CP, Han CS, Ho I, Larrondo LF, Leon ALd, Magnuson JK, Merino S, Misra M, Nelson B, Putnam N, Robbertse B, Salamov AA, Schmoll M, Terry A, Thayer N, Westerholm-Parvinen A, et al. (2008) Genome sequencing and analysis of the biomass-degrading fungus *Trichoderma reesei* (syn. *Hypocrea jecorina*). Nat Biotechnol 26(5):553–560

Montenecourt BS, Eveleigh DE (1979) Selective screening methods for the isolation of high yielding cellulase mutants of *Trichoderma reesei*. In: Brown RD, Jurasek L (eds) Hydrolysis of cellulose: Mechanisms of enzymatic and acid catalysis. American Chemical Society, Washington, D. C., pp 289–301. doi:10.1021/ba-1979-0181.ch014

Moon HJ, Jeya M, Kim IW, Lee IK (2010) Biotechnological production of erythritol and its applications. Appl Microbiol Biotechnol 86(4):1017–1025. doi:10.1007/s00253-010-2496-4

Munro I, Berndt W, Borzelleca J, Flamm G, Lynch B, Kennepohl E, Br E, Modderman J (1998) Erythritol: an interpretive summary of biochemical, metabolic, toxicological and clinical data. Food Chem Toxicol 36(12):1139–1174. http://dx.doi.org/10.1016/S0278-6915(98)00091-X

Noé P, Chevalier J, Mors F, Comtat J (1986) Action of xylanases on chemical pulp fibers part II : Enzymatic beating. J Wood Chem Technol 6(2):167–184. doi:10.1080/02773818608085222

Pedersen GP, Screws GA, Cereoni DA (1992) Biopolishing of cellulosic fabrics. Can Text J Dec: 31–35

Peterson R, Nevalainen H (2012) *Trichoderma reesei* RUT-C30 thirty years of strain improvement. Microbiology 158(1):58–68. doi:10.1099/mic.0.054031-0

Pfaffl MW (2001) A new mathematical model for relative quantification in real-time RT-PCR. Nucleic Acids Res 29(9):45

Punt PJ, Oliver RP, Dingemanse MA, Pouwels PH, van den Hondel CA (1987) Transformation of *Aspergillus* based on the hygromycin B resistance marker from *Escherichia coli*. Gene 56(1):117–124. doi:10.1016/0378-1119(87)90164-8

Ragauskas AJ, Williams CK, Davison BH, Britovsek G, Cairney J, Eckert CA, Frederick WJ, Hallett JP, Leak DJ, Liotta CL, Mielenz JR, Murphy R, Templer R, Tschaplinski T (2006) The path forward for biofuels and biomaterials. Science 311(5760):484–489. doi:10.1126/science.1114736

Rymowicz W, Rywińska A, Marcinkiewicz M (2009) High-yield production of erythritol from raw glycerol in fed-batch cultures of *Yarrowia lipolytica*. Biotechnol Lett 31(3):377–380. doi:10.1007/s10529-008-9884-1

Ryu YW, Park CY, Park JB, Kim SY, Seo JH (2000) Optimization of erythritol production by *Candida magnoliae* in fed-batch culture. J Ind Microbiol Biotechnol 25(2):100–103. doi:10.1038/sj.jim.7000039

Savergave LS, Gadre RV, Vaidya BK, Narayanan K (2011) Strain improvement and statistical media optimization for enhanced erythritol production with minimal by-products from *Candida magnoliae* mutant R23. Biochem Eng J 55(2):92–100. http://dx.doi.org/10.1016/j.bej.2011.03.009

Sawada K, Taki A, Yamakawa T, Seki M (2009) Key role for transketolase activity in erythritol production by *Trichosporonoides megachiliensis* SN-G42. J Biosci Bioeng 108(5):385–390. http://dx.doi.org/10.1016/j.jbiosc.2009.05.008

Steiger M, Vitikainen M, Uskonen P, Brunner K, Adam G, Pakula T, Penttilä M, Saloheimo M, Mach R, Mach-Aigner AR (2011) Transformation system for *Hypocrea jecorina* (*Trichoderma reesei*) that favors homologous integration and employs reusable bidirectionally selectable markers. Appl Environ Microbiol 77(1):114–121. doi:10.1128/AEM.02100-10

Steiger MG, Mach RL, Mach-Aigner AR (2010) An accurate normalization strategy for RT-qPCR in *Hypocrea jecorina* (*Trichoderma reesei*). J Biotech 145(1):30–37. doi:10.1016/j.jbiotec.2009.10.012

Te'o V, Bergquist P, Nevalainen K (2002) Biolistic transformation of *Trichoderma reesei* using the bio-rad seven barrels hepta adaptor system. J Microbiol Meth 51(3):393–399. http://dx.doi.org/10.1016/S0167-7012(02)00126-4

Walsh G, Power R, Headon D (1993) Enzymes in the animal-feed industry. Trends Biotechnol 11(10):424–430

Welt T, Dinus R (1995) Enzymatic deinking - a review. Prog Paper Recycl 4(2):36–47

Yang SW, Park JB, Han NS, Ryu YW, Seo JH (1999) Production of erythritol from glucose by an osmophilic mutant of *Candida magnoliae*. Biotechnol Lett 21:887–890

2

The effect on growth of *Chlamydomonas reinhardtii* of flue gas from a power plant based on waste combustion

Leiv M Mortensen[*] and Hans R Gislerød

Abstract

Flue gases from a power plant based on waste combustion were tested as a carbon dioxide (CO_2) source for growing *Chlamydomonas reinhardtii*. To achieve recognition as an environmentally friendly hydrogen production method, waste gases should be used to grow this hydrogen-producing microalgae. The algae were grown in undiluted flue gas containing $11.4\pm0.2\%$ CO_2 by volume, in diluted flue gas containing $6.7\pm0.1\%$ or $2.5\pm0.0\%$ CO_2, and in pure liquid CO_2 at a concentration of $2.7\pm0.2\%$. The NO_x concentration was 45 ± 16 mg m^{-3}, the SO_2 concentration was 36 ± 19 mg m^{-3}, the HCl concentration 4.1 ± 1.0 mg m^{-3} and the O_2 concentration $7.9\pm0.2\%$ in the undiluted flue gas. Undiluted flue gas reduced the dry weight production by around 20-25% when grown at a photon flux density (PFD) of 300 μmol m^{-2} s^{-1} artificial light and at 24 or 33°C, compared with the other treatments. A less negative effect was found at the highest flue gas concentration when the algae were grown at 75 μmol m^{-2} s^{-1} PFD. Growing the algae outdoors at a day length of 12.5 h and a temperature of around 24°C, the dry weight production was higher (about 15%) in the 2.6% CO_2 flue gas treatment compared with all other treatments. Reducing the light level by 30% through shading did not affect the dry weight production. Calculated on aerial basis the productivity reached approximately 70 g m^{-2} day^{-1} in the 300 μmol m^{-2} s^{-1} PFD treatment (corresponding to 25 mol m^{-2} day^{-1}) and approximately 17 g m^{-2} day^{-1} in the 75μmol m^{-2} s^{-1} PFD treatment (corresponding to 6.5 mol m^{-2} day^{-1}). The outdoor production reached around 14 g m^{-2} day^{-1}. It was concluded that the negative effect of the undiluted flue gas was attributable to the high CO_2 concentration and not to the other pollutants.

Keywords: Carbon dioxide concentration; *Chlamydomonas* reinhardtii; Flue gas; Photosynthetic active radiation

Introduction

The single-celled green alga *Chlamydomonas reinhardtii* is known to produce hydrogen when starved of sulphur under anaerobic conditions (Skjånes et al. 2007; Nguyen et al. 2011; Geier et al. 2012). At present, conventional hydrogen production is energy-intensive, and more environmentally friendly production based on biological processes is therefore of great interest (Jo et al. 2006). Today, the atmospheric CO_2 concentration of about 400 μmol mol^{-1} strongly limits the algal growth, and additional CO_2 gas has to be supplied throughout the production phase (Geier et al. 2012). Waste CO_2 from industrial flue gases should be used in order to make the production environmentally friendly. This will also contribute to reducing

CO_2 emissions that are important to the environment (IPCC 2013). Several studies have been carried out on the effect of flue gases on the growth of microalgae (Douskova et al. 2009; Kastanek et al. 2010; Borkenstein et al. 2011). *Chlamydomonas reinhardtii* seems to have been little studied, however (see review by van den Hende et al. 2012). Flue gases contain pollutants such as NO_x and SO_2 that can reach harmful levels depending on the species (van den Hende et al. 2012). However, few studies have devoted attention to whether the harmful effects depend on environmental factors such as irradiance level and temperature. In tomato plants, it is known that susceptibility to NO_x is much higher in low-light as opposed to high-light conditions (Mortensen 1986). For microalgae, and particularly for *C. reinhardtii*, little is known about the modifying effects of climate factors. Therefore, in this work the effect of flue gas was studied on *C. reinhardtii* at

* Correspondence: lei-mo@online.no
Department of Plant Science, The University of Life Sciences, Ås NO-1432, Norway

different levels of artificial light and in outdoor conditions with and without shade, as well as at two temperature levels.

Material and methods

Chlamydomonas reinhardtii strain SAG 34.89 from SAG (Göttingen, Germany) obtained from the NIVA culture collection, Norway, was used in the experiments. The algae were stored on Petri dishes covered with TAP medium 1.5% agar (Gorman and Levine 1965). The algae were grown in the high-salt Sueoka medium (Sueoka 1960). Sodium bicarbonate was used in the medium to buffer the culture at 10 mM. The microalgae were grown in 1.0 l clear plastic bottles (80 mm inner and 82 mm outer diameter) filled with 0.85 l of growing medium (filled up to 17 cm). Tubes with these dimensions have a volume of approximately 60 l per m^2 surface area when placed closely together, as the bottles were in the present experiments. The light was supplied by cool white fluorescence tubes (Osram L58W/840) 24 h day^{-1} placed about 10 cm in front of the row of bottles. The photon flux density (PFD) of the artificial light was measured by a LI-COR Model Li-250 instrument with quantum sensor (400-700 nm). The light was supplied from one side and was measured at the surface of the bottles. However, inside the culture the light level strongly decreased from the light exposed side to the opposite side of the bottles, as well as with increasing cell concentration during growth. Typically, the light level decreased by about 70% through the 8.0 cm diameter bottle at start of the experiment and by more than 99.9% at the end of the experiment, due to the increase in the algae concentration.

Two experiments were carried out indoor with artificial light, while a third experiment was carried out outdoor in daylight. The daylight was measured by a Delta-T Devices PAR sensor (cosine corrected within ±5% up to 70° incidence). The temperature was controlled by placing the bottles with the microalgae culture in water baths controlled by aquarium heaters. A circulation pump ensured a homogenous temperature in the water baths. The temperature was measured by cupper-constantan thermocouples. The CO_2 concentration was measured by a Vaisala CO_2 transmitter (Type GMT221, range 0-5%). The CO_2 concentration as well as the temperatures and the daylight PAR were recorded as hourly means by a Campbell CR10X logger with an AM25T thermocouple multiplexer. In addition a Vaisala GMP instrument was used to measure the CO_2 concentrations between 0 and 20%, and the measurements were recorded as hourly means.

The flue gas

The flue gas was provided by 'Borregaard Waste to Energy' located in Sarpsborg, Norway (www.hafslund.no). This modern fuel-flexible energy recovery plant burns approximately 80,000 tonnes of waste-based fuel and produces approximately 230 GWh per year. It has a high environmental standard. The CO_2, O_2, NO_x, NO, NO_2, SO_2, HCl, CO and TOC concentrations in the flue gas were measured at 10-minute intervals by an ABB Advance Cemas FTIR NT continuous monitoring system with extra modules for O_2 and TOC measurements (Figure 1, Table 1). NO constituted the main part of the NO_x, while NO_2 contributed only 3.4±1.4% of the total NO+NO_2 (data not presented). The mean O_2 concentration in the flue gas was 7.9±0.2%. In addition, license measurements on a series of heavy metals and dioxins in the flue gas were performed 2-4 times per year since the start of the power plant in 2010 (Table 1).

Flue gas from the chimney was sucked by pumps through two 100 l plastic tubs connected in series for condensation of water vapour. The microalgae were grown in undiluted flue gas (11.4% CO_2) or mixed with fresh air in a constant ratio using air pumps (Resun ACO-008A) to yield 6.7% and 2.5% CO_2, respectively (Figure 1, Table 1). One CO_2 concentration (2.66±0.16%) was established by mixing pure CO_2 (food quality) from bottles with fresh air. The CO_2 gas flow was determined by a capillary with a defined resistance, while the gas pressure was defined by the height of a water column. In this way, a very accurate CO_2 flow could be added to a constant rate of fresh air supplied by air pumps (Resun ACO-001, ACO-004).

The different gas mixtures were bubbled through plastic tubes with 0.3 cm inner diameter to the bottom of the bottles at a rate of approximately 100 l h^{-1}. All treatments in all experiments included three parallel bottles containing 0.85 l of culture. Three independent experiments (including a total of 60 bottles) were carried out during the same time period, all of which started with the same algae concentration of 0.20 g dry weight per litre culture. This concentration was established by adding algae from a start culture. Two of the experiments were conducted indoor with artificial lighting while the third was conducted outdoor in daylight.

Dissolved CO_2 in the growth medium

For algal growth, the concentration of dissolved CO_2 in the nutrient medium is important and not the concentration of CO_2 in the air bubbled into the culture, although a close relationship should be expected. In order to document this relationship a test with different concentrations of pure CO_2 mixed with air were bubbled through the bottles filled with nutrient medium. The concentration of dissolved CO_2 was measured using hand-held titration cells for titrimetric analysis (CHEMetrics Inc., USA, www.chemetric.com). The results showed that a progressive increase in the dissolved CO_2 concentration from about 100 to about 500 mg l^{-1} with increasing CO_2 concentration from about 1% up to about 20% (Figure 2). Parallel to

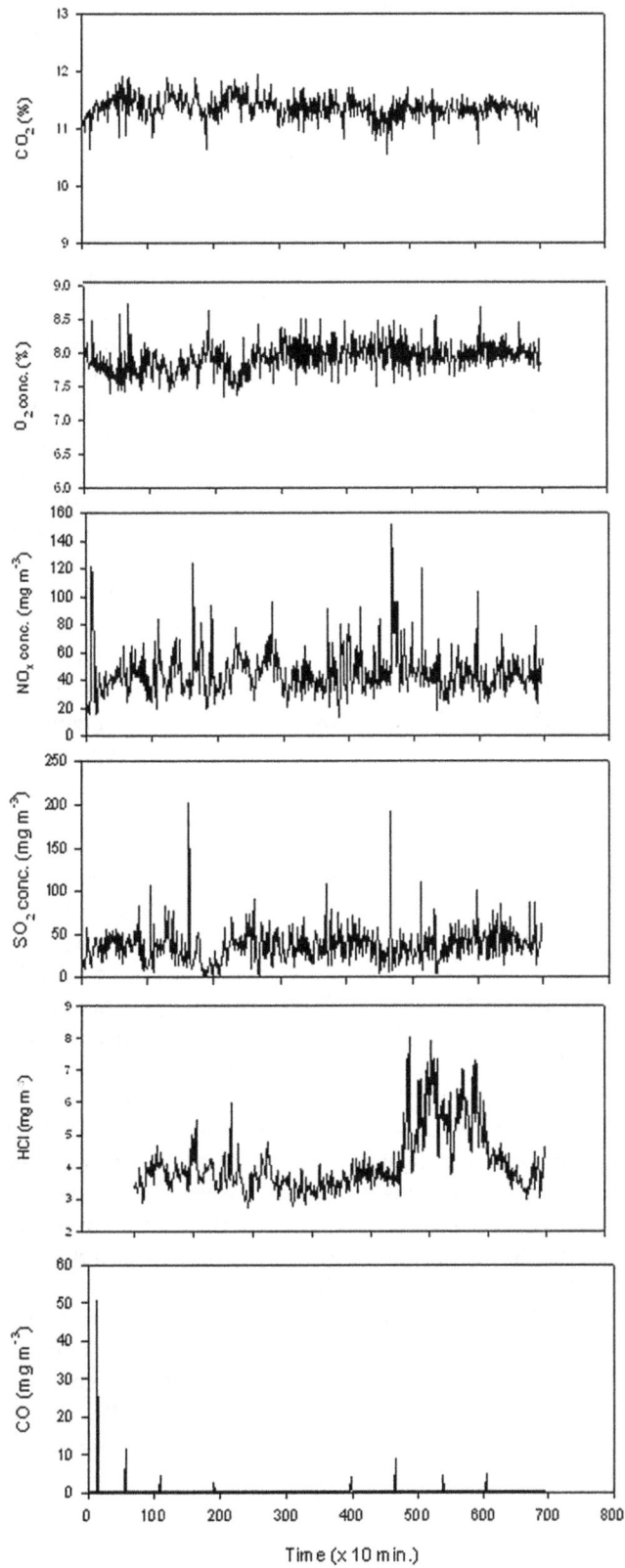

Figure 1 The concentration of different gases in undiluted flue gas.

Table 1 Mean concentrations (±SD) of different pollutants as measured in the different flue gas concentrations

	CO$_2$ conc. (%)		
	11.37±0.19	6.71±0.11	2.50±0.04
NO$_x$ (mg m^{-3})	45.0±15.8	26.6±9.3	9.9±3.5
SO$_2$ (mg m^{-3})	36.1±19.0	21.3±11.2	7.9±4.2
HCl (mg m^{-3})	4.11±0.95	2.43±0.56	0.90±0.2
CO (mg m^{-3})	0.45±2.00	0.27±1.18	0.10±0.4
TOC (mg m^{-3})	0.714±0.436	0.421±0.257	0.157±0.096
*Hg (µg m^{-3})	0.28±0.35	0.17±0.21	0.062±0.046
*HF (mg m^{-3})	0.063±0.020	0.037±0.012	0.014±0.004
*Dioxins (ng m^{-3})	0.00151±0.00134	0.00089±0.00079	0.00033±00029
*As+Co+Cr+Cu+Mn +Ni+Pb+Sb+V (mg m^{-3})	0.0211±0.0545	0.0124±0.0322	0.0046±0.0120

The concentrations were measured continuously in the undiluted flue gas (11.37%) and the concentrations in the diluted flue gases were reduced to the same extent as the CO$_2$ concentration. *These concentrations were measured 2-4 times per year in the period 2010-2013 (n=10, ±SD).

this increase the pH decreased from 7.6 to about 6.5. The measurements were done at 23°C. Dissolved CO$_2$ as measured at 7.0% CO$_2$ in the air was 311±12, 297±12 and 297±12 mg l^{-1} (n=3, ±SE) at 23, 28 and 33°C, respectively.

The experiments
Experiment 1

The microalgae were grown at the three flue gas concentrations and one concentration with pure CO$_2$ from bottles (Figure 1, Table 1). Two photon flux densities (PFD) were continuously applied, 75 and 300 µmol m^{-2} s^{-1}, corresponding to 6.5 and 25.9 mol m^{-2} day^{-1} PAR, respectively. Two rows of twelve bottles with algae culture were placed closely adjacent to each other in a water bath. One row along one side of the water bath was exposed to 300 µmol m^{-2} s^{-1} PFD, and the other row along the opposite side was exposed to 75 µmol m^{-2}s^{-1} PFD. A black sheet across the water bath eliminated any light pollution between the two light treatments. The water bath was made of transparent plexiglass, and one and four fluorescent tubes placed 10-15 cm from the bottles (outside the water bath) produced the low and high PFD, respectively. The temperature was 33±2°C. The dry weight (mg l^{-1} culture), pH and O$_2$ concentration in the culture were measured after three and five days, and the production per m^2 and day was calculated using the vertical projected area of the bottles.

Experiment 2

The same flue gas and pure CO$_2$ gas treatments were applied in this experiment as in Experiment 1. In this experiment a PFD of 300 µmol m^{-2} s^{-1} given continuously was used. The temperature was 19±2°C during the first day, and was thereafter increased to 24±2°C. The

temperature was controlled as in Experiment 1. Twelve bottles were included in the experiment, and the dry weight concentration and pH were measured four and five days after the start.

Experiment 3

In this experiment the microalgae were grown outdoors during four days under the different CO$_2$ treatments in full daylight and in 70% daylight by shading with white plastic (Figure 3). The bottles were closely placed adjacent to each other in water baths in rows with six bottles facing to the south. In the forefront row the culture received full daylight while the shade was given on the back row placed about 30 cm behind. Two water baths were needed for the 24 bottles including four CO$_2$ and two light treatments. The temperature was as a mean 24°C, varying from a peak of around 30°C at midday down to around 22°C during the night. The experiment was carried out in mid-September and the day length was 12.5 h (06.50 – 19.30 h). The building of the power station was located a few meters north of the experiment. The PFD varied from 0 to a maximum of about 1600 µmol m^{-2} s^{-1} in full daylight and up to about 1100 µmol m^{-2} s^{-1} in shaded conditions (Figure 2). The mean PAR was 17.1 and 12.0 mol m^{-2} day^{-1} in full daylight and in shaded conditions, respectively. At the Meterological station 5 km from the experimental site (Østad, Sarpsborg, 59°N, 11°E) the corresponding daylight was measured to 19.7 mol m^{-2} day^{-1} when converted from global radiation to PAR (www.bioforsk.no, Agricultural Meteorological service). The higher measured value here was probably due to the light sensor with 180° view (Kipp & Zonen, CM11 pyranometer) and more diffuse light from the north since the building shaded for the light from this direction in the experiment. Mean effective PFD in the experiment was calculated by assuming that PFD above different threshold values (100, 200 µmol m^{-2} s^{-1} etc.) has no effect on the growth (has reached the light saturation level) of the algae (Figure 4).

The dry weight was measured by vacuum filtering 10 or 20 ml of culture through a 90 mm filter (Whatman GF/B, cat. No. 1821-090) and drying it in an oven for four hours at 100°C. No pore size of this filter is given, however, all algal cells remained on the filter since no colouration of the filtered water was observed. The data were analysed using the SAS-GLM procedure (SAS institute Inc., Cary, USA) based on the bottles as replicates (n=3).

Results
Experiment 1

From an initial concentration of 0. 2 g l^{-1}, the dry weight reached its maximum level after three days at the highest PFD, since no further increase was found on the fifth

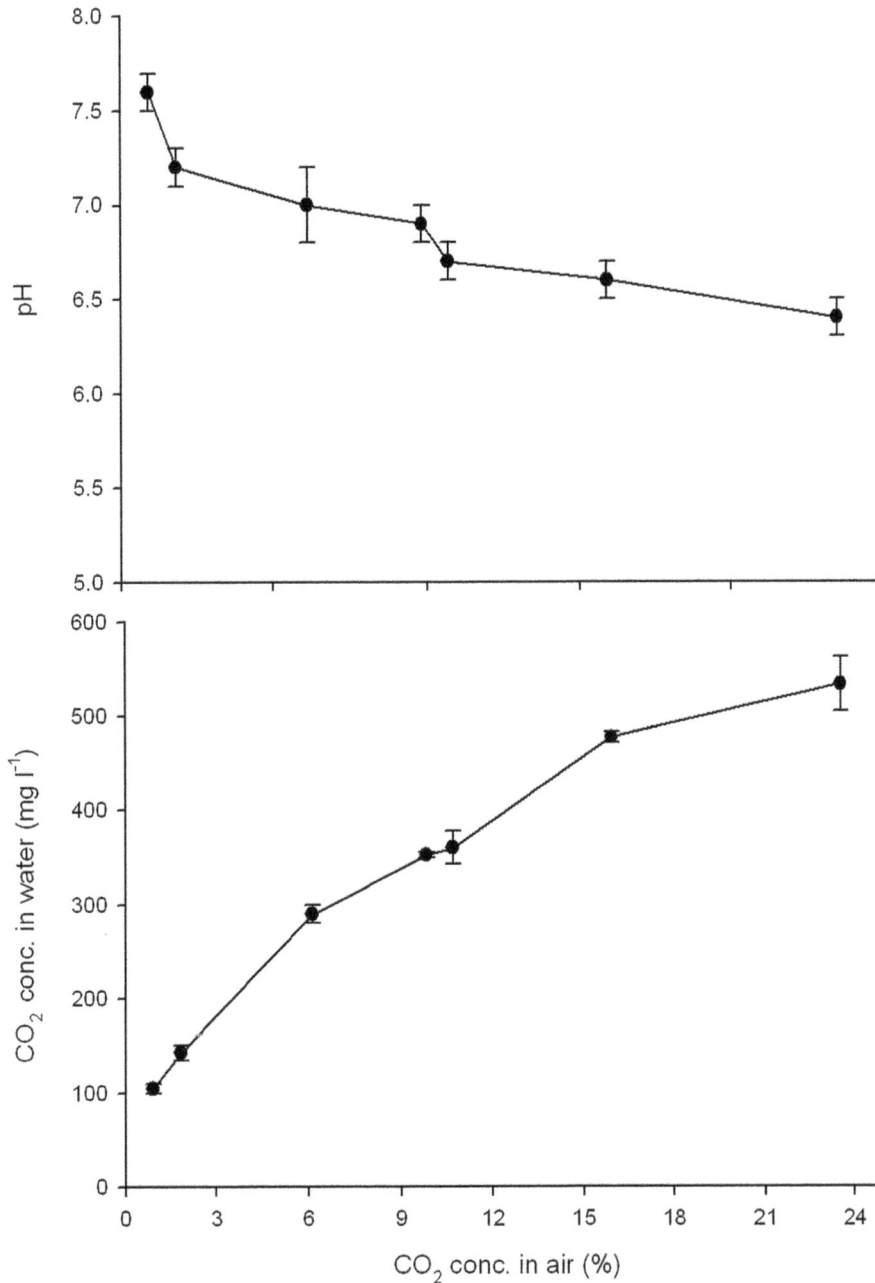

Figure 2 The concentration of CO_2 dissolved in the culture and pH of the growth medium as influenced by the CO_2 concentration in the air bubbled through the bottles (means, ±SE).

day (Table 2). At low PFD, the algae concentration continued to grow until the fifth day. The dry weight production during the first three days was significantly lower in the 11.4% CO_2 flue gas treatment (about 20%) than in the 6.7% and 2.5% CO_2 flue gas treatments at 300 µmol m^{-2} s^{-1} PFD, while the decrease was less (5-10%) in the low PFD treatment. The dry weights at the two lower flue gas concentrations were slightly higher compared with the 2.6% control CO_2 treatment using pure liquid gas. The dry weight production per day was about four times higher at 300 than at 75 µmol m^{-2} s^{-1} PFD, reaching about 70 g m^{-2} day^{-1}. The dry weight produced per mol photosynthetic active photons was the same at both PFD levels. Increasing the flue gas concentration slightly decreased the O_2 content and decreased the pH in the algae culture. Increasing the light level slightly increased the O_2 content and decreased the pH.

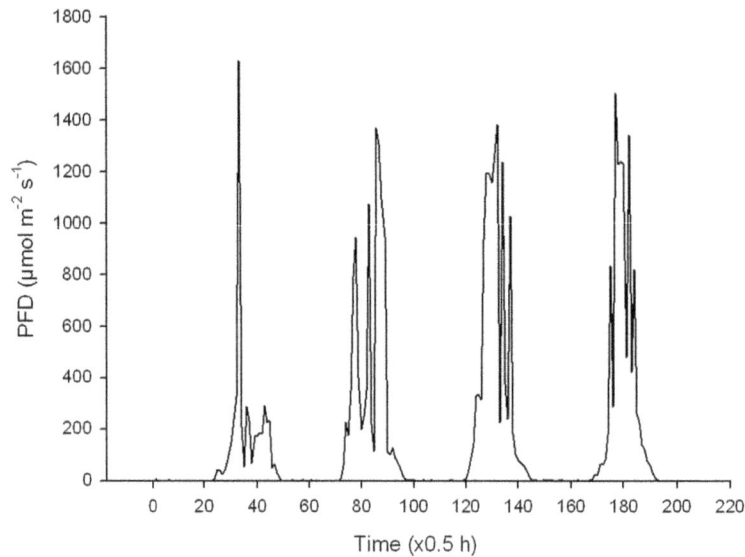

Figure 3 The photon flux density (PFD) of daylight during the experimental period.

Experiment 2

The dry weight concentration increased significantly from the fourth to the fifth day in this experiment when the algae were grown at 300 μmol m^{-2} s^{-1} PFD and 24±2°C (Table 3). The dry weight production was significantly lower (about 25%) at the highest flue gas concentration compared with the other treatments. The increase in algae concentration from 1.6 to 2.9 g l^{-1} from the fourth to the fifth day resulted in an algal production of around 80 g m^{-2} day^{-1} in the different treatments, except in the 11.4%

CO$_2$ flue gas treatment, where the production was around 60 g m^{-2} day^{-1}. At the end of the experiment, the pH decreased from 6.8 to 6.0 when the flue gas concentration was increased from the lowest to the highest level.

Experiment 3

Reducing daylight by 30% shade had no significant effect on the growth of the algae (Table 4). The dry weight production was 12-14 g m^{-2} day^{-1} as a mean during four days. In this experiment, the dry weight production was

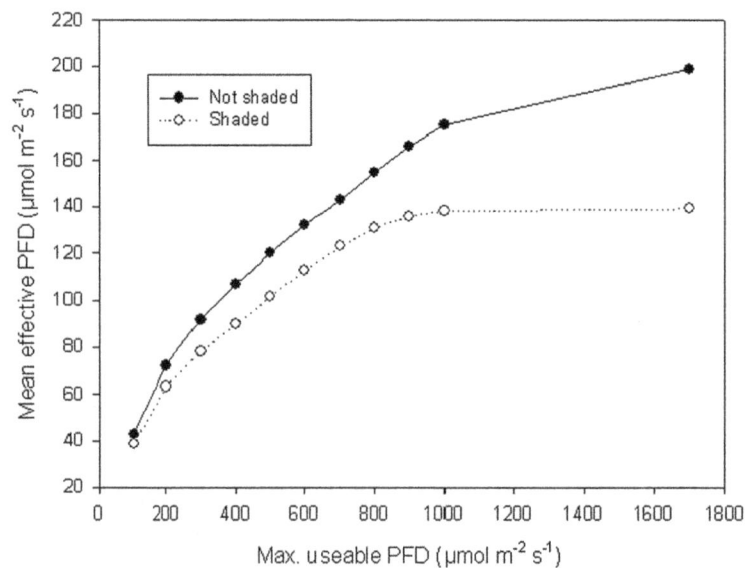

Figure 4 Mean effective PFD at different threshold values of daylight useable for the growth of the microalgae. PFD values above the threshold value were set to the threshold value, i.e. if PFD values above 300 μmol m^{-2} s^{-1} are recognised as having no effect, the effective PFD was set to 300 μmol m^{-2} s^{-1}.

Table 2 The effect of different CO_2 concentrations supplied by flue gas (Fl) and one concentration supplied by pure liquid CO_2 gas (C) on pH, O_2 concentration in the culture and dry weight concentration (n=3, ±SE) after 3 and 5 days of C. reinhardtii grown at 75 and 300 µmol m^{-2} s^{-1} PFD

| CO_2 treatment | PFD | Day 3 | | | Day 5 | | Mean dry weight production | | |
| | | | | | | | Day 0 - 3 | | |
		pH	O_2 (mg l^{-1})	Dry w. (mg l^{-1})	pH	Dry w. (mg l^{-1})	mg l^{-1} day^{-1}	g m^{-2} day^{-1}	g mol^{-1}
2.6% C	75	7.1±0.0	7.5±0.2	798±27	6.9±0.1	1360±149	266±9	16.2±0.6	2.50±0.09
2.6% C	300	7.0±0.1	8.3±0.1	3297±46	6.8±0.1	3040±94	1099±15	67.0±0.9	2.59±0.04
2.5% Fl	75	7.3±0.1	7.8±0.4	877±84	7.1±0.1	1338±88	292±28	17.8±1.7	2.74±0.26
2.5% Fl	300	7.0±0.0	8.5±0.1	3527±48	6.8±0.1	3423±91	1175±16	71.7±0.9	2.77±0.04
6.7% Fl	75	7.0±0.1	7.0±0.3	845±36	6.6±0.1	1343±70	282±12	17.2±0.8	2.64±0.12
6.7% Fl	300	6.7±0.0	7.5±0.2	3663±124	6.4±0.2	3343±91	1221±41	74.5±2.5	2.89±0.10
11.4% Fl	75	6.8±0.1	6.4±0.2	777±14	6.5±0.1	1147±23	259±5	15.8±0.3	2.43±0.04
11.4% Fl	300	6.7±0.1	7.3±0.2	2933±36	6.5±0.1	2843±45	978±12	59.6±0.7	2.30±0.03
F-value and significance level:									
CO_2		64.5***	50.9***	17.6***	25.6***	3.54*	17.6***	17.6***	5.73**
PFD		91.1***	77.4***	3512***	10.7**	403***	3511***	3511***	0.51
CO_2 x PFD		3.79*	0.59	10.9***	1.32	1.25	10.9***	10.9***	0.89

The productivity during the first three days was calculated as the increase in culture concentration, in g dry weight production per m^2 and day, as well as in g dry weight produced per mol of photosynthetic active radiation. F-values and significance levels are stated as follows: *p<0.05; **p<0.01; ***p<0.001.

10-20% higher in the 2.5% CO_2 flue gas treatment than in the other treatments. The production per mol photons was increased 40-50% by 30% shading.

Discussion

The undiluted flue gas containing 11.4% CO_2 caused a decrease in the dry weight production compared with lower flue gas concentrations (2.5 and 6.7%). This was particularly the case when the dry weight production was very high (up to 70-80 g m^{-2} day^{-1}), obtained at 300 µmol m^{-2} s^{-1} PFD continuously applied (25.9 mol m^{-2} day^{-1} PAR). In low-light conditions, (continuously 75 µmol m^{-2} s^{-1} PFD or 6.5 mol m^{-2} day^{-1} PAR) or in sunny daylight with a day length of 12.5 h (17.2 mol m^{-2} day^{-1} PAR) when the growth rate was much lower, less or no negative effect was found of the undiluted flue gas. The

question was whether the negative effect was related to the high CO_2 concentration itself or to the accompanying air pollutants. Separate measurements indicated that the dissolved CO_2 concentration in the culture with undiluted flue gas might be about 400 mg l^{-1} as compared with about 150 mg l^{-1} in diluted flue gas with a concentration of 2.5% CO_2. This is far below the saturating level of CO_2 in water that is about 1500 mg l^{-1} at 23°C and 1200 mg l^{-1} at 33°C. The present pollutant levels of NO_x and SO_2 below about 50 mg m^{-3} in the flue gas seldom seem to cause growth reduction in microalgae (Matsumoto et al. 1997; Douskova et al. 2010; van den Hende et al. 2012; Farrelly et al. 2013; Jiang et al. 2013). Other flue gas compounds such as CO, HCl, HF and heavy metals such as Hg have received little attention so far (van den Hende et al. 2012). Probably the concentrations

Table 3 The effect of different CO_2 concentrations supplied by flue gas (Fl) and one concentration supplied by pure liquid CO_2 gas (C) on pH and dry weight concentration (n=3, ±SE) after 4 and 5 days of C. reinhardtii grown at 300 µmol m^{-2} s^{-1} PFD

| | Day 4 | | Day 5 | | Dry weight increase | | |
	pH	Dry w. (mg l^{-1})	pH	Dry w. (mg l^{-1})	(mg l^{-1} day^{-1})	g m^{-2} day^{-1}	g mol^{-1}
2.6% C	7.1±0.1	1457±56	6.9±0.1	2810±141	1353±92	82.6±5.6	3.19±0.22
2.5% Fl	7.0±0.0	1622±21	6.8±0.1	2943±48	1321±51	80.6±3.1	3.11±0.12
6.7 Fl	6.5±0.1	1635±21	6.3±0.1	2917±100	1282±79	78.2±4.9	3.02±0.19
11.4% Fl	6.4±0.1	1377±42	6.0±0.3	2380±81	1003±54	61.2±3.3	2.36±0.13
F-value and significance level:							
CO_2	49.8***	17.6***	20.4***	7.31*	5.24*	5.24*	5.24*

The productivity from day four to five was calculated as the increase in culture concentration, in g dry weight production per m^2 and day, as well as in g dry weight produced per mol of photosynthetic active radiation.
For significance levels see Table 2 footnote.

Table 4 The effect of different CO_2 concentrations supplied by flue gas (Fl) and one concentration supplied by pure liquid CO_2 gas (C) on pH and dry weight concentration (n=3, ±SE) after four days of *C. reinhardtii* grown in daylight or 70% daylight (shaded)

CO_2 treatment	Light	pH	Dry w. (mg l^{-1})	Dry weight increase	
				g m^{-2} day^{-1}	g mol^{-1}
2.6% C	Shaded	7.19±0.1	815±45	11.6±0.6	1.00±0.06
2.6% C	Daylight	7.1±0.0	762±43	12.4±0.7	0.72±0.04
2.5% Fl	Shaded	7.1±0.1	960±18	14.3±0.7	1.23±0.06
2.5% Fl	Daylight	7.1±0.1	935±45	14.6±0.3	0.85±0.02
6.7% Fl	Shaded	6.8±0.1	867±83	11.9±0.7	1.03±0.06
6.7% Fl	Daylight	6.7±0.1	780±46	13.2±1.3	0.76±0.07
11.4% Fl	Shaded	6.5±0.1	792±32	12.1±0.6	1.05±0.05
11.4% Fl	Daylight	6.4±0.2	797±41	12.1±0.5	0.70±0.03
F-value and significance level:					
CO_2		110***	5.14*	5.14*	5.31**
Light		3.00	1.49	1.49	79.9***
CO_2 x Light		1.00	0.36	0.36	0.63

The mean productivity during the four days of the experimental period was calculated as g dry weight production per m^2 and day and as g dry weight produced per mol of photosynthetic active radiation.
For significance levels see Table 2 footnote.

in the present flue gas were so low that they would have no effect on the growth. However, microalgae possess very high metal uptake capacities and accumulation in the cells will therefore take place (de-Bashan and Bashan 2010). High CO_2 concentrations (18-19%) from pure liquid CO_2 gas, however, have recently been found to decrease the dry weight production in the same *C. reinhardtii* strain (Mortensen and Gislerød 2014). Fischer et al. (2006) showed that cells of the same species were more susceptible to high-light stress under high CO_2 concentrations than under low concentrations. In the present study, however, the negative effect of the high concentrations seemed to be more related to a high growth rate than to high-light conditions. It can also be noted that the maximum dry weight concentration reached in the algae culture in the flue gas decreased to the same extent (in percentage) as the dry weight production, indicating higher respiration or lower photosynthetic activity in the algae. The negative effect of the 11.4% flue gas in the present experiment was in contrast to the stimulating effect of flue gas, probably due to lower O_2 content, found in some studies on microalgae (Vance and Spalding 2005; Douskova et al. 2009; Kliphuis et al. 2011). Growing *Chlorella sp.* at 2-20% CO_2 (v/v) simulating flue gas from biogas gave the same effect as growing the algae in food grade CO_2 at the same concentrations (Douskova et al. 2010). The environmental conditions could play a role here, and they might also be the reason for the positive effect of the moderate flue gas concentration with 2.5% CO_2 in the present experiment in daylight.

The production at low-level light 24 h day^{-1} (6.5 mol m^{-2} day^{-1} PAR) was at the same level (around 14 g m^{-2} da^{-1}) as at about a three times higher PAR in daylight, which demonstrates the limitation of the algae as regards utilising the high irradiance level. The productivity in daylight was typical of outdoor production systems and the high productivity was typical of controlled environmental conditions in laboratories (Grobbelaar 2012). The light use efficiency in the present study was found to be the same in the range 75-300 μmol m^{-2} s^{-1} PFD. If we assume that all daylight above 300 μmol m^{-2} s^{-1} PFD has a value of 300 μmol m^{-2} s^{-1}, the mean PFD of the daylight will decrease from 199 to about 90 μmol m^{-2} s^{-1} or 7.8 mol m^{-2} day^{-1} PAR. This level is comparable to the low-light level with artificial light applied 24 h day^{-1}. In addition to the constraint caused by light saturation, the presence of a dark period is known to decrease algae growth much more than would be expected from the reduction in PAR (Jacob-Lopez et al. 2009). This means that long day lengths and lower maximum irradiance levels at high latitudes would be beneficial for algae production during the summer months. However, short days and low PAR during large parts of the year make the production of algae impractical in such locations. Growing *C. reinhardtii* with the aim of using it to produce hydrogen should be based on using daylight in combination with flue gas in order to ensure a positive energy balance (Lam et al. 2012). However, large-scale systems that can utilise the high irradiance levels of daylight much better than today (Slegers et al. 2013) are a prerequisite for future energy-efficient hydrogen production using microalgae. Flue gas is an important CO_2 source. However, while care should be taken to ensure a CO_2 concentration that is optimal, the presence of pollutants in the flue gas in today's industrial emissions seems to be less of a problem in relation to the growth of the algae.

Competing interests

The authors declare that they have no competing interests.

Acknowledgements

The authors thank 'Borregaard Waste to Energy' with Jørgen Karlsen and staff in Sarpsborg for their hospitality, excellent assistance and for providing the records of the flue gas measurements. This work was done as a part of the project 'Use of solar energy for CO_2 capture, algae cultivation and hydrogen production' headed by Dr Stig Borgvang (Bioforsk). It was financed by the Research Council of Norway.

References

Borkenstein CG, Knoblechner J, Frühwirth H, Schagert M (2011) Cultivation of *Chlorella emersonii* with flue gas derived from cement plant. J Appl Phycol 23:131–135

de-Bashan LE, Bashan Y (2010) Immobilized microalgae for removing pollutants. Revew of practical aspects. Bioresour Technol 101:1611–1627

Douskova I, Doucha J, Livansky K, Machat J, Novak P, Umysova D, Zachleder V, Vitova M (2009) Simultaneous flue gas bioremediation and reduction of microalgal biomass production costs. Appl Microbiol Biotechnol 82:179–185

Douskova I, Kastanek F, Maleterova Y, Kastanek P, Doucha ZV (2010) Utilization of distillery stilage for energy generation and concurrent production of valuable microalgal biomass in the sequence: Biogas-cogeneration-microalge-products. Energy Conversion Management 51:606–611

Farrelly DJ, Everard CD, Fagan CC, McDonnell KP (2013) Carbon sequestration and the role of biological carbon mitigation: a review. Renew Sust Energ Rev 21:712–727

Fischer BB, Wiesendanger M, Eggen RIL (2006) Growth condition-dependent sensitivity, photodamage and stress response of *Chlamydomonas reinhardtii* exposed to high light conditions. Plant Cell Physiol 47:1135–1145

Geier SC, Huyer S, Praebst K, Husmann M, Walter C, Buchholz R (2012) Outdoor cultivation of *Chlamydomonas reinhardtii* for photobiological hydrogen production. J Appl Phycol 24:319–327

Gorman DS, Levine RP (1965) Cytochrome f and plastocyanin: their sequence in the photosynthetic electron transport chain of *Chlamydomonas reinhardtii*. Proc Natl Acad Sci U S A 54(6):1665–1669

Grobbelaar JU (2012) Microalgae mass culture: the constraints of scaling-up. J Appl Phycol 24:315–318

IPCC (2013) Climate Change 2013: The Physical Science Basis. Contribution of Working Group I to the Fifth Assessment Report of the Intergovernmental Panel on Climate Change. In: Stocker TF, Qin D, Plattner G-K, Tignor M, Allen SK, Boschung J, Nauels A, Xia Y, Bex V, Midgley PM (eds). Cambridge University Press, Cambridge, United Kingdom and New York, NY, USA, www.climate change2013.org

Jacob-Lopez AJ, Scoparo CHG, Lacerda LMCF, Franco TT (2009) Effect of light cycles (night/day) on CO_2 fixation and biomass production by microalgae in photobioreactors. Chem Eng Process 48:306–310

Jiang Y, Zhang W, Wang J, Chen Y, Shen S, Liu T (2013) Utilization of simulated flue gas for cultivation of *Scenedesmus dimorphus*. Bioresour Technol 128:359–364

Jo HJ, Lee DS, Park JM (2006) Modeling and optimization of photosynthetic hydrogen gas Production by green alga Chlamydomonas reinhardtii in sulphur-deprived circumstance. Biotechnol Prog 22:431–437

Kastanek F, Sabata S, Solcova O, Maleterova Y, Kastanek P, Branyikova I, Kuthan K, Zachleder V (2010) In-field experimental verification of cultivation of microalgae *Chlorella* sp. using flue gas from cogeneration unit as a source of carbon dioxide. Waste Management and Research 28(Kastanek F, Sabata S, Solcova O, Maleterova Y, Kastanek P, Branyikova I, Kuthan K, Zachleder V):961–966

Kliphuis AMJ, Martens DE, Janssen M, Wijffels RH (2011) Effect of O_2:CO_2 ratio on the primary metabolism of *Chlamydomonas reinhardtii*. Biotechnol And Bioengineering 108:2390–2402

Lam MK, Lee KT, Mohamed AR (2012) Current status and challenges on microalgae-based capture. Int J Greenhouse Gas Contr 10:456–469

Matsumoto H, Hamasaki A, Sioji NIkuta Y (1997) Influence of CO_2, SO_2 and NO in flue gas on microalgae productivity. J Chem Eng Japan 30:620–624

Mortensen LM (1986) Nitrogen oxides produced during CO2 enrichment III. Effects on tomato at different photon flux densities. New Phytol 104:653–660

Mortensen LM, Gislerød HR (2014) The growth of *Chlamydomonas reinhardtii* as influenced by high CO_2 and low O_2 in flue gas from a silicomanganese smelter. J Appl Phycol (in press)

Nguyen AV, Toepel J, Burgess S, Uhmeyer A, Bilfernez O, Doebbe A, Hankamer B, Nixon P, Wobbe L, Kruse O (2011) Time-course global expression profiles of *Chlamydomonas reinhardtii* during photo-biological H_2 production. PLoS ONE 6(12):e29364, doi:10.1371

Skjånes K, Lindblad P, Muller J (2007) BioCO$_2$ – a multidisciplinary, biological approach using solar energy to capture CO_2 while producing H_2 and high value products. Biomol Eng 24:405–413

Slegers PM, van Beveren PJM, Wijffels RH, van Straten G, Boxtel AJB (2013) Scenario analysis of large scale algae production in tubular photobioreactors. Appl Energy 105:395–406

Sueoka N (1960) Mitotic replication of deoxyribonucleic acid in Chlamydomonas reinhardii. Proc Natl Acad Sci U S A 46:83–91

van den Hende S, Vervaeren H, Boon N (2012) Flue gas compounds and microalgae: (Bio-) chemical interactions leading to biotechnological opportunities. Biotechn Advances 30:1405–1424

Vance P, Spalding MH (2005) Growth, photosynthesis, and gene expression in *Chlamydomonas* over a range of CO_2 concentrations and CO_2/O_2 ratios: CO_2 regulates multiple acclimation states. Can J Botany 83:796–809

3

Re-assessment of *YAP1* and *MCR1* contributions to inhibitor tolerance in robust engineered *Saccharomyces cerevisiae* fermenting undetoxified lignocellulosic hydrolysate

Valeria Wallace-Salinas[1†], Lorenzo Signori[2†], Ying-Ying Li[3,4], Magnus Ask[5], Maurizio Bettiga[5], Danilo Porro[2], Johan M Thevelein[3,4], Paola Branduardi[2], María R Foulquié-Moreno[3,4] and Marie Gorwa-Grauslund[1*]

Abstract

Development of robust yeast strains that can efficiently ferment lignocellulose-based feedstocks is one of the requirements for achieving economically feasible bioethanol production processes. With this goal, several genes have been identified as promising candidates to confer improved tolerance to *S. cerevisiae*. In most of the cases, however, the evaluation of the genetic modification was performed only in laboratory strains, that is, in strains that are known to be quite sensitive to various types of stresses. In the present study, we evaluated the effects of overexpressing genes encoding the transcription factor (*YAP1*) and the mitochondrial NADH-cytochrome b5 reductase (*MCR1*), either alone or in combination, in an already robust and xylose-consuming industrial strain of *S. cerevisiae* and evaluated the effect during the fermentation of undiluted and undetoxified spruce hydrolysate. Overexpression of either gene resulted in faster hexose catabolism, but no cumulative effect was observed with the simultaneous overexpression. The improved phenotype of *MCR1* overexpression appeared to be related, at least in part, to a faster furaldehyde reduction capacity, indicating that this reductase may have a wider substrate range than previously reported. Unexpectedly a decreased xylose fermentation rate was also observed in *YAP1* overexpressing strains and possible reasons behind this phenotype are discussed.

Keywords: *Saccharomyces cerevisiae*; Hydrolysate; Inhibitors; *YAP1*; *MCR1*; Ethanol

Introduction

Production of second-generation bioethanol from lignocellulosic biomass requires robust *Saccharomyces cerevisiae* strains with improved capacity to cope with the toxic compounds formed during the biomass pre-treatment, among which are 5-hydroxymethylfurfural (HMF), furfural, weak organic acids and phenolic compounds (Parawira and Tekere 2011). This has led to extensive studies to decipher mechanisms behind the compounds toxicity and the yeast natural tolerance responses to them and, among others, genes involved in detoxification and yeast tolerance to individual inhibitors have been identified, such as *ADH6*,

HAA1 or *PMA1* (Haitani et al. 2012; Mira et al. 2010; Petersson et al. 2006); for a more exhaustive review, see (Almeida et al. 2009a; Liu 2011). *YAP1* is another interesting candidate for industrial strain engineering because it encodes a transcription factor (Yap1p) that simultaneously controls a wide range of stress-related targets (Toone and Jones 1999). Notably, its overexpression has a beneficial role in the response of laboratory *S. cerevisiae* towards HMF, furfural, and different concentrations of hydrolysate (Alriksson et al. 2010; Kim and Hahn 2013; Ma and Liu 2010; Sundström et al. 2010). Another interesting and complementary candidate for gene overexpression is *MCR1* that encodes the mitochondrial NADH-cytochrome b5 reductase (Hahne et al. 1994; Meineke et al. 2008). Previous experiments performed in our group revealed that overexpression of *MCR1* in *S. cerevisiae* resulted in a reduced lag

* Correspondence: Marie-Francoise.Gorwa@tmb.lth.se
†Equal contributors
[1]Applied Microbiology, Department of Chemistry, Lund University, P.O. Box 124, SE-22100 Lund, Sweden
Full list of author information is available at the end of the article

phase and faster growth rate when the yeast was grown with high concentrations of acetic acid (Signori et al., personal communication). This weak acid is one of the inhibitors directly affecting xylose metabolism in *S. cerevisiae* (Almeida et al. 2011; Bellissimi et al. 2009; Casey et al. 2010; Helle et al. 2003). Still, considering that many of the studies about strain improvement towards hydrolysate-derived inhibitors concern laboratory strains, it is difficult to predict the real effect of these changes in an industrial, and more robust, strain background.

The objective of the present study was to evaluate the effect of overexpressing *YAP1* and *MCR1*, either alone or in combination, in process-like conditions, that is using a robust industrial *S. cerevisiae* strain and undetoxified lignocellulosic hydrolysate. For this, the strain GSE16 was chosen as background strain for engineering since it combines a robust industrial background with the ability to ferment xylose, using the xylose isomerase pathway (Demeke et al. 2013b). This robust strain was developed by a combination of different strategies including rational metabolic engineering, mutagenesis, evolutionary engineering, genome shuffling and meiotic recombination (Demeke et al. 2013a,b). Moreover, and as part of the genetic engineering strategy used in the current study, the overexpression of *YAP1* was combined with the deletion of the chaperone-encoding gene *APJ1* since deletion of this gene has been previously reported to enable growth on rich medium at inhibitory ethanol concentrations for the parental strain (Swinnen et al. 2012). The relevance of overexpressing *YAP1* and *MCR1* in an industrial background was confirmed for undetoxified spruce hydrolysate fermentation. We also uncovered unexpected interactions between *YAP1* overexpression and xylose metabolism.

Materials and methods

Strains

S. cerevisiae strains utilized in this study are presented in Table 1. *Escherichia coli* DH5α and *E. coli* NEB 5-alpha were used for sub-cloning and were grown on Luria-Bertani (LB) medium supplemented with 100 mg.L^{-1} ampicillin, when required. Plasmids utilized in the study are described in Table 2.

Molecular biology methods

Standard molecular biology methods were used for all cloning procedures. (Sambrook and Russel 2001). Thermo-Scientific GeneJET plasmid miniprep kit (ThermoScientific, Lithuania) was used for plasmid extraction. E.Z.N.A Cycle-Pure Kit (Omega Biotek, USA) was used for purification of polymerase chain reaction (PCR) products. Qiagen Qiaquick gel extraction kit (Qiagen GmbH, Germany) was used to extract DNA from agarose gels. All DNA-modifying enzymes were purchased from ThermoScientific. Primers for PCR and sequencing of DNA constructs were ordered from and performed by MWG (MWG-Biotech AG, Germany). All primers are shown in Additional file 1: Table S1. Transformation of *E. coli* was performed using the Inoue Method (Sambrook and Russel 2001). Either the lithium acetate (LiAc) method (Gietz et al. 1995), a modified version of it that uses dimethyl sulfoxide (DMSO) (Hill et al. 1991) or electroporation (Benatuil et al. 2010), were used as transformation methods of *S. cerevisiae*.

Construction of plasmids pJET1,2-attB-KanMX-attP, pJET1,2-attB-hph-attP and pBEVY-Nat-PhiC31

The antibiotic markers KanMX and hph expressed under the TEF promoter and terminator were amplified with the primers Fw-A1-attP-tefpr and Rv-A2-attB-teft and cloned into pJET1,2 (Thermo Scientific, Belgium) following the protocol of the kit. The resulting plasmids were called pJET1,2-attB-KanMX-attP and pJET1,2-attB-hph-attP respectively. The PhiC31 integrase was amplified from pCMVInt (Addgene, Cambridge, Massachusetts, USA) using the primers Fw-PstI-PhiC31 and Rv-BamHI-PhiC31 and cloned into the *Pst*I and *Bam*HI sites of pBEVY-Nat giving the pBEVY-Nat-PhiC31.

Table 1 *Saccharomyces cerevisiae* strains used in the current work

S. cerevisiae strains	Genotype	Plasmid	Source
CEN.PK 102-5b	*Mat a, ura 3–52, his2Δ1, leu 2-3/112*		(van Dijken et al. 2000)
CEN.PKc	CEN.PK 102-5b	[pYX012; pYX022; pYX042]	This work
TMB3400	*S. cerevisiae* industrial strain		(Almeida et al. 2009b)
GSE16	GS1.11-26 + backcrossing with a segregant of Ethanol Red that is tolerant towards acetic acid; *MATα/α*		(Demeke et al. 2013b)
GSE16 - YAP1	GSE16-*APJ1-1:: TDH3p-YAP1-CYC1*t		This work
GSE16 - MCR1	GSE16- *YLR446W-1::TPIp-MCR1*		This work
GSE16 - MCR1-YAP1	GSE16- *YLR446W-1::TPIp-MCR1, APJ1-1:: TDH3p-YAP1-CYC1*t		This work
GSE16 - ΔΔAPJ1	GSE16-APJ1/APJ1::attL/attL		This work

Table 2 Plasmids used in the current work

Plasmids	Relevant features	Origin
p426GPD	*Multicopy URA3* 2 μm *TDH3*p-*CYC1*t	(Mumberg et al. 1995)
p426GPD –YAP1	*TDH3*p-*YAP1*-*CYC1*t	This work
pUG6	*kanMX* flanked by *loxP* sites	(Güldener et al. 1996) (EUROSCARF; accession number P30114)
YE-plac 112 KanR	Multicopy, KanMX	(Jeppsson et al. 2003)
YE-plac 112 KanR-YAP1	*TDH3*p-*YAP1*-*CYC1*t	This work
pJET1,2	Multicopy	Thermo Scientific, Belgium
pJET1,2-attB-KanMX-attP	Multicopy, KanMX under TEFp	This work
p-intYAP1	*TDH3*p-*YAP1*-*CYC1*t – KanMX attB/attP system; integrative	This work
pSTBlue-1	multi-purpose cloning vector with dual kanamycin/ampicillin resistance.	Novagen (EMD Millipore)
pSTBlue-YLR446W	pSTBlue-1 with the *YLR446W* gene cloned into the multiple cloning region	This work
pYX012	Integrative; *URA3*; TPI1p	R&D System, Inc., Wiesbaden, D
pYX012-LoxPkanMXLoxP	pYX012, with the KanMX cassette flanked by loxP sites, deriving from pUG6	This work
pYX012-LoxPKanMXLoxP-MCR1	pYX012- LoxPKanMXLoxP with the *MCR1* gene inserted in the MCS under the control of the TPI1 promoter	This work
pSTBlue-YLR446WΔ- LoxPKanMXLoxP-MCR1	pSTBlue-YLR446W with the LoxPKanMXLoxP-MCR1 cassette, deriving from pYX012 LoxPKanMXLoxP-MCR1, inserted into the YLR446W sequence	This work
pSH65	Centromeric plasmid, GAL1p-cre, ble^r	(Gueldener et al. 2002) (EUROSCARF; accession number P30122)
pJET1,2- attB-hph- attP	Multicopy, hph under TEFp	This work
pBEVY-Nat-phiC31	Multicopy, PhiC31 integrase	This work

Construction of S. cerevisiae strains

S. cerevisiae GSE16-YAP1

The open reading frame of the *YAP1* gene was amplified from *S. cerevisiae* strain TMB3500 (Almeida et al. 2009b), using primers YAP1-F and YAP1-R. The amplicon was ligated into p426GPD (Mumberg et al. 1995) resulting in p426GPD-YAP1 used for transformation of *E. coli* DH5α cells and followed by sequence verification. The cassette from p426GPD-YAP1 was amplified using primers GPD-YAP1-F and CYC1t-YAP1-R. The amplified cassette was ligated into pUG6 after restriction with *Aat*II and *Pvu*II, resulting in pUG6-YAP1. Amplification of the two homologous regions (HR) used for integration of the *YAP1* cassette into the *APJ1* locus were performed from *S. cerevisiae* GSE16 (Demeke et al. 2013b) with primers HR1-F/HR1-R and HR2-F/HR2-R. Amplicon for RH2 was first ligated into pUG6-YAP1 and transformed into *E. coli* DH5α cells resulting in pYAP1-HR2. Amplicon for HR1 was ligated into pYAP1-HR2 and transformed into *E. coli* DH5α cells resulting in pYAP1-HR2-HR1. The selection cassette (KanMX) flanked by the attB and attP sites was amplified from pJET1,2-attB-KanMX-attP using attBP-F and attBP-R. The amplicon was ligated into pYAP1-HR2-HR1 and transformed into *E. coli* NEB 5-alpha resulting in p-intYAP1. The nucleotide sequence of each amplicon

was verified after every subsequent cloning step. *S. cerevisiae* strain GSE16 was used for expression of the *YAP1* construct. In this strain, the expression cassette containing the transcription factor *YAP1* was integrated by linearization of the integrative cassette using *Aat*II and *Not*I, followed by transformation using a DMSO-modified version of the LiAc method (Hill et al. 1991). The selection of colonies was done on YNB plates with 150 μg.mL^{-1} geneticin G418 (Sigma). Verification of the correct insertion (*APJ1* locus) was done by sequencing using primers Ver. ins1-F/Ver.ins1-R and Ver.ins2-F/Ver.ins2-R. The resulting strain was named GSE16-YAP1.

S. cerevisiae GSE16-MCR1

The open reading frame of *MCR1* was amplified from CEN.PK 102-5b using primers MCR1-F and MCR1-R. The amplified DNA fragment (1544 bp) was cloned into pSTBlue (*Eco*RV site) and used to transform *E. coli* DH5 α cells. A 1.5 Kb *Eco*RI fragment containing *MCR1* was isolated from pSTBlue-MCR1 and cloned into the MCS of the yeast integrative plasmid pYX012, resulting in pYX012-MCR1. In parallel, the ORF YLR446W was amplified from *S. cerevisiae* CEN.PK 102-5b using primers YLR446W-F and YLR446W-R and cloned into pSTBlue (*Eco*RV site) resulting in pSTBlue-YLR446W. The selection cassette (KanMX) flanked by two LoxP sites was amplified from pUG6 using primers YLR446W Lox-F and Lox-R, and cloned into pYX012-MCR1 plasmid (*Kpn*I site). The expression cassette (LoxPKanMXLoxP + promoter + *MCR1* ORF + terminator) was amplified using primers YLR446W Lox-F and YLR446W TER-R and cloned into pSTBlue-YLR446W after restriction with *Btg*I and *Bsr*GI (this double digestion allowed the removal of the inner part of the *YLR446W* gene (~601 bp)). Each amplicon was verified by sequencing analysis after every subsequent cloning step. *S. cerevisiae* strain GSE16 was used for expression of the *MCR1* construct. The expression cassette containing *MCR1* was integrated into GSE16 after PCR amplification using primers YLR446W-F and YLR446W-R. Transformation was carried out using a DMSO-modified version of the LiAc method (Hill et al. 1991). Correct insertion in the *YLR446W* locus was verified by PCR. The resulting strain was named GSE16-MCR1. The removal of the dominant marker (KanMX) was obtained by transforming GSE16-MCR1 with pSH65.

S. cerevisiae GSE16-MCR1-YAP1

The strain was constructed from GSE16-MCR1 by integration of p-intYAP1 previously digested with *Aat*II and *Not*I (see *S. cerevisiae* strain overexpressing *YAP1*). The correct integration in the *APJ1* locus was verified by PCR.

S. cerevisiae GSE16- ΔΔAPJ1

The selection cassettes (KanMX and hph) flanked by the attB and attP sites were amplified by PCR from pJET1,2-attB-KanMX-attP and pJET1,2- attB-hph- attP using primers Fw-APJ1-A1 and Rv-APJ1-A2 with 50 bp homologues regions. Deletion of the two *APJ1* alleles was carried out by integrating both selection cassettes into the *APJ1* loci of GSE16 (Demeke et al. 2013b) using an adapted electroporation method (Benatuil et al. 2010). *APJ1* double deletion colonies were selected from YPD plates with hygromycin (300 μg.mL^{-1}) and geneticin (G418) (200 μg.mL^{-1}), and checked by PCR using the primers Fw-APJ1-check and Rv-APJ1-check. To loop out the markers, the colonies were transformed with pBevy-Nat-phi31 and selected in YPD plates with 100 μg.mL^{-1} of nourseotricin. These colonies were also checked on YPD hygromycin and YPD geneticin plates. The plasmids containing the integrase were lost by growing the colonies in YPD liquid medium overnight and transferred twice. The resulting strain was named GSE16-ΔΔAPJ1.

Spruce hydrolysate Spruce hydrolysate was obtained from SEKAB E-Technology AB (Örnsköldsvik, Sweden), and consisted of the non-detoxified liquid fraction of spruce after a pretreatment by SO_2 catalyzed steam explosion. It is referred in this work as spruce hydrolysate and had the following sugar composition: 11 g.L^{-1} glucose,

17 g.L^{-1} mannose, 4 g.L^{-1} galactose, and 10 g.L^{-1} xylose. All the batch fermentations were carried out with the same batch of spruce hydrolysate which was kept at 4°C.

Anaerobic batch fermentations of spruce hydrolysate Inoculum was prepared by growing the cells overnight in 1 L shake flasks containing 100 mL defined mineral medium with vitamins (Verduyn et al. 1992) with glucose as carbon source (20 g.L^{-1}) buffered with phthalate buffer (50 mM, pH 5.0). After centrifugation and a washing step with deionized water, the pellet was resuspended with 20 mL of fermentation medium, and immediately used for inoculation of the fermenter. The fermentation medium consisted of 100% (v/v) spruce hydrolysate supplemented with 10 g.L^{-1} of xylose, 1 g.L^{-1} yeast extract, 0.5 g.L^{-1} $(NH_4)_2HPO_4$ and 0.025 g.L^{-1} $MgSO_4$. $7H_2O$. The pH of the hydrolysate was adjusted to 5.0 with 8 M KOH prior to supplementation. The fermentations were carried out in 1.2 L MultiFors fermenters with 0.5 L working volume. Temperature was maintained at 30°C and the pH was kept at 5.0 by addition of 3 M KOH. Oxygen free conditions were maintained by sparging N_2 at 0.2 L.min^{-1}, and the agitation was set to 600 rpm. Reactors were inoculated to a biomass concentration of 1 g.L^{-1} (cdw). Cultivations were performed in biological duplicates for each investigated strain. Specific conversion rates of furfural and HMF were calculated assuming a pseudo-steady state during the exponential growth on glucose.

Anaerobic batch fermentations in mineral medium Inoculum was prepared by growing the cells overnight in 100 mL shake flasks containing 20 mL defined mineral medium with vitamins (Verduyn et al. 1992) with glucose as carbon source (20 g.L^{-1}) buffered with phthalate buffer (50 mM, pH 5.0). Fermentations were carried out in 100 mL glass bottles with 40 mL working volume of the same mineral medium and vitamins supplemented with 5 g.L^{-1} of glucose and 20 g.L^{-1} of xylose as carbon and energy sources. Temperature was maintained at 30°C and stirring was set at ca. 160 rpm. Oxygen-limited conditions were obtained by sealing the bottles and sparging N_2 at 0.2 L.min^{-1} for at least 5 minutes before inoculation. A cotton-filled syringe was inserted through the rubber stopper using a needle to avoid accumulation of gas inside the bottles. Bottles were inoculated to a biomass concentration of ca. 1 g.L^{-1} (cdw). Cultivations were performed in biological duplicates for each investigated strain.

Analysis of substrates and products Controlled volumes of samples were taken regularly for analysis. Biomass was followed by OD620 measurements during the length of the fermentations and determination of cell dry weight measurements were also done at time zero (just after inoculation) and at times 42 h and 92 h. For biomass determination, the cell pellet from 5 mL culture was washed with distilled water and dried on Gelman filters (ø 47 mm Supor-450, 0.45 μm) in a microwave oven (350 W) for 8 minutes. Ethanol, glycerol, acetic acid, HMF and furfural were analysed by high performance liquid chromatography (HPLC; Waters Corporation, MA, USA) using an Aminex HPX-87H column (Bio-Rad, CA, USA) at 65°C. The mobile phase was 5 mM sulphuric acid with a flow rate of 0.6 mL.min^{-1}. Analysis of glucose, mannose, xylose, galactose and xylitol was performed on a Shodex™ SP-0810 sugar column (Showa Denko K.K, Japan) at 85°C with water as mobile phase and 0.6 mL.min^{-1} flow rate. All compounds were detected with a refractive index detector (Shimadzu, Tokyo, Japan, and Waters 2414, MA, USA respectively). Yields were calculated based on HPLC measurements.

Results

Genetic engineering of the xylose consuming industrial strain GSE16

It has previously been reported that, whereas overexpression of the transcription factor YAP1 resulted in a strain with a faster sugar consumption rate when fermenting 60% (v/v) of spruce hydrolysate, the same strain was severely inhibited at higher concentrations of the substrate (Alriksson et al. 2010). However the study was performed using laboratory strains with a sensitive genetic background towards hydrolysate-derived inhibitors (Martin and Jönsson 2003). From these results and further studies carried out in our group (data not shown), the importance of the genetic background and the initial inhibitors concentrations on the effect of YAP1 overexpression was further highlighted, and stressed the necessity of assessing the effect of gene modification under more industrial relevant conditions, that is using up-to-date engineered industrial strains and process-like fermentation conditions.

In this study, overexpression cassettes containing the genes that encode the transcription factor YAP1 and the mitochondrial NADH-cytochrome b5 reductase MCR1 were designed and integrated in the genome of the robust industrial strain GSE16, either alone or in combination, generating strains GSE16-YAP1, GSE16-MCR1 and GSE16-YAP1-MCR1 (Table 1). Simultaneous YAP1 integration and APJ1 deletion were obtained by inserting the YAP1 overexpression cassette into the APJ1 locus. The MCR1 cassette was targeted to the YLR446W locus, a gene whose deletion has not affected the yeast performance during fermentation (Subtil and Boles 2012). A control strain in which both alleles of APJ1 were deleted (GSE16-ΔΔAPJ1) was included during the fermentations (Table 1).

The effect of each genetic modification was assessed during anaerobic batch fermentations of undiluted spruce hydrolysate. In order to have a medium composition similar

to large scale lignocellulose-based fermentations, spruce hydrolysate with limited nutrient supplementation was used (See Materials and methods). The xylose concentration was increased from 10 g.L^{-1} to 20 g.L^{-1} to allow a longer period for analysis of the xylose consumption phase. The initial composition of the fermentation medium consisted therefore of 11 g.L^{-1} glucose, 17 g.L^{-1} mannose, 4 g.L^{-1} galactose, and 20 g.L^{-1} xylose. The following inhibitors could also be identified: acetate (3.7 g.L^{-1}), HMF (0.96 g.L^{-1}) and furfural (0.78 g.L^{-1}). The strain robustness was first assessed by comparing lag phase duration, specific growth rate, and specific HMF and furfural conversion rates. Next, we evaluated the fermentation performance of the different strains in terms of glucose and xylose consumption rates, ethanol production rate, and product distribution. The carbon dioxide profile of all strains during the batch fermentations is presented in Additional file 1: Figure S1.

Impact of strain engineering on growth and furaldehyde conversion

During the anaerobic batch fermentations of 100% spruce hydrolysate, and with an initial biomass of 1 g.L^{-1}cell dry weight (cdw), the control strain GSE16 showed a lag phase of 9 hours (Table 3). The strains overexpressing YAP1 showed a consistent decrease in the duration of the lag phase, which lasted for 5.3 ± 0.6 hours. The lag phase duration was not significantly affected by overexpression of MCR1 while it was increased (by ~3 h) for the strain ΔΔAPJ1 (p =0.01). The specific growth rate was also altered in all the strains with the exception of ΔΔAPJ1. GSE16-YAP1 showed a specific growth rate of 0.21 h^{-1}, that is an increase of around 60% when compared to the control strain (p = 0.08). A similar increase was also displayed by GSE16-MCR1 (p = 0.15) (Table 3). No additional increase was observed by the strain combining both modifications.

As for furaldehyde conversion, all strains exhibited a higher specific conversion rate (g.g cells^{-1}.h^{-1}) for furfural than for HMF (Figure 1). The control strain displayed a

Table 3 Lag phase duration and maximum specific growth rate during batch fermentation of spruce hydrolysate

Strain	Lag phase[a] (h)	Growth rate, μ[b] (h^{-1})
GSE16	8.96 ± 0.08	0.13 ± 0.02
GSE16 – YAP1	5.29 ± 0.59	0.21 ± 0.03
GSE16 – MCR1	7.65 ± 1.06	0.20 ± 0.04
GSE16 – YAP1- MCR1	5.49 ± 0.83	0.21 ± 0.02
GSE16 – ΔΔAPJ1	12.04 ± 0.48	0.12 ± 0.02

Initial biomass concentration was 1 g.L^{-1} cdw. The values show the mean and standard deviation of two biological replicates.
[a]The lag phase is defined as the time between inoculation and the onset of the increase on carbon dioxide production. [b]Growth rates were calculated from the carbon dioxide production rates during the exponential phase on glucose.

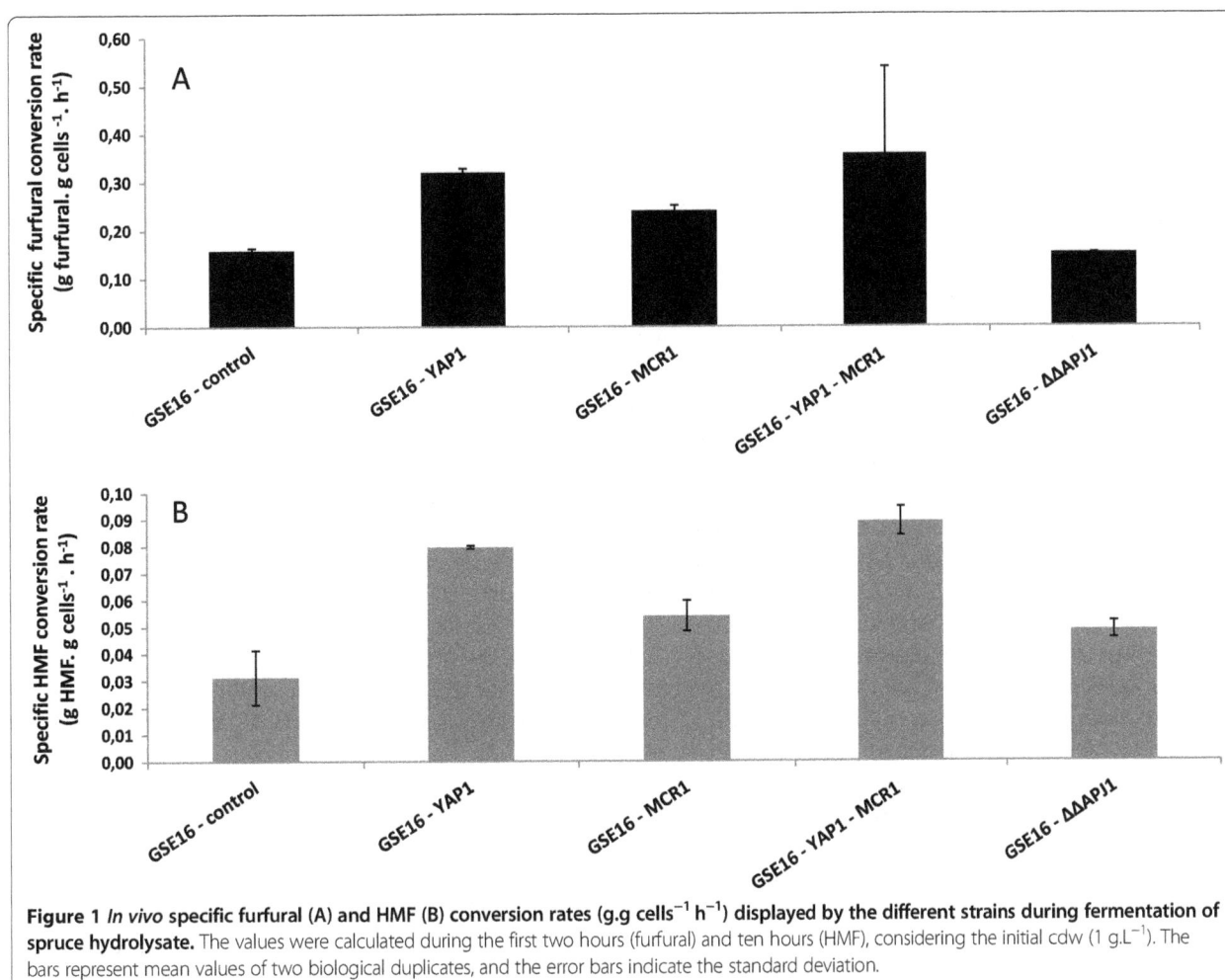

Figure 1 *In vivo* specific furfural (A) and HMF (B) conversion rates (g.g cells^{-1} h^{-1}) displayed by the different strains during fermentation of spruce hydrolysate. The values were calculated during the first two hours (furfural) and ten hours (HMF), considering the initial cdw (1 g.L^{-1}). The bars represent mean values of two biological duplicates, and the error bars indicate the standard deviation.

furfural specific conversion rate of 0.16 g.g cells^{-1}.h^{-1}. GSE16-YAP1 showed the highest furfural specific conversion rate among all the strains (two times higher than the control strain, p = 0.002). Conversion of furfural was also improved when *MCR1* was overexpressed (Figure 1). Likewise, in vivo conversion of HMF was enhanced by overexpression of *YAP1* or *MCR1*. GSE16-YAP1 was able to convert the inhibitor around two times faster (0.08 g.g cells^{-1}.h^{-1}) than the parental strain (p = 0.02), and a 67% increase (p = 0.11) was observed for GSE16-MCR1 (Figure 1). As observed for the growth rate, overexpression of *YAP1* in combination with *MCR1* did not result in additional improvements neither in furfural nor in HMF conversion capacity. Neither was any significant difference in reduction capacity observed between the control strain and the *APJ1* double deletion mutant.

Impact of strain engineering on sugar metabolism

Cell dry weight (cdw) measured after 48 and 92 h of fermentation, suggested that all strains had a similar increase in biomass (data not shown). However, accurate determination of biomass concentration in the presence

of lignocellulosic hydrolysate was made difficult by the presence of solid particles in suspension in the medium or because of variations in the broth color (in the case of optical density measurement) over time. Therefore the strains were compared in terms of volumetric rates (of sugar and product formation) instead of specific rates.

The volumetric consumption rate of glucose was around 0.81 g.L^{-1}.h^{-1} for the strains overexpressing *YAP1* or *MCR1*, which was at least 65% higher than the rate observed for the control strain (Table 4). And again, no additive effect was seen in the strains combining the overexpression of both genes. Deletion of *APJ1* resulted in a marginal increase in glucose utilization rate. Our analytical set-up did not allow accurate quantification of the mannose and galactose consumption rates. However, approximate determinations showed that mannose was consumed shortly after glucose was depleted from the medium. Co-consumption of mannose and xylose was observed during this phase. After mannose was exhausted from the medium, the relatively low amount of galactose (4 g.L^{-1}) was co-consumed together with xylose (data not shown). In contrast to glucose, the xylose consumption

Table 4 Volumetric consumption rate of glucose and xylose and volumetric production rate of ethanol during fermentation of spruce hydrolysate

Strain	Volumetric consumption and production rates (g.L^{-1}.h^{-1})				
	Glucose	Xylose	Ethanol (glucose phase)	Ethanol (xylose phase[+])	Ethanol (overall)
GSE16	0.49 ± 0.01	0.22 ± 0.01	0.60 ± 0.13	0.13 ± 0.00	0.27 ± 0.04
GSE16 – YAP1	0.81 ± 0.15	0.13 ± 0.00	0.90 ± 0.00	0.09 ± 0.02	0.21 ± 0.00
GSE16 – MCR1	0.81 ± 0.12	0.24 ± 0.02	0.84 ± 0.08	0.13 ± 0.00	0.27 ± 0.01
GSE16 – YAP1- MCR1	0.84 ± 0.12	0.13 ± 0.00	0.90 ± 0.08	0.08 ± 0.00	0.20 ± 0.01
GSE16 – ΔΔAPJ1	0.64 ± 0.03	0.19 ± 0.02	0.57 ± 0.05	0.11 ± 0.01	0.28 ± 0.01

[+]During this phase galactose is also consumed.

rate was found to be negatively affected by *YAP1* overexpression. While the control strain consumed 0.22 g xylose. L^{-1}.h^{-1}, a 40% reduction was observed for the two strains overexpressing the transcription factor (p ≤ 0.03). Overexpression of *MCR1* alone and deletion of *APJ1* did not have a significant effect on xylose utilization (Table 4). Since the deletion of both alleles of *APJ1* in GSE16-ΔΔAPJ1 did not affect xylose consumption, it is very unlikely that the effect observed in the *YAP1*-overexpressing strains (carrying only one deleted allele) could be responsible for the lower xylose consumption rate observed in these strains.

The volumetric ethanol production rate was calculated both during the glucose and the xylose consumption phases (Table 4). As expected from the differences between the rates of consumption of the two sugars, the ethanol production rate measured during the xylose phase was considerably lower than the one observed during the glucose phase for all strains. For the control strain, for example, the ethanol production rate on xylose decreased by 78% as compared to the one measured during the glucose phase. When comparing the different strains on glucose, GSE16-YAP1 showed one of the fastest ethanol production rates (0.90 g ethanol.L^{-1}.h^{-1}, p = 0.08). More generally, the ethanol production rate calculated for the new constructs was between 18% and 53% higher than the one obtained for the control strain, thereby matching the increase in the glucose consumption rate. An exception to this correlation was GSE16- ΔΔAPJ1, for which the ethanol rate during the glucose phase was the same as for the control strain (Table 4). On xylose, both *YAP1*-overexpressing strains exhibited a 40% reduction in the ethanol production rate when compared to GSE16. For GSE16-MCR1 and GSE16- ΔΔAPJ1, the ethanol rate in the xylose phase was comparable with that of the control strain. When considering the total length of the fermentations (92 h), no substantial differences were observed between the strains in terms of the volumetric production rate of ethanol (p>0.1) (Table 4).

All strains showed similar metabolite distributions, with the exception of xylitol which was around 2.4 times higher for the strains overexpressing *YAP1* (p ≤ 0.03) (Table 5). The ethanol yields at the end of the fermentation

were very similar between all strains, and accounted for about 50-58% of the maximum theoretical yield (Table 5). These low ethanol yields are very likely a result of a significant rate of evaporation in the fermenter. As previously reported (Bengtsson et al. 2009; Wahlbom et al. 2001), the low boiling point of ethanol and the continuous sparging of the fermenters with nitrogen importantly affect the ethanol yields during prolonged fermentations. As mentioned earlier, the biomass yield was also similar for all strains as was the glycerol yield. The acetate yield was very low for all the strains and less than 0.01 g/g in all the cases (data not shown).

In summary, the deletion of *APJ1* in GSE16 was not detrimental for the strain performance when fermenting spruce hydrolysate. While it was not possible to evaluate the positive benefit of this deletion with the set-up used for the fermentations (expected concentrations of ethanol lower than 3% v/v), *APJ1* can be considered as a good candidate for integration of expression cassettes in strains to be used during very high gravity fermentations. In contrast, *YAP1* and *MCR1* had a positive effect on glucose fermentation in undetoxified hydrolysate but the effect was not cumulative. Finally, *YAP1* overexpression had an unexpected negative impact on xylose utilization.

YAP1 and xylose consumption in mineral medium

The positive effect of *YAP1* in relation to resistance to inhibitors was clearly shown by the improved fermentation rate of glucose during the anaerobic batch fermentations of spruce hydrolysate. The effect during the xylose phase, on the other hand, was not clear. In order to further investigate the effect of *YAP1* overexpression on xylose consumption and to prevent any interference caused by the complex matrix of spruce hydrolysate, the xylose consumption rate of strain GSE16-YAP1 was evaluated during anaerobic fermentation of mineral medium containing 5 g.L^{-1} of glucose and 20 g.L^{-1} xylose. Strain GSE16 was included as a control. With an initial biomass of 1 g.L^{-1} (cdw), both strains consumed all the glucose within 5 hours. However, the negative effect of *YAP1* overexpression on xylose utilization was also observed in mineral medium, indicating that such effect was not

Table 5 Ethanol, glycerol, biomass and xylitol yields per gram of consumed sugars in anaerobic batch fermentation of spruce hydrolysate

Strain	Yields [1] (g.g^{-1})			
	$Y_{ethanol}$	$Y_{glycerol}$	$Y_{biomass}$	$Y_{xylitol}$
GSE16	0.27 ± 0.08	0.05 ± 0.00	0.04 ± 0.01	0.05 ± 0.01
GSE16 – YAP1	0.28 ± 0.05	0.05 ± 0.01	0.04 ± 0.00	0.12 ± 0.00
GSE16 – MCR1	0.26 ± 0.00	0.06 ± 0.00	0.04 ± 0.00	0.05 ± 0.01
GSE16 – YAP1- MCR1	0.25 ± 0.00	0.06 ± 0.00	0.04 ± 0.00	0.12 ± 0.01
GSE16 – ΔΔAPJ1	0.30 ± 0.00	0.04 ± 0.00	0.04 ± 0.00	0.07 ± 0.02

Yields of ethanol, glycerol and biomass were calculated based on the total consumed sugars, while xylitol yields were calculated based on the consumed xylose only. The numbers reported are means ± standard deviation (n = 2).

related to the presence of inhibitors. After 54 h, the control strain GSE16 consumed 13.06 ± 0.46 g of xylose, while GSE16-YAP1 only consumed 3.17 ± 0.1 g xylose (Figure 2). As observed during the fermentations of hydrolysate, the final yields of biomass and extracellular metabolites were similar between the strains, except for xylitol. After 138 h, the xylitol yield of GSE16-YAP1 was 93% higher than that of the control strain (Table 6).

Discussion

In the present study two different genetic modifications that have been reported to improve strain tolerance in a laboratory strain background were introduced in the industrial *S. cerevisiae* strain GSE16 and assessed during

fermentation of undetoxified spruce hydrolysate. The objective was to test whether the overexpression of the selected genes would still be relevant in a strain with a high robust genetic background, to identify the most promising gene among the two candidates, and to assess any putative additive or synergistic effects.

During the glucose consumption phase, the overexpression of *YAP1* was shown to be relevant for inhibitor conversion, even in a robust genetic background. Similarly, the strain robustness in relation to spruce inhibitors was further increased when overexpressing *MCR1*, but no cumulative effect of combined overexpression was revealed. Slightly increased inhibitor tolerance was observed when overexpressing *YAP1* as compared to *MCR1*, which might be explained by the diversity of genes controlled by this transcription factor with relevant functions for detoxification (reductases, transporters, and oxidative stress-related enzymes among others) (Alriksson et al. 2010; Herrero et al. 2008; Toone and Jones 1999). Still, the explanation for the almost equivalent effect obtained by *MCR1* overexpression is less evident. Previous results showed that overexpression of *MCR1* resulted in better growth in mineral medium supplemented with 12 g.L^{-1} acetic acid. (Signori et al., personal communication). Considering the role of the enzyme encoded by *MCR1* in maintaining the antioxidant D-erythroascorbic acid (EASC) in its reduced form (Lee et al. 2001), it is plausible that higher concentrations of this antioxidant may help counteracting the oxidative

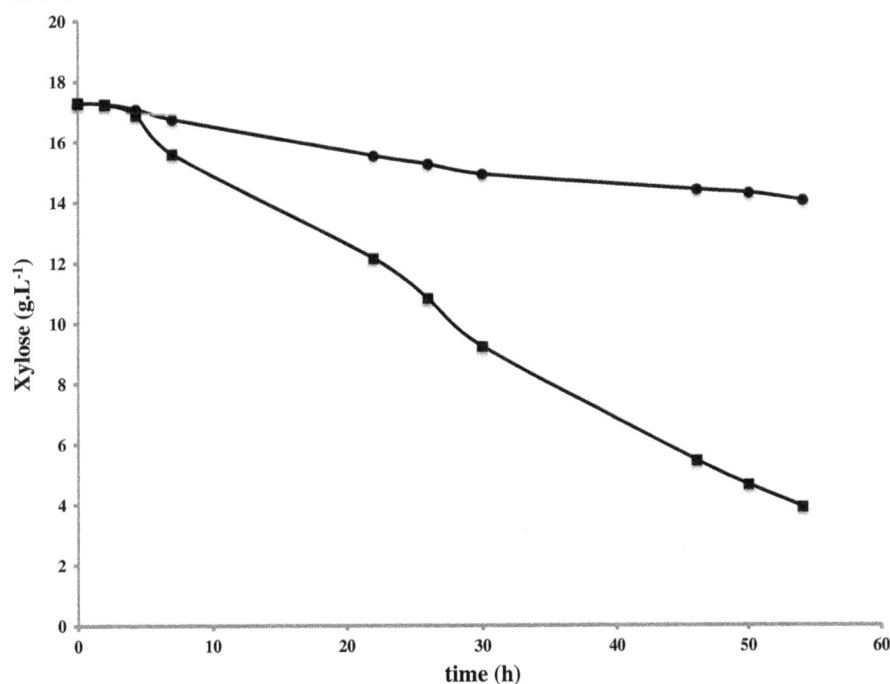

Figure 2 Profile of xylose consumption for GSE16 (squares) and GSE16-YAP1 (circles) in small vials with mineral medium and 5 g.L^{-1} glucose + 20 g.L^{-1} xylose. Initial biomass was 1 g.L^{-1} cdw. The experiment was carried out in biological duplicates. The figure shows the data of a representative profile for each strain with deviation <10%.

Table 6 Ethanol, glycerol, biomass, xylitol and acetate yields obtained during fermentations on mineral medium with 5 g.L^{-1} glucose and 20 g.L^{-1} xylose

Strain	Yield (g.g^{-1})				
	Y$_{Ethanol}$	Y$_{Glycerol}$	Y$_{Biomass}$	Y$_{Xylitol}$	Y$_{Acetate}$
GSE16	0.45 ± 0.03	0.04 ± 0.00	0.09 ± 0.00	0.03 ± 0.00	0.01 ± 0.00
GSE16-YAP1	0.45 ± 0.02	0.04 ± 0.00	0.10 ± 0.00	0.06 ± 0.00	0.01 ± 0.00

Yields of ethanol, glycerol, biomass and acetate were calculated based on the total sugar consumed, while xylitol yields were calculated based on the consumed xylose only. The numbers reported are means ± standard deviation (n = 2).

effect exerted by the acetic acid present in the hydrolysate (Semchyshyn et al. 2011). In fact, the effect seen by *MCR1* overexpression in the presence of acetic acid could be similar to the effect obtained with the biosynthesis of ascorbic acid (ASC) reported by Branduardi (Branduardi et al. 2007). In this study the authors showed that in yeasts, biosynthesis of ASC -a molecule very similar in structure and properties to D-erythroascorbic acid (EASC) - conferred increased resistance to H_2O_2, low pH and organic acids. Moreover, the higher in vivo conversion rates of furaldehydes displayed by GSE16-MCR1 also points towards an involvement of the enzyme in the reduction (by a yet unknown mechanism) of HMF and furfural probably into their less inhibitory alcohol forms (Liu et al. 2005). Given the specific location of this enzyme, this result supports previous observations that indicated a damaging effect of hydrolysate-derived inhibitors to the mitochondria (Allen et al. 2010; Nguyen et al. 2014). To the best of our knowledge this is the first report of a positive effect of *MCR1* overexpression on hydrolysate detoxification; and all together the results suggest that increased concentrations of NADH-cytochrome b5 reductase can improve the resistance of yeast to hydrolysate inhibitors and therefore increase the ethanol production rate on glucose.

On xylose, the much slower sugar consumption rate (as compared to glucose) in one of the best reported pentose fermenting industrial yeasts (Demeke et al. 2013b) and its derivatives emphasized that anaerobic pentose metabolism still requires further improvement, especially in the presence of lignocellulosic inhibitors. Recent comparisons of metabolic profiles between the glucose and xylose consumption phases of *S. cerevisiae* with (Wang et al. 2014; Ask et al. 2013) and without inhibitors (Bergdahl et al. 2012; Matsushika et al. 2013), highlighted the significant perturbations in the metabolic capacities of the yeast caused by xylose. In these metabolomics studies, such perturbations were ascribed to the depletion of key metabolites in glycolysis and cofactors (among other important metabolic variations) suggesting inefficient metabolic states such as carbon starvation, and diminished biosynthetic capacities (Bergdahl et al. 2012; Matsushika et al. 2013).

Our results also revealed a far more complex set of cellular responses deriving from the interactions between xylose metabolism and *YAP1* overexpression. And although the evaluated physiological responses do not provide enough information for explaining the decrease in xylose consumption in *YAP1*-overexpressing strains, the integration of these observations with the results of previous studies could give some hints for further analysis.

First, higher concentrations of xylitol were obtained for *YAP1*-overexpressing strains, which may result from the unspecific reduction of xylose by the different reductases whose transcription is under the control of *YAP1* (Toone and Jones 1999) In fact, *YAP1*-overexpressing strains showed a higher xylose reductase activity than the control strain when cells extracts were used for reduction of xylose (data not shown). Possible inhibition of xylose isomerase (XI) by xylitol was considered, although the used XI originates from *C. phytofermentans*, and has been shown to be much less inhibited by xylitol than other previously expressed XIs in *S. cerevisiae* (Brat et al. 2009). Moreover, the lower xylose consumption for GSE16-YAP1 was observed from the beginning of the fermentation, i.e. when negligible levels of xylitol had been formed. Besides, the rate of consumption was almost constant during most of the process (except at the end of the fermentation when the low concentrations of xylose probably reduce the conversion rate for all strains). This implies that the consumption of xylose did not vary in relation to the increase of xylitol in the medium, i.e. xylitol was not inhibiting the conversion of xylose to xylulose by XI.

A second aspect was connected to the "history" of GSE16. Demeke and co-workers (Demeke et al. 2013b) reported that the diverse and complex paths followed during the development of GSE16 included unknown mutations that appeared to be linked to the high capacity of the strain to ferment xylose, but that such mutations could also be correlated with a possible detrimental effect in terms of inhibitor tolerance (Demeke et al. 2013b). The authors suggested that this mutually exclusive phenotype (good xylose fermentation – bad inhibitor tolerance and vice versa) could be either causally or structurally linked, i. e. genes responsible for improved xylose utilization could be functionally connected with genes responsible for the slower growth; or in the other case, such genes may be located close to each other in the genome (Demeke et al. 2013b). The results obtained in the current study with the *YAP1* overexpressing strains also point towards a mutually exclusive phenotype, but in this case a structural link does not seem likely since the deletion of the locus used for *YAP1* integration (*APJ1*) did not cause a detrimental effect on xylose utilization.

With the exception of xylitol, comparable values were obtained for biomass and extracellular metabolites yields between the different strains in hydrolysate fermentations, which suggested that the overexpression of *YAP1* affected the rate of xylose metabolism but not the product distribution. As similar lower xylose uptake was obtained in mineral medium we could conclude that the negative interactions between xylose metabolism and overexpression of the Yap1 transcription factor were not dependent on the presence of inhibitors and their associated cell responses. When considering available omics data about the effects of overexpressing *YAP1* in *S. cerevisiae* in mineral medium, 17 transcripts were up-regulated by overexpression of the transcription factor (DeRisi et al. 1997) while around 55 proteins were present in higher concentrations (Jun et al. 2012). Although the differences in cultivation conditions do not permit to make any conclusion towards particular genes of interest, these two studies suggest that the overexpression of *YAP1* imposes a higher demand to the cells biosynthetic capacity. This, together with an impairment of biosynthetic capabilities on xylose (as presented in the metabolomics studies previously commented (Bergdahl et al. 2012; Matsushika et al. 2013)), and the high cell maintenance energy required during growth on this sugar (Feng and Zhao 2013) could explain the slower growth on xylose in *YAP1* overexpressing strains.

Nevertheless, the unexpected results seen with *YAP1*-overexpression during xylose assimilation require deeper analysis to further understand the biological responses that limit the development of robust xylose consuming strains. In this respect, we consider that the study at the molecular level of the cellular responses of *YAP1*-overexpression in GSE16 and other xylose consuming strains (for example expressing the fungal redox pathway) during glucose and xylose utilization would reveal important insights about limiting steps for xylose metabolism.

In conclusion, overexpression of the transcription factor *YAP1* and the mitochondrial reductase *MCR1* in the already robust strain GSE16 resulted in an even faster hexose catabolism in the presence of spruce hydrolysate-derived inhibitors, but the effect was not cumulative. The improved phenotype of *MCR1* overexpression seems to be related, at least in part, to a faster furaldehyde reduction, indicating that this reductase may have a wider substrate range than previously reported. Unexpected reduced xylose fermentation rate was observed in *YAP1* overexpressing strains and further studies are needed to elucidate the mechanisms behind this observation.

Additional file

> **Additional file 1: Table S1.** Oligonucleotides used in the current work. Shows the sequence of the primers used during the study. **Figure S1.** Carbon dioxide profile of the strains GSE16 (parental), GSE16-YAP1, GSE16-MCR1, GSE16-YAP1-MCR1 and GSE16- ΔΔAPJ1 during anaerobic batch fermentations of spruce hydrolysate (only shown for the first 35 h). The figure shows the carbon dioxide profile of the different strains evaluated in the study during anaerobic batch fermentations of spruce hydrolysate.

Competing interests
The authors declare that they have no competing interests.

Authors' contributions
VWS participated in the design of the study, performed the experiments and wrote the manuscript. LS, YL, MA performed the experiments and commented on the manuscript. MB participated in the initial design of the study and commented the manuscript. DP and PB contributed in conceiving the study and revised the manuscript. MFM and JT participated in design of the study and revised the manuscript. MGG conceived the study and revised the manuscript. All authors read and approved the final manuscript.

Acknowledgements
This project was financed by the 7th European Commission Framework Project 222699 NEMO (Novel High performance Enzymes and Micro-organisms for conversion of lignocellulosic biomass to bioethanol) and SBO grants (IWT 90043) from IWT-Flanders. SEKAB, Sweden is gratefully acknowledged for the provision of the hydrolysate. VWS was co-financed by the Swedish Energy Agency (Energimyndigheten). LS acknowledges the doctoral fellowship of the University of Milano Bicocca. YL was supported by the China Scholarship Council. MB is financed by the Chalmers Energy Initiative.

Author details
[1]Applied Microbiology, Department of Chemistry, Lund University, P.O. Box 124, SE-22100 Lund, Sweden. [2]University of Milano Bicocca, Piazza della Scienza 2, 20126 Milan, Italy. [3]Laboratory of Molecular Cell Biology, Institute of Botany and Microbiology, Leuven, KU, Belgium. [4]Department of Molecular Microbiology, VIB, Kasteelpark Arenberg 31, Leuven, B-3001 Heverlee, Flanders, Belgium. [5]Department of Chemical and Biological Engineering, Industrial Biotechnology, Chalmers University of Technology, SE-41296 Gothenburg, Sweden.

References
Allen SA, Clark W, McCaffery JM, Cai Z, Lanctot A, Slininger PJ, Liu ZL, Gorsich SW (2010) Furfural induces reactive oxygen species accumulation and cellular damage in *Saccharomyces cerevisiae*. Biotechnology for Biofuels 3:2. doi:10.1186/1754-6834-3-2

Almeida JRM, Bertilsson M, Gorwa-Grauslund MF, Gorsich S, Lidén G (2009a) Metabolic effects of furaldehydes and impacts on biotechnological processes. Appl Microbiol Biotechnol 82(4):625–638. doi:10.1007/s00253-009-1875-1

Almeida JRM, Karhumaa K, Bengtsson O, Gorwa-Grauslund M-F (2009b) Screening of *Saccharomyces cerevisiae* strains with respect to anaerobic growth in non-detoxified lignocellulose hydrolysate. Bioresour Technol 100(14):3674–3677. doi:http://dx.doi.org/10.1016/j.biortech.2009.02.057

Almeida JRM, Runquist D, Sànchez Nogué V, Lidén G, Gorwa-Grauslund MF (2011) Stress-related challenges in pentose fermentation to ethanol by the yeast *Saccharomyces cerevisiae*. Biotechnol J 6(3):286–299. doi:10.1002/biot.201000301

Alriksson B, Horváth IS, Jönsson LJ (2010) Overexpression of *Saccharomyces cerevisiae* transcription factor and multidrug resistance genes conveys enhanced resistance to lignocellulose-derived fermentation inhibitors. Process Biochem 45(2):264–271. doi:10.1016/j.procbio.2009.09.016

Ask M, Bettiga M, Duraiswamy V, Olsson L (2013) Pulsed addition of HMF and furfural to batch-grown xylose-utilizing *Saccharomyces cerevisiae* results in different physiological responses in glucose and xylose consumption phase. Biotechnology for Biofuels 6(1):181

Bellissimi E, van Dijken JP, Pronk JT, van Maris AJA (2009) Effects of acetic acid on the kinetics of xylose fermentation by an engineered, xylose-isomerase-based *Saccharomyces cerevisiae* strain. FEMS Yeast Res 9(3):358–364. doi:10.1111/j.1567-1364.2009.00487.x

Benatuil L, Perez JM, Belk J, Hsieh CM (2010) An improved yeast transformation method for the generation of very large human antibody libraries. Protein Eng Des Sel 23(4):155–159. doi:10.1093/protein/gzq002

Bengtsson O, Hahn-Hagerdal B, Gorwa-Grauslund M (2009) Xylose reductase from *Pichia stipitis* with altered coenzyme preference improves ethanolic xylose fermentation by recombinant *Saccharomyces cerevisiae*. Biotechnology for Biofuels 2(1):9

Bergdahl B, Heer D, Sauer U, Hahn-Hägerdal B, van Niel EW (2012) Dynamic metabolomics differentiates between carbon and energy starvation in recombinant *Saccharomyces cerevisiae* fermenting xylose. Biotechnology for Biofuels 5(1):34

Branduardi P, Fossati T, Sauer M, Pagani R, Mattanovich D, Porro D (2007) Biosynthesis of vitamin C by yeast leads to increased stress resistance. PLoS One 2(10):e1092. doi:10.1371/journal.pone.0001092

Brat D, Boles E, Wiedemann B (2009) Functional expression of a bacterial xylose isomerase in *Saccharomyces cerevisiae*. Appl Environ Microbiol 75(8):2304–2311. doi:AEM.02522-08

Casey E, Sedlak M, Ho NWY, Mosier NS (2010) Effect of acetic acid and pH on the cofermentation of glucose and xylose to ethanol by a genetically engineered strain of *Saccharomyces cerevisiae*. FEMS Yeast Res 10(4):385–393. doi:10.1111/j.1567-1364.2010.00623.x

Demeke MM, Dietz H, Li Y, Foulquie-Moreno MR, Mutturi S, Deprez S, Den Abt T, Bonini BM, Lidén G, Dumortier F, Verplaetse A, Boles E, Thevelein JM (2013a) Development of a D-xylose fermenting and inhibitor tolerant industrial *Saccharomyces cerevisiae* strain with high performance in lignocellulose hydrolysates using metabolic and evolutionary engineering. Biotechnology for Biofuels 6(1):89. doi:10.1186/1754-6834-6-89

Demeke MM, Dumortier F, Li Y, Broeckx T, Foulquie-Moreno MR, Thevelein JM (2013b) Combining inhibitor tolerance and D-xylose fermentation in industrial *Saccharomyces cerevisiae* for efficient lignocellulose-based bioethanol production. Biotechnology for Biofuels 6(1):120. doi:10.1186/1754-6834-6-120

DeRisi JL, Iyer VR, Brown PO (1997) Exploring the metabolic and genetic control of gene expression on a genomic scale. Science 278(5338):680–686

Feng X, Zhao H (2013) Investigating xylose metabolism in recombinant *Saccharomyces cerevisiae* via 13C metabolic flux analysis. Microb Cell Fact 12(1):114

Gietz RD, Schiestl RH, Willems AR, Woods RA (1995) Studies on the transformation of intact yeast cells by the LiAc/s-DNA/PEG procedure. Yeast 11(4):355–360

Gueldener U, Heinisch J, Koehler GJ, Voss D, Hegemann JH (2002) A second set of loxP marker cassettes for Cre-mediated multiple gene knockouts in budding yeast. Nucleic Acids Res 30(6):e23. doi:10.1093/nar/30.6.e23

Güldener U, Heck S, Fiedler T, Beinhauer J, Hegemann JH (1996) A New efficient gene disruption cassette for repeated Use in budding yeast. Nucleic Acids Res 24(13):2519–2524. doi:10.1093/nar/24.13.2519

Hahne K, Haucke V, Ramage L, Schatz G (1994) Incomplete arrest in the outer membrane sorts NADH-cytochrome b5 reductase to two different submitochondrial compartments. Cell 79(5):829–839. doi:http://dx.doi.org/10.1016/0092-8674(94)90072-8

Haitani Y, Tanaka K, Yamamoto M, Nakamura T, Ando A, Ogawa J, Shima J (2012) Identification of an acetate-tolerant strain of *Saccharomyces cerevisiae* and characterization by gene expression analysis. J Biosci Bioeng 114(6):648–651. doi:10.1016/j.jbiosc.2012.07.002

Helle S, Cameron D, Lam J, White B, Duff S (2003) Effect of inhibitory compounds found in biomass hydrolysates on growth and xylose fermentation by a genetically engineered strain of *S. cerevisiae*. Enzyme Microb Technol 33(6):786–792. doi:10.1016/S0141-0229(03)00214-X

Herrero E, Ros J, Belli G, Cabiscol E (2008) Redox control and oxidative stress in yeast cells. Biochim Biophys Acta 1780(11):1217–1235. doi:10.1016/j.bbagen.2007.12.004

Hill J, Donald KA, Griffiths DE (1991) DMSO-enhanced whole cell yeast transformation. Nucleic Acids Res 19(20):5791

Jeppsson M, Johansson B, Jensen PR, Hahn-Hägerdal B, Gorwa-Grauslund MF (2003) The level of glucose-6-phosphate dehydrogenase activity strongly influences xylose fermentation and inhibitor sensitivity in recombinant *Saccharomyces cerevisiae* strains. Yeast 20(15):1263–1272. doi:10.1002/yea.1043

Jun H, Kieselbach T, Jonsson L (2012) Comparative proteome analysis of *Saccharomyces cerevisiae*: a global overview of in vivo targets of the yeast activator protein 1. BMC Genomics 13(1):230

Kim D, Hahn J-S (2013) Roles of Yap1 transcription factor and antioxidants in yeast tolerance to furfural and 5-hydroxymethylfurfural that function as thiol-reactive electrophiles generating oxidative stress. Appl Environ Microbiol 79(16):5069–5077. doi:10.1128/aem.00643-13

Lee J-S, Huh W-K, Lee B-H, Baek Y-U, Hwang C-S, Kim S-T, Kim Y-R, Kang S-O (2001) Mitochondrial NADH-cytochrome b5 reductase plays a crucial role in the reduction of d-erythroascorbyl free radical in *Saccharomyces cerevisiae*. Biochim Biophys Acta Gen Subj 1527(1–2):31–38. doi:http://dx.doi.org/10.1016/S0304-4165(01)00134-9

Liu Z (2011) Molecular mechanisms of yeast tolerance and in situ detoxification of lignocellulose hydrolysates. Appl Microbiol Biotechnol 90(3):809–825. doi:10.1007/s00253-011-3167-9

Liu ZL, Slininger P, Gorsich S (2005) Enhanced biotransformation of furfural and hydroxymethylfurfural by newly developed ethanologenic yeast strains. Appl Biochem Biotechnol 121(1–3):451–460. doi:10.1385/abab:121:1-3:0451

Ma M, Liu ZL (2010) Comparative transcriptome profiling analyses during the lag phase uncover *YAP1*, *PDR1*, *PDR3*, *RPN4*, and *HSF1* as key regulatory genes in genomic adaptation to the lignocellulose derived inhibitor HMF for *Saccharomyces cerevisiae*. BMC Genomics 11(1):660

Martin C, Jönsson L (2003) Comparison of the resistance of industrial and laboratory strains of *Saccharomyces* and *Zygosaccharomyces* to lignocellulose-derived fermentation inhibitors. Enzyme Microb Technol 32(3–4):386–395. doi:http://dx.doi.org/10.1016/S0141-0229(02)00310-1

Matsushika A, Nagashima A, Goshima T, Hoshino T (2013) Fermentation of xylose causes inefficient metabolic state Due to carbon/energy starvation and reduced glycolytic flux in recombinant industrial *Saccharomyces cerevisiae*. PLoS One 8(7):e69005. doi:10.1371/journal.pone.0069005

Meineke B, Engl G, Kemper C, Vasiljev-Neumeyer A, Paulitschke H, Rapaport D (2008) The outer membrane form of the mitochondrial protein Mcr1 follows a TOM-independent membrane insertion pathway. FEBS Lett 582(6):855–860. doi:http://dx.doi.org/10.1016/j.febslet.2008.02.009

Mira N, Palma M, Guerreiro J, Sa-Correia I (2010) Genome-wide identification of *Saccharomyces cerevisiae* genes required for tolerance to acetic acid. Microb Cell Fact 9(1):79

Mumberg D, Muller R, Funk M (1995) Yeast vectors for the controlled expression of heterologous proteins in different genetic backgrounds. Gene 156(1):119–122

Nguyen TTM, Iwaki A, Ohya Y, Izawa S (2014) Vanillin causes the activation of Yap1 and mitochondrial fragmentation in *Saccharomyces cerevisiae*. J Biosci Bioeng 117(1):33–38. doi:http://dx.doi.org/10.1016/j.jbiosc.2013.06.008

Parawira W, Tekere M (2011) Biotechnological strategies to overcome inhibitors in lignocellulose hydrolysates for ethanol production: review. Crit Rev Biotechnol 31(1):20–31. doi:10.3109/07388551003757816

Petersson A, Almeida JRM, Modig T, Karhumaa K, Hahn-Hägerdal B, Gorwa-Grauslund MF, Lidén G (2006) A 5-hydroxymethyl furfural reducing enzyme encoded by the *Saccharomyces cerevisiae* ADH6 gene conveys HMF tolerance. Yeast 23(6):455–464. doi:10.1002/yea.1370

Sambrook J, Russel D (2001) Molecular Cloning: A Laboratory Manual. Cold Spring Harbor Laboratory Press, Cold Spring Harbor, NY, USA

Semchyshyn HM, Abrat OB, Miedzobrodzki J, Inoue Y, Lushchak V (2011) Acetate but not propionate induces oxidative stress in bakers' yeast *Saccharomyces cerevisiae*. Redox Rep 16(1):15–23

Subtil T, Boles E (2012) Competition between pentoses and glucose during uptake and catabolism in recombinant *Saccharomyces cerevisiae*. Biotechnology for Biofuels 5(1):14

Sundström L, Larsson S, Jönsson L (2010) Identification of *Saccharomyces cerevisiae* genes involved in the resistance to Phenolic fermentation inhibitors. Appl Biochem Biotechnol 161(1–8):106–115. doi:10.1007/s12010-009-8811-9

Swinnen S, Schaerlaekens K, Pais T, Claesen J, Hubmann G, Yang Y, Demeke M, Foulquié-Moreno MR, Goovaerts A, Souvereyns K, Clement L, Dumortier F, Thevelein JM (2012) Identification of novel causative genes determining the complex trait of high ethanol tolerance in yeast using pooled-segregant whole-genome sequence analysis. Genome Res 22(5):975–984. doi:10.1101/gr.131698.111

Toone WM, Jones N (1999) AP-1 transcription factors in yeast. Curr Opin Genet Dev 9(1):55–61. doi:http://dx.doi.org/10.1016/S0959-437X(99)80008-2

van Dijken JP, Bauer J, Brambilla L, Duboc P, Francois JM, Gancedo C, Giuseppin MLF, Heijnen JJ, Hoare M, Lange HC, Madden EA, Niederberger P, Nielsen J, Parrou JL, Petit T, Porro D, Reuss M, van Riel N, Rizzi M, Steensma HY, Verrips CT, Vindeløv J, Pronk JT (2000) An interlaboratory comparison of physiological and genetic properties of four *Saccharomyces cerevisiae* strains. Enzyme Microb Technol 26(9–10):706–714. doi:http://dx.doi.org/10.1016/S0141-0229(00)00162-9

Verduyn C, Postma E, Scheffers WA, van Dijken JP (1992) Effect of benzoic acid on metabolic fluxes in yeasts - a continuous culture study on the regulation of respiration and alcoholic fermentation. Yeast 8(7):501–517

Wahlbom CF, Eliasson A, Hahn-Hägerdal B (2001) Intracellular fluxes in a recombinant xylose-utilizing Saccharomyces cerevisiae cultivated anaerobically at different dilution rates and feed concentrations. Biotechnol Bioeng 72(3):289–296. doi:10.1002/1097-0290(20010205) 72:3<289::aid-bit5>3.0.co;2-9

Wang X, Jin M, Balan V, Jones AD, Li X, Li B-Z, Dale BE, Yuan Y-J (2014) Comparative metabolic profiling revealed limitations in xylose-fermenting yeast during co-fermentation of glucose and xylose in the presence of inhibitors. Biotechnol Bioeng 111(1):152–164. doi:10.1002/bit.24992

4

Functional characterization and structural modeling of synthetic polyester-degrading hydrolases from *Thermomonospora curvata*

Ren Wei, Thorsten Oeser, Johannes Then, Nancy Kühn, Markus Barth, Juliane Schmidt and Wolfgang Zimmermann[*]

Abstract

Thermomonospora curvata is a thermophilic actinomycete phylogenetically related to *Thermobifida fusca* that produces extracellular hydrolases capable of degrading synthetic polyesters. Analysis of the genome of *T. curvata* DSM43183 revealed two genes coding for putative polyester hydrolases Tcur1278 and Tcur0390 sharing 61% sequence identity with the *T. fusca* enzymes. Mature proteins of Tcur1278 and Tcur0390 were cloned and expressed in *Escherichia coli* TOP10. Tcur1278 and Tcur0390 exhibited an optimal reaction temperature against p-nitrophenyl butyrate at 60°C and 55°C, respectively. The optimal pH for both enzymes was determined at pH 8.5. Tcur1278 retained more than 80% and Tcur0390 less than 10% of their initial activity following incubation for 60 min at 55°C. Tcur0390 showed a higher hydrolytic activity against poly(ε-caprolactone) and polyethylene terephthalate (PET) nanoparticles compared to Tcur1278 at reaction temperatures up to 50°C. At 55°C and 60°C, hydrolytic activity against PET nanoparticles was only detected with Tcur1278. *In silico* modeling of the polyester hydrolases and docking with a model substrate composed of two repeating units of PET revealed the typical fold of α/β serine hydrolases with an exposed catalytic triad. Molecular dynamics simulations confirmed the superior thermal stability of Tcur1278 considered as the main reason for its higher hydrolytic activity on PET.

Keywords: Polyester hydrolase; Synthetic polyester; Polyethylene terephthalate (PET); *Thermomonospora curvata*

Introduction

The widespread use of synthetic polyesters such as polyethylene terephthalate (PET) in industry and daily life has resulted in serious environmental pollution over the last decades. However, the recycling of PET by chemical methods performed under extreme temperature and pH conditions is an energy-consuming process (Paszun and Spychaj 1997). Recently, alternative processes using biocatalysis have been proposed for recycling and surface functionalization of PET (Müller et al. 2005; Zimmermann and Billig 2011). Microbial enzymes capable of degrading PET have been previously described from various fungal (Egmond and de Vlieg 2000; Alisch et al. 2004; Alisch-Mark et al. 2006; Nimchua et al. 2007; Ronkvist et al. 2009) and bacterial (Müller et al. 2005; Eberl et al. 2009; Herrero Acero et al. 2011; Ribitsch et al. 2012a; Ribitsch et al. 2012b; Sulaiman et al. 2012; Kitadokoro et al. 2012; Chen et al. 2010; Oeser et al. 2010) sources. Enzymes with high PET-hydrolyzing activity are mostly extracellular proteins secreted by thermophilic microorganisms such as *Thermomyces insolens* (Ronkvist et al. 2009) and several *Thermobifida* species (Müller et al. 2005; Eberl et al. 2009; Herrero Acero et al. 2011; Ribitsch et al. 2012a; Ribitsch et al. 2012b; Kitadokoro et al. 2012; Chen et al. 2010; Oeser et al. 2010). The biodegradability of PET by these enzymes has been shown to strongly depend on the flexibility of polymer chains that is directly influenced by the hydrolysis reaction temperatures (Ronkvist et al. 2009; Wei et al. 2013).

Thermomonospora curvata DSM 43183, a facultative aerobic thermophilic actinomycete, has been isolated from composts containing plant materials (Henssen 1957; Henssen and Schnepf 1967; Chertkov et al. 2011). The optimal growth temperature of *T. curvata* is 50°C (Henssen and Schnepf 1967) at a wide range of pH from 7.5 to 11 (Chertkov et al. 2011). Weak growth of *T. curvata* has

[*] Correspondence: wolfgang.zimmermann@uni-leipzig.de
Department of Microbiology and Bioprocess Technology, Institute of Biochemistry, University of Leipzig, Johannisallee 21-23, D-04103 Leipzig, Germany

been also observed at higher temperatures up to 65°C (Henssen and Schnepf 1967). The phylogenetic analysis of *T. curvata* revealed a distant relationship to other thermophilic actinomycetes isolated from a similar habitat including *Thermobifida fusca* and *Thermobifida alba*, as indicated by a lower level of 16S rRNA sequence similarity between 89% and 90% (Henssen 1957; Zhang et al. 1998; Chertkov et al. 2011). Several of these bacteria have been shown to express extracellular enzymes with polyester-hydrolyzing activity (Kleeberg et al. 1998; Alisch et al. 2004; Herrero Acero et al. 2011; Thumarat et al. 2012; Ribitsch et al. 2012b).

In this study, we report the identification of two genes coding for the polyester hydrolases Tcur1278 and Tcur0390 by genome mining of *T. curvata* DSM43183 (Chertkov et al. 2011), the characterization of their catalytic properties and thermal stability, as well as the modeling and analysis of their three-dimensional structures.

Materials and methods

Cloning, expression and purification of Tcur1278 and Tcur0390

The genes encoding Tcur1278 and Tcur0390 without the Gram-positive secretion signal peptides were selected from the annotated genome sequences of *T. curvata* DSM43183 (Chertkov et al. 2011). Synthetic gene constructs with adapted codon usage to *E. coli* (Geneart GmbH, Regensburg, Germany) for Tcur1278 [EMBL: HG939554] and Tcur0390 [EMBL: HG939555] were applied for direct cloning into the pBAD TOPO expression vector (Invitrogen, Life Technologies, Carlsbad, USA). The recombinant expression of *T. curvata* hydrolases was carried out in One Shot *E. coli* TOP10 (Invitrogen) at room temperature for 14 h in lysogeny broth (LB) containing 0.2% (m/v) of L-arabinose as inducer as described previously (Oeser et al. 2010). Bacterial cells were harvested by centrifugation and resuspended in a lysis buffer containing 50 mM phosphate (pH 8) and 300 mM NaCl. After sonication, the soluble cell extracts were subjected to immobilized metal ion affinity chromatography (IMAC) using Ni-NTA columns (Qiagen, Hilden, Germany). The protein elutions containing the recombinant hydrolases were separated by SDS PAGE and analyzed by esterase activity-staining with 1-naphthyl acetate and Fast Red dye (Sztajer et al. 1992) as well as by staining with Coomassie Brilliant Blue.

Determination of esterase activity

Esterase activity was determined with p-nitrophenyl butyrate (pNPB) as a substrate in a microplate format (BioTek PowerWave XS, BioTek Instruments Inc., Winooski, USA) (Billig et al. 2010). To avoid the adsorption of proteins to the plastic vials, the dilution was carried out in the presence of 15% poly(ethylene glycol) (PEG$_{6000}$, Sigma-Aldrich

Co., St. Louis, USA) in Davies buffer (Davies 1959) between pH 6.5 and 9.5 or in 100 mM Tris-HCl. One unit of esterase activity was defined as the amount of enzyme required to hydrolyze 1 µmol pNPB per min (Alisch et al. 2004). To investigate their thermal stability, 250 µg/mL of enzymes were incubated in 100 mM Tris buffer (pH 8.5) at 50°C, 55°C and 60°C for up to 1 h. Residual esterase activity against pNPB was determined at 25°C in triplicate. The Michaelis-Menten kinetic constants for the hydrolysis of pNPB by Tcur1278 and Tcur0390 were determined at 25°C and pH 8.5.

Enzymatic hydrolysis of polyester nanoparticles

The enzymatic hydrolysis of polyesters was analyzed by monitoring the change of turbidity of a polyester nanoparticle suspension at 600 nm (Wei et al. 2013). Poly(ε-caprolactone) (PCL) and PET nanoparticles were prepared by a precipitation and solvent evaporation technique from amorphous PCL (Sigma-Aldrich Co., St. Louis, USA) and low-crystallinity PET film (Goodfellow GmbH, Bad Nauheim, Germany) dissolved in acetone and 1,1,1,3,3,3-hexafluoro-2-propanol, respectively. The enzymatic hydrolysis of PCL was performed in a microplate format (BioTek PowerWave XS) at 49°C with 0.22 mg/mL of PCL nanoparticles in each well, whereas the enzymatic hydrolysis of PET was performed at 50°C to 60°C in cuvettes containing 0.25 mg/mL of PET nanoparticles immobilized in 0.9% agarose gel. The change of turbidity was monitored over an incubation period of 15 min at 1 min intervals for PCL hydrolysis, whereas PET hydrolysis was determined for 60 min at 5 min intervals. The initial degradation rates were defined as the square roots of turbidity decrease during the initial linear phase of the hydrolysis, and plotted as a function of enzyme concentration using a kinetic model (Eq. 1) modified from Wei et al. (2013)

$$\frac{d(\sqrt{\tau})}{dt} = \frac{k_\tau K_A[E]}{1 + K_A[E]} \tag{1}$$

where τ is the turbidity of a nanoparticle suspension, t, the reaction time, k_τ, the hydrolysis rate constant based on the turbidity change, K_A, the adsorption equilibrium constant, and $[E]$, the enzyme concentration.

Homology modeling and molecular docking

Homology modeling of *T. curvata* polyester hydrolases was carried out using the Phyre2 web server (Kelley and Sternberg 2009) based on the crystal structure of *T. alba* AHK 119 (Est119, PDB ID: 3VIS) (Kitadokoro et al. 2012). The sequence identity of *T. curvata* polyester hydrolases with the corresponding template structure is summarized in Table 1 and Additional file 1: Figure S1.

Table 1 Sequence identity (in percent, upper right part) and the root-mean-square deviation (RMSD) of C$_\alpha$ atoms (in Å, lower left part) of *T. curvata* polyester hydrolases in comparison with the crystal structure of homologous *T. alba* Est119 (PDB ID: 3VIS)

RMSD (Å)	Identity (%)		
	Tcur1278	Tcur0390	*T. alba* Est119
Tcur1278		82.2	61.7
Tcur0390	0.11		61.7
T. alba Est119	0.85	0.83	

The molecular docking program GOLD version 5.1 (Cambridge Crystallographic Data Centre, Cambridge, UK) (Jones et al. 1997) was used to study the substrate-binding pocket of *T. curvata* polyester hydrolases. The polyester model substrate 2PET composed of 2 repeating units of PET (*bis* 2-hydroxyethyl terephthalate, BHET) was constructed with the software MOE (Chemical Computing Group, Montreal, Canada). The central ester bond of 2PET was constrained in the oxyanion hole composed of the main chain NH groups of amino acid residues F62 and M131 with the correct orientation to form a tetrahedral intermediate based on the catalytic mechanism of ester hydrolases (Jaeger et al. 1999). The other atoms of 2PET were allowed to be flexible for a conformation to be docked to the rigid protein structural model by a genetic algorithm (Jones et al. 1997). Based on the default scoring function of GOLD, the top-ranked productive docking conformations in accordance with the catalytic mechanism of ester hydrolases (Jaeger et al. 1999) were selected for the illustrations generated by the MOE software.

Molecular dynamics simulations

The molecular dynamics (MD) simulations were carried out using GROMACS 4.6 (Groningen University, The Netherlands) (Hess et al. 2008) in the Amber99SB force field (Hornak et al. 2006) in explicit solvent. Protein structural models of both *T. curvata* polyester hydrolases were centered in a cube with a distance of ≥1.0 nm from each edge as the starting structures. The steepest descent method was applied to perform energy minimization until a maximum force (F$_{max}$) of less than 1000 kJ/mol/nm was reached. The system was equilibrated for 100 ps by a position-restrained simulation at the desired temperatures in the isothermal-isobaric (NPT) ensemble. The isotropic pressure coupling using the Berendsen algorithm was applied with a reference pressure of 1.0 bar (Berendsen et al. 1984). For each protein structure, three independent simulations were performed under equilibration conditions for 50 ns in 2 fs steps at 298 K (25°C) and 353 K (80°C), respectively. To analyze the thermal stability of the polyester hydrolases, the time course of the root-mean-square

deviation (RMSD) of backbone structures and the root-mean-square fluctuation (RMSF) of C$_\alpha$ atoms of each amino acid residue over the complete 50 ns simulation were calculated using GROMACS 4.6 (Hess et al. 2008).

Results

Cloning, expression and purification of Tcur1278 and Tcur0390

Synthetic genes encoding Tcur1278 and Tcur0390 were amplified in the pBAD-TOPO expression vector (Invitrogen) for recombinant expression in One Shot *E. coli* TOP10 (Invitrogen). Following an expression period of 14 h at 25°C and the subsequent IMAC purification, 2.5 mg of Tcur1278 and 2.9 mg of Tcur0390 were obtained from a 500 mL culture with a specific activity of 3.0 U/mg and 17.9 U/mg against pNPB, respectively. By SDS PAGE analysis, both *T. curvata* hydrolases were obtained as single bands with esterase activity against 1-naphthyl acetate, corresponding to an apparent molecular mass of approximately 35 kDa (Additional file 1: Figure S2).

Effect of pH and temperature on the hydrolytic activity of Tcur1278 and Tcur0390

The effect of pH on the hydrolytic activity of Tcur1278 and Tcur0390 was investigated against pNPB in a pH range from 6.5 to 9.5 (Figure 1A). Both enzymes displayed an optimal pH at pH 8.5 and still retained more than 60% of their maximum activity at pH 9.5.

The effect of temperature on the hydrolytic activity of both enzymes was assayed against pNPB in a temperature range from 30°C to 70°C (Figure 1B). Tcur1278 and Tcur0390 showed an optimal temperature at 60°C and 55°C, respectively.

Thermal stability of Tcur1278 and Tcur0390

The stability of Tcur1278 and Tcur0390 at 50°C, 55°C and 60°C was investigated at pH 8.5 over a period of 60 min by monitoring the residual activities against pNPB (Figure 2A-B). Tcur1278 showed a higher thermal stability compared to Tcur0390 retaining more than 80% of its initial activity following incubation for 60 min at 50°C and 55°C. At 60°C, approximately 65% loss of its initial activity was detected following incubation for 10 min. In contrast, Tcur0390 showed a residual activity of only 40% following incubation for 60 min at 50°C and of 15% following incubation for 10 min at 55°C and 60°C.

Kinetic analysis of the hydrolysis of pNPB by Tcur1278 and Tcur0390

Based on the Michaelis-Menten kinetic model, Tcur0390 revealed an almost 6-fold higher k_{cat} and no significantly lower K_m than Tcur1278 for pNPB hydrolysis indicating a higher hydrolytic activity of Tcur0390 against the soluble pNPB compared to Tcur1278 (Table 2).

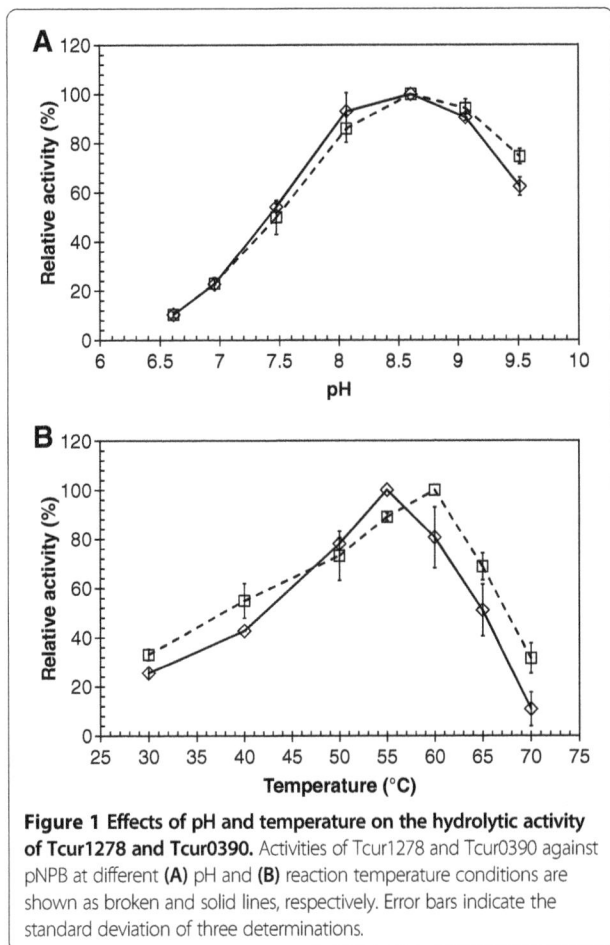

Figure 1 Effects of pH and temperature on the hydrolytic activity of Tcur1278 and Tcur0390. Activities of Tcur1278 and Tcur0390 against pNPB at different **(A)** pH and **(B)** reaction temperature conditions are shown as broken and solid lines, respectively. Error bars indicate the standard deviation of three determinations.

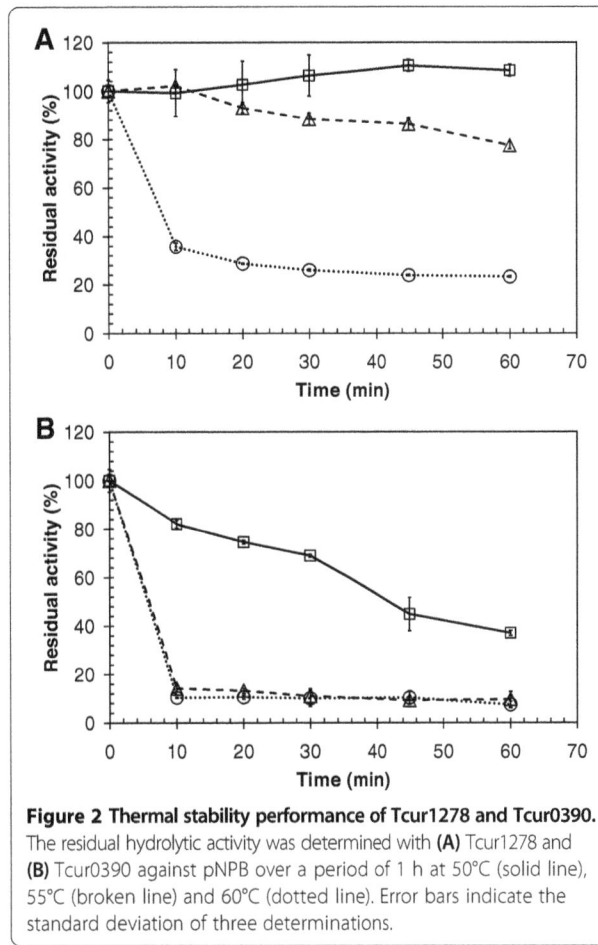

Figure 2 Thermal stability performance of Tcur1278 and Tcur0390. The residual hydrolytic activity was determined with **(A)** Tcur1278 and **(B)** Tcur0390 against pNPB over a period of 1 h at 50°C (solid line), 55°C (broken line) and 60°C (dotted line). Error bars indicate the standard deviation of three determinations.

Kinetic analysis of the hydrolysis of PCL nanoparticles by Tcur1278 and Tcur0390

A PCL nanoparticle suspension with a concentration of 0.22 mg/mL was completely hydrolyzed following incubation for 15 min at 49°C with 20 µg/mL of Tcur0390 or 30 µg/mL of Tcur1278 (data not shown). The kinetic analysis of the hydrolysis of PCL nanoparticles was therefore performed at enzyme concentrations up to 20 µg/mL and 30 µg/mL for Tcur0390 and Tcur1278, respectively. The hydrolysis rates of PCL nanoparticles calculated from the square roots of turbidity decrease are shown as a function of enzyme concentration (Figure 3A-B). By fitting the experimental data to Eq. (1), the kinetic constants for the PCL nanoparticle hydrolysis by the two enzymes were determined (Table 3). Compared to Tcur1278, Tcur0390 showed a 2.3-fold higher adsorption equilibrium constant (K_A) and no significantly higher hydrolysis rate constant (k_r).

Kinetic analysis of the hydrolysis of PET nanoparticles by Tcur1278 and Tcur0390

The enzymatic hydrolysis of PET nanoparticles by Tcur1278 and Tcur0390 was investigated at pH 8.5 and temperatures of 50°C, 55°C and 60°C (Figure 3C-F). Due to the lower

thermal stability of Tcur0390 (Figure 2B), a hydrolytic activity at 55°C and 60°C was not detected. At 50°C, a maximum hydrolysis rate ($d(\sqrt{\tau})/dt$) of 3.3×10^{-3} min^{-1} and 5.9×10^{-3} min^{-1} was determined with 80 µg/mL of Tcur1278 and 20 µg/mL of Tcur0390, respectively. With 50 µg/mL of Tcur1278, the hydrolysis rate was increased 1.8-fold at 55°C and 2.6-fold at 60°C. Higher enzyme concentrations exceeding the amount required for the maximum reaction rate resulted in lower hydrolysis rates. This effect has also been observed in the hydrolysis of PET nanoparticles by TfCut2, a polyester hydrolase from *T. fusca*, and has been attributed to the adsorption of catalytically inactive enzyme in excess to the monolayer coverage of the PET surface (Wei et al. 2013).

Table 3 summarizes the kinetic constants for the enzymatic PET nanoparticle hydrolysis by fitting the

Table 2 Kinetic parameters for pNPB hydrolysis by Tcur1278 and Tcur0390 at 25°C and pH 8.5

	Tcur1278	Tcur0390
K_m (µM)	88.8 ± 12.8	83.1 ± 11.1
k_{cat} (1/s)	2.3 ± 0.1	12.4 ± 0.4

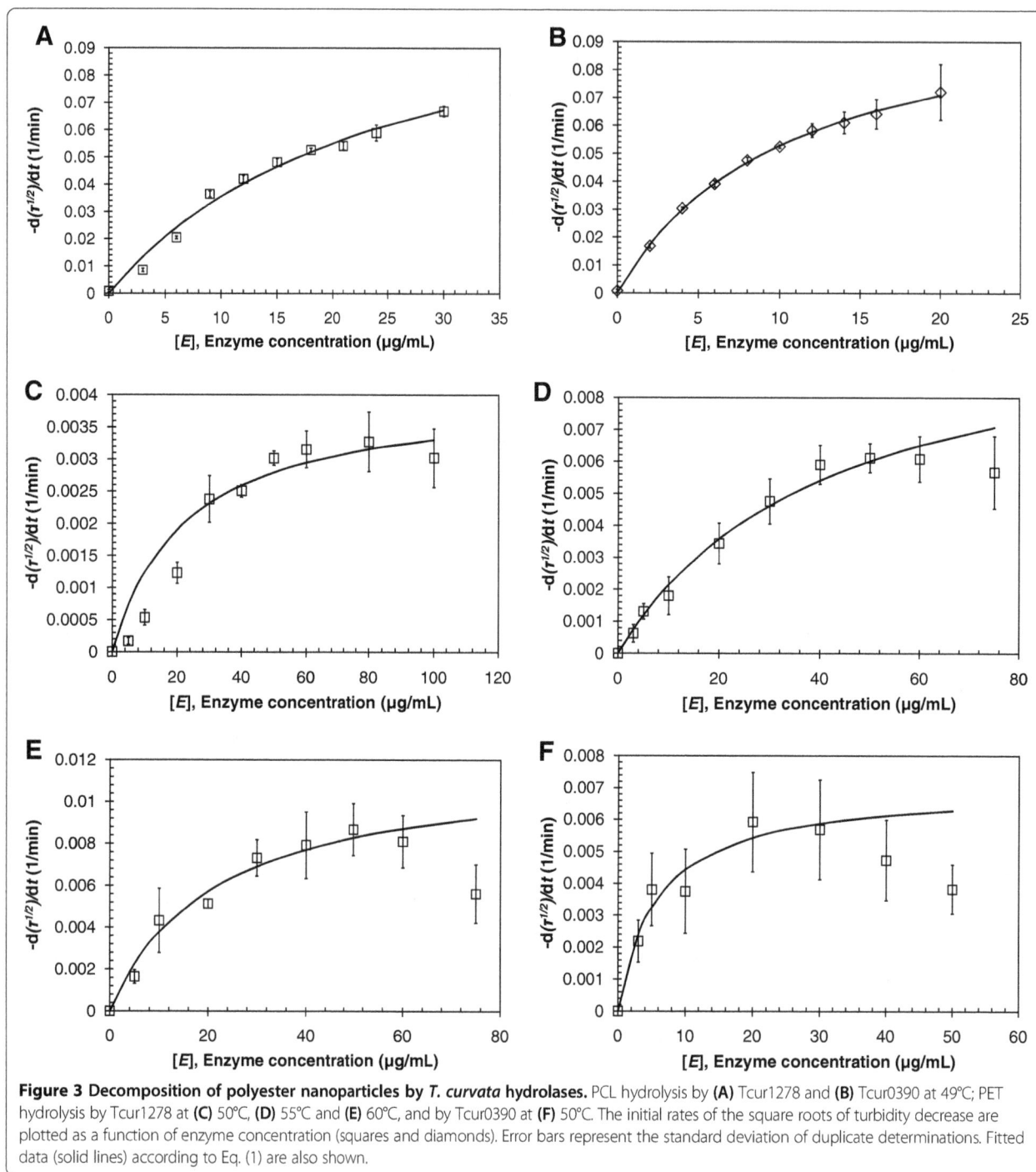

Figure 3 Decomposition of polyester nanoparticles by *T. curvata* hydrolases. PCL hydrolysis by **(A)** Tcur1278 and **(B)** Tcur0390 at 49℃; PET hydrolysis by Tcur1278 at **(C)** 50℃, **(D)** 55℃ and **(E)** 60℃, and by Tcur0390 at **(F)** 50℃. The initial rates of the square roots of turbidity decrease are plotted as a function of enzyme concentration (squares and diamonds). Error bars represent the standard deviation of duplicate determinations. Fitted data (solid lines) according to Eq. (1) are also shown.

experimental data to Eq. (1). Compared to Tcur1278, a 1.7-fold higher k_r and a 3.9-fold higher K_A were obtained with Tcur0390 at 50℃. With Tcur1278, the highest values of both kinetic constants were determined at 60℃.

In silico modeling of Tcur1278 and Tcur0390
Structural models of Tcur1278 and Tcur0390 were generated on the basis of the crystal structure of Est119 from *T. alba* AHK119 (PDB ID: 3VIS) (Kitadokoro et al.

2012). Homology models of *T. curvata* polyester hydrolases revealed a typical α/β hydrolase fold (Ollis et al. 1992; Carr and Ollis 2009) with low RMSD values of C_α atomic coordinates of less than 1 Å in comparison with the template crystal structure (Table 1).

Similar to TfCut2 from *T. fusca* KW3 (Wei 2011; Herrero Acero et al. 2011), the catalytic triad of *T. curvata* polyester hydrolases formed by S130, D176 and H208 was found to be exposed to the solvent

Table 3 Kinetic parameters for polyester nanoparticle hydrolysis by Tcur1278 and Tcur0390 at 49°C to 60°C and pH 8.5

Polyester nanoparticles		PCL	PET		
Temperature (°C)		49	50	55	60
Tcur1278	K_A (mL/mg)	41.1 ± 4.5	44.4 ± 8.6	24.3 ± 5.3	46.9 ± 12.1
	k_τ (10^{-3}/min)	122.2 ± 11.9	4.1 ± 0.5	11.0 ± 1.2	11.8 ± 1.2
Tcur0390	K_A (mL/mg)	96.0 ± 9.8	172.7 ± 30.7	n. d.	n. d.
	k_τ (10^{-3}/min)	108.3 ± 5.7	7.0 ± 0.8	n. d.	n. d.

n. d. = not detectable.

(Figure 4A-B). By docking of the 2PET model substrate, the substrate-binding pocket could be identified as a large groove on the surface of Tcur1278 and Tcur0390 (Figure 4C-F). The negative charge buried in this major groove was contributed by the deprotonated S130 (Figure 4C-D). As shown in Figure 4C-F, Tcur1278 and Tcur0390 displayed similar surface properties in the vicinity of the active site with extended hydrophobic regions around the substrate-binding groove.

Molecular dynamics simulations of Tcur1278 and Tcur0390
The overall C_α RMSD values for *T. curvata* polyester hydrolases obtained by MD simulations at 298 K and 353 K are shown as a function of simulation time (Figure 5A-B). In all simulations, the RMSD values for both proteins stabilized rapidly within 0.02 ns. At 298 K, the RMSD for Tcur1278 showed values below 0.1 nm, slightly lower than the corresponding values for Tcur0390. At 353 K, the RMSD values for Tcur0390 fluctuated more strongly compared to Tcur1278. This effect was most pronounced after 15 ns simulation time. The RMSD values for Tcur0390 were almost doubled at 353 K compared to those obtained at 298 K. In contrast, Tcur1278 exhibited only a slight increase in backbone structure deviations at 353 K further confirming its superior thermal stability properties.

The corresponding RMSF plots revealed the flexibility profiles for the complete protein sequence (Figure 5C-D). Both the enzymes displayed a flexible N-terminus with the highest deviation of C_α atoms contributed mainly by a short helical part of the molecule (12-17) embedded in loop structures (2-11, 18-23). The high RMSF observed in this part of the protein also affected the neighboring beta-sheet structure (24-30) and furthermore the whole enzyme. Compared to Tcur1278, Tcur0390 showed generally a larger difference in the flexibility profiles obtained by the MD simulations performed at temperatures from 298 K to 353 K. A significant increase of the RMSF values at higher temperatures was observed in the neighborhood of D176 (172-180) with both enzymes. This resulted also in a higher flexibility of H208 at 353 K due to its interaction with D176 according to the catalytic mechanism of ester hydrolases (Jaeger et al. 1999). As a consequence, the distance between the catalytic H208 and S130 (H-S) increased from 0.3 nm to 0.5 nm after 34 ns and 28 ns of

the MD simulation at 353 K with Tcur1278 and Tcur0390, respectively (Figure 5E-F). Compared to the RMSD plot shown in Figure 5A that suggested a relatively stable backbone structure of Tcur1278 during the complete 50 ns simulation at 353 K, the permanent change of the H-S distance occurred at an earlier stage prior to the denaturation of its other temperature-labile parts.

Discussion
Bacterial polyester hydrolases have been previously described mainly from thermophilic *Thermobifida* species (Kleeberg et al. 1998; Alisch et al. 2004; Herrero Acero et al. 2011; Thumarat et al. 2012; Ribitsch et al. 2012b; Oeser et al. 2010). *T. curvata* is a phylogenetically related actinomycete that has been isolated from a similar habitat (Henssen 1957; Zhang et al. 1998; Chertkov et al. 2011). By genome mining of *T. curvata* DSM 43183 (Chertkov et al. 2011), we identified two genes encoding the proteins Tcur1278 and Tcur0390 with sequences similar to TfCut2 from *T. fusca* KW3 (Wei 2011; Herrero Acero et al. 2011). As shown in the protein sequence alignment (Simossis and Heringa 2005), Tcur1278 and Tcur0390 share a sequence identity of about 82% and both enzymes show a sequence identity of about 61% with TfCut2 (Additional file 1: Figure S1).

Codon-optimized genes of Tcur1278 and Tcur0390 were synthesized for cloning and recombinant expression in *E. coli*. When the pET-20b(+) vector (Novagen) was used for the recombinant expression of the complete proteins of Tcur1278 and Tcur0390 in *E. coli* BL21(DE3), no active proteins could be detected suggesting an interference of the original Gram-positive signal peptides with the recombinant system. With the pBAD expression vector and *E. coli* TOP10 (Invitrogen) for shorter mature proteins, both *T. curvata* polyester hydrolases could be expressed as active enzymes fused with a C-terminal His-tag and purified by affinity chromatography (Additional file 1: Figure S2).

Similar to homologous polyester hydrolases from *T. fusca* (Chen et al. 2008; Wei 2011; Herrero Acero et al. 2011), Tcur1278 and Tcur0390 showed their highest activity against pNPB between pH 8 and pH 9 in a temperature range from 50°C to 60°C (Figure 1). Compared to Tcur1278, Tcur0390 revealed a significantly higher

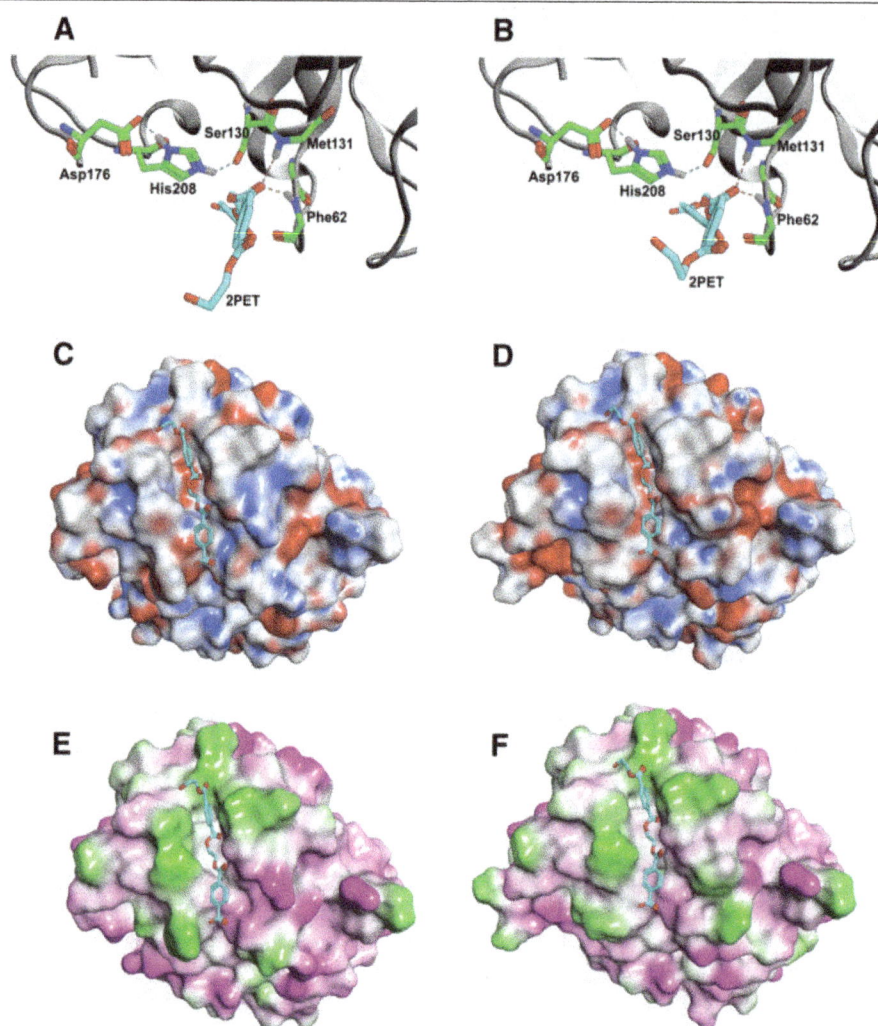

Figure 4 Structural modeling of Tcur1278 and Tcur0390 polyester hydrolases. Homology modeling was performed with the Phyre2 web server (Kelley and Sternberg 2009). The catalytic triad of **(A)** Tcur1278 and **(B)** Tcur0390 is formed by S130, D176 and H208. The 2PET model substrate was docked using GOLD 5.1 with its central ester bond constrained between 2.7 and 3.1 Å in the oxyanion hole formed by the main chain NH groups of F62 and M131 (broken yellow lines). The hydrogen bonds stabilizing the tetrahedral intermediate formed during the catalytic reaction are shown as broken lines in blue. The backbone structures are shown as gray cartoons. The electrostatic surface properties of Tcur1278 **(C)** and Tcur0390 **(D)** are shown with negatively charged residues in red, positively charged residues in blue and neutral residues in white/gray, respectively. The lipophilic surface properties of Tcur1278 **(E)** and Tcur0390 **(F)** are shown with hydrophilic residues in pink and hydrophobic residues in bright green, respectively. The docked 2PET model substrate is shown in cyan.

hydrolytic activity against both soluble (pNPB) and insoluble substrates (polyester nanoparticles) at reaction temperatures up to 50°C (Figure 3, Tables 2 and 3). This higher hydrolytic activity could be attributed to the stronger substrate affinity of Tcur0390 (Table 3). Molecular docking experiments with the model substrate 2PET to the structural models of *T. curvata* polyester hydrolases confirmed the presence of extended hydrophobic regions in close vicinity to the catalytic triad (Figure 4E-F). The hydrophobic character of the regions near the substrate-binding groove may facilitate the binding of hydrophobic polymeric substrates. Compared to Tcur1278, the hydrophobic properties were

more pronounced in Tcur0390 and may account for its observed higher substrate affinity (Figure 4E-F). This result is confirming earlier observations that more hydrophobic and less charged amino acid residues clustered in the neighborhood of the substrate-binding groove of Thc_Cut1 compared to Thc_Cut2 and a concomitantly higher hydrolytic activity of the former isoenzyme from *T. cellulosilytica* (Herrero Acero et al. 2011; Herrero Acero et al. 2013).

The optimal temperatures for pNPB hydrolysis by Tcur1278 and Tcur0390 were 60°C and 55°C, respectively (Figure 1B). However, both enzymes showed poor thermal stability at their optimal temperature, as indicated by an

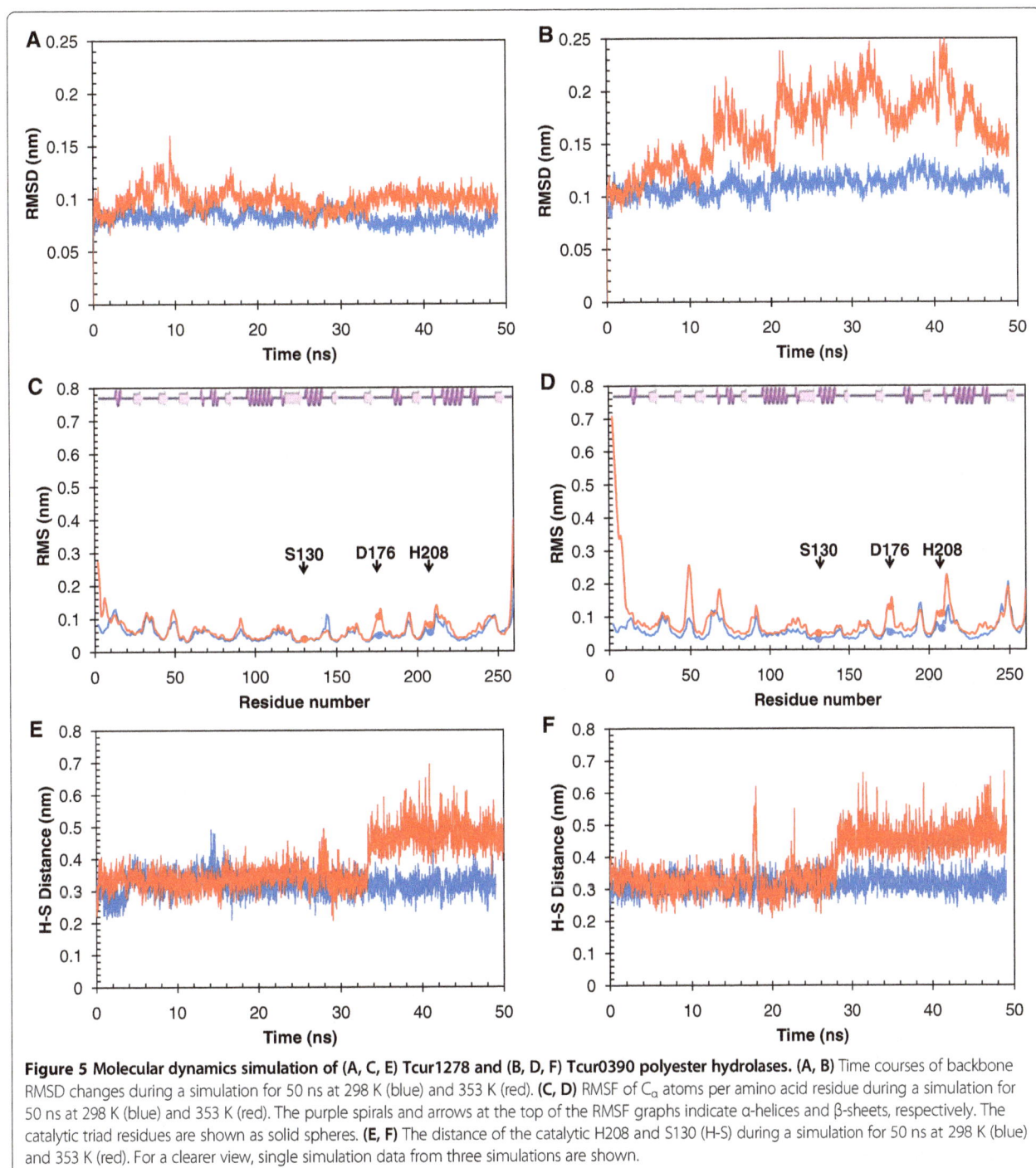

Figure 5 Molecular dynamics simulation of (A, C, E) Tcur1278 and (B, D, F) Tcur0390 polyester hydrolases. (A, B) Time courses of backbone RMSD changes during a simulation for 50 ns at 298 K (blue) and 353 K (red). **(C, D)** RMSF of C_α atoms per amino acid residue during a simulation for 50 ns at 298 K (blue) and 353 K (red). The purple spirals and arrows at the top of the RMSF graphs indicate α-helices and β-sheets, respectively. The catalytic triad residues are shown as solid spheres. **(E, F)** The distance of the catalytic H208 and S130 (H-S) during a simulation for 50 ns at 298 K (blue) and 353 K (red). For a clearer view, single simulation data from three simulations are shown.

irreversible loss of more than 65% of their initial activities following incubation for 10 min (Figure 2A-B). Tcur1278 maintained its maximum activity against PET nanoparticles for approximately 15 min at 60°C (data not shown). This suggests that the thermal stability of Tcur1278 was improved in the presence of the insoluble polymeric substrate. In contrast, a significant improvement of the thermal stability was not detected with Tcur0390 in the presence of PET nanoparticles. The backbone RMSD

plots obtained by MD simulations also indicated a more rigid structure of Tcur1278 thus verifying its superior thermal stability compared to Tcur0390 (Figure 5A-B). The RMSF profiles that describe the deviation of C_α atoms of single amino acid residues from the averaged position over the simulation period showed highly flexible regions clustered in the neighborhood of the catalytic residues H208 and D176 (Figure 5C-D). These regions may enable some induced fit motions necessary for the catalytic reaction at the active site. A comparison of the backbone RMSD plots (Figure 5A) and the H-S distance of Tcur1278 (Figure 5E) over the complete MD simulation period indicated that the exposed flexible catalytic triad was also prone to local unfolding prior to the denaturation of the overall structure. By contrast, the permanent increase of the H-S distance in Tcur0390 was accompanied by the unfolding of the overall structure and occurred at an earlier stage of MD simulations compared to Tcur1278 (Figure 5B, F).

A reaction temperature close to the glass transition temperature of PET at approximately 75°C (Alves et al. 2002) is required for an optimal performance of the enzymatic hydrolysis due to the restricted mobility of polymer chains at temperatures below (Marten et al. 2003, 2005; Herzog et al. 2006; Ronkvist et al. 2009; Wei et al. 2013). The thermal stability of Tcur1278 needs therefore be further improved for an efficient degradation of PET. Protein engineering in regions near the catalytic triad as well as at the flexible N-terminus could be a useful approach for further optimizations to overcome the limited thermal stability of these polyester hydrolases.

In summary, the polyester hydrolases Tcur1278 and Tcur0390 from *T. curvata* have been shown to exhibit catalytic and structural features similar to enzymes from *T. fusca* and *T. cellulosilytica*. The comparison of the catalytic characteristics of Tcur1278 and Tcur0390 revealed a correlation between their hydrolytic activity and their surface properties in the vicinity of the catalytic triad. However, a comparison of the thermal stability of the two enzymes provided evidence that their ability to hydrolyze PET is predominately limited by their stability at higher reaction temperatures.

Additional file

Additional file 1: Figure S1. Alignment of the mature protein sequences of Tcur1278, Tcur0390, TfCut2 and Est119 polyester hydrolases. The regions of similarity of individual amino acid residues are indicated with colors from blue, unconserved to red, conserved. The multiple sequence alignment was performed with the PRALINE web server (Simossis and Heringa 2005). **Figure S2.** SDS PAGE analysis of Tcur0390 (lanes 1-2) and Tcur1278 (lanes 3-4). 10 µg of crude cell lysate (1, 3) and eluate obtained after IMAC purification (2, 4) were loaded in each lane; protein size markers (M). The gel was first stained with Fast Red dye for esterase activity against 1-naphthyl acetate (purple bands) followed by staining with Coomassie Brilliant Blue (blue bands).

Competing interests

The authors declare that they have no competing interests.

Authors' contributions

RW and NK carried out the recombinant cloning of genes coding for the polyester hydrolases. MB and JS participated in the expression and purification of the recombinant enzymes. TO and RW carried out the biochemical characterization of the polyester hydrolases. JT performed the molecular dynamics simulations. RW and TO analyzed the experimental and simulation data and prepared the manuscript. WZ conceived the study and contributed to manuscript writing. All authors read and approved the final manuscript.

Acknowledgements

Dr. René Meier, Institute of Biochemistry, University of Leipzig is acknowledged for his assistance in the MD simulations. This work was supported by the Deutsche Bundesstiftung Umwelt (AZ 13267; AZ 2012/202).

References

Alisch-Mark M, Herrmann A, Zimmermann W (2006) Increase of the hydrophilicity of polyethylene terephthalate fibres by hydrolases from *Thermomonospora fusca* and *Fusarium solani f. sp. pisi*. Biotechnol Lett 28(10):681–685. doi:10.1007/s10529-006-9041-7

Alisch M, Feuerhack A, Müller H, Mensak B, Andreaus J, Zimmermann W (2004) Biocatalytic modification of polyethylene terephthalate fibres by esterases from actinomycete isolates. Biocatal Biotransform 22(5-6):347–351. doi:10.1080/10242420400025877

Alves NM, Mano JF, Balaguer E, Meseguer Duenas JM, Gomez Ribelles JL (2002) Glass transition and structural relaxation in semi-crystalline poly(ethylene terephthalate): a DSC study. Polymer 43(15):4111–4122. doi:10.1016/S0032-3861(02)00236-7

Berendsen HJC, Postma JPM, van Gunsteren WF, DiNola A, Haak JR (1984) Molecular dynamics with coupling to an external bath. J Chem Phys 81 (8):3684–3690

Billig S, Oeser T, Birkemeyer C, Zimmermann W (2010) Hydrolysis of cyclic poly (ethylene terephthalate) trimers by a carboxylesterase from *Thermobifida fusca* KW3. Appl Microbiol Biotechnol 87(5):1753–1764. doi:10.1007/s00253-010-2635-y

Carr PD, Ollis DL (2009) Alpha/beta hydrolase fold: an update. Protein Pept Lett 16(10):1137–1148. doi:10.2174/092986609789071298

Chen S, Su L, Billig S, Zimmermann W, Chen J, Wu J (2010) Biochemical characterization of the cutinases from *Thermobifida fusca*. J Mol Catal B-Enzym 63(3–4):121–127. doi:10.1016/j.molcatb.2010.01.001

Chen S, Tong X, Woodard RW, Du G, Wu J, Chen J (2008) Identification and characterization of bacterial cutinase. J Biol Chem 283(38):25854–25862. doi:10.1074/jbc.M800848200

Chertkov O, Sikorski J, Nolan M, Lapidus A, Lucas S, Del Rio TG, Tice H, Cheng JF, Goodwin L, Pitluck S, Liolios K, Ivanova N, Mavromatis K, Mikhailova N, Ovchinnikova G, Pati A, Chen A, Palaniappan K, Djao OD, Land M, Hauser L, Chang YJ, Jeffries CD, Brettin T, Han C, Detter JC, Rohde M, Goker M, Woyke T, Bristow J, Eisen JA, Markowitz V, Hugenholtz P, Klenk HP, Kyrpides NC (2011) Complete genome sequence of *Thermomonospora curvata* type strain (B9). Stand Genomic Sci 4(1):13–22. doi:10.4056/sigs.1453580

Davies MT (1959) A universal buffer solution for use in ultra-violet spectrophotometry. Analyst 84:248–251. doi:10.1039/AN9598400248

Eberl A, Heumann S, Bruckner T, Araujo R, Cavaco-Paulo A, Kaufmann F, Kroutil W, Guebitz GM (2009) Enzymatic surface hydrolysis of poly(ethylene terephthalate) and bis(benzoyloxyethyl) terephthalate by lipase and cutinase in the presence of surface active molecules. J Biotechnol 143(3):207–212. doi:10.1016/j.jbiotec.2009.07.008

Egmond MR, de Vlieg J (2000) *Fusarium solani pisi* cutinase. Biochimie 82(11):1015–1021. doi:10.1016/S0300-9084(00)01183-4

Henssen A (1957) Beitraege zur Morphologie und Systematik der thermophilen Actinomyceten. Arch Mikrobiol 26(4):373–414. doi:10.1007/bf00407588

Henssen A, Schnepf E (1967) Zur Kenntnis thermophiler Actinomyceten. Arch Mikrobiol 57(3):214–231. doi:10.1007/bf00405948

Herrero Acero E, Ribitsch D, Dellacher A, Zitzenbacher S, Marold A, Steinkellner G, Gruber K, Schwab H, Guebitz GM (2013) Surface engineering of a cutinase from *Thermobifida cellulosilytica* for improved polyester hydrolysis. Biotechnol Bioeng 110(10):2581–2590. doi:10.1002/bit.24930

Herrero Acero E, Ribitsch D, Steinkellner G, Gruber K, Greimel K, Eiteljoerg I, Trotscha E, Wei R, Zimmermann W, Zinn M, Cavaco-Paulo A, Freddi G, Schwab H, Guebitz G (2011) Enzymatic surface hydrolysis of PET: Effect of structural diversity on kinetic properties of cutinases from *Thermobifida*. Macromolecules 44(12):4632–4640. doi:10.1021/ma200949p

Herzog K, Müller RJ, Deckwer WD (2006) Mechanism and kinetics of the enzymatic hydrolysis of polyester nanoparticles by lipases. Polym Degrad Stab 91(10):2486–2498. doi:10.1016/j.polymdegradstab.2006.03.005

Hess B, Kutzner C, van der Spoel D, Lindahl E (2008) GROMACS 4: Algorithms for highly efficient, load-balanced, and scalable molecular simulation. J Chem Theory Comput 4(3):435–447. doi:10.1021/ct700301q

Hornak V, Abel R, Okur A, Strockbine B, Roitberg A, Simmerling C (2006) Comparison of multiple Amber force fields and development of improved protein backbone parameters. Proteins 65(3):712–725. doi:10.1002/prot.21123

Jaeger KE, Dijkstra BW, Reetz MT (1999) Bacterial biocatalysts: molecular biology, three-dimensional structures, and biotechnological applications of lipases. Annu Rev Microbiol 53:315–351. doi:10.1146/annurev.micro.53.1.315

Jones G, Willett P, Glen RC, Leach AR, Taylor R (1997) Development and validation of a genetic algorithm for flexible docking. J Mol Biol 267(3):727–748. doi:10.1006/jmbi.1996.0897

Kelley LA, Sternberg MJ (2009) Protein structure prediction on the Web: a case study using the Phyre server. Nat Protoc 4(3):363–371. doi:10.1038/nprot.2009.2

Kitadokoro K, Thumarat U, Nakamura R, Nishimura K, Karatani H, Suzuki H, Kawai F (2012) Crystal structure of cutinase Est119 from *Thermobifida alba* AHK119 that can degrade modified polyethylene terephthalate at 1.76 Å resolution. Polym Degrad Stab 97(5):771–775. doi:10.1016/j.polymdegradstab.2012.02.003

Kleeberg I, Hetz C, Kroppenstedt RM, Muller RJ, Deckwer WD (1998) Biodegradation of aliphatic-aromatic copolyesters by *Thermomonospora fusca* and other thermophilic compost isolates. Appl Environ Microbiol 64(5):1731–1735

Müller R-J, Schrader H, Profe J, Dresler K, Deckwer W-D (2005) Enzymatic degradation of poly(ethylene terephthalate): rapid hydrolyse using a hydrolase from *T. fusca*. Macromol Rapid Commun 26(17):1400–1405. doi:10.1002/marc.200500410

Marten E, Müller R-J, Deckwer W-D (2003) Studies on the enzymatic hydrolysis of polyesters I. Low molecular mass model esters and aliphatic polyesters. Polym Degrad Stab 80(3):485–501. doi:10.1016/S0141-3910(03)00032-6

Marten E, Müller R-J, Deckwer W-D (2005) Studies on the enzymatic hydrolysis of polyesters. II Aliphatic-aromatic copolyesters. Polym Degrad Stab 88(3):371–381. doi:10.1016/j.polymdegradstab.2004.12.001

Nimchua T, Punnapayak H, Zimmermann W (2007) Comparison of the hydrolysis of polyethylene terephthalate fibers by a hydrolase from *Fusarium oxysporum* LCH I and *Fusarium solani* f. sp. *pisi*. Biotechnol J 2(3):361–364. doi:10.1002/biot.200600095

Oeser T, Wei R, Baumgarten T, Billig S, Follner C, Zimmermann W (2010) High level expression of a hydrophobic poly(ethylene terephthalate)-hydrolyzing carboxylesterase from *Thermobifida fusca* KW3 in *Escherichia coli* BL21(DE3). J Biotechnol 146(3):100–104. doi:10.1016/j.jbiotec.2010.02.006

Ollis DL, Cheah E, Cygler M, Dijkstra B, Frolow F, Franken SM, Harel M, Remington SJ, Silman I, Schrag J, Sussman JL, Verschueren KHG, Goldman A (1992) The alpha/beta hydrolase fold. Protein Eng 5(3):197–211. doi:10.1093/protein/5.3.197

Paszun D, Spychaj T (1997) Chemical recycling of poly(ethylene terephthalate). Ind Eng Chem Res 36(4):1373–1383. doi:10.1021/ie960563c

Ribitsch D, Herrero Acero E, Greimel K, Dellacher A, Zitzenbacher S, Marold A, Rodriguez RD, Steinkellner G, Gruber K, Schwab H, Guebitz GM (2012a) A new esterase from *Thermobifida halotolerans* hydrolyses polyethylene terephthalate (PET) and polylactic acid (PLA). Polymers 4(1):617–629

Ribitsch D, Herrero Acero E, Greimel K, Eiteljoerg I, Trotscha E, Freddi G, Schwab H, Guebitz GM (2012b) Characterization of a new cutinase from *Thermobifida alba* for PET-surface hydrolysis. Biocatal Biotransform 30(1):2–9. doi:10.3109/10242422.2012.644435

Ronkvist ÅM, Xie W, Lu W, Gross RA (2009) Cutinase-catalyzed hydrolysis of poly (ethylene terephthalate). Macromolecules 42(14):5128–5138. doi:10.1021/ma9005318

Simossis VA, Heringa J (2005) PRALINE: a multiple sequence alignment toolbox that integrates homology-extended and secondary structure information. Nucleic Acids Res 33(Web Server issue):W289–W294. doi:10.1093/nar/gki390

Sulaiman S, Yamato S, Kanaya E, Kim JJ, Koga Y, Takano K, Kanaya S (2012) Isolation of a novel cutinase homolog with polyethylene terephthalate-degrading activity from leaf-branch compost by using a metagenomic approach. Appl Environ Microbiol 78(5):1556–1562. doi:10.1128/AEM.06725-11

Sztajer H, Lunsdorf H, Erdmann H, Menge U, Schmid R (1992) Purification and properties of lipase from *Penicillium simplicissimum*. Biochim Biophys Acta 1124(3):253–261

Thumarat U, Nakamura R, Kawabata T, Suzuki H, Kawai F (2012) Biochemical and genetic analysis of a cutinase-type polyesterase from a thermophilic *Thermobifida alba* AHK119. Appl Microbiol Biotechnol 95(2):419–430. doi:10.1007/s00253-011-3781-6

Wei R (2011) Hydrolysis of polyethylene terephthalate by cutinases from *Thermobifida fusca* KW3. PhD Thesis. Universität Leipzig, Leipzig

Wei R, Oeser T, Barth M, Weigl N, Lübs A, Schulz-Siegmund M, Hacker M, Zimmermann W (2013) Turbidimetric analysis of the enzymatic hydrolysis of polyethylene terephthalate nanoparticles. J Mol Catal B-Enzym. doi:10.1016/j.molcatb.2013.08.010

Zhang Z, Wang Y, Ruan J (1998) Reclassification of *Thermomonospora* and *Microtetraspora*. Int J Syst Bacteriol 48(2):411–422. doi:10.1099/00207713-48-2-411

Zimmermann W, Billig S (2011) Enzymes for the biofunctionalization of poly (ethylene terephthalate). In: Nyanhongo GS, Steiner W, Guebitz G (ed) Biofunctionalization of polymers and their applications. Adv Biochem Eng/Biotechnol, vol 125. Springer, Berlin/Heidelberg, pp 97–120. doi:10.1007/10_2010_87

5

Foam-free production of Surfactin via anaerobic fermentation of *Bacillus subtilis* DSM 10^T

Judit Willenbacher[1*], Jens-Tilman Rau[1], Jonas Rogalla[1], Christoph Syldatk[1] and Rudolf Hausmann[2]

Abstract

Surfactin is one of the most popular biosurfactants due to its numerous potential applications. The usually aerobic production via fermentation of *Bacillus subtilis* is accompanied by vigorous foaming which leads to complex constructions and great expense. Therefore it is reasonable to search for alternative foam-free production processes. The current study introduces a novel approach to produce Surfactin in a foam-free process applying a strictly anaerobic bioreactor cultivation. The process was performed several times with different glucose concentrations in mineral salt medium. The fermentations were analyzed regarding specific ($q_{Surfactin}$, vol. $q_{Surfactin}$) and overall product yields ($Y_{P/X}$, $Y_{P/S}$) as well as substrate utilization ($Y_{X/S}$). Fermentations in which 2.5 g/L glucose were employed proofed to be the most effective, reaching product yields of $Y_{P/X} = 0.278$ g/g. Most interesting, the product yields exceeded classical aerobic fermentations, in which foam fractionation was applied. Additionally, values for specific production rate $q_{Surfactin}$ (0.005 g/(g·h)) and product yield per consumed substrate ($Y_{P/S} = 0.033$ g/g) surpass results of comparable foam-free processes. The current study introduces an alternative to produce a biosurfactant that overcomes the challenges of severe foaming and need for additional constructions.

Keywords: Surfactin; Anaerobic fermentation; *Bacillus subtilis*; Foam-free

Introduction

Biosurfactants become increasingly attractive based on their biodegradability and production on the basis of renewable resources (Banat et al. 2010). Surfactin is one of the most popular biosurfactants and was already discovered in 1968 by Arima et al. (1968). The lipopeptide consists of a seven amino acid peptide ring comprising a β-hydroxy fatty acid. Surfactin is produced by *Bacillus subtilis*, a gram positive bacterium known for its application in several industrial processes, for instance the production of detergent enzymes and others (Schallmey et al. 2004). The molecule exhibits various different characteristics, which might lead to several applications. For instance, Surfactin was shown to improve plant self-resistance mechanism against soil bacteria (Ongena et al. 2007) or vigorously affects mycoplasma cells (Vollenbroich et al. 1997). Therefore, next to an application in detergents, washing agents or food products, a usage in agriculture or pharmaceutical products is also imaginable.

Naturally, amphiphile molecules produced by bacteria in cultivation processes accumulate at gas–liquid interfaces and lead to massive foam formation. The main challenge in cultivating microorganisms producing biosurfactants is to overcome this severe foam production. In the majority of cases foaming is handled by the addition of antifoam agents. Unfortunately, this strategy harbors several disadvantages, as antifoam agents are expensive and very hard to remove in downstream processes. The second most common method to cope with foam formation is to disrupt the foam by shear stress or pressure using foam breakers. However, this method is often insufficient and increases the overall costs for the production of biosurfactants. Another, more elegant, way to manage foaming in biosurfactant production processes is to apply foam fractionation, which was already shown by Cooper et al. 1981 (Cooper et al. 1981). This technique inverts the disadvantage into an advantage by using the accumulation of biosurfactants in the foam for *in situ* product enrichment and recovery. The Surfactin producer *Bacillus subtilis* is especially suited for the

* Correspondence: judit.willenbacher@kit.edu
[1]Institute of Process Engineering in Life Sciences, Section II: Technical Biology, Karlsruhe Institute of Technology (KIT), Engler-Bunte-Ring 1, 76131, Karlsruhe, Germany
Full list of author information is available at the end of the article

employment of foam fractionation, yielding high values in product recovery and enrichment (Willenbacher et al. 2014). Although this is a possible way to handle foam and to improve product yields, a realization in industrial scale is probably unrealistic in the near future.

Another artful approach is to avoid foaming at all instead of dealing with it. Several attempts have been made to establish foam-free fermentation processes. Ohno et al. for instance employed a solid state fermentation of recombinant *Bacillus subtilis* MI113 (pC112), using soybean curd residue as solid substrate (Ohno et al. 1995), which led to a yield of 2.0 g/kg (Surfactin per wet weight). Another attempt to produce Surfactin in a foam-free fashion implemented a membrane bioreactor (Coutte et al. 2010). A culture of *Bacillus subtilis* ATCC 21332 obtained a maximal Surfactin concentration of 0.242 g/L. However, a significant amount of Surfactin was adsorbed at the membranes and oxygen transfer was reduced significantly. In contrast, Chtioui et al. focused on a rotating disc bioreactor for the production of Surfactin, allowing *Bacillus subtilis* ATCC 21332 to grow free and immobilized in a biofilm at the same time (Chtioui et al. 2012). Aeration was realized above the fluid level, when the overgrown discs arose from the liquid. Maximal Surfactin concentrations of 0.212 g/L were obtained, but oxygen supply was limited and Fengycin concentrations surpassed Surfactin concentrations by far. While all these studies implemented innovative ideas to circumvent foaming, those processes are either difficult to scale up or lack high specificity.

Bacillus subtilis was for a long time believed to be a strict aerobic bacterium. Since 1995 research on the anaerobic growth behavior of *Bacillus subtilis* increased dramatically (Hoffmann et al. 1995; Nakano et al. 1997). By using nitrate as the terminal electron acceptor, *Bacillus subtilis* is able to perform anaerobic respiration via a nitrate reductase encoded by operon *narGHJI* (Ramos et al. 1995). In this manner nitrate is reduced to nitrite, which thereafter is transformed to ammonium via a nitrite reductase encoded by *nasDEF* (Nakano et al. 1998).

The production of biosurfactants under anaerobic conditions was already shown in 1985. The study presents the production of an undefined biosurfactant by *Bacillus licheniformis* in glucose mineral salt medium (Javaheri et al. 1985). The cultivation was performed in shake flasks, in the course of which the decreasing surface tension (from 70 mN/m to 28 mN/m) was measured. Although the characterization of the biosurfactant was only performed by thin layer chromatography and no high pressure liquid chromatography (HPLC) was applied, Javaheri et al. laid the foundation of anaerobic biosurfactant production. Subsequently, Davis et al. investigated the impact of nitrogen, carbon and oxygen conditions on Surfactin production of *Bacillus subtilis*

ATCC 21332 (Davis et al. 1999). Interestingly, maximal product yields were obtained under nitrate-limited and oxygen-depleted conditions ($Y_{P/X} = 0.075$), which gives a further impulse to examine anaerobic Surfactin production. The proof of concept was provided by Zhang et al., who produced Surfactin with *Bacillus subtilis* ATCC 21332 strictly anaerobic for the first time (Zhang et al. 2007). The investigation focused on a connected shake flask system, introducing a nitrogen flow to induce vigorous foaming. The foam was channeled through several flasks with distilled water to collect the produced biosurfactant. While these studies demonstrate that anaerobic production of Surfactin is possible, none of them propose a solution to overcome foaming.

The aim of the current study is to combine the relatively new research field of anaerobic biosurfactant production with a foam-free bioprocess strategy (Figure 1 B). Therefore the anaerobic growth behavior of *Bacillus subtilis* DSM 10[T] was investigated in a 2.5 L benchtop bioreactor without any gas flow through the liquid phase. Four different glucose concentrations were tested and evaluated regarding their influence on Surfactin production. The processes were analyzed focusing on maximal Surfactin concentrations ($c_{Surfactin}$), growth rates (μ_{max}), product and substrate yields ($Y_{P/X}$, $Y_{X/S}$, $Y_{P/S}$), specific production rates ($q_{Surfactin}$) and specific volumetric production rates (vol. $q_{Surfactin}$).

Materials and methods
Chemicals
All chemicals applied in the current study were of analytical grade and purchased from Carl Roth GmbH (Karlsruhe, Germany). The Surfactin standards for HPLC analysis were obtained from Sigma-Aldrich Laborchemikalien GmbH (Seelze, Germany).

Microorganism and strain maintenance
The wildtype strain *Bacillus subtilis* DSM 10[T] was used for all experiments during this study. The microorganism was obtained from the DSMZ (Deutsche Sammlung von Mikroorganismen und Zellkulturen GmbH, Braunschweig, Germany) and stored as glycerol stocks, prepared from a culture in Lysogeny Broth (Bertani 1951) from the exponential growth phase, at –80°C.

Culture conditions
Media
The employed mineral salt medium was based on the fermentation medium of Cooper (Cooper et al. 1981): 8.0×10^{-4} M $MgSO_4$, 7.0×10^{-6} M $CaCl_2$, 4.0×10^{-6} M $FeSO_4$, 4.0×10^{-6} M Na_2EDTA, 1×10^{-6} M $MnSO_4$. In contrast to the original medium (40 g/L glucose) the concentration of glucose was altered to 2.5 g/L, 5 g/L, 7.5 g/L and 10 g/L, during various cultivations.

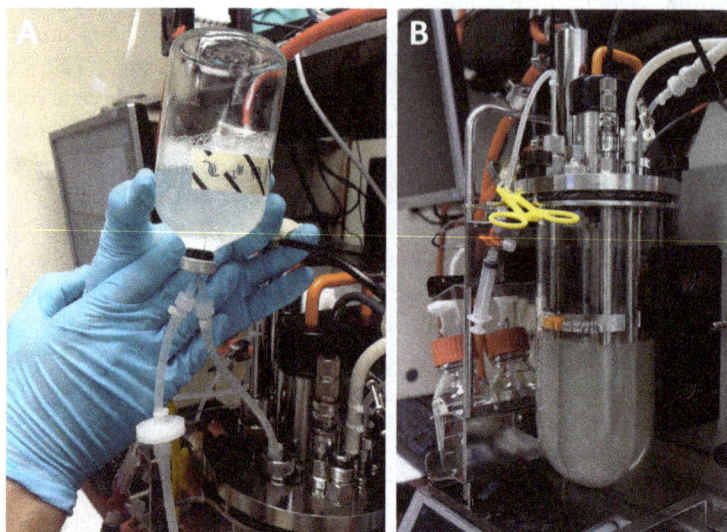

Figure 1 Inoculation and fermentation of *Bacillus subtilis* DSM 10^T in 2.5 L benchtop bioreactor. **A**. Direct inoculation of the benchtop fermenter using a serum bottle with preculture. Nitrogen was introduced into the serum bottle via a small filter creating excess pressure inside the bottle. A second tube was used to channel the preculture directly into the inoculum device. **B**. Foam-free cultivation of *Bacillus subtilis* DSM 10^T applying an anaerobic fermentation process.

Furthermore, the former nitrogen source 0.05 M NH$_4$NO$_3$ was replaced with 0.1 M NH$_4$Cl and 0.1177 M NaNO$_3$. The deployed concentration of the phosphate buffer demanded slight changes depending on its usage for inoculum cultures or fermentation processes. For the cultivation in serum bottles the original 0.07 M phosphate buffer (0.03 M KH$_2$PO$_4$ and 0.04 M Na$_2$HPO$_4$) was used, whereas for the cultivation in benchtop bioreactors a 0.01 M phosphate buffer was employed (4.29×10^{-3} M KH$_2$PO$_4$ and 5.71×10^{-3} M Na$_2$HPO$_4$).

The preparation of medium for the cultivation in serum bottles demanded a different approach compared to the preparation of medium for the cultivation in benchtop bioreactors. Four different stock solutions were prepared for the cultivation in serum bottles. One stock solution contained the salt compounds (NH$_4$Cl, NaNO$_3$, KH$_2$PO$_4$, Na$_2$HPO$_4$) and was later completed to the final volume of 50 or 100 mL, respectively. The second stock solution included a 5.56-fold glucose solution of the final glucose concentration. In comparison, the third and fourth stock solution contained a 50-fold MgSO$_4$ solution and a 1000-fold solution of the trace elements (CaCl$_2$, FeSO$_4$, Na$_2$EDTA, MnSO$_4$). All solutions were filled into separate serum bottles and anaerobic conditions were adjusted by 20 alternating cycles of purging with gas (20 vol.-% CO$_2$ in N$_2$, 45 s) and evacuating (70 mbar, 45 s). Subsequently the bottles were autoclaved and the salt stock solution was completed under anaerobic conditions to receive the final concentrations of glucose, MgSO$_4$ and trace elements.

For the preparation of the bioreactor medium four stock solutions were prepared in a similar fashion. However, the first stock solution (NH$_4$Cl, NaNO$_3$, KH$_2$PO$_4$, Na$_2$HPO$_4$) was autoclaved inside the bioreactor. Whereas the glucose, MgSO$_4$ and trace elements stock solutions were prepared and autoclaved in separate vessels. The medium was completed inside the bioreactor after sterilization and thereafter anaerobic conditions were reached by purging the bioreactor with N$_2$ (4 Lpm, 1050 rpm, 20 min, Figure 2: valve 1 open).

Preparation of inoculum cultures

For the preparation of the first seed culture a loop of *B. subtilis* DSM 10^T from the glycerol stock solution was inoculated in 20 mL of Lysogeny Broth (inside a 100 mL baffled shake flask) and incubated in a shake incubator chamber (Multitron II, HT Infors, Bottmingen, Switzerland) at 30°C and 120 rpm for 24 h. The second seed culture was inoculated with a resulting OD$_{600}$ of 0.05 under anaerobic conditions in prepared serum bottles with 50 or 100 mL of mineral salt medium, respectively. The serum bottles were incubated in a horizontal position but otherwise in the same manner as the first seed culture. After 24 h of incubation approximately 200 mL of the second seed culture were used to inoculate the aqueous phase of the bioreactor (Figure 1 A). The initial OD$_{600}$ inside the bioreactor fluctuated between 0.03 and 0.07, depending on bacterial growth of the second seed culture.

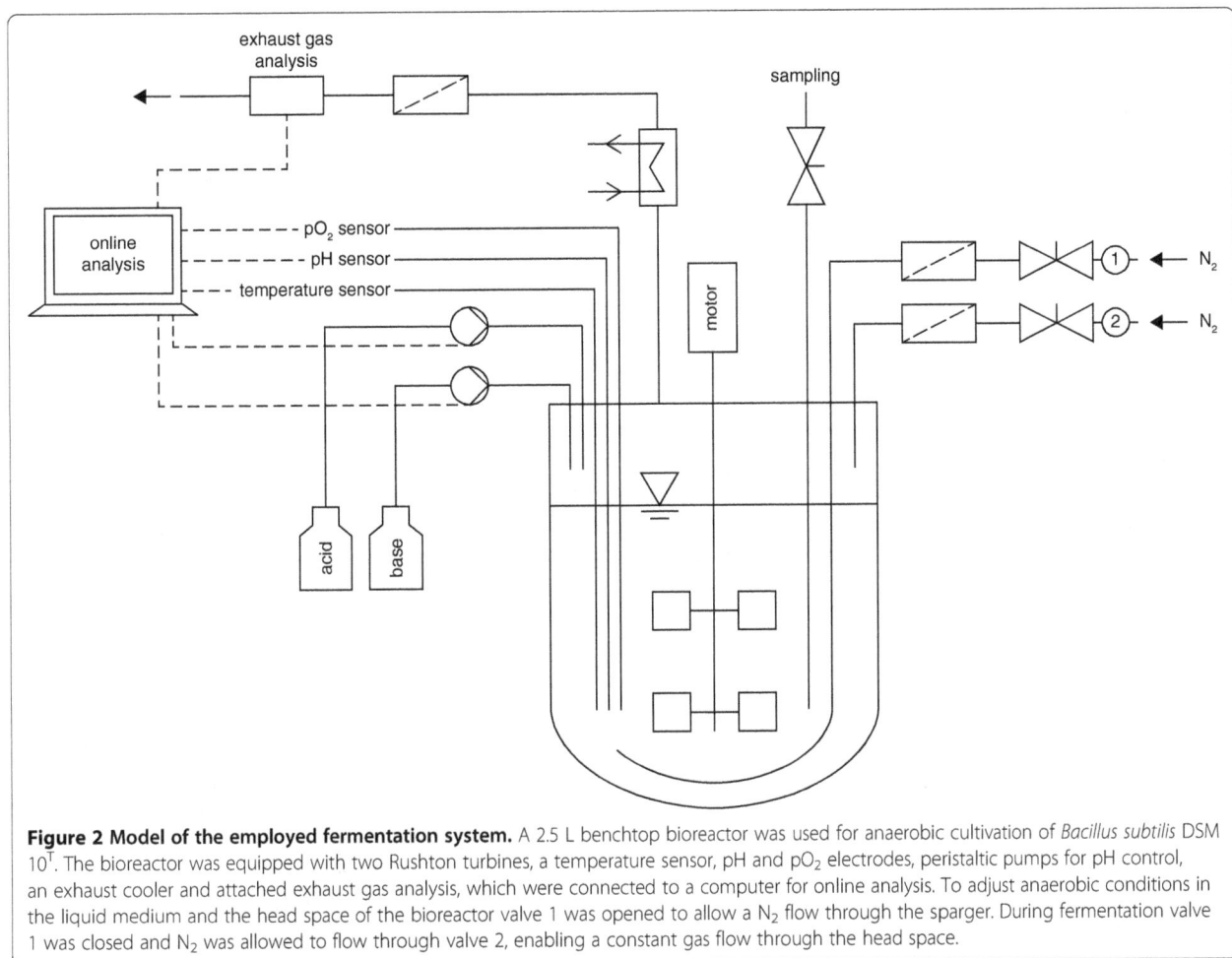

Figure 2 Model of the employed fermentation system. A 2.5 L benchtop bioreactor was used for anaerobic cultivation of *Bacillus subtilis* DSM 10[T]. The bioreactor was equipped with two Rushton turbines, a temperature sensor, pH and pO_2 electrodes, peristaltic pumps for pH control, an exhaust cooler and attached exhaust gas analysis, which were connected to a computer for online analysis. To adjust anaerobic conditions in the liquid medium and the head space of the bioreactor valve 1 was opened to allow a N_2 flow through the sparger. During fermentation valve 1 was closed and N_2 was allowed to flow through valve 2, enabling a constant gas flow through the head space.

Cultivation in a 2.5 L benchtop bioreactor

All cultivations were carried out in 2.5 L benchtop bioreactors (Minifors, HT Infors, Bottmingen, Switzerland) with 1.0 L mineral salt medium. The bioreactors were equipped with pH (Mettler-Toledo International Inc., Greifensee, Switzerland) and pO_2 electrodes (Oxyferm, Hamilton Bonaduz AG, Bonaduz, Switzerland), a temperature sensor and Rushton turbines. The temperature was adjusted to 30°C and the pH was controlled to a value of 7.0 by the addition of 4 M NaOH or 4 M H_3PO_4 (Figure 2). The stirrer was adjusted to 300 rpm the entire time of cultivation. The medium was not exposed to gas flow throughout the whole fermentation process to guarantee an absolutely foam-free cultivation. However, to avoid reflux of air through the exhaust cooler and to allow the measurement of CO_2 through the exhaust gas analysis, a constant N_2 gas flow through the headspace of the bioreactor with 0.1 Lpm (1.5 L headspace volume) was adjusted (Figure 2: valve 2 open).

The fermentation process was started with 1.0 L of the described mineral salt medium and the additional volume of the inoculated seed culture (200 mL). Since the bioreactor cultivation was realized as a batch cultivation, no further medium components were added. During the cultivation pH, pO_2, CO_2 exhaust, temperature, stirrer speed and addition of acid and base were consistently monitored (Figure 2). Samples were taken from the cultivation broth (4 mL) without allowing any air flow inside the bioreactor. All fermentations were performed as duplicates.

Analytical methods
Sampling and sample processing

By day samples were taken every three hours, whereas during nights the intervals were between five and seven hours. The sampling was designed to prevent air from entering the bioreactor system to guarantee anaerobic conditions inside. The offline analysis of the cultivation broth samples included the determination of the OD_{600} and the glucose, nitrate and Surfactin concentration of the supernatant. The concentration of glucose and nitrate was analyzed using a glucose assay kit (Cat. No. 10 716 251 035, R-Biopharm AG, Darmstadt, Germany) and a nitrate assay kit (1.09713.0001, Merck KGaA, Darmstadt, Germany). The concentrations were determined according to the manufacturer instructions, utilizing a spectrophotometric method (Ultrospec 2100 pro, General Electric Deutschland Holding GmbH, Frankfurt, Germany). The concentration of Surfactin was determined by analyzing the sample supernatant using HPLC (Willenbacher et al. 2014).

Data analysis

To enable the evaluation of the fermentation processes, several values were calculated to compare the different experiments. Using the results of CDW, glucose and Surfactin mass the values of $Y_{X/S}$ [g/g], $Y_{P/X}$ [g/g], $Y_{P/S}$ [g/g], μ [h^{-1}], $q_{Surfactin}$ [g/(g·h)], volumetric $q_{Surfactin}$ [g/(L·h)], were determined.

The biomass yield $Y_{X/S}$ was defined in an integral manner, using the maximal mass of produced CDW (m_{Xmax}) and the corresponding mass of depleted glucose (m_S; Eq. 1).

$$Y_{X/S} = \frac{\Delta m_{Xmax}}{\Delta m_S} \qquad (1)$$

The product yield $Y_{P/X}$ was calculated in the same manner as $Y_{X/S}$ using the maximal mass of produced product ($m_{Surfactin\ max}$) and the corresponding CDW over the whole fermentation process (Eq. 2).

$$Y_{P/X} = \frac{\Delta m_{Surfactin\ max}}{\Delta m_{Xmax}} \qquad (2)$$

The product yield $Y_{P/S}$ was calculated dividing the maximal produced mass of Surfactin by the corresponding mass of consumed glucose during the entire fermentation process (Eq. 3).

$$Y_{P/S} = \frac{\Delta m_{Surfactin\ max}}{\Delta m_S} \qquad (3)$$

The specific growth rate μ was determined in a differential manner using Eq. 4.

$$\mu = \frac{ln\frac{m_{x_2}}{m_{x_1}}}{t_2 - t_1} \qquad (4)$$

The specific productivity $q_{Surfactin}$ was calculated in an integral manner using the maximal mass of produced Surfactin, the corresponding mass of CDW and cultivation time (Eq. 5).

$$q_{Surfactin} = \frac{\Delta m_{Surfactin\ max}}{\Delta m_{Xmax} \cdot \Delta t} \qquad (5)$$

The integral volumetric specific productivity $q_{Surfactin}$ was determined using the maximal mass of Surfactin, the average medium volume and cultivation time (Eq. 6).

$$volumetric\ q_{Surfactin\ max} = \frac{\Delta m_{Surfactin\ max}}{V_{Reactor} \cdot \Delta t} \qquad (6)$$

Results
Anaerobic growth

Altogether eight fermentations were performed testing four different glucose concentrations as duplicates. An example is given in Figure 3, showing a fermentation of *Bacillus subtilis* DSM 10^T using 2.5 g/L glucose as carbon source (for exemplary fermentations using 5 g/L, 7.5 g/L and 10 g/L glucose see Additional file 1: Figure S2-S4, the fermentation employing 2.5 g/L glucose is additionally presented with matching axis scaling in Additional file 1: Figure S1). All figures present the course of the CDW, CO$_2$, phosphoric acid, nitrate, glucose and Surfactin concentrations with time. The fermentation shown in Figure 3 endured 55 h. The process was terminated because the levels of CO$_2$ and CDW were drastically decreasing and the glucose was completely consumed. During the fermentation the CDW continually increased reaching 0.320 g/L at its maximum. The amount of CO$_2$ (no longer solved in the medium and therefore carried on within the N$_2$ stream in the headspace) increased simultaneously with the CDW. Meanwhile the glucose concentration consistently decreased until its depletion. In contrast, only 1 g/L nitrate was consumed during this fermentation (during fermentations with 10 g/L glucose about 5 g/L nitrate were used up, Additional file 1: Figure S4). The concentration of Surfactin in the fermentation medium started to increase after 24 h of incubation. It reached its maximum at the end of the fermentation yielding 0.09 g/L Surfactin. The amount of added phosphoric acid to adjust the mediums pH level increased significantly after 34 h of cultivation. The demand for pH regulation is caused by *Bacillus subtilis* anaerobic metabolism. In this pathway nitrate is used as terminal electron acceptor. The reduction of nitrate to nitrite via a nitrate reductase and the additional conversion of nitrite to ammonia via a nitrite reductase results in the production of an alkaline end product. In contrast to conventional aerobic cultivations of *Bacillus subtilis*, where the addition of base marks cell growth, the addition of acid represents vivid cell growth under anaerobic conditions. The amount of dissolved oxygen was monitored throughout the fermentation processes but is not shown in the figures, because values were below detection limit.

Comparison of process parameters during anaerobic fermentation with different glucose concentrations

The fermentations of *Bacillus subtilis* DSM 10^T with various glucose concentrations were analyzed regarding product yields and substrate utilization. Table 1 presents an overview of the most interesting process parameters,

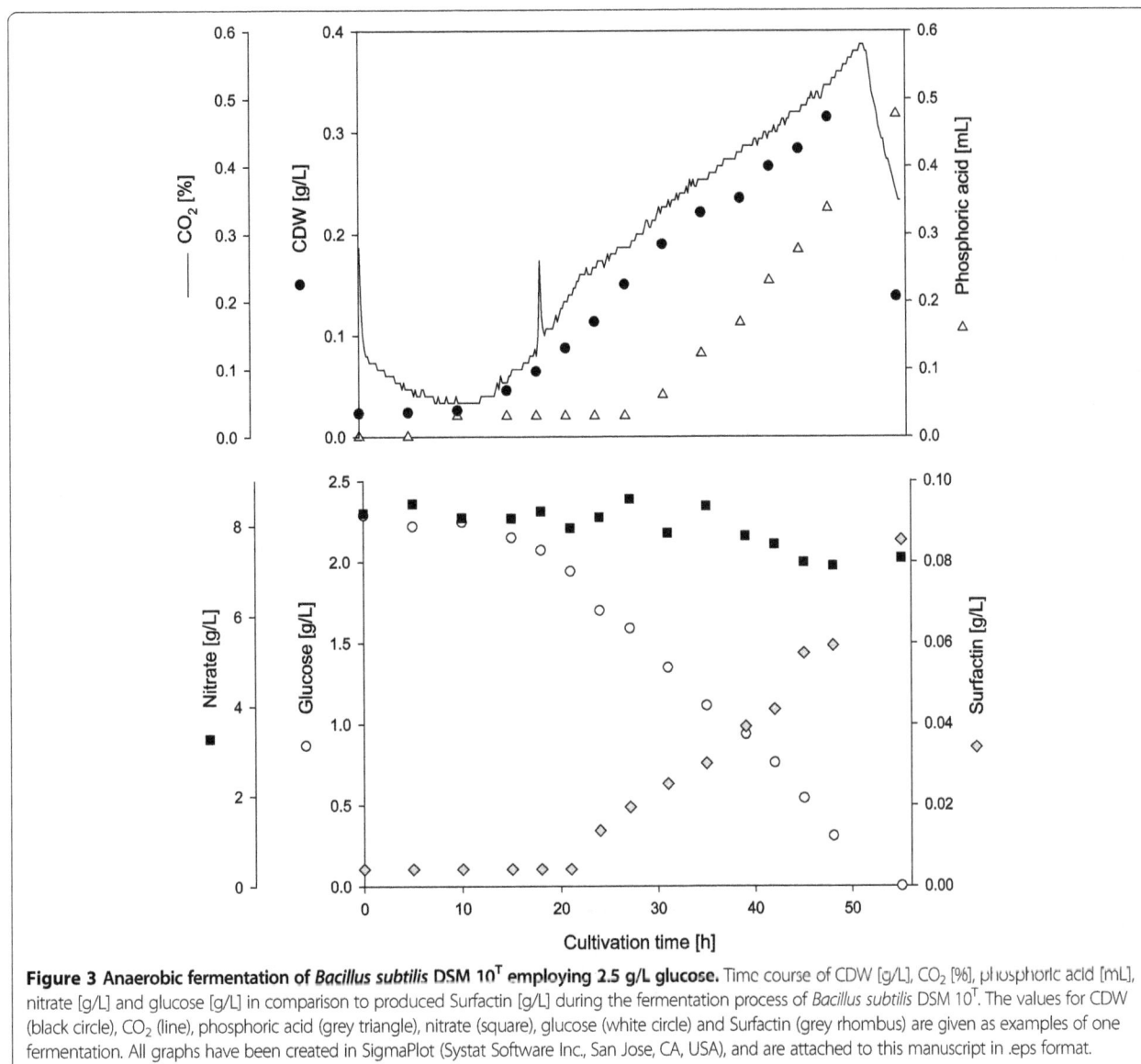

Figure 3 Anaerobic fermentation of *Bacillus subtilis* DSM 10^{T} employing 2.5 g/L glucose. Time course of CDW [g/L], CO_2 [%], phosphoric acid [mL], nitrate [g/L] and glucose [g/L] in comparison to produced Surfactin [g/L] during the fermentation process of *Bacillus subtilis* DSM 10^{T}. The values for CDW (black circle), CO_2 (line), phosphoric acid (grey triangle), nitrate (square), glucose (white circle) and Surfactin (grey rhombus) are given as examples of one fermentation. All graphs have been created in SigmaPlot (Systat Software Inc., San Jose, CA, USA), and are attached to this manuscript in .eps format.

Table 1 Summary of the process parameters during various fermentations

Glucose concentration [g/L]	2.5	5	7.5	10
Cultivation time [h]	55	102	108	161
Max. CDW [g/L]	0.320	0.612	0.856	0.586
Max. $c_{Surfactin}$ [g/L]	0.087	0.105	0.150	0.158
μ_{max} [h^{-1}]	0.105	0.114	0.118	0.074
$Y_{P/X}$ [g/g]	0.278	0.169	0.179	0.259
$Y_{X/S}$ [g/g]	0.120	0.105	0.119	0.049
$Y_{P/S}$ [g/g]	0.033	0.018	0.022	0.011
$q_{Surfactin}$ [g/(g·h)]	0.005	0.002	0.002	0.002
vol. $q_{Surfactin}$ [g/(L·h)]	0.002	0.001	0.002	0.001

All values are mean values of two fermentations.
Comparison of process parameters during anaerobic fermentation of *Bacillus subtilis* DSM 10^{T} with different glucose concentrations.

such as cultivation time, maximal CDW, maximal Surfactin concentration, maximal growth rate, product yields ($Y_{P/X}$, $Y_{P/S}$, $q_{Surfactin}$, vol. $q_{Surfactin}$) and substrate utilization ($Y_{X/S}$). All illustrated values are mean values of two fermentations. The duration of the fermentation depended on the starting glucose concentration. Fermentations with 2.5 g/L glucose lasted for 55 h, whereas fermentations with 10 g/L glucose averagely endured 161 h. Fermentations with 5 g/L and 7.5 g/L glucose ran for approximately 100 h. The maximal CDW was reached during fermentations with 7.5 g/L glucose (0.856 g/L). In contrast, only 0.320 g/L CDW were yielded in fermentations with 2.5 g/L glucose. Fermentations with 5 g/L glucose or more reached at least 0.105 g/L Surfactin as maximal concentration. Fermentations with 2.5 g/L glucose earned 0.087 g/L Surfactin.

The highest maximal growth rate μ_{max} was reached by fermentations with 7.5 g/L glucose (0.118 h^{-1}), whereas fermentations with 10 g/L glucose only reached maximal growth rates of 0.074 h^{-1}. The values of overall $Y_{P/X}$ differed widely between the fermentations with different glucose concentrations. Fermentations with 5 g/L or 7.5 g/L glucose earned product yields around 0.17 g/g. In contrast, fermentations with 2.5 g/L and 10 g/L reached $Y_{P/X}$ values of 0.278 g/g and 0.259 g/g, respectively. Overall values of $Y_{X/S}$ varied around 0.1 g/g except for fermentations with 10 g/L glucose. These cultivations led to $Y_{X/S}$ values of 0.049 g/g. The results for $Y_{P/S}$ show much higher values for fermentations with low glucose concentrations. Fermentations with 2.5 g/L glucose reached 0.033 g/g instead of 0.011 g/g with 10 g/L glucose in mineral salt medium. Additionally, cultivations using 2.5 g/L glucose yielded high specific production rates of 0.005 g/(g·h). Interestingly, all other fermentations reached only 0.002 g/(g·h). Volumetric specific production rates varied for all fermentations between 0.001 g/(L·h) and 0.002 g/(L·h).

Although cultivations with 2.5 g/L glucose reached only small amounts of CDW and Surfactin, these fermentations are comparably efficient. The cultivation time is much shorter and values for μ_{max}, $Y_{X/S}$ and vol. $q_{Surfactin}$ are comparatively high. Moreover, fermentations with 2.5 g/L glucose reached excellent values for $Y_{P/X}$, $Y_{P/S}$ and specific production rate $q_{Surfactin}$ emphasizing an outstanding conversion of substrate into product. Nevertheless, fermentations with 2.5 g/L glucose yielded only small amounts of Surfactin, due to the short cultivation time. As a consequence it would be interesting to test whether higher overall amounts of Surfactin can be reached by applying a repeated fed-batch process. Interestingly, on closer inspections Surfactin concentrations did increase simultaneously to rising initial glucose concentrations possibly due to longer cultivation times. Surprisingly, fermentations employing 10 g/L did also

achieve an almost equal value for $Y_{P/X}$ in comparison to fermentations with 2.5 g/L glucose. But this positive result is misleading as overflow metabolism (as a result of the high initial glucose concentration) leads to low values of CDW, μ_{max} and $Y_{X/S}$. This means that the bacterial growth is already strongly restricted under the employment of 10 g/L glucose. As a result data for $Y_{P/S}$ and $q_{Surfactin}$ are comparably low. These findings support the usage of lower initial glucose concentrations for the anaerobe fermentation of *B. subtilis* DSM 10^T for the production of Surfactin to avoid overflow metabolism.

Discussion

Comparison with other foam-free cultivation systems and aerobic fermentation with foam fractionation

The aim of the current study was to introduce a new approach for a foam-free biosurfactant production process. The results shown in Figure 3 and Table 1 demonstrate a high efficiency for anaerobic cultivations with low glucose concentrations. There are only three other fermentation processes described for the foam-free production of Surfactin. The solid state fermentation analyzed by Ohno et al. is incomparable with aqueous fermentations (Ohno et al. 1995), hence these data are not further discussed in comparison to the current study. However, Chtioui et al. established a rotating disc bioreactor allowing air flow only above the liquid phase. The growth of a *Bacillus subtilis* ATCC 21332 biofilm led to the production of Surfactin and Fengycin (Chtioui et al. 2012). Chtioui et al. provided several results about product yields and substrate utilization. On basis of these findings further process parameters were calculated (see Table 2) to achieve a more complete comparison with the results of the current study (Table 2). Coutte et al. introduced a novel membrane bioreactor for the production of biosurfactants (Coutte et al. 2010). The data of the *Bacillus subtilis* ATCC 21332 cultivation were also

Table 2 Summary of the process parameters of different foam-free processes

	Chtioui et al. 2012	Coutte et al. 2010	The current study	Willenbacher et al. 2014
Surfactin producer	*Bacillus subtilis* ATCC 21332	*Bacillus subtilis* ATCC 21332	*Bacillus subtilis* DSM 10^T	*Bacillus subtilis* DSM 10^T
Fermentation approach	Rotating discs	Membrane bioreactor	Anaerobic, no gas flow	Foam fractionation
Cultivation time [h]	72	72	55	30
Max. $c_{Surfactin}$ [g/L]	0.212*	0.242*	0.087	3.995 (foam)
$Y_{P/X}$ [g/g]	0.068	0.078*	0.278	0.192
$Y_{X/S}$ [g/g]	0.189	0.164*	0.120	0.268
$Y_{P/S}$ [g/g]	0.013	0.013	0.033	0.052
$q_{Surfactin}$ [g/(g·h)]	0.001*	0.001*	0.005	0.006
vol. $q_{Surfactin}$ [g/(L·h)]	0.003*	0.003*	0.002	0.018

*the values were calculated during the current study, using data of Chtioui et al. 2012 and Coutte et al. 2010 ($m_{Surfactin}$, CDW, cultivation time and cultivation volume).

Comparison of different foam-free Surfactin production processes regarding their process parameters and collation with a fermentation process applying foam fractionation [2].

used for the calculation of additional process parameters (Table 2). Therefore, Table 2 compares the data of three different foam-free fermentation processes for the production of Surfactin. To outline the differences between these methods and a traditional aerobic cultivation for the production of Surfactin, these results are additionally collated with a fermentation process applying foam fractionation (Willenbacher et al. 2014).

The processes of Chtioui et al. and Coutte et al. each yielded above 0.2 g/L Surfactin. Whereas only 0.087 g/L Surfactin were reached in the current study (with 2.5 g/L glucose in mineral salt medium). However, the fermentations of Chtioui et al. and Coutte et al. lasted comparatively longer (72 h instead of 55 h). Aerobic fermentations with *Bacillus subtilis* using foam fractionation take much shorter time (30 h) and yield much higher concentrations in foam (3.995 g/L). Values for $Y_{X/S}$ differ only slightly between the foam-free processes (0.120 g/g – 0.189 g/g), but are relatively low compared to cultivations applying foam fractionation (0.268 g/g). The results for volumetric production rates are very similar, too, between the foam-free fermentations (0.002 g/(L·h) – 0.003 g/(L·h)). The foam fractionation fermentation reached a much higher value for vol. $q_{Surfactin}$ in comparison (0.018 g/(L·h)). The product yield in contrast to substrate utilization is given by the parameter $Y_{P/S}$. The values for cultivations of Chtioui et al. and Coutte et al. are both 0.013 g/g. The current study reached a much higher value of 0.033 g/g for $Y_{P/S}$. However, fermentations employing foam fractionation still yield higher $Y_{P/S}$ values (0.052 g/g). The specific production rate $q_{Surfactin}$ is five-times higher in anaerobic fermentations using 2.5 g/L glucose (0.005 g/(g·h)) in comparison to other foam-free fermentations (0.001 g/(g·h)). Aerobic processes applying foam fractionation yield rather similar results for $q_{Surfactin}$ (0.006 g(g·h)). Most surprising are the results for $Y_{P/X}$. Fermentations of Chtioui et al. and Coutte et al. reached 0.068 g/g and 0.078 g/g, respectively. In contrast, anaerobic fermentations of the current study employing 2.5 g/L glucose yielded 0.278 g/g. These findings surpass even $Y_{P/X}$ values of aerobic fermentations employing foam fractionation (0.192 g/g).

Interestingly, the results of Chtioui et al. and Coutte et al. show very similar values for efficiency, product yields and substrate utilization although completely different fermentation approaches were applied. This similarity was revealed only after calculating some additional process parameters from the original data of these publications (Table 2). While rotating disc bioreactors or membrane reactors seem very attractive alternatives to common foam fractionation processes the presented data in Table 2 expose their low yields in comparison to the results of a classic foam fractionation process. The comparison of the results of Chtioui et al. and Coutte et al. with data of the current study displays a much

higher effectiveness of the anaerobic fermentation approach. Although overall less Surfactin was produced, much more Surfactin was produced per CDW. This implies that the bacterial growth is probably lower compared to the rotating discs or membrane bioreactors, but single cells produce more Surfactin under completely anaerobic conditions. These findings explain the much higher values for $Y_{P/X}$, $Y_{P/S}$ and $q_{Surfactin}$. In comparison to an aerobic fermentation process with foam fractionation some process parameters are lower (e.g., vol. $q_{Surfactin}$ and $Y_{X/S}$), but values for $Y_{P/S}$ and $q_{Surfactin}$ are at the same level. Most important is the much higher value for $Y_{P/X}$ under anaerobic conditions. This implies a much better production of Surfactin per CDW not only in comparison to other foam-free processes, but even in comparison to aerobic foam fractionation processes.

The current study demonstrates a new approach to produce Surfactin without any foam formation. Moreover, anaerobic cultivation and foam-free biosurfactant production are combined in one process for the first time. The anaerobic production of Surfactin was shown before, but never analyzed for product yields and substrate utilization. The comparison of different fermentations with various glucose concentrations displayed great efficiency for processes applying low glucose concentrations. Furthermore, the confrontation with other foam-free processes revealed a much higher effectiveness of the anaerobic fermentation process of the current study.

Additional file

Additional file 1: Exemplary anaerobic fermentations employing 2.5 g/L, 5 g/L, 7.5 g/L and 10 g/l glucose.

Competing interests
The authors declare that they have no competing interests.

Authors' contributions
JW collected all data, created the graphs and figures and calculated additional data from comparable studies for further analysis. Furthermore, JW drafted this manuscript. J-TR conducted his Bachelor thesis under the supervision of JW and contributed the data of fermentations with 2.5 g/L, 5.0 g/L and 7.5 g/L glucose. JR conducted his Bachelor thesis under the supervision of JW and provided the data of fermentations with 10.0 g/L glucose. CS and RH substantially contributed to conception and design of the conducted experiments. All authors read and approved the final version of this manuscript.

Acknowledgements
The authors would like to thank Florian Oswald, a colleague at the Karlsruhe Institute of technology (KIT), for his helpful tips during anaerobic cultivations and contribution to scientific discussions. We acknowledge the support by Deutsche Forschungsgemeinschaft and Open Access Publishing Fund of Karlsruhe Institute of Technology.

Author details
[1]Institute of Process Engineering in Life Sciences, Section II: Technical Biology, Karlsruhe Institute of Technology (KIT), Engler-Bunte-Ring 1, 76131, Karlsruhe, Germany. [2]Institute of Food Science and Biotechnology (150), Section Bioprocess Engineering (150 k), University of Hohenheim, Garbenstr. 25, 70599 Stuttgart, Germany.

References

Arima K, Kakinuma A, Tamura G (1968) Surfactin, a crystalline peptidelipid surfactant produced by *Bacillus subtilis*: Isolation, characterization and its inhibition of fibrin clot formation. Biochem Bioph Res Co 31:488–494

Banat IM, Franzetti A, Gandolfi I, Bestetti G, Martinotti MG, Fracchia L, Smyth TJ, Marchant R (2010) Microbial biosurfactants production, applications and future potential. Appl Microbiol Biot 87:427–444

Bertani G (1951) STUDIES ON LYSOGENESIS I.: The Mode of Phage Liberation by Lysogenic *Escherichia coli* 1. J Bacteriol 62:293

Chtioui O, Dimitrov K, Gancel F, Dhulster P, Nikov I (2012) Rotating discs bioreactor, a new tool for lipopeptides production. Process Biochem 47:2020–2024

Cooper D, Macdonald C, Duff S, Kosaric N (1981) Enhanced production of surfactin from *Bacillus subtilis* by continuous product removal and metal cation additions. Appl Environ Microb 42:408–412

Coutte F, Lecouturier D, Ait Yahia S, Leclère V, Béchet M, Jacques P, Dhulster P (2010) Production of surfactin and fengycin by *Bacillus subtilis* in a bubbleless membrane bioreactor. Appl Microbiol Biot 87:499–507

Davis D, Lynch H, Varley J (1999) The production of Surfactin in batch culture by *Bacillus subtilis* ATCC 21332 is strongly influenced by the conditions of nitrogen metabolism. Enzyme Microbial Tech 25:322–329

Hoffmann T, Troup B, Szabo A, Hungerer C, Jahn D (1995) The anaerobic life of *Bacillus subtilis*: Cloning of the genes encoding the respiratory nitrate reductase system. FEMS Microbiol Lett 131:219–225

Javaheri M, Jenneman GE, McInerney MJ, Knapp RM (1985) Anaerobic Production of a Biosurfactant by *Bacillus licheniformis* JF-2. Appl Environ Microb 50:698–700

Nakano MM, Dailly YP, Zuber P, Clark DP (1997) Characterization of anaerobic fermentative growth of *Bacillus subtilis*: identification of fermentation end products and genes required for growth. J Bacteriol 179:6749–6755

Nakano MM, Hoffmann T, Zhu Y, Jahn D (1998) Nitrogen and Oxygen Regulation of *Bacillus subtilis* nasDEF Encoding NADH-Dependent Nitrite Reductase by TnrA and ResDE. J Bacteriol 180:5344–5350

Ohno A, Ano T, Shoda M (1995) Production of a lipopeptide antibiotic, surfactin, by recombinant *Bacillus subtilis* in solid state fermentation. Biotechnol Bioeng 47:209–214

Ongena M, Jourdan E, Adam A, Paquot M, Brans A, Joris B, Arpigny J-L, Thonart P (2007) Surfactin and fengycin lipopeptides of *Bacillus subtilis* as elicitors of induced systemic resistance in plants. Environ Microbiol 9:1084–1090

Ramos HC, Boursier L, Moszer I, Kunst F, Danchin A, Glaser P (1995) Anaerobic transcription activation in *Bacillus subtilis*: identification of distinct FNR-dependent and-independent regulatory mechanisms. EMBO J 14:5984

Schallmey M, Singh A, Ward OP (2004) Developments in the use of *Bacillus* species for industrial production. Can J Microbiol 50:1–17

Vollenbroich D, Pauli G, Ozel M, Vater J (1997) Antimycoplasma properties and application in cell culture of surfactin, a lipopeptide antibiotic from *Bacillus subtilis*. Appl Environ Microb 63:44–49

Willenbacher J, Zwick M, Mohr T, Schmid F, Syldatk C, Hausmann R (2014) Evaluation of different *Bacillus* strains in respect of their ability to produce Surfactin in a model fermentation process with integrated foam fractionation. Appl Microbiol Biot 98:9623–9632

Zhang G, Rogers RE, French WT, Lao W (2007) Investigation of microbial influences on seafloor gas-hydrate formations. Mar Chem 103:359–369

6

Influence of operational parameters on the fluid-side mass transfer resistance observed in a packed bed bioreactor

Amir Hussain[1], Martin Kangwa[1], Ahmed Gad Abo-Elwafa[2] and Marcelo Fernandez-Lahore[1*]

Abstract

The influence of mass transfer on productivity as well as the performance of packed bed bioreactor was determined by varying a number of parameters; flow rate, glucose concentration and polymers (chitosan). *Saccharomyces cerevisiae* cells were immobilized in chitosan and non-chitosan coated alginate beads to demonstrate the effect on external mass transfer by substrate consumption time, lag phase and ethanol production. The results indicate that coating has a significant effect on the lag phase duration, being 30–40 min higher than non-coated beads. After lag phase, no significant change was observed in both types of beads on consumption of glucose with the same flow rate. It was observed that by increasing flow rates; lag phase and glucose consumption time decreased. The reason is due to the reduction of external mass transfer as a result of increase in flow rate as glucose is easily transported to and from the beads surface by diffusion. It is observed that chitosan acts as barrier for transfer of substrate and products, in and out of beads, at initial time of fermentation as it shows longer lag phase for chitosan coated beads than non-coated. Glucose consumption at low flow rate was lower as compared to higher flow rates. The optimum combination of parameters consisting of higher flow rates 30–90 ml/min and between 10 and 20 g/l of glucose was found for maximum production of ethanol.

Keywords: Packed bed reactor, External mass transfer, *Saccharomyces cerevisiae*, Alginate, Chitosan, Glucose

Introduction

The increase in fossil fuel usage has resulted in both environmental and health problems due to pollutants produced (Shafiee and Topal 2008). This effect has encouraged researchers in finding alternative, less/non-pollutant cheaper fuel sources like ethanol. Therefore, the use of (bio) ethanol as a fuel has been widely encouraged. The most favored method in ethanol production is through the use of yeast *Saccharomyces cerevisiae* fermentation process in bioreactors (Pscheidt and Glieder 2008; de Jong et al. 2012; Cha et al. 2014; Djordjevic et al. 2014). For many centuries, yeast wholecells have profoundly been used as a work horse in the production of bioethanol and it is currently the most used microorganism due to its extensively high rate of fermentation of sugars and its high tolerance to by-products produced during fermentation (Matsushika et al. 2009; Hasunuma and Kondo 2012; De Bari et al. 2013; Borovikova et al. 2014). However, as the demand in biofuel increases, there is need in finding both the best bioreactor and fermentation conditions that favor's higher production and quality. Bioreactors have found their extensive usage in biotechnology and are assimilated in the heart of biotechnological process, being the equipment in which the substrate is effectively bio-converted to the desired products under the microbial cells or enzyme activity (Pilkington et al. 1998; Yu et al. 2007; Crespo et al. 2012; de Jong et al. 2012; Lee et al. 2012; Mathew et al. 2014).

For the past decades researchers have focused both on selecting the best favorable strains for bioconversion as while as in the design of the best bioreactors. To achieve high, effective and economically commercialized

*Correspondence: m.fernandez-lahore@jacobs-university.de
[1] Downstream Bioprocessing Laboratory, School of Engineering and Science, Jacobs University, Campus Ring 1, 28759 Bremen, Germany
Full list of author information is available at the end of the article

industrial production of bioethanol and other bioproducts, there is need to use a bioreactor with immobilized cells and, having an enhanced flow regime that, in turn, will minimize mass transfer limitations. Therefore, the study on the influence of mass transfer on productivity as well as the performance of bioreactor is still needed. These factors are severely affected by both external mass transfer limitations (transfer of reactants to and products from immobilized cell system) and internal mass transfer limitations (rate of transport inside the system (Saini and Vieth 1975; Converti et al. 1985; Anselme and Tedder 1987; Galaction et al. 2011). Cell immobilization technology, the localization of intact cells to a defined region of space with the preservation of catalytic activity presents for the biochemical process industry a radical advance, similar to the introduction of heterogeneous catalysis in the petrochemical and heavy chemical industries (Yu et al. 2007; Willaert and Flickinger 2009; Duarte et al. 2013). This justifies the interest in the research and development advanced materials for biotechnology with the combined effort of scientists from various fields to obtain polymers with well-defined structures and specific chemical, physicochemical, mechanical and biological properties which are used in cell enzyme entrapments (Terada et al. 2006; Duarte et al. 2013). The immobilization technique has found numerous advantages over free cells such as easiness of product separation, reutilization of entrapped cells, maintaining of specific growth, high cells densities and lack of contamination. Additionally, immobilized cells are less susceptible than free cells to the effect of substrate inhibition and pH variations, all these help to improve the overall process. Presently, natural and synthetic polymers such as cellulose, alginate, chitosan, agarose polyurethane, and polyacrylate are being used for cell immobilization with calcium alginate beads being widely used in immobilization of bacteria, yeast, fungi and algae for different bioprocesses (Gòdia et al. 1987; Pacheco et al. 2010; Galaction et al. 2012; Duarte et al. 2013). These polymers have potential application in bioethanol production, vinegar production, and wastewater treatment due to its simplicity, cheap, non-toxic to cells and good mechanical properties. However, there are some disadvantages with their use, such as gel degradation, severe mass transfer limitations, low mechanical strength as it can cause the release of cells from the support and large pore size. To overcome this, a combination of chitosan, a polycationic polymer and alginate, a polyanionic polymer is diffused into the alginate beads to provide a strong ionic interaction between chitosan amino groups and carboxyl groups of alginate which forms a polyelectrolyte complex (PEC) that gives more mechanical support to cells (Yu et al. 2007; Galaction et al. 2012; Duarte et al. 2013).

For several decades, traditional setups like membrane, air lift and stirrer tank bioreactors have been used in bioethanol production. However, some drawbacks like, less product yield due to low mass and heat transfer, inefficient conversion of substrate, uneven mixing and shear stress on biocatalysts have been observed. Therefore, there is need in utilizing a reactor that is able to sustain an excellent hydrodynamic regime coupled to reduced overall mass transfer limitations (Saini and Vieth 1975; Pilkington et al. 1998; Karagoz and Ozkan 2014). In this article, we used the packed bed bioreactor (PBR) with one bed containing immobilized beads and a vessel for culture medium, in which the culture medium is circulated from the vessel through the fixed bed and back (Figure 1).

Medium enriched with glucose re-enter the packed bed where it can be re-utilized to convert glucose into ethanol. Toxic metabolites and other by-products are diluted; oxygen and pH can be adjusted to optimal levels. This reactor has several advantages over other bioreactors like, low manufacturing and operating cost, automation process and facility to operate at low temperatures. The preference for fixed bed bioreactor has increased due to its higher sensitivity/effectiveness of immobilized cells or enzymes (Cascaval et al. 2012).

In this article we focused on the operational performance of the immobilized packed-bed bioreactor in the course of physiological and biochemical studies on the substrate uptake of immobilized yeast cells. The reactor was operated in batch mode fermentation; yeast physiology and mass transfer behavior in packed bed reactor were monitored in close relation to parameters such as glucose concentration, medium flow rate and different support materials like alginate beads with and without chitosan coating.

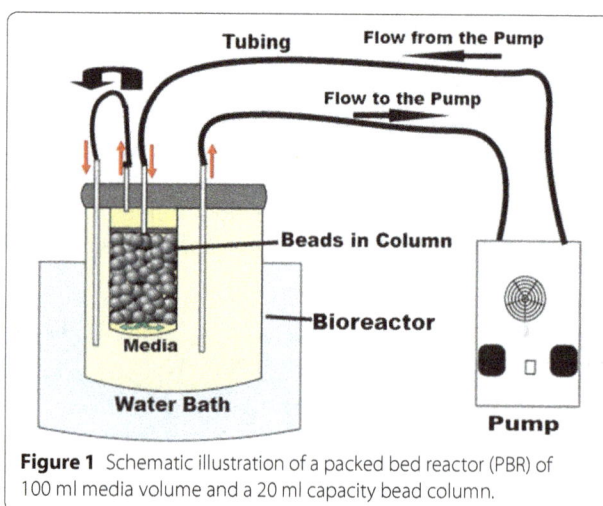

Figure 1 Schematic illustration of a packed bed reactor (PBR) of 100 ml media volume and a 20 ml capacity bead column.

Materials and methods

Materials

Microorganism

The yeast *S. cerevisiae* (baker yeast) was obtained from DHW Vital Gold, Nürnberg, Germany, while the *S. cerevisiae* Ethanol Red 11 strain was purchased from Fermentis Inc, Marcq-en-Baroeul, France and were stored at 4 and $-80°C$, respectively.

Fermentation medium and cultivation

Minimal media was prepared with 6.7 g/l yeast extract nitrogen base without amino acid, 1.7 g/l ammonium acetate and glucose (2, 4, 10, 20 and 40 g/l) were prepared separately and mixed after sterilizing (121°C, 20 min). Amino acid mixture (100×) was prepared by mixing the following different amino acids; 200 mg L-arginine, 1,000 mg L-aspartic acid, 1,000 mg L-glutamic acid, 300 mg L-lysine, 500 mg L-phenylalanine, 4,000 mg L-serine, 2,000 mg L-threonine, 300 mg L-tyrosine, 1,500 mg L-valine, dissolved in water by adjusting pH 10 with 0.1 N NaOH and used 0.2 μm filter for sterilization. During culturing, 10 ml of amino acids solution was added to a final 1 l media.

Ethanol Red 11 strain was refreshed by streaked onto YPD agar plate (1% yeast extract, 2% peptone and 2% glucose, 2% agar), incubated for 2 days at 35°C. The resulting single colonies were used to start a fresh culture. Twenty milliliters of YPD media (1% yeast extract, 2% peptone and 10% D-glucose) in a 100 ml flask was inoculated with a single colony of Yeast Ethanol Red 11 grown overnight at 35°C with vigorous shaking at 250 rpm. One percent of the pre-culture was used to inoculate 2 l Erlenmeyer baffled flask containing 1,000 ml YPD media final volume. The inoculated flask was incubated on a rotary shaker at 200 rpm and 35°C for 24 h. Furthermore, the cells were collected by centrifugation at 4,000 rpm for 15 min, washed twice with sterile distilled water, centrifuged and re-suspended in sterile water to obtain a dense cell suspension.

Calcium alginate beads preparation and yeast immobilization

A sterile sodium alginate solution (2.5% w/v, autoclaved at 121°C, for 15 min, was prepared in 50 mM phosphate buffer at pH 7. For yeast immobilization, 3% final amount of the above obtained cell suspension were mixed with alginate solution. For beads preparation, alginate-yeast solution was drop by drop allowed to dip using 1 ml pipette tip into 200 ml, 180 mM $CaCl_2$. Beads were let to harden in this solution for 1 h. Beads were further rinsed three times with sterile 2% NaCl solution and then with sterile water. The alginate beads with diameters between 3 and 4 mm were used in experiments. For the preparation of alginate beads with chitosan coating, the above prepared beads were dipped in sterilized chitosan solution (3% chitosan, 0.1 N HCl, pH 5) for 10 min and later washed 3 times with sterile water.

Packed bed reactor and beads packaging

A packed bed bioreactor (100 ml) was purchase from Medorex GmbH, Noerden-Hardenberg, Germany. The bioreactor column has a 2 cm diameter glass vessel for beads package, with one end close and other closed by rubber plug (Figure 1). The reactor was 2/3 filled with beads and temperature was kept at 35°C using a water bath. The immobilized yeast was grown on minimal media with varying factors: glucose (2, 4, 10, 20 and 40 g/l), flow rate (1, 4, 12, 30 and 90 ml/min.) and alginate bead with and without chitosan coating while factors like initial cells amount (3%) and temperature (35°C) were kept constant.

Glucose consumption measurements

For immobilized yeast glucose consumption measurements, the DNS method was used. For each measurement, 0.5 ml sample and 0.5 ml DNS solution were mixed in a 1.5 ml Eppendorf tube, vortex for 10 s, and incubated for 10 min at 90°C. After incubation, 40% 0.16 ml potassium sodium tartrate was added, mixed by vortex and placed on ice for 3 min. Two hundred microliter of each sample was measured at 575 nm. The obtained results were compared with calibration curve of different glucose concentration to get actual concentration.

Ethanol production measurements

For measurement of ethanol concentration produced in fermentation broth as well as calibration curve preparation, the underlined method was used. Six hundred microliter of fermentation broth samples were collected, transferred to an Eppendorf tube and centrifuged at 9,000 rpm for a min to pellet the cells. Later, 500 μl of the clear supernatant were transferred into a new tube without disturbing the cell pellet, and 5 μl of 1% *n*-butanol was added as an internal standard. The samples were vortexed for 30 s and 1 ml of 25% ethyl acetate was added with a further 5 min vortexing. For phase separation, the samples were centrifuged at 5,000 rpm and the organic phase was used for gas chromatography (GC). For sample measurements, gas chromatograph equipped with flame ionization detector (FID) was used. The columns used were the 30 and 0.25 mm CP-WAX-57CB (Santa Clara, CA, USA). During liquid analysis, temperature programming was employed and the column temperature was initially maintained at 120°C for 2 min and later the oven temperature was increased at a rate of 10°C/min until it reached 150°C. The injector and detector temperature were kept at 150 and 200°C, respectively. The flow rate for carrier gas (helium) was set at 30 ml/min. The injection sample volume was 2 µl. Each set of the experiment and data points were repeated thrice and the reported value was the mean average.

Results

Effect of chitosan coating on lag phase and glucose consumption

Sequential fermentation experiments with two parameters; flow rate and glucose concentration were varied to understand the effect on lag phase and glucose consumption rate till C/C_0 of 0.1 in both chitosan and non-chitosan coated calcium alginate beads, where C_0 represent the initial glucose concentration at time zero, C is the concentration at a particular time and 0.1 (10%) is the remaining glucose in the media. Figure 2a–d shows the flow rates used to determine the effect of chitosan coating on glucose consumption and from the curves we observed two phases: lag and exponential phase. After lag phase, no significant change was observed in both types of beads on glucose consumption with the same flow rate. Additionally, it was also observed that by increasing flow rates; lag phase and glucose consumption time decreased (Figure 2c, d).

Effect of flow rate and glucose concentration on lag phase

The results in Figure 3a, b shows two parameters; flow rate and glucose concentration, varied from 1 to 90 ml/min and 2 to 40 g/l, respectively, having a tremendous effect on lag phase. In the study, it was observed that lag phase of both types of beads decreases by increasing flow rate, moreover longer lag phase was found at higher glucose medium concentration. The maximum time of lag phase was found to be 290 min at lower flow rate of 1 ml/min and 190 min at higher flow rate 90 ml/min when using 40 g/l of glucose. It was also observed that by decreasing glucose concentration from 40 to 10 g/l, lag phase decreased too. Furthermore, no lag phase was found at glucose concentration of 4 and 2 g/l (Figures 2, 3). As shown in Figure 3, non-chitosan coated beads have shorter lag phase as compared to coated beads, indicating an improved mass transfer effect observed at higher flow rate and less inhibition of glucose transfer. While higher flow rate was shown to have a major effect in reducing time on lag phases in both types of beads (Figures 2, 3). To support the above data, fermentation results of the Ethanol Red 11 yeast strain was compared with Baker's yeast using flow rates of 4 and 90 ml/min and glucose concentration of 4 and 10 g/l. The results show that there

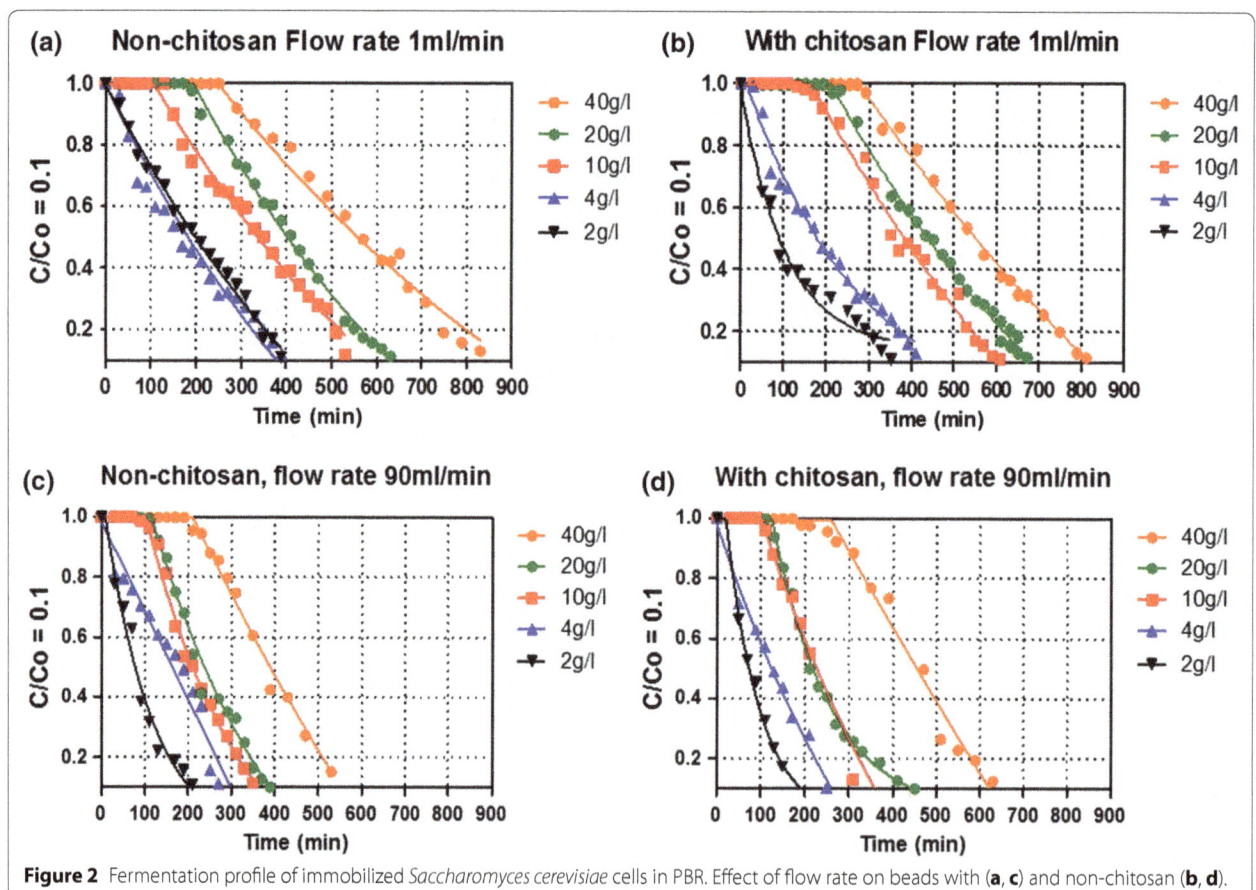

Figure 2 Fermentation profile of immobilized *Saccharomyces cerevisiae* cells in PBR. Effect of flow rate on beads with (**a**, **c**) and non-chitosan (**b**, **d**).

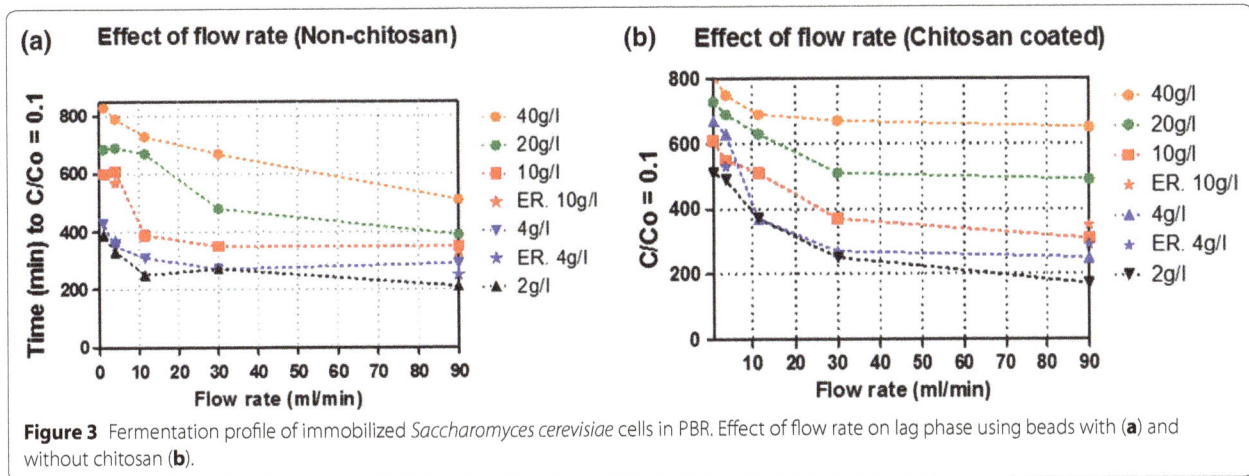

Figure 3 Fermentation profile of immobilized *Saccharomyces cerevisiae* cells in PBR. Effect of flow rate on lag phase using beads with (**a**) and without chitosan (**b**).

is no significant difference in lag phase of two types of yeast.

Effect of flow rate and glucose concentration on glucose consumption

In this study, glucose consumption of up to the level of $C/C_0 = 0.1$ was measured so as to understand the performance of the bioreactor and mass transfer properties regarding chitosan and non-chitosan coated beads. Figure 4 shows that by varying the flow rate from 1 to 90 ml/min, time for glucose consumption decreased. The major difference in glucose consumption behavior was observed when using both types of beads at higher flow rate like 90 ml/min. Time for glucose consumption by chitosan coated beads, at 30 and 90 ml/min is rather equal when using higher glucose concentration i.e. 40 and 20 g/l as compared to lower glucose concentration 10, 4 and 2 g/l. Moreover, beads' having no layer of chitosan, glucose consumption time tends to decrease by increasing flow rate. Further experiments have been performed to compare the *S. cerevisiae* Ethanol Red strain and wild type Baker's yeast using parameter glucose consumption time. Both strains performance were observed to be relatively equal at 4 and 10 g/l glucose.

Effect of flow rate and glucose concentration on ethanol productivity and yield

The minimal medium was used in all experiments so that yeast growth rate was at its minimal and cells inside the beads were assumed to be uniform. Experiments were conducted using the above mentioned yeast strains having initial glucose concentrations 4 and 10 g/l and flow rate 4 and 90 ml/min with dilution rate of 0.2 and 4.5 h^{-1}, respectively. The effect of flow rate and dilution rate at different glucose concentration on ethanol productivity as well as on ethanol yield is presented in Table 1. It can be observed that when the initial glucose concentration was 4 and 10 g/l, the ethanol productivity increase linearly with the dilution rate from 0.2 to 4.5 h^{-1}. An optimal ethanol productivity of 21.9 g/(g h) was obtained when using Ethanol Red strain at D of 4.5 h^{-1} with glucose concentration of 10 g/l. It was also observed that there was

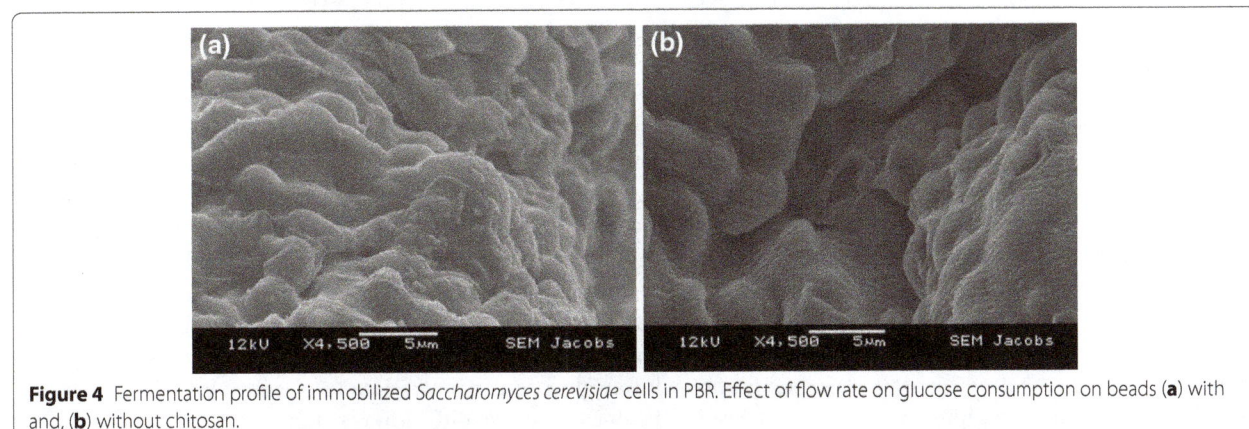

Figure 4 Fermentation profile of immobilized *Saccharomyces cerevisiae* cells in PBR. Effect of flow rate on glucose consumption on beads (**a**) with and, (**b**) without chitosan.

Table 1 Ethanol productivity and yield by yeast strains

Flow rate (ml/min)	Dilution rate = flow rate/bed volume	Glucose conc. (g/l)	Ethanol productivity		Ethanol yield	
			$D \times P = $ (g/g yeast h) at: 300 min		$Y(p/s) = Pl - Po/So - Sl$ at: 300 min	
			B.Yeast	ER.Yeast	B.Yeast	ER.Yeast
4	0.2	4	0.38	0.4	1.11	1.2
4	0.2	10	0.56	0.64	1.0	1.2
90	4.5	4	10.8	12.6	0.63	0.73
90	4.5	10	17.1	19.8	0.48	0.55

D dilution rate, *P* product concentration.

no significant difference in ethanol productivity for both *S. cerevisiae* strains at lower flow rate i.e. 4 ml/min, while higher productivity was obtained at higher flow rate (90 ml/min).

Discussion

In recyclable biocatalyst, the mechanical strength of calcium alginate beads had not fully been found to effectively support entrapped cells. To solve this problem, we focused on using Baker's yeast immobilized in chitosan coated alginate beads of 4 mm in diameter to facilitate the needed mechanical support. However, the chitosan coating may cause resistance in external mass transfer. The results in Figure 2 indicates that coating has a significant effect on lag phase duration, as it was observed with chitosan coated beads being 30–40 min higher than non-coated beads. The reason is due to the reduction of external mass transfer as a result of increase in flow rate as glucose is easily transported to and from the beads surface by diffusion (Willaert and Flickinger 2009; Galaction et al. 2012; Karagoz and Ozkan 2014).

Our results show an improvement over some literature data, were it was observed that chitosan-covered alginate beads have longer glucose conversion time when compared to alginate beads (Duarte et al. 2013). From the results it can be observed that chitosan acts as barrier for transfer of substrate and products, in and out of beads, at initial time of fermentation as it shows longer lag phase for chitosan coated beads than non-coated. This study gives the significant understanding of both alginate beads with and without chitosan coating as indicated in the differences in lag phases. A number of researchers have been using chitosan coating on alginate beads in order to reduce cell and enzyme release but it has disadvantage on mass transfer and may have an impact on the metabolic activity of cells in beads due to limited substrate supply that ultimately may have an effect on product formation.

Lag phase is considered as the adaptation time of yeast within new environment before the start of fermentation process. The similar effect in Figure 3 was also observed by Irfan et al. (2014) indicating that sugar concentration is critical in fermentation process as it has influence on yeast physiological, growth, rate of production and yield.

The dependence of lag phase on glucose concentration (Figures 2, 3) might be as a result of substrate diffusion and increase in concentration gradient between surface and inner regions of beads (Galaction et al. 2012). The observed prolonged lag phase might be due to higher accumulation of cAMP level stimulated by the effect of glucose on cAMP synthesis as the level of cAMP is higher during initial fermentation time (Ma et al. 1997) i.e. lag phase time and decreased on initiation of exponential growth in yeast cells (Duarte et al. 2013; Djordjevic et al. 2014; Mukherjee et al. 2014).

This summarizes the fact that inter-particle diffusional resistance reduces by increasing velocity around beads (Saini and Vieth 1975; Zhao and Delancey 2000; Galaction et al. 2012). Consequently, chitosan coated beds have more inter-particle diffusional resistance i.e. longer lag phase at early times of fermentation as compare to non-coated at lower flow rate. At this point it can be concluded that lag phase is not due to the physiology of yeast, but it may be due to the resistance in internal diffusion of glucose. This could be due to the fact that lag phase is directly depending on the glucose concentration as well as on flow rate.

Glucose is the most fundamental carbon source playing a central role in metabolic pathways providing energy to living organisms, and for product synthesis. Yeast metabolize glucose via the Embden–Mereyhof Parnas metabolic pathway (Galaction et al. 2012) there-by producing energy necessary for it survival. Furthermore, the efficiency of ethanol production can be affected by glucose concentration and flow rate. From the ethanol production experiments, it was observed that the time for glucose consumption by chitosan coated beads, at 30 and 90 ml/min was rather equal when using higher glucose concentration i.e. 40 and 20 g/l as compared to lower glucose concentration 10, 4 and 2 g/l. This might be due to glucose diffusion resistance that did not reduce even when using higher flow rate. However, this result indicates that chitosan coating characteristics influences glucose internal diffusion at higher glucose concentration. In literature, it was also observed that magnitude of glucose diffusion resistance is directly related to glucose concentration gradient created in and outside of beads (Galaction et al. 2011), indicating substrate inhibition phenomenon, affecting the fermentation performance. In non-chitosan coated beads' when compared with coated beads, glucose consumption time tends to decrease by increasing flow rate. The reason might be due to the fact that these types of beads did not pose any significant barrier (Figure 5) for glucose diffusion to metabolically active cells. This result was supported by Chen et al. (2012) observation that under scanning electron microscopy (SEM), surface of chitosan-coated beads was rough and compact compared to the non-coated alginate beads, due to strong electrostatic interaction between chitosan and alginate. The interpretation of these results indicates that glucose consumption behavior was not due to the yeast strain, but to mass transfer barrier that might have occurred by layer of chitosan coating on alginate beads and glucose concentration inhibition phenomenon.

As indicated in Table 1, higher ethanol productivity was observed on increasing flow rate and glucose concentration. A higher productivity can be attributed to the improved mass transfer properties when using higher flow rate that might be due to reduced substrate diffusional resistance (Anselme and Tedder 1987; Yu et al. 2007; Matsushika et al. 2009; Pacheco et al. 2010; Bangrak et al. 2011; Mathew et al. 2014). Although higher glucose concentration can give higher productivity, it can also facilitate increase in inter-particle diffusional resistance that enhances the lag phase as shown in Figures 2 and 3. It was also found the enhancement of ethanol production on increasing liquid velocity decrease mass

transfer resistance and substrate inhibitory effect (Bangrak et al. 2011).

Duarte et al. (2013) found that the maximum ethanol production during fermentation was after 4 h for non-chitosan coated alginate beads while for coated ones was after 6 h. While it was also reported that hydrodynamics of medium exhibits an important influence on glucose conversion and transfer processes (Cascaval et al. 2012; Galaction et al. 2012; Mathew et al. 2014).

Furthermore, in the case of ethanol yield, the industrial strain Ethanol Red 11 strain has higher yield than Baker's yeast at all flow rate and glucose concentration. On the other hand at 4 ml/min flow rate and 10 g/l glucose, ethanol yield of both yeast strains was observed to be high as compare to flow rate of 90 ml/min with same glucose. This result is due to higher residence time up to which yield is high (Singh et al. 2009).

Ethanol yield for both strains has been observed to decrease on addition of glucose that might be due to increased substrate diffusional resistance. The magnitude of resistance is directly related to the glucose concentration gradient between the inner and outer regions of beads, consequently concentration gradient can induce substrate inhibition and it was found that there was significant decrease in ethanol yield on addition of sugar concentration in fermentation medium. (Bangrak et al. 2011; Galaction et al. 2011, 2012 Rotaru et al. 2011; Cascaval et al. 2012).

It was also reported that in batch fermentation of *S. cerevisiae*, the ethanol yield was significantly depended on initial glucose concentration and substrate inhibition was notices at high initial glucose concentration (Wendhausen et al. 2001).

Sequential experiments on varying flow rates and glucose in a packed bed bioreactor with immobilized *S. cerevisiae* cells shades significant understanding on mass

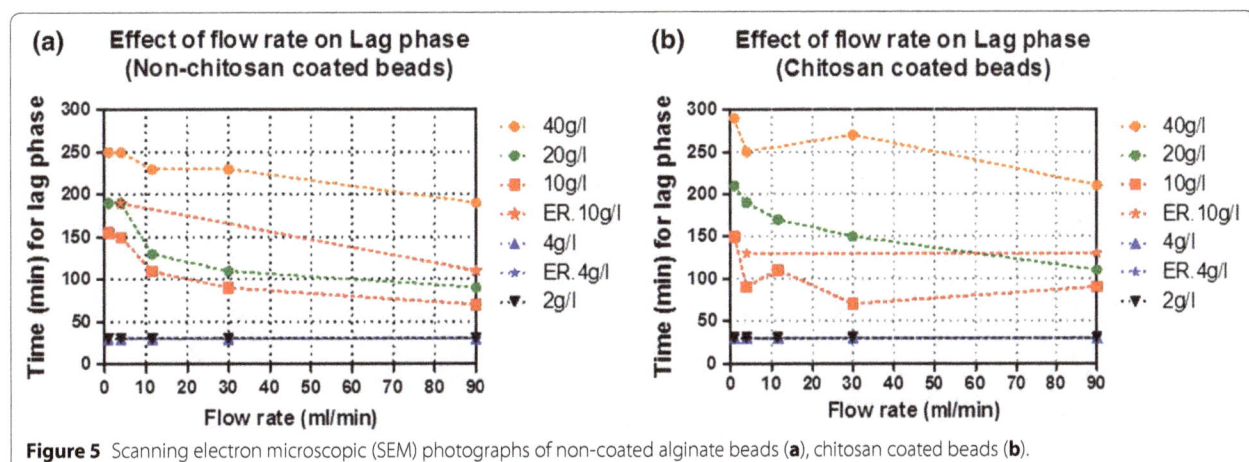

Figure 5 Scanning electron microscopic (SEM) photographs of non-coated alginate beads (**a**), chitosan coated beads (**b**).

transfer. Moreover, glucose consumption at low flow rate was lower when compared to when higher flow rates were used. By means of the analysis of the influence of different concentration of glucose and varying flow rate, the optimum combination was found to be that consisting of higher flow rates and between 10 and 20 g/l of glucose. This combination leads to the optimum glucose consumption rate and maximum product formation. The selected system for mixing as well as glucose concentration will be used in the further experiments for internal mass transfer or active pharmaceutical ingredient production in this basket bioreactor.

Authors' contributions

AH, MK and MFL have designed the work. AH, AG carried out the experiment. AH, AG, MK and MFL analyzed the data and contributed for the statistical analysis. AH, MK and MFL wrote the manuscript and reviewed the manuscript critically. All the authors have read the article and approved the final manuscript.

Author details

[1] Downstream Bioprocessing Laboratory, School of Engineering and Science, Jacobs University, Campus Ring 1, 28759 Bremen, Germany. [2] Department of Biotechnology, Faculty of Agriculture, Al-Azhar University, Naser City, Cairo 11884, Egypt.

Acknowledgements

Partial support for this study was provided from Project PGSYS-EXCHANGE EU-PIRSES#269211, ERA Net Euro TransBio-3, PGYSYS and Jacobs University Bremen.

Compliance with ethical guidelines

Competing interests

The authors declare that they have no competing interests.

References

Anselme MJ, Tedder DW (1987) Characteristics of immobilized yeast reactors producing ethanol from glucose. Biotechnol Bioeng 30(6):736–745

Bangrak P, Limtong S, Phisalaphong M (2011) Continuous ethanol production using immobilized yeast cells entrapped in loofa-reinforced alginate carriers. Braz J Microbiol 42(2):676–684

Borovikova D, Scherbaka R, Patmalnieks A, Rapoport A (2014) Effects of yeast immobilization on bioethanol production. Biotechnol Appl Biochem 61(1):33–39

Cascaval D, Galaction AI, Turnea M (2012) Influences of internal diffusion on the lipids bio-degradation with immobilized Bacillus sp. cells in fixed bed of basket type. Rev Med Chir Soc Med Nat Iasi 116(1):228–232

Cha HG, Kim YO, Lee HY, Choi WY, Kang DH, Jung KH (2014) Ethanol production from glycerol by the yeast Pachysolen tannophilus immobilized on celite during repeated-batch flask culture. Mycobiology 42(3):305–309

Chen XH, Wang XT, Lou WY, Li Y, Wu H, Zong MH, Chen XD (2012) Immobilization of Acetobacter sp. CCTCC M209061 for efficient asymmetric reduction of ketones and biocatalyst recycling. Microb Cell Fact 11:119

Converti A, Perego P, Lodi A, Parisi F, del Borghi M (1985) A kinetic study of Saccharomyces strains: performance at high sugar concentrations. Biotechnol Bioeng 27(8):1108–1114

Crespo CF, Badshah M, Alvarez MT, Mattiasson B (2012) Ethanol production by continuous fermentation of D-(+)-cellobiose, D-(+)-xylose and sugarcane bagasse hydrolysate using the thermoanaerobe Caloramator boliviensis. Bioresour Technol 103(1):186–191

De Bari I, De Canio P, Cuna D, Liuzzi F, Capece A, Romano P (2013) Bioethanol production from mixed sugars by Scheffersomyces stipitis free and immobilized cells, and co-cultures with Saccharomyces cerevisiae. N Biotechnol 30(6):591–597

de Jong B, Siewers V, Nielsen J (2012) Systems biology of yeast: enabling technology for development of cell factories for production of advanced biofuels. Curr Opin Biotechnol 23(4):624–630

Djordjevic R, Gibson B, Sandell M, De Billerbeck GM, Bugarski B, Leskosek-Cukalovic I et al (2014) Raspberry wine fermentation with suspended and immobilized yeast cells of two strains of Saccharomyces cerevisiae. Yeast 32(1):271–279

Duarte JC, Rodrigues JA, Moran PJ, Valenca GP, Nunhez JR (2013) Effect of immobilized cells in calcium alginate beads in alcoholic fermentation. AMB Express 3(1):31

Galaction AI, Rotaru R, Kloetzer L, Vlysidis A, Webb C, Turnea M et al (2011) External and internal glucose mass transfers in succinic acid fermentation with stirred bed of immobilized Actinobacillus succinogenes under substrate and product inhibitions. J Microbiol Biotechnol 21(12):1257–1263

Galaction AI, Kloetzer L, Turnea M, Webb C, Vlysidis A, Cascaval D (2012) Succinic acid fermentation in a stationary-basket bioreactor with a packed bed of immobilized Actinobacillus succinogenes: 1. Influence of internal diffusion on substrate mass transfer and consumption rate. J Ind Microbiol Biotechnol 39(6):877–888

Gòdia F, Casas C, Castellano B, Solà C (1987) Immobilized cells: behaviour of carrageenan entrapped yeast during continuous ethanol fermentation. Appl Microbiol Biotechnol 26(4):342–346

Hasunuma T, Kondo A (2012) Development of yeast cell factories for consolidated bioprocessing of lignocellulose to bioethanol through cell surface engineering. Biotechnol Adv 30(6):1207–1218

Irfan M, Nadeem M, Syed Q (2014) Ethanol production from agricultural wastes using Sacchromyces cervisae. Braz J Microbiol 45(2):457–465

Karagoz P, Ozkan M (2014) Ethanol production from wheat straw by Saccharomyces cerevisiae and Scheffersomyces stipitis co-culture in batch and continuous system. Bioresour Technol 158:286–293

Lee SE, Lee CG, Kangdo H, Lee HY, Jung KH (2012) Preparation of corncob grits as a carrier for immobilizing yeast cells for ethanol production. J Microbiol Biotechnol 22(12):1673–1680

Ma P, Goncalves T, Maretzek A, Dias MC, Thevelein JM (1997) The lag phase rather than the exponential-growth phase on glucose is associated with a higher cAMP level in wild-type and cAPK-attenuated strains of the yeast Saccharomyces cerevisiae. Microbiology 143(Pt 11):3451–3459

Mathew AK, Crook M, Chaney K, Humphries AC (2014) Continuous bioethanol production from oilseed rape straw hydrosylate using immobilised Saccharomyces cerevisiae cells. Bioresour Technol 154:248–253

Matsushika A, Inoue H, Kodaki T, Sawayama S (2009) Ethanol production from xylose in engineered Saccharomyces cerevisiae strains: current state and perspectives. Appl Microbiol Biotechnol 84(1):37–53

Mukherjee V, Steensels J, Lievens B, Van de Voorde I, Verplaetse A, Aerts G et al (2014) Phenotypic evaluation of natural and industrial Saccharomyces yeasts for different traits desirable in industrial bioethanol production. Appl Microbiol Biotechnol 98(22):9483–9498

Pacheco AM, Gondim DR, Goncalves LR (2010) Ethanol production by fermentation using immobilized cells of Saccharomyces cerevisiae in cashew apple bagasse. Appl Biochem Biotechnol 161(1–8):209–217

Pilkington PH, Margaritis A, Mensour NA, Russell I (1998) Fundamentals of immobilised yeast cells for continuous beer fermentation: a review. J Inst Brew 104(1):19–31

Pscheidt B, Glieder A (2008) Yeast cell factories for fine chemical and API production. Microb Cell Fact 7:25

Rotaru R, Kloetzer L, Galaction AI, Cascaval D (2011) Succinic acid production using mobile bed of immobilized Actinobacillus succinogenes in alginate. Rev Med Chir Soc Med Nat Iasi 115(1):264–268

Saini R, Vieth WR (1975) Reaction kinetics and mass transfer in glucose isomerisation with collagen-immobilised whole microbial cells. J Appl Chem Biotech 25(2):115–141

Shafiee S, Topal E (2008) An econometrics view of worldwide fossil fuel consumption and the role of US. Energy Policy 36(2):775–786

Singh NL, Srivastava P, Mishra PK (2009) Studies on ethanol production using immobilized cells of Kluyveromyces thermotolerans in a packed bed reactor. J Sci Ind Res 68:617–623

Terada A, Yuasa A, Kushimoto T, Tsuneda S, Katakai A, Tamada M (2006) Bacterial adhesion to and viability on positively charged polymer surfaces. Microbiology 152(Pt 12):3575–3583

Wendhausen R, Fregonesi A, Moran PJ, Joekes I, Rodrigues JA, Tonella E et al (2001) Continuous fermentation of sugar cane syrup using immobilized yeast cells. J Biosci Bioeng 91(1):48–52

Willaert R, Flickinger MC (2009) Cell immobilization: engineering aspects encyclopedia of industrial biotechnology. Wiley, New York

Yu J, Zhang X, Tan T (2007) An novel immobilization method of Saccharomyces cerevisiae to sorghum bagasse for ethanol production. J Biotechnol 129(3):415–420

Zhao Y, Delancey GB (2000) A diffusion model and optimal cell loading for immobilized cell biocatalysts. Biotechnol Bioeng 69(6):639–647

7

Proteomics analysis of metabolically engineered yeast cells and medium-chained hydrocarbon biofuel precursors synthesis

Xiang Li and Wei Ning Chen[*]

Abstract

Recently, various biofuels have been synthesized through metabolic engineering approaches to meet the exploding energy demands. Hydrocarbon biofuels, energy-equivalent to petroleum-based fuels, are identified as promising replacements for petroleum. Metabolically engineered *Saccharomyces cerevisiae* capable of synthesize precursors of medium-chained hydrocarbons is proposed in this study. The hydroperoxide pathway introduced in *S. cerevisiae* consisted of lipoxygenase (LOX) and hydroperoxide lyase (HPL) from almond, which catalyzes linoleic acid to 3(*Z*)-nonenal, the precursor for medium-chained hydrocarbon biofuels. Proteomics study showed that 31 proteins displayed different expression levels among four functional strains and most of them were related to carbohydrate metabolism and protein synthesis, suggested prospective capabilities of energy generation and exogenous protein synthesis. Biotransformation efficiency studies carried out by GC-FID were in accordance with the expectations. The highest yield of 3(*Z*)-nonenal was up to 1.21 ± 0.05 mg/L with the carbon recovery of up to 12.4%.

Keywords: *Saccharomyces cerevisiae*; Lipoxygenase; Hydroperoxide lyase; Medium-chained biofuel precursors; Proteomics

Introduction

Recently, petroleum shortage and environmental concerns have emphasized the synthesis and utilization of renewable fuels (Atsumi et al. 2008; Chang and Keasling 2006; Lennen et al. 2010). Biomass-derived ethanol as drop-in fuel is currently in use, while hydrocarbons which eliminate the drawbacks of ethanol are also promising biofuels (Regalbuto 2009).

Hydrocarbons are highly compatible with existing energy infrastructure due to its chemical resemblance to traditional petroleum-based fuels. Besides, hydrocarbons are energy-equivalent to petroleum-based fuels and render no mileage penalty in the procedure of usage. Moreover, being immiscible in water eliminate the additional effort required for water separation and distillation step (Boundy et al. 2011), further makes hydrocarbons promising diesel substitutes.

Recent research has identified various genetically-engineered micro-organisms capable of producing hydrocarbons (Steen et al. 2010; Rutherford et al. 2010). Fatty aldehydes derived from lipid biosynthesis were identified to be metabolically flexible precursors for a diversity of biofuels, including alkanes, free fatty acids and wax esters (Kaiser et al. 2013). In this study, we will therefore explore the biosynthesis capabilities of medium-chained aldehydes through metabolic engineering approaches.

The aldehyde-producing hydroperoxide pathway in plants has been studied and the corresponding genetic information has been elucidated (Mita et al. 2001; Santino et al. 2005; Tijet et al. 2001; Mita et al. 2005). Hydroperoxide pathway starts with hydroperoxidation of polyunsaturated fatty acid, linoleic acid. With LOX catalyzing, one peroxy is inserted onto the backbone of linoleic acid and yield one unsaturated acid hydroperoxide (HPOD). HPOD can subsequently be metabolized via a number of secondary reactions while one of them is to be cleaved by HPL and yield one aldehyde and one oxo-acid (Feussner and Wasternack 2002). Linoleic acid can be oxygenated either at carbon atom 9(9LOX) or 13(LOX) of the backbone. In the case of oxygenating at carbon atom position 9, linoleic

* Correspondence: WNChen@ntu.edu.sg
School of Chemical and Biomedical Engineering, Nanyang Technological University, 62 Nanyang Drive, Singapore 637459, Singapore

acid will be diverted to 3(*Z*)-nonenal and 9-oxo-nonanoic acid, as shown in Figure 1. 3(*Z*)-nonenal is our target medium-chained biofuel precursor in this study. We used *S. cerevisiae* to construct whole-cell based catalyst which was capable of synthesize 3(*Z*)-nonenal through exogenous expressing of 9LOX and 9HPL from almond (*Prunus dulcis*).

It has been reported that, after absorption into *S. cerevisiae* from media, the degradations of long-chained fatty acids (LCFAs) are confined in peroxisomes (Hiltunen et al. 2003; Hettema and Tabak 2000). The protein complex, Pxa1p-Pxa2p, which embeds in the peroxisomal membrane, functions as transporter and translocates activated fatty acids into peroxisomes for beta-oxidation, utilizing the energy of ATP hydrolysis. Furthermore, previous results have confirmed that LCFAs cannot enter peroxisomes in Δ*pxa1* and Δ*pxa2* mutant and disruption of either pxa1 or pxa2 leads to latency of LCFA β-oxidation while disrupting both genes exhibited similar phenotype (Hettema et al. 1996).

In this study, exogenous genes 9LOX and 9HPL were expressed in *S. cerevisiae*. Apart from wild type as control, we used Δ*pxa1*, Δ*pxa2* and Δ*pxa1&2* mutants as the hosts to block the translocations of absorbed LCFAs into peroxisomes and divert them to the exogenous hydroperoxide pathway. Proteomics analysis using 2D LC-MS/MS approaches provided us a global overview of protein expression levels to determine the potentials of the constructed whole-cell based catalysts (Wiese 2007). The biotransformation efficiencies of the functional strains were also characterized by GC-FID approach and the highest yield of 3(*Z*)-nonenal we achieved was up to 1.21 mg/L.

Materials and methods
Strains and culture media
E.coli strain Top10 was used for cloning and plasmid propagation and cultured at 37°C with constant shaking

at 250 rpm. LB broth contained 10 g/L bacto-tryptone (Fluka), 5 g/L yeast extract and 5 g/L NaCl (Sigma). The *S. cerevisiae* strains (Table 1) were cultured at 30°C with constant shaking at 250 rpm. YPD medium consisted of 10 g/L yeast extract (Fluka), 20 g/L peptone (Bacto) and 20 g/L dextrose (Sigma). YNB-LEU selective media consisted of 6.7 g/L yeast nitrogen base without amino acids (Sigma), 0.69 g/L DO Supplement-LEU (Clontech) and 20/L dextrose or galactose (Sigma). YNB-HIS selective media contained 6.7 g/L yeast nitrogen base without amino acids (Sigma), 0.69 g/L DO Supplement-HIS (Clontech) and 20/L dextrose or galactose (Sigma).

Recombinant plasmid construction
All the oligonucleotide primers in Table 1 were synthesized by Integrated DNA Technologies. All restriction enzymes used in this study were purchased from New England Biolabs. Ligation reactions were performed using T4 ligase (Fermentas). PCR reactions were carried out with HotStarTaq *Plus* Master Mix Kit (Qiagen) according to standard protocols. Gel extractions were carried out using QIAquick Gel Extraction Kit (Qiagen). *E.coli* minipreps were performed with QIAprep Spin Miniprep Kit (Qiagen).

Codon optimized genes 9LOX and 9HPL were generated by Geneart (acc. No. KC920894 and KC920895). The cloning vector pESC-LEU (Agilent) was adopted, which contains the *GAL1* and *GAL10* yeast promoters in opposing orientations, capable of introducing two genes into one strain under the control of a repressible promoter.

Primers F-*BamHI* and R-*SalI* in Table 1 were used to introduce *BamHI* and *SalI* into 9LOX. Flanked by 5′ *BamHI* restriction enzyme site and 3′ *SalI* site, 9LOX gene was then inserted into pESC plasmid to obtain 9LOX-pESC recombinant plasmid. *SacI* and *NotI* restriction endonucleases were adopted to double digest 9HPL gene from default plasmid pMK-RQ. The DNA fragment is flanked by 5′ *SacI* restriction enzyme site and 3′ *NotI* site, and inserted into 9LOX-pESC recombinant plasmid to obtain the recombinant plasmid 9LOX-9HPL-pESC (9LHP), shown in Additional file 1: Figure AF2.

Double deletion strain construction
The pUG plasmid carrying gene disruption cassettes containing HIS5 heterologous marker genes with *loxP* sites was selected for gene disruption (Gueldener et al. 2002). The target genes in *S. cerevisiae* were pxa1 and pxa2, the heterodimers of peroxisomal membrane transporter Pxa1p-Pxa2p. The sequences flanking the target genes were added to the 5′ end of OL3′ and OL3′ sequences: 40 nucleotide stretches that are homologous to sequences upstream of the ATG start codon, and down-stream of the stop codon of the targeted gene respectively. Primer sequences are shown in Table 1.

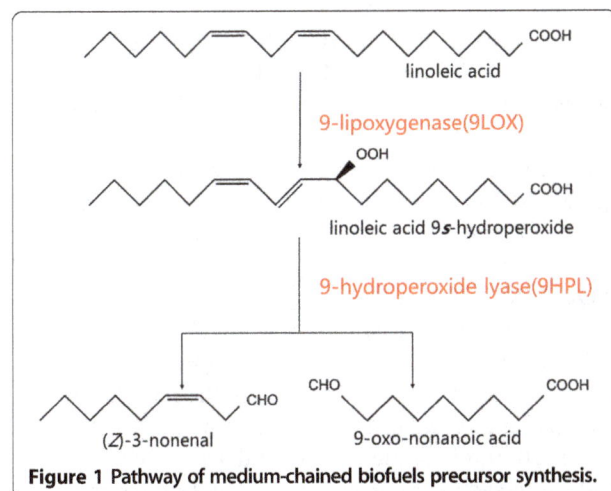

Figure 1 Pathway of medium-chained biofuels precursor synthesis.

Table 1 Primers, plasmids and strains used in this work

Name	Description	Reference
E. coli strain		
Top 10	F-*mcrA* Δ(*mrr-hsd*RMS-*mcr*BC) Φ80*lacZ*ΔM15 Δ*lacX74 recA1 araD139* Δ(*ara leu*) 7697 *galU galKrps*L (StrR) *endA1 nupG*	Life Technology
Primers for gene disruption		
Δ*pxa2*-F	5'- ATAATAATAC AATTAAAAGT TACCGAAGAA AGATTTATA CAGCTGAAGC TTCGTACGC-3'	This work
Δ*pxa2*-R	5'- CAATTTATAC ATGATTTGGA TCCTCCTTTG GCTATGTATG GCATAGGCCA CTAGTGGATC TG-3'	This work
Primers for restriction endonuclease		
F-*Bam*HI	5'-CG*GGATCC*AT GTTGCATAAC TTGTTCGACA AGA-3'	This work
R-*Sal*I	5'-GC*GTCGAC*AG TAGAATCCAA ACCCAACAAT GGA-3'	This work
Plasmids		
PUG27	*loxP-kanMX-loxP* disruption module plasmid	EUROSCARF
pESC-*leu*	With *GAL1* and *GAL10* yeast promoters in opposite orientation, CYC1 and ADH1 terminator respectively	Agilent
S. cerevisiae strains		
Wild type	*MATa; his3Δ 1; leu2Δ 0; met15Δ 0; ura3Δ 0*	EUROSCARF
Δ*pxa1*	*BY4741; Mat a; his3D1; leu2D0; met15D0; ura3D0; YPL147w::kanMX4*	EUROSCARF
Δ*pxa2*	*BY4741; Mat a; his3D1; leu2D0; met15D0; ura3D0; YKL188c::kanMX4*	EUROSCARF
Δ*pxa1&2*	*BY4741; Mat a; his3D1; leu2D0; met15D0; ura3D0; YPL147w::kanMX4; YKL188c::his*	This work
WT-pESC	wild type carring pESC	This work
Δ*pxa1*-pESC	Δ*pxa1* carrying pESC	This work
Δ*pxa2*-pESC	Δ*pxa2* carrying pESC	This work
Δ*pxa1&2*-pESC	Δ*pxa1&2* carrying pESC	This work
WT-9LHP	wild type carring 9LHP	This work
Δ*pxa1*-9LHP	Δ*pxa1* carrying 9LHP	This work
Δ*pxa2*-9LHP	Δ*pxa2* carrying 9LHP	This work
Δ*pxa1&2*-9LHP	Δ*pxa1&2* carrying 9LHP	This work

S. cerevisiae wild type, Δ*pxa1* and Δ*pxa2* strains were purchased from EUROSCARF. The Δ*pxa1* strain was transformed with the *pxa2::his* using PEG-LiAc method (Gietz and Schiestl 2007) to construct double deletion strain Δ*pxa1&2*. Transformed deletion stains Δ*pxa1&2* strain were selected via histidin prototroph by growing on synthetic complete minimal medium deficient in histidine. Yeast colony PCR was carried out to further confirm the gene disruption.

Functional strains construction

Recombinant plasmid 9LHP was transformed into *S. cerevisiae* wild type, Δ*pxa1*, Δ*pxa2* and Δ*pxa1&2*, to obtain WT-9LHP, Δ*pxa1*-9LHP, Δ*pxa2*-9LHP and Δ*pxa1&2*-9LHP functional strains. Corresponding controls, WT-pESC, Δ*pxa1*-pESC, Δ*pxa2*-pESC and Δ*pxa1&2*-pESC were constructed by transforming empty pESC plasmid into the four strains (Table 1).

Protein extraction and labeling

The *S. cerevisiae* functional strains WT-9LHP, Δ*pxa1*-9LHP, Δ*pxa2*-9LHP and Δ*pxa1&2*-9LHP were cultured at 30°C with constant shaking at 250 rpm using 50mLYPD-

LEU selective media containing galactose to induce the promoter. After 3 days' culture, *S. cerevisiae* cells were collected. For cell lysis and protein extraction, all steps were carried out on ice to avoid denaturation of proteins. Same amount ($OD_{600} = 20$) units of yeast cells were pelleted at 13,000 rpm, 4°C for 5 min. The cell pellets were washed twice by distilled water and re-suspended in 300 μL of yeast lysis buffer which consisted of: 8 M Urea, 50 mM DTT, 50 mM Tris-Cl (pH7.6), 100 mM NaCl, 0.1% Triton X-100, 1 mM EDTA and 1 mM PMSF. Equal volumes of acid-washed glass beads were added and the mixtures were performed in the bead mill by 5 cycles of 30s of vortex at 4.0 m/s with 30s of cooling on ice. Lysates were centrifuged at 10,000 rpm for 10 min at 4°C and supernatants were collected and stored at −80°C. The protein concentrations were determined following the standard protocol of 2D Quant Kit (GE Healthcare).

A total of 100 μg proteins from functional strains WT-9LHP, Δ*pxa1*-9LHP, Δ*pxa2*-9LHP and Δ*pxa1&2*-9LHP were collected and labeled by iTRAQ Reagent Multi-Plex Kit (AB Sciex) according to the standard protocol as follows: 20 μL dissolution buffer and 1 μL denaturant were added to each sample; vortex to mix; 2 μL reducing reagent was added to each sample; incubation at 60°C for 1 h; 1 µL cysteine-blocking reagent was added to each sample; vortex to mix; incubate 10 min at room temperature; 20 µL of 0.25 µg/µL sequence grade modified trypsin (Promega, US) was added to each sample to digest the protein overnight at 37°C; amino-modifying labeling reagent 114, 115, 116 and 117 were used to label four samples respectively: WT-9LHP protein sample was labeled with iTRAQ tag 114; Δpxa1-9LHP protein protein sample was labeled with iTRAQ tag 115; Δpxa2-9LHP protein sample was labeled with iTRAQ tag 116; Δpxa1&2-9LHP protein sample was labeled with iTRAQ tag 117. The labeled samples were then combined together and condensed to roughly 100 µL using a thermal shaker at 30°C.

LC-MS analysis

The labeled samples were analyzed by online 2D Nano-LC-MS/MS 1200 series nanoflow liquid chromatography system (Agilent Technologies) interfaced with 6500 Q-TOF mass-spectrometer with HPLC-Chip Cube (Agilent Technologies). The HPLC-Chip was a combination of Zorbax 300SB C_{18} reversed-phase column (75 µm × 50 mm, 3.5 µm) packing with Zorbax 300SB C_{18} enrichment column (0.3 × 5 mm, 5 µm).

In the first dimension, 4 µL of sample was loaded onto the polysulfoethyl, a strong cation-exchange (SCX) column (0.32 × 50 mm, 5 µm). The retained peptides were then eluted by injecting 8 µL ammonium formate solutions in concentration gradient of 20, 40, 60, 80, 100, 200, 500 and 1000 mM. In the second dimension, the effluent was trapped onto Zorbax 300SB C_{18} enrichment column during the enrichment mode by buffer A (5% acetonitrile and 0.1% formic acid) with a flow rate of 4 µL/min. Then the peptides trapped on enrichment column were eluted for 60 min by buffer B (0.1% formic acid) and buffer C (0.1% formic acid + acetonitrile nanoflow gradient from 5% to 80% in 60 min) at a flow rate of 300 µL/min. Subsequently, the effluent flowed through the analytical Zorbax 300SB C_{18} reversed-phase column for separation with the HPLC-Chip on analytical mode. The analysis was accomplished by 6500 Q-TOF mass spectrometer with a capillary voltage of 1950 V for 10 runs in total. For MS analysis, positive ionization mode was used. Survey scans were from m/z 300 to 2000 with an acquisition rate of 4 spectra per second.

LC-MS/MS data analysis

Peptide quantification and protein identification were performed with Spectrum Mill MS Proteomics Workbench (Agilent Technologies). Each MS/MS spectrum was searched for the species of S. cerevisiae against the UniProt-Swiss-Prot database. Methyl-methane-thiosulfate-labeled cysteine and iTRAQ modification of free amine in the amino terminus and lysine were set as fixed modification.

Protein relative quantification using iTRAQ was performed on the MS/MS scans. Protein quantification data with two or more unique peptides identified with confidence > 99% and the p value < 0.05 were selective for further statistical analysis. Three independent batches were performed to increase statistically evidence of protein expression. The overlapping isotopic contributions were used to correct the calculated peak area ratios and to estimate the relative abundances of a specific peptide.

Biotransformation product detection

Functional strains were cultured in YNB-LEU containing galactose to induce the promoters. After reaching an $OD_{600} = 1$, 20 mL of the culture was collected and pelleted and washed twice with 100 mM potassium phosphate buffer at pH = 6.5 to prepare resting cells and then transferred to 250 mL GL-45 Erlenmeyer flask (Chemglass Life Sciences). Biotransformation buffer used was 20 mL potassium phosphate buffer with 100 µL linoleic acid solution (5% v/v with 0.2% tween-80). The flasks were sealed with GL-45 open top cap and parafilms (Chemglass Life Sciences) and incubated at 30°C on an orbital shaker (250 rpm) for 3 days.

Headspace samples of the cultures were determined by Agilent 6890 N GC-FID system (Agilent) equipped with Agilent J&W DB-WAX column (30 m × 0.25 mm × 0.25 µm, Agilent). 1 mL SampleLock syringe (Hamilton) was used to draw out the headspaces of the 20 mL cultures to inject into GC system. GC settings were: carrier gas: helium; column flow: 2.0 ml/min; splitless; inject temperature: 230°C The analyzing temperature program used was: 50-230°C in 18 min; 230°C for 2 min. Product identification was carried out by comparing with authentic standards and benzoaldehyde was used as internal control for quantification.

Results
Strains construction

Recombinant plasmid 9LHP (shown in Additional file 1: Figure AF2) was constructed according to the procedure described in "Materials and methods". The size of the recombinant plasmid 9LHP was 11820 bp and gene-sequencing results proved that no site mutation in the recombinant plasmid.

We constructed double mutant Δpxa1&2 using PUG27 plasmid carrying lox-his5⁺-lox. Then recombinant plasmid 9LHP or empty vector pESC was transformed into S. cerevisiae strains wild type, single mutants Δpxa1-9LHP and Δpxa2-9LHP and double mutant Δpxa1&2 to obtain functional strains WT-9LHP, Δpxa1-9LHP, Δpxa2-9LHP, Δpxa1&2-9LHP and control strains WT-pESC, Δpxa1-pESC, Δpxa2-pESC, Δpxa1&2-pESC. All the functional strains and control strains grown well on YPD and YNB-LEU selective minimal media..

Proteomic analysis

The proteomic profiling of four functional strains was carried out by On-line 2D LC-MS/MS system (Additional file 1: Figure AF1). The Spectrum Mill system was used for peptides identification. Figure 2 and Additional file 1: Figure AF4 showed the representative peptide fragmentation spectrum of glucose-6-phosphate isomerase: (R)AVYHVALR(N). Basing on the analytical conditions, more than 200 proteins were detected while 31 showed different levels among the four functional strains, as shown in Table 2 with WT-9LHP as reference. The average of protein expression levels in WT-9LHP strain was taken as 1. The "average of B/A", "average of C/A" and "average of D/A" refer to the average ratios of protein expression levels in $\Delta pxa1$-9LHP, $\Delta pxa2$-9LHP and $\Delta pxa1\&2$-9LHP strains over those in WT-9LHP strain.

We classified the 31 protein of interest into eight categories according to their functions: GAL1 and GAL7 from galactose metabolism; HXK1, GI1, PFK2, FBA1, TPI1, TDH, PGK1, GPM1, ENO and PYK1 from glycolysis; CIT1 and ACO1 from TCA cycle; ATP1 and ATP2 from ATP synthesis; LEU1, LEU2, MET6 and PDC from amino-acid metabolism; TIF, TEF1, RPL4 and RPL19 from protein biosynthesis; HSP12, HSP26 from heat shock protein; POR1, SAM2, YMR226C and SOD1 involved in other bioprocess.

It is noteworthy that in the $\Delta pxa1\&2$-9LHP strain, all the proteins listed showed higher levels than strain WT-LHP to different extents. The levels of the listed proteins in strains $\Delta pxa1$-9LHP and $\Delta pxa2$-9LHP were mostly equivalent to strain WT-9LHP.

Biotransformation

Functional strains and control strains were cultured in YNB-LEU selective media with galactose inducing the promoters of heterologous genes. As the produced 3(Z)-nonenal would be secreted outside and then vaporize into the headspace for being insoluble in water phase and volatile, 1 mL of the headspace of the cultures were extracted and injected into GC-FID for qualification and quantification.

Preliminary results have shown that, when linoleic acid was added to cultures of the growing cells, no detectable targeting volatile compounds was produced (Julsing et al. 2012). Thus non-growing but metabolically-active resting cells with higher specific catalyzing activities were obtained in this study.

Functional strains WT-9LHP, $\Delta pxa1$-9LHP, $\Delta pxa2$-9LHP, $\Delta pxa1\&2$-9LHP and corresponding control strains WT-pESC, $\Delta pxa1$-pESC, $\Delta pxa2$-pESC, $\Delta pxa1\&2$-pESC were cultured, collected and prepared as resting cells for 3 days' biotransformation. Gas samples of the headspaces of the cultures were determined with GC-FID system.

In the GC spectra, peaks at 8.82 min were identified as 3(Z)-nonenal (Additional file 1: Figure AF3). The characterizations of catalyzing activities of functional strains were repeated for 5 times, and the control strains were repeated for 3 times. Figure 3 showed the 3(Z)-nonenal production level. All the control strains produced non-detectable levels of 3(Z)-nonenal. WT-9LHP, $\Delta pxa1$-9LHP and $\Delta pxa2$-9LHP strains displayed similar producing capabilities, which were up to 0.57 ± 0.09 mg/L, 0.50 ± 0.07 mg/L and 0.48 ± 0.02 mg/L respectively. While $\Delta pxa1\&2$-9LHP produced twofold higher level of 3(Z)-nonenal, up to 1.21 ± 0.05 mg/L. The catalyzing efficiencies of the functional strains were also calculated. WT-9LHP, $\Delta pxa1$-9LHP and $\Delta pxa2$-9LHP have biotransformed 5.8%, 5.11% and 4.95% of linoleic acid into 3(Z)-nonenal respectively while $\Delta pxa1\&2$-9LHP performed the highest carbon recovery rate of up to 12.4% (Table 3).

Discussion

Since the emergence of "metabolic engineering", the potentials in producing unnatural specialty chemicals through

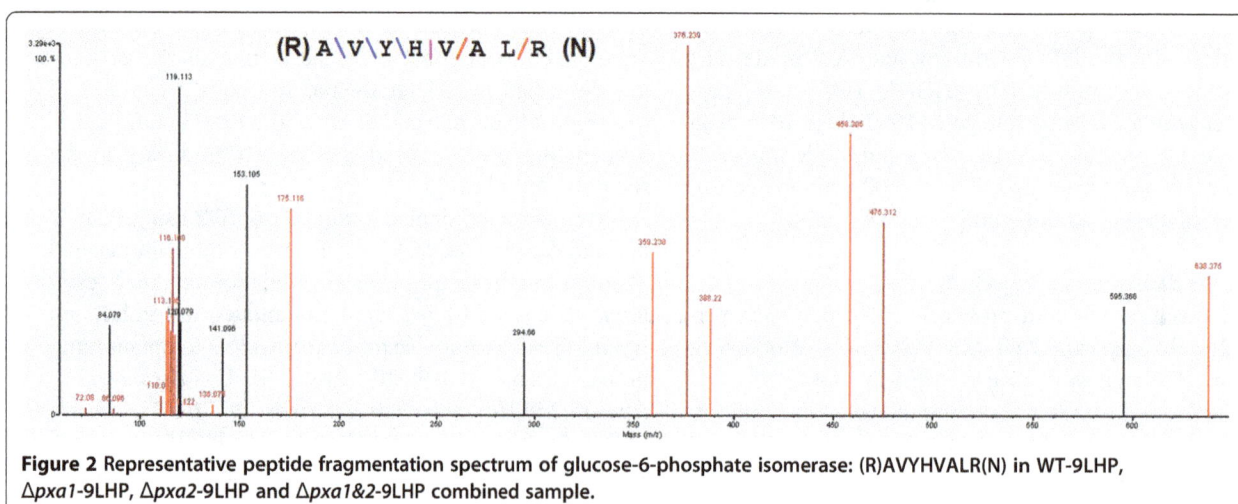

Figure 2 Representative peptide fragmentation spectrum of glucose-6-phosphate isomerase: (R)AVYHVALR(N) in WT-9LHP, $\Delta pxa1$-9LHP, $\Delta pxa2$-9LHP and $\Delta pxa1\&2$-9LHP combined sample.

Table 2 Relative changes in protein expression between *S. cerevisiae* wild type and engineered strains

Protein	Description	No. of peptides	Average of B/A	Average of C/A	Average of D/A
Galactose metabolism					
GAL1	Galactokinase	9	0.870 ± 0.340	0.587 ± 0.225	3.783 ± 0.215
GAL7	Galactose-1-phosphate uridylyltransferase	2	1.230 ± 0.005	1.387 ± 0.965	1.008 ± 0.021
Glycolysis					
HXK1	Hexokinase-1	2	0.723 ± 0.259	0.691 ± 0.104	3.489 ± 0.368
PGI1	Glucose-6-phosphate isomerase	3	1.032 ± 0.460	1.330 ± 0.180	9.891 ± 0.251
PFK2	Phosphofructokinase	3	1.236 ± 0.155	1.149 ± 0.100	1.426 ± 0.197
FBA1	Fructose-biophosphate aldolase	8	1.094 ± 0.155	0.871 ± 0.235	2.304 ± 0.942
TPI1	Triosephosphate isomerase	6	1.123 ± 0.380	0.849 ± 0.070	1.038 ± 0.357
TDH	Glyceraldehyde 3-phosphate dehydrogenase	15	1.117 ± 0.305	0.925 ± 0.220	1.664 ± 0.541
PGK1	Phosphoglycerate kinase	19	1.066 ± 0.225	0.798 ± 0.120	1.286 ± 0.076
GPM1	Phosphoglycerate mutase 1	14	1.047 ± 0.210	0.750 ± 0.145	1.500 ± 0.457
ENO	Enolase	19	1.185 ± 0.160	0.912 ± 0.155	2.176 ± 0.478
PYK1	Pyruvate kinase	6	1.186 ± 0.200	0.831 ± 0.195	1.831 ± 0.147
TCA cycle					
CIT1	Citrate synthase, mitochondrial	3	1.256 ± 0.820	1.982 ± 0.230	5.580 ± 0.248
ACO1	Aconitate hydratase, mitochondrial	3	0.961 ± 0.400	1.033 ± 0.075	1.737 ± 0.128
ATP synthesis					
ATP1	ATP synthase subunit alpha, mitochondrial	6	1.222 ± 0.260	1.005 ± 0.030	1.157 ± 0.160
ATP2	ATP synthase subunit beta, mitochondrial	8	1.176 ± 0.040	1.123 ± 0.085	3.216 ± 0.205
Amino-acid metabolism					
LEU1	3-isopropylmalate dehydratase	3	0.883 ± 0.075	1.023 ± 0.335	1.424 ± 0.200
LEU2	3-isopropylmalate dehydrogenase	17	3.477 ± 0.630	1.070 ± 0.335	2.570 ± 0.254
MET6	5-methyltetrahydropteroyltriglutamate–homocysteine methyltransferase	10	1.050 ± 0.125	0.881 ± 0.430	2.018 ± 0.121
PDC	Pyruvate decarboxylase isozyme	12	1.305 ± 0.355	1.118 ± 0.115	1.894 ± 0.218
Protein biosynthesis					
TIF	ATP-dependent RNA helicase eIF4A	3	1.408 ± 0.025	0.752 ± 0.295	1.655 ± 0.245
TEF1	Elongation factor 1-alpha	8	0.910 ± 0.375	0.758 ± 0.185	1.507 ± 0.110
RPL4	60s ribosomal protein L4	9	1.245 ± 0.255	0.778 ± 0.035	1.418 ± 0.068
RPL19	60s ribosomal protein L19	2	1.218 ± 0.285	1.114 ± 0.805	2.995 ± 0.197
Heat shock proteins					
HSP 12	12 kDa Heat shock protein	2	2.199 ± 0.640	0.882 ± 0.135	2.308 ± 0.219
HSP 26	Heat shock protein 26	3	2.281 ± 0.675	1.823 ± 0.360	2.453 ± 0.195
STI1	Heat shock protein STI1	2	1.363 ± 0.665	0.485 ± 0.035	3.450 ± 0.377
Unknown					
POR1	Mitochondrial outer membrane protein porin 1	4	1.033 ± 0.395	0.808 ± 0.445	2.785 ± 0.066
SAM2	S-adenosylmethionine synthetase 2	2	1.271 ± 0.300	0.624 ± 0.085	2.125 ± 0.151
YMR226C	Uncharacterized oxidoreductase YMR226C	2	1.856 ± 0.375	1.260 ± 0.110	3.051 ± 0.265
SOD1	Superoxide dismutase [Cu-Zn]	2	6.360 ± 0.420	3.942 ± 1.400	7.910 ± 0.330

Average of protein expression levels in WT-9LHP strain was taken as 1 and the deviation was calculated from three independent LC-MS/MS analysis results. The "Average of B/A" refers to the average ratio of protein expression level in Δpxa1-9LHP strain over that in WT-9LHP strain. "Average of C/A"refers to the average ratio of protein expression level in Δpxa2-9LHP strain over that in WT-9LHP strain. The "Average of D/A" refers to the average ratio of protein expression level in Δpxa1&2-9LHP strain over that in WT-9LHP strain.

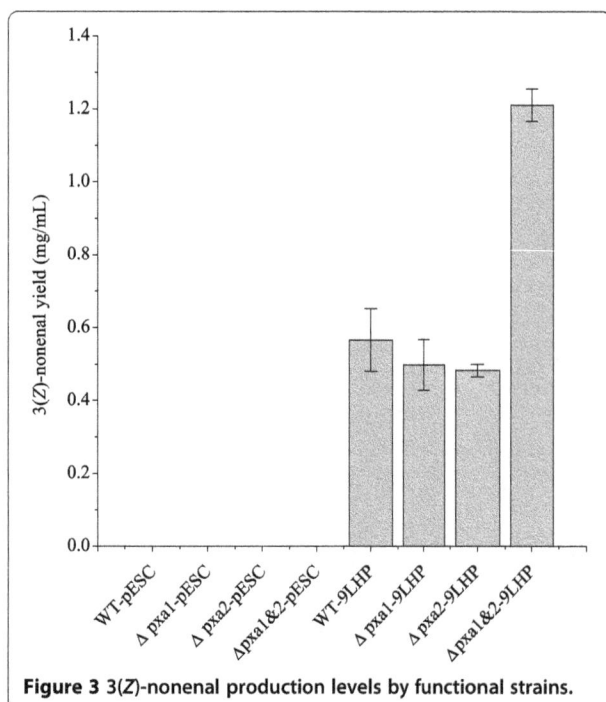

Figure 3 3(*Z*)-nonenal production levels by functional strains.

genetic and metabolic modifications have been extensively explored, especially for the discovery of petroleum-replacing biofuels (Keasling 2012). Among various reported biofuels, hydrocarbons, with high energy density and compatibility with current energy storage, transportation and utilization system, outstood as promising petroleum substitutes. While productions of short-chained (Atsumi et al. 2008; Steen et al. 2008; Santiago-Gomez et al. 2009) and long-chained hydrocarbons (Schirmer et al. 2010; Blazeck et al. 2013) have been widely explored, biofuels and precursors in medium-chained range were seldom reported. In this study, we introduced hydroperoxide pathway to convert linoleic acid to 3(*Z*)-nonenal, one promising medium-chained hydrocarbon precursor.

The 2D LC-MS/MS approach was adopted to analyze the relative protein expression levels among four functional strains, WT-9LHP, Δ*pxa1*-9LHP, Δ*pxa2*-9LHP and Δ*pxa1&2*-9LHP as shown in Table 2. We classified the 31 protein of interest into eight categories according to their functions.

In order to induce the promoters on the recombinant plasmid, galactose was added to the culture medium as the sole carbon source. After absorption, galactose would be converted to glucose-1-phosphate to enter glycolysis

through Leloir pathway. Two proteins, GAL1 and GAL7, which catalyze the two irreversible steps in Leloir pathway, showed comparable levels among WT-9LHP strain, Δ*pxa1*-9LHP strain and Δ*pxa2*-9LHP strain while up-regulated in Δ*pxa1&2*-9LHP strain.

Glycolysis is the metabolic pathway that converts glucose to pyruvate with the production of two molecules of ATP. Glycolysis pathway consists of ten enzymes: HXK1, PGI1, PFK2, FBA1, TPI1, TDH, PGK1, GPM1, ENO and PYK1. Our results showed that the levels of the ten enzymes in Δ*pxa1*-9LHP strain and Δ*pxa2*-9LHP strain were comparable to WT-9LHP strain. However, the levels of all the ten proteins were much higher (even 9.891 folds for PGI1) in Δ*pxa1&2*-9LHP strain which suggested that the up-regulated activity of glycolysis.

Pyruvate, the end product of glycolysis, can be used in aerobic respiration via TCA cycle. Pyruvate decarboxylated by pyruvate dehydrogenase catalyzing was converted into acetyl-CoA, the starting point of TCA cycle. Our results showed that, the mitochondrial enzymes CIT1 and ACO1 involved in TCA cycle and the enzymes ATP1 and ATP2 involved in ATP synthesis displayed equivalent or higher expression levels in Δ*pxa1*-9LHP strain, Δ*pxa2*-9LHP strain and Δ*pxa1&2*-9LHP strain comparing to WT-9LHP strain, suggesting the up-regulated activities.

The metabolic pathways mentioned above, galactose metabolism, glycolysis, TCA cycle an ATP synthesis, are steps in carbohydrate catabolism which breaks down carbohydrates and release energy in the form of ATP. It is noteworthy that the enzymes involved in these pathways were most notably up-regulated in the strain Δ*pxa1&2*-9LHP which suggested the most active metabolism and energy provision.

With galactose inducing the promoters, exogenous genes 9LOX and 9LHP carried by high copy number vector pESC were expressed. In the procedure of synthesis of unique proteins and peptides, amino acids metabolism was also important. LC-MS/MS results showed that four enzymes involved in the amino acid metabolism LEU1, LEU2, MET6 and PDC, along with four enzymes involved in protein biosynthesis TIF, TEF1, RPL4 and RPL19 were also significantly up-regulated in the strain Δ*pxa1&2*-9LHP. The up-regulations of these enzymes supported the exogenous genes expression well.

Proteins mentioned above were involved in energy-generation and protein synthesis. The significant higher levels in strain Δ*pxa1&2*-9LHP suggested possibly highest

Table 3 Production of functional strains and carbon recovery rates

Functional strains	WT-9LHP	Δ*pxa1*-9LHP	Δ*pxa2*-9LHP	Δ*pxa1&2*-9LHP
Yield (mg/L)	0.57 ± 0.09	0.50 ± 0.07	0.48 ± 0.02	1.21 ± 0.05
Carbon recovery (%)	5.80 ± 2.01	5.11 ± 1.63	4.95 ± 0.41	12.40 ± 0.05

biotransformation efficiency. While for the strains WT-LHP, Δpxa1-9LHP and Δpxa2-9LHP, the expression level differences were slighter which suggested comparable biotransformation efficiencies. Our expectations would be tested in subsequent biotransformation studies performed on the functional strains and control strains.

Furthermore, three heat shock proteins related to stress response, HSP12, HSP26 and STI1 showed significantly different levels among the four functional strain. The introduction of exogenous genes and the folding of the proteins may well be stress to the yeast cells and the up-regulations of these proteins were to keep the balance of the intracellular metabolism. In addition, the levels of POR1, SAM2, YMR226C and SOD1 were also found different among the four functional strain which remained to be studied. The mechanism details of the above 31 proteins level differences however still need to be further investigated.

While resting cells showed good biotransformation activities, growing cells did not produce detectable amount of 3(Z)-nonenal. Possible explanation is that growing cells were more active in cell divisions rather than performing catalyzing reactions. Furthermore, with the presence of galactose in the culture, which is the preferred carbon source, growing cells would less likely to take linoleic acid from the medium.

In this study, single deletion strains displayed comparable biotransformation efficiency as the wild type strain, as shown in Figure 3. The significantly higher biotransformation efficiency of functional strain Δpxa1&2-9LHP indicates that the combination of the two mutations would influence the flux of absorbed linoleic and further retain the absorbed linoleic acid in cytosol to be degraded through the introduced hydroperoxide pathway.

As in the biotransformation cultures, linoleic acid was the sole carbon source. Certain flux ratio would be degraded and generate energy to support the living activities of the cells apart from as substrate for 3(Z)-nonenal biotransformation,. Functional strains WT-9LHP, Δpxa1-9LHP and Δpxa2-9LHP performed equivalent biotransformation efficiencies, while Δpxa1&2-9LHP strain showed two-fold higher biotransformation with an efficiency of up to 12.1%. This biotransformation results were consistent with our expectations from the proteomics analysis results.

In conclusion, we have demonstrated a yeast-based whole-cell biocatalyst capable of transforming polyunsaturated fatty acids into medium-chained aldehyde, the medium-chained biofuel precursor. The comparative proteomics analysis offered an approach to study the overall protein in the cells and potentials as catalyst. This study lay foundation in our future direction to synthesize medium-chained hydrocarbons through metabolic engineering approaches.

Additional file

Additional file 1: Figure AF1. Total intensity chromatogram results of peptides eluted by gradient concentrations of ammonium formate. **Figure AF2** Scheme of recombinant plasmid 9LHP. **Figure AF3** GC-FID spectra of biotransformation detection: retention time at 8.82 min was identified as 3(Z)-nonenal. Blue:3(Z)-nonenal standard; red: Δpxa1&2-9LHP strain; green: Δpxa1&2-pESC strain. **Figure AF4** LC-MS qualification results of representative peptide fragmentation spectrum of glucose-6-phosphate isomerase. **Table AF1** Heat map of proteomics results in Table.

Competing interests
The authors declare that they have no competing interests.

Acknowledgements
This work was supported by funding from Competitive Research Programme Funding Scheme. X. Li was a recipient of a research scholarship from Nanyang Technological University.

References
Atsumi S, Hanai T, Liao JC (2008) Non-fermentative pathways for synthesis of branched-chain higher alcohols as biofuels. Nature 451(7174):86–89, 10.1038/nature06450

Blazeck J, Liu LQ, Knight R, Alper HS (2013) Heterologous production of pentane in the oleaginous yeast Yarrowia lipolytica. J Biotechnol 165(3–4):184–194, 10.1016/j.jbiotec.2013.04.003

Boundy B, Diegel SW, Wright L, Davis SC (2011) Biomass Energy Data Book, 4th edn, Oak Ridge National Laboratory. Department of energy. cta.ornl.gov/bedb, U.S, Accessed September 2011

Chang MCY, Keasling JD (2006) Production of isoprenoid pharmaceuticals by engineered microbes. Nat Chem Biol 2(12):674–681, 10.1038/Nchembio836

Feussner I, Wasternack C (2002) The lipoxygenase pathway. Annu Rev Plant Biol 53:275–297, 10.1146/annurev.arplant.53.100301.135248

Gietz RD, Schiestl RH (2007) Frozen competent yeast cells that can be transformed with high efficiency using the LiAc/SS carrier DNA/PEG method. Nat Protoc 2(1):1–4, 10.1038/nprot.2007.17

Gueldener U, Heinisch J, Koehler GJ, Voss D, Hegemann JH (2002) A second set of loxP marker cassettes for Cre-mediated multiple gene knockouts in budding yeast. Nucleic Acids Res 30(6):e23, 10.1093/nar/30.6.e23

Hettema EH, Tabak HF (2000) Transport of fatty acids and metabolites across the peroxisomal membrane. Bba-Mol Cell Biol L 1486(1):18–27, 10.1016/S1388-1981(00)00045-7

Hettema EH, van Roermund CWT, Distel B, vanden Berg M, Vilela C, RodriguesPousada C, Wanders RJA, Tabak HF (1996) The ABC transporter proteins Pat1 and Pat2 are required for import of long-chain fatty acids into peroxisomes of Saccharomyces cerevisiae. EMBO J 15(15):3813–3822

Hiltunen JK, Mursula AM, Rottensteiner H, Wierenga RK, Kastaniotis AJ, Gurvitz A (2003) The biochemistry of peroxisomal beta-oxidation in the yeast Saccharomyces cerevisiae. Fems Microbiol Rev 27(1):35–64, 10.1016/S0168-6445(03)00017-2

Julsing MK, Kuhn D, Schmid A, Buhler B (2012) Resting cells of recombinant E. coli show high epoxidation yields on energy source and high sensitivity to product inhibition. Biotechnol Bioeng 109(5):1109–1119

Kaiser BK, Carleton M, Hickman JW, Miller C, Lawson D, Budde M, Warrener P, Paredes A, Mullapudi S, Navarro P, Cross F, Roberts JM (2013) Fatty aldehydes in cyanobacteria are a metabolically flexible precursor for a diversity of biofuel products. Plos One 8(3):1–11, 10.1371/journal.pone.0058307

Keasling JD (2012) Synthetic biology and the development of tools for metabolic engineering. Metab Eng 14(3):189–195, 10.1016/j.ymben.2012.01.004

Lennen RM, Braden DJ, West RM, Dumesic JA, Pfleger BF (2010) A process for microbial hydrocarbon synthesis: overproduction of fatty acids in Escherichia coli and catalytic conversion to alkanes. Biotechnol Bioeng 106(2):193–202, 10.1002/Bit.22660

Mita G, Gallo A, Greco V, Zasiura C, Casey R, Zacheo G, Santino A (2001) Molecular cloning and biochemical characterization of a lipoxygenase in almond (Prunus dulcis) seed. Eur J Biochem 268(5):1500–1507

Mita G, Quarta A, Fasano P, De Paolis A, Di Sansebastiano GP, Perrotta C, Iannacone R, Belfield E, Hughes R, Tsesmetzis N, Casey R, Santino A (2005) Molecular cloning and characterization of an almond 9-hydroperoxide lyase, a new CYP74 targeted to lipid bodies. J Exp Bot 56(419):2321–2333, 10.1093/Jxb/Eri225

Regalbuto JR (2009) Cellulosic biofuels-got gasoline? Science 325(5942):822–824, 10.1126/science.1174581

Rutherford BJ, Dahl RH, Price RE, Szmidt HL, Benke PI, Mukhopadhyay A, Keasling JD (2010) Functional genomic study of exogenous n-butanol stress in *Escherichia coli*. Appl Environ Microb 76(6):1935–1945, 10.1128/Aem.02323-09

Santiago-Gomez MP, Thanh HT, De Coninck J, Cachon R, Kermasha S, Belin JM, Gervais P, Husson F (2009) Modeling hexanal production in oxido-reducing conditions by the yeast *Yarrowia lipolytica*. Process Biochem 44(9):1013–1018, 10.1016/j.procbio.2009.04.028

Santino A, Iannacone R, Hughes R, Casey R, Mita G (2005) Cloning and characterisation of an almond 9-lipoxygenase expressed early during seed development. Plant Sci 168(3):699–706, 10.1016/j.plantsci.2004.10.001

Schirmer A, Rude MA, Li XZ, Popova E, del Cardayre SB (2010) Microbial biosynthesis of alkanes. Science 329(5991):559–562

Steen E, Chan R, Prasad N, Myers S, Petzold CJ, Redding A, Ouellet M, Keasling J (2008) Metabolic engineering of *Saccharomyces cerevisiae* for the production of n-butanol. Microb Cell Fact 7(1):36, 10.1186/1475-2859-7-36

Steen EJ, Yisheng K, Bokinsky G, Zhihao H, Schirmer A, McClure A, del Cardayre SB, Keasling JD (2010) Microbial production of fatty-acid-derived fuels and chemicals from plant biomass. Nature 463(7280):559–562, 10.1038/nature08721

Tijet N, Schneider C, Muller BL, Brash AR (2001) Biogenesis of volatile aldehydes from fatty acid hydroperoxides: Molecular cloning of a hydroperoxide lyase (CYP74C) with specificity for both the 9-and 13-hydroperoxides of linoleic and linolenic acids. Arch Biochem Biophys 386(2):281–289

Wiese S (2007) Protein labeling by iTRAQ: A new tool for quantitative mass spectrometry in proteome research (vol 7, pg 340, 2007). Proteomics 7(6):1004–1004

8

Improved production of poly(lactic acid)-like polyester based on metabolite analysis to address the rate-limiting step

Ken'ichiro Matsumoto[1,2]*, Kota Tobitani[1], Shunsuke Aoki[1], Yuyang Song[1,3], Toshihiko Ooi[1,4] and Seiichi Taguchi[1,4]*

Abstract

The biosynthesis of poly(lactic acid) (PLA)-like polymers, composed of >99 mol% lactate and a trace amount of 3-hydroxybutyrate, in engineered *Corynebacterium glutamicum* consists of two steps; the generation of the monomer substrate lactyl-coenzyme A (CoA) and the polyhydroxyalkanoate (PHA) synthase-catalyzed polymerization of lactyl-CoA. In order to increase polymer productivity, we explored the rate-limiting step in PLA-like polymer synthesis based on quantitative metabolite analysis using liquid chromatography mass spectroscopy (LC-MS). A significant pool of lactyl-CoA was found during polymer synthesis. This result suggested that the rate-limitation occurred at the polymerization step. Accordingly, the expression level of PHA synthase was increased by means of codon-optimization of the corresponding gene that consequently led to an increase in polymer content by 4.4-fold compared to the control. Notably, the codon-optimization did not significantly affect the concentration of lactyl-CoA, suggesting that the polymerization reaction was still the rate-limiting step upon the overexpression of PHA synthase. Another important finding was that the generation of lactyl-CoA was concomitant with a decrease in the acetyl-CoA level, indicating that acetyl-CoA served as a CoA donor for lactyl-CoA synthesis. These results show that obtaining information on the metabolite concentrations is highly useful for improving PLA-like polymer production. This strategy should be applicable to a wide range of PHA-producing systems.

Keywords: Biobased plastic; P(LA-*co*-3HB); Polyhydroxybutyrate; Metabolome analysis

Introduction

Bacterial polyesters polyhydroxyalkanoates (PHAs) are synthesized via the supply of monomer hydroxyacyl-CoA molecules and polymerization of the monomers catalyzed by PHA synthases (Rehm 2003; Matsumoto and Taguchi 2013b; Lu et al. 2009). The rate-limiting step in PHA synthesis may be either the monomer supply or polymerization, which can vary depending on the combination of the relevant enzymes, production hosts and carbon source. To date, the rate-limiting step has been estimated by modulating the activity of each step to see its effect on polymer productivity (Jung et al. 2000; Kichise et al. 1999; Taguchi et al. 2001). This indirect approach, however, was unable to provide quantitative information on the

metabolic pathways. The aim of this study was to determine the intracellular concentration of the metabolic intermediates in the PHA biosynthetic pathways in order to explore the rate-limiting step.

In this study, we chose the PLA-like polymer-producing *Corynebacterium glutamicum* as the target (Song et al. 2012). This microorganism, which is known as an industrial amino acid producer with GRAS (Generally Regarded As Safe) status, has been engineered to express three exogenous genes encoding D-lactate dehydrogenase, propionyl-CoA transferase (PCT) and LA-polymerizing PHA synthase (PhaC1$_{Ps}$STQK) (Taguchi and Doi 2004; Taguchi et al. 2008) (see pathway in Figure 1). This unique bacterial system was shown to be capable of producing PLA-like polymer directly from glucose via one-pot fermentation. It should be noted that the PLA-like polymer consists of >99 mol% LA and a trace amount of 3-hydroxybutyrate (3HB) (Song et al. 2012), and thus is here referred to as PLA'. The challenge of the system was that

* Correspondence: mken@eng.hokudai.ac.jp; staguchi@eng.hokudai.ac.jp
[1]Division of Biotechnology and Macromolecular Chemistry, Graduate School of Engineering, Hokkaido University, N13-W8, Kita-ku, Sapporo 060-8628, Japan
[4]CREST, JST, 4-1-8 Honcho, Kawaguchi, Saitama 332-0012, Japan
Full list of author information is available at the end of the article

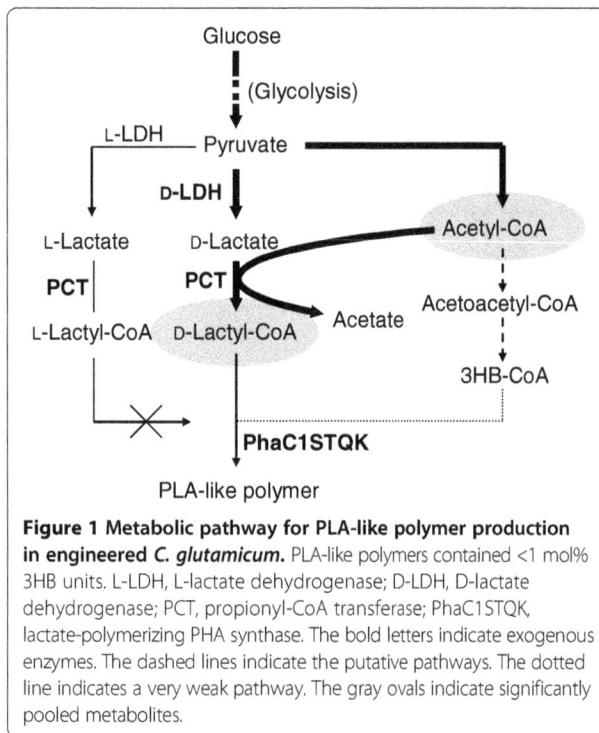

Figure 1 Metabolic pathway for PLA-like polymer production in engineered _C. glutamicum_. PLA-like polymers contained <1 mol% 3HB units. L-LDH, L-lactate dehydrogenase; D-LDH, D-lactate dehydrogenase; PCT, propionyl-CoA transferase; PhaC1STQK, lactate-polymerizing PHA synthase. The bold letters indicate exogenous enzymes. The dashed lines indicate the putative pathways. The dotted line indicates a very weak pathway. The gray ovals indicate significantly pooled metabolites.

had to be overcome was the low productivity of the polymer (0.03 g/L) compared to typical bacterial PHAs.

To meet this challenge, we attempted to identify the rate-limiting step in PLA′ synthesis by means of quantitative metabolite analysis using liquid chromatography mass spectroscopy (LC-MS) (Zhou et al. 2012). This method was reported to be suitable for measuring the derivatives of CoA having a relatively high molecular weight and polarity, while gas chromatography-MS has been used for detecting relatively low-polarity and small molecules (Poblete-Castro et al. 2012). In this study, the concentrations of the important intermediates for PLA′ production were determined, i.e. lactyl-CoA, acetyl-CoA, acetoacetyl-CoA and 3HB-CoA (Figure 1). To the best of our knowledge this was the first successful monitoring of the intracellular CoA-derivatives during PLA′ production, which was realized by a rational metabolic engineering approach.

Materials and methods

Plasmid construction

The four oligonucleotides 5′-gtcaccggatcccggttaactctag-3′, 5′-actcgagcctgcaggagatcttcgatatca-3′, 5′-ctagtgatatcgaagat ctcctgc-3′ and 5′-aggctcgagtctagagttaaccgggatccg-3′ were inserted into the _Xba_I/_Bst_EII sites of pPSPTG1 (Song et al. 2012) so as to create _Bam_HI, _Hpa_I, _Xba_I, _Xho_I, _Sbf_I, _Bgl_II and _Eco_RV sites (pPSDCP). Fragments of the _ldhA_ gene from _Escherichia coli_ (gene ID: 12930508), the _pct_ gene from _Megasphaera elsdenii_ and the _phaC1STQK_ gene (Taguchi et al. 2008) were inserted into pPSDCP at

_Bam_HI/_Hpa_I, _Bgl_II/_Eco_RV and _Xba_I/_Bgl_II sites, respectively, to yield pPS_ldhAC1STQKpct_. The codon-optimized _phaC1STQK_ gene [_ephaC1STQK_, accession No. AB983346 (DDBJ)] was chemically synthesized (Eurofins Genomics) for the expression in _C. glutamicum_ and its _Xba_I/_Sbf_I fragment was inserted in a similar manner so as to yield pPS_ldhAeC1STQKpct_ (Additional file 1: Table S1).

Strain, culture conditions and polymer analysis

C. glutamicum ATCC13803 was transformed by electroporation, as described previously (Liebl et al. 1989). For polymer production, the engineered strains were grown in 2 ml nutrient-rich CM2G medium (Kikuchi et al. 2003) at 30°C for 24 h with reciprocal shaking at 180 strokes/ min. Two hundred microliters of the preculture were then transferred into 2 mL minimal MMTG medium (Kikuchi et al. 2003) containing 60 g/L glucose and 0.45 mg/L of biotin, and further cultivated for 72 h at 30°C. When needed, kanamycin (50 μg/mL) was added to the medium. After cultivation, cells were lyophilized for polymer extraction. The polymer content was determined using gas chromatography as described previously (Takase et al. 2004). Based on this analytical method, 3HB units in the polymer were below the detection limit.

LC-MS analysis

The cell extract was prepared using a method modified from a previous report (Kiefer et al. 2011). The cells were cultivated on 2 mL MMTG medium as described above, then harvested at 18 h. The cells were resuspended in 200 μL of chilled water, then combined with 1 mL chilled acetonitrile containing 0.1 M formic acid and treated with sonication for 5 sec × 5 times. The supernatant was transferred to a new microtube and evaporated _in vacuo_ at 4°C. The sample was dissolved in 200 μL of chilled water. LC-MS analysis was performed using an LCMS-8030 (Shimadzu) equipped with a Mastro C18 column (150 mm), electrospray ionization (ESI) and triple quadrupole mass spectroscopy. Carrier A: 5 mM ammonium acetate (pH 5.6) containing 5 mM dimethylbutylamine (Gao et al. 2007) and carrier B: methanol were used with a flow rate of 0.2 mL/min in gradient mode, as follows: 0 min, 10% B; 3 min, 10% B; 15 min, 95% B; 18 min, 95% B; 23 min, 10% B. The ESI voltage was 3.5 kV in the negative mode. Nitrogen was used as a nebulizer (3.0 mL/ min) and drying gas (15.0 mL/min). $[M-H]^-$ ions from acetyl-CoA (m/z = 808, retention time: 9.1 min), acetoacetyl-CoA (m/z = 850, rt: 8.9 min), 3HB-CoA (m/z = 852, rt: 9.1 min) and lactyl-CoA (m/z = 838, rt: 8.7 min) were monitored using the selected ion monitoring mode. Acetyl-CoA, acetoacetyl-CoA and 3HB-CoA, used as standards, were purchased from Sigma Aldrich. Lactyl-CoA was synthesized via CoA-transferring reaction by PCT, as follows. The reaction mixture containing 100 mM Tris–

HCl (pH 7.4), 0.4 mM acetyl-CoA, 12.5 mM sodium lactate and 0.1 mg/mL purified His-tagged PCT (Additional file 1) was incubated at 30°C for 30 min. Then lactyl-CoA was purified using a preparative HPLC equipped with a C18 reverse phase column.

SDS-PAGE analysis

The cells cultivated under the polymer producing conditions were harvested at 18 h. The cells were resuspended in 25 mM Tris–HCl (pH 7.5) buffer and treated with sonication. The whole cell extracts were subjected to SDS-PAGE.

Results

The use of a single plasmid system increased transformation efficiency

In a previous study, a dual plasmid system using pPS and pVC vectors was used for PLA′ production (Song et al. 2012). To reduce the use of antibiotics, the expression vector was reconstructed so that the recombinant form was maintained only in the presence of kanamycin. In addition, the single plasmid system improved the transformation efficiency compared to the dual plasmid system (from 3×10^2 to 2×10^4 colonies/μg DNA). Therefore, the single plasmid system was used for further study.

Determination of the CoA intermediates in the engineered *C. glutamicum*

The wild-type *C. glutamicum* and the recombinant cells harboring pPS*ldhAC1STQKpct* were cultivated under the polymer-producing conditions. The cells were harvested in the logarithmic growth phase (18 h) and subjected to metabolite analysis. The concentrations of aectyl-CoA, lactyl-CoA, acetoacetyl-CoA and 3HB-CoA in the cells were determined. The key points for the successful measurement of the CoA derivatives were shown to be a rapid extraction of the cells along with the an appropriate ion pair reagent. For the wild-type cells, the acetyl-CoA concentration was 89 nmol/g-dry cells, and unexpectedly, a small amount of lactyl-CoA was also observed (Figure 2). In contrast, the recombinant cells exhibited an elevated concentration of lactyl-CoA, which was concomitant with a significant reduction in the concentration of acetyl-CoA. This result suggests that PCT promoted lactyl-CoA synthesis, and more importantly, acetyl-CoA is able to serve as a CoA donor for the generation of lactyl-CoA (Figure 1). The concentrations of acetoacetyl-CoA and 3HB-CoA were below the detection limit for all of the conditions tested (data not shown). From these observations, the interconversion of lactate + acetyl-CoA ↔ lactyl-CoA + acetate presumably achieved an equilibrium state, namely, the polymerization of lactyl-CoA would be a retarded step.

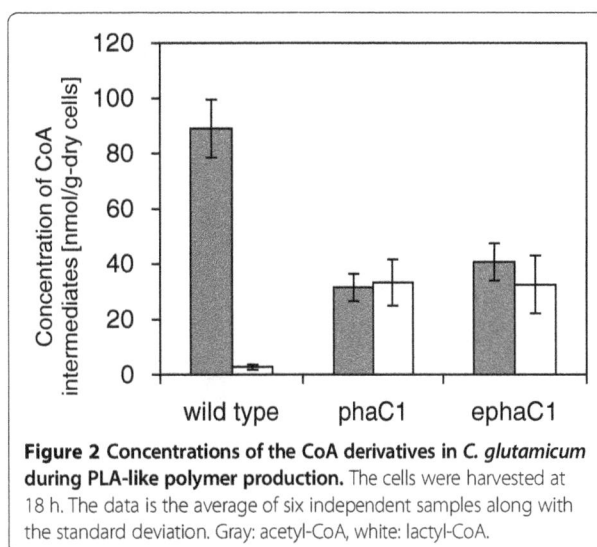

Figure 2 Concentrations of the CoA derivatives in *C. glutamicum* during PLA-like polymer production. The cells were harvested at 18 h. The data is the average of six independent samples along with the standard deviation. Gray: acetyl-CoA, white: lactyl-CoA.

Enhanced expression of PHA synthase elevated PLA-like polymer production

In order to evaluate the aforementioned hypothesis and to overcome the existing limitation, we attempted to improve the expression level of PHA synthase. For this purpose, the codon-optimized PHA synthase gene *ephaC1*$_{Ps}$*STQK*, which was supposed to be efficiently translated in *C. glutamicum*, was synthesized. First, the effect of the codon-optimization on the expression of the enzyme was evaluated. As shown in Figure 3, the cells harboring the codon-optimized gene had an increase in the expression level of PHA synthase.

Next, the PLA′ production was investigated for the cells harboring parent and codon-optimized PHA synthase genes. As expected, the cells expressing a higher level of PHA synthase accumulated more PLA′ (4.4-fold) than the control (Table 1). Thus, the metabolic engineering, which was designed based on the metabolite analysis, did successfully improve the polymer production. In addition, this result supported the validity of the method for the determination of the metabolite concentrations.

Metabolite analysis of the modified cells indicated the rate-limiting step remained

The metabolite concentrations in the cells harboring the codon-optimized gene were measured. Despite the increase in the polymer content (Table 1), there was no significant difference between the metabolite levels in the two types of recombinant cells (Figure 2). This result indicated that the rate-limiting at polymerization step remained even with the reinforced expression of PHA synthase.

Discussion

In this study, metabolite analysis was shown to provide quantitative information that was very useful for addressing the rate-limiting step in the PLA′ production. In the

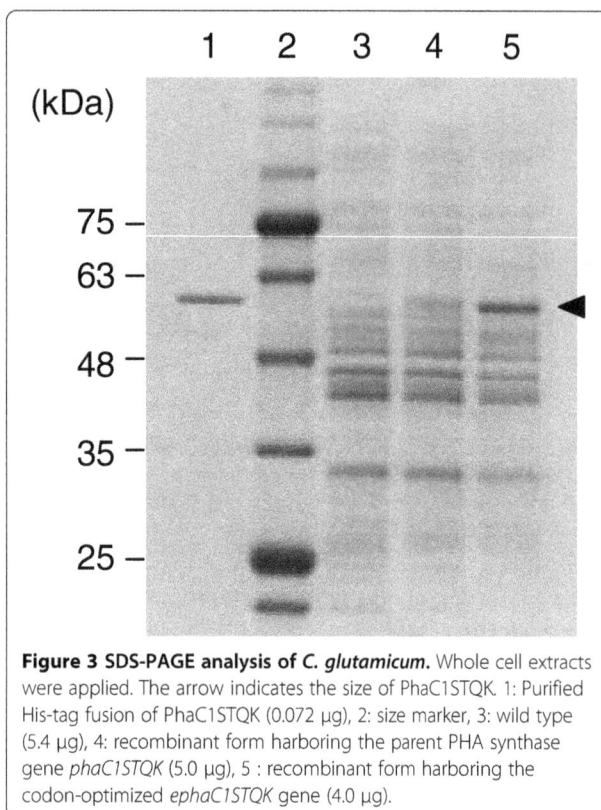

Figure 3 SDS-PAGE analysis of *C. glutamicum*. Whole cell extracts were applied. The arrow indicates the size of PhaC1STQK. 1: Purified His-tag fusion of PhaC1STQK (0.072 μg), 2: size marker, 3: wild type (5.4 μg), 4: recombinant form harboring the parent PHA synthase gene *phaC1STQK* (5.0 μg), 5 : recombinant form harboring the codon-optimized *ephaC1STQK* gene (4.0 μg).

' synthesis at polymerization step was shown to remain after PHA synthase was overexpressed. Therefore, it was expected that an additional enhancement in polymerizing activity would be needed to increase PLA' production.

It was a chance discovery that the engineered *C. glutamicum* expressing PhaC1$_{Ps}$STQK synthesized PLA'. On the other hand, it has been reportedly demonstrated that *E. coli* engineered to express the same set of enzymes did not produce PLA' (Nduko et al. 2013; Shozui et al. 2011; Yamada et al. 2009; Matsumoto and Taguchi 2013a). Thus, the mechanism for PLA' synthesis in *C. glutamicum* has been an important issue. A clue for answering this question might be the small amount of 3HB units (<1 mol %) incorporated into the polymer (Song et al. 2012). The presence of 3HB units in the polymer suggests that this organism is likely to possess intrinsic 3HB-CoA (Figure 1). The result of the present study, however, demonstrated that the concentration of 3HB-CoA was below the detection limit. Thus, 3HB-CoA may be synthesized via an unidentified, very weak route and/or rapidly metabolized. In comparison, in *R. eutropha*, which is an efficient P(3HB) producer, a much higher concentration of 3HB-CoA (0.1-1 nmol/g-dry cells) has been reportedly observed (Fukui et al. 2014). The very low 3HB-CoA level in *C. glutamicum* might account for the capacity of this organism to produce the copolymer with an extremely high LA fraction. Further investigation of LA-based polymer-producing *E. coli* is needed to clarify this issue.

The wild-type *C. glutamicum* unexpectedly synthesized lactyl-CoA or another compound having the same m/z and retention time. The metabolite level was too low to be detected using the MS/MS mode. Thus, currently the molecule is not confidently identified. However, if the existence of lactyl-CoA is postulated, this molecule should be L-lactyl-CoA, which cannot be incorporated into the polymer due to the strict stereospecificity of PHA synthase (Tajima et al. 2009), because the wild-type *C. glutamicum* possesses no D-LDH gene (Kalinowski et al. 2003). To date, the presence of lactyl-CoA in *C. glutamicum* has not been reported and its physiological role is unknown. Thus it may be an interesting research target.

Although PLA' production was successfully increased by the overexpression of PHA synthase (Table 1), the polymer content was lower than previously reported results (up to 1.4 wt%) obtained using100 mL-scale flask cultures (Song et al. 2012). This difference was probably due to the aeration efficiencies in the test tubes and flasks. Because the lactic acid production in *C. glutamicum* is influenced by the oxygen supply (Inui et al. 2004), the aeration rate would be expected to have an impact on PLA' production. More detailed analysis and fine-tuning of the aeration using a jar-fermentor is needed to optimize the culture conditions.

In summary, the levels of CoA derivatives in engineered *C. glutamicum* during PLA' production were determined,

previous studies, the rate-limiting step in PHA production has been only qualitatively explored based on the polymer production *in vivo*. For examples, in the case of P(3HB) production in *C. glutamicum*, the expression of the highly active acetoacetyl-CoA reductase (PhaB) mutant from *Ralstonia eutropha* improved P(3HB) production (Matsumoto et al. 2013), suggesting that monomer supply is a rate-limiting step in P(3HB) synthesis. A similar result was obtained in transgenic P(3HB)-producing tobacco (Matsumoto et al. 2011). Here it should be noted that in these studies it was impossible to determine whether the improved the PhaB activity was sufficient or not unless the PhaB activity was further increased. In other words, P(3HB) production should reach a plateau if the PhaB activity was sufficient under the original conditions. In contrast, based on the metabolite analysis, the rate-limitation of PLA

Table 1 PLA-like polymer production in engineered *C. glutamicum*[a]

Relevant genes	Cell dry weight (g/L)	Polymer content (wt%)
None (wild-type control)	11.4 ± 1.7	ND[b]
phaC1STQK, pct, ldhA	10.7 ± 2.0	0.27 ± 0.01
ephaC1STQK, pct, ldhA	9.3 ± 0.5	1.19 ± 0.01

[a]Cells were grown on 2 mL MMTG medium containing glucose for 72 h at 30°C. The data is the average of triplicate samples along with the standard deviation.
[b]ND: not detected.

which allowed us to identify a rate-limitation at the polymerization step. In fact, overexpression of PHA synthase successfully increased the polymer production. In addition, acetyl-CoA probably served as a CoA donor for supplying lactyl-CoA in *C. glutamicum*.

Additional file

Additional file 1: Protocol for purification of PCT.

Competing interests
The authors declare that they have no competing interests.

Author contributions
KM designed the study. KM wrote the manuscript TO and ST participated herein. KT performed the experiment work and developed LC-MS method. SA synthesized lactyl-CoA. YS developed protocols to manipulate *C. glutamicum*. All authors read and approved the submission of the manuscript.

Acknowledgements
This work was supported by Japans Society for the Promotion of Science (JSPS) Kakenhi (No. 26281043 to K.M., Nos. 23310059 and 26660080 to S.T. and No. 23580452 to T.O.), Japan Science and Technology Agency (JST), Precursory Research for Embryonic Science and Technology (PRESTO) and JST, CREST. Pacific Edit reviewed the manuscript prior to submission.

Author details
[1]Division of Biotechnology and Macromolecular Chemistry, Graduate School of Engineering, Hokkaido University, N13-W8, Kita-ku, Sapporo 060-8628, Japan. [2]PRESTO-JST, K's Gobancho, Building 7, Gobancho Chiyoda-ku, Tokyo 102-0076, Japan. [3]College of Enology, Northwest A&F University, 22 Xinong Road, Yangling, 712100 Shaanxi, China. [4]CREST, JST, 4-1-8 Honcho, Kawaguchi, Saitama 332-0012, Japan.

References
Fukui T, Chou K, Harada K, Orita I, Nakayama Y, Bamba T, Nakamura S, Fukusaki E (2014) Metabolite profiles of polyhydroxyalkanoate-producing *Ralstonia eutropha* H16. Metabolomics 10(2):190–202. doi:10.1007/s11306-013-0567-0

Gao L, Chiou W, Tang H, Cheng XH, Camp HS, Burns DJ (2007) Simultaneous quantification of malonyl-CoA and several other short-chain acyl-CoAs in animal tissues by ion-pairing reversed-phase HPLC/MS. J Chromatogr B 853(1–2):303–313. doi:10.1016/j.jchromb.2007.03.029

Inui M, Murakami S, Okino S, Kawaguchi H, Vertes AA, Yukawa H (2004) Metabolic analysis of *Corynebacterium glutamicum* during lactate and succinate productions under oxygen deprivation conditions. J Mol Microb Biotech 7(4):182–196. doi:10.1159/000079827

Jung Y, Park J, Lee Y (2000) Metabolic engineering of *Alcaligenes eutrophus* through the transformation of cloned *phbCAB* genes for the investigation of the regulatory mechanism of polyhydroxyalkanoate biosynthesis. Enzyme Microb Technol 26(2–4):201–208

Kalinowski J, Bathe B, Bartels D, Bischoff N, Bott M, Burkovski A, Dusch N, Eggeling L, Eikmanns BJ, Gaigalat L, Goesmann A, Hartmann M, Huthmacher K, Kramer R, Linke B, McHardy AC, Meyer F, Mockel B, Pfefferle W, Puhler A, Rey DA, Ruckert C, Rupp O, Sahm H, Wendisch VF, Wiegrabe I, Tauch A (2003) The complete *Corynebacterium glutamicum* ATCC 13032 genome sequence and its impact on the production of L-aspartate-derived amino acids and vitamins. J Biotechnol 104(1–3):5–25. doi:10.1016/S0168-1656(03)00154-8

Kichise T, Fukui T, Yoshida Y, Doi Y (1999) Biosynthesis of polyhydroxyalkanoates (PHA) by recombinant *Ralstonia eutropha* and effects of PHA synthase activity on *in vivo* PHA biosynthesis. Int J Biol Macromol 25(1–3):69–77

Kiefer P, Delmotte N, Vorholt JA (2011) Nanoscale ion-pair reversed-phase HPLC-MS for sensitive metabolome analysis. Anal Chem 83(3):850–855. doi:10.1021/Ac102445r

Kikuchi Y, Date M, Yokoyama K, Umezawa Y, Matsui H (2003) Secretion of active-form *Streptoverticillium mobaraense* transglutaminase by *Corynebacterium*

glutamicum: processing of the pro-transglutaminase by a cosecreted subtilisin-Like protease from *Streptomyces albogriseolus*. Appl Environ Microbiol 69(1):358–366

Liebl W, Bayerl A, Schein B, Stillner U, Schleifer KH (1989) High efficiency electroporation of intact *Corynebacterium glutamicum* cells. FEMS Microbiol Lett 53(3):299–303

Lu JN, Tappel RC, Nomura CT (2009) Mini-review: Biosynthesis of poly(hydroxyalkanoates). Polym Rev 49(3):226–248. doi:10.1080/15583720903048243

Matsumoto K, Taguchi S (2013a) Biosynthetic polyesters consisting of 2-hydroxyalkanoic acids: current challenges and unresolved questions. Appl Microbiol Biotechnol 97(18):8011–8021. doi:10.1007/s00253-013-5120-6

Matsumoto K, Taguchi S (2013b) Enzyme and metabolic engineering for the production of novel biopolymers: crossover of biological and chemical processes. Current Opinion in Biotechnology 24(6):1054–1060. doi:10.1016/j.copbio.2013.02.021

Matsumoto K, Morimoto K, Gohda A, Shimada H, Taguchi S (2011) Improved polyhydroxybutyrate (PHB) production in transgenic tobacco by enhancing translation efficiency of bacterial PHB biosynthetic genes. J Biosci Bioeng 111(4):485–488. doi:10.1016/j.jbiosc.2010.11.020

Matsumoto K, Tanaka Y, Watanabe T, Motohashi R, Ikeda K, Tobitani K, Yao M, Tanaka I, Taguchi S (2013) Directed evolution and structural analysis of NADPH-dependent acetoacetyl Coenzyme A (acetoacetyl-CoA) reductase from *Ralstonia eutropha* reveals two mutations responsible for enhanced kinetics. Appl Environ Microbiol 79(19):6134–6139. doi:10.1128/Aem. 01768-13

Nduko JM, Matsumoto K, Ooi T, Taguchi S (2013) Effectiveness of xylose utilization for high yield production of lactate-enriched P(lactate-*co*-3-hydroxybutyrate) using a lactate-overproducing strain of *Escherichia coli* and an evolved lactate-polymerizing enzyme. Metab Eng 15:159–166. doi:10.1016/j.ymben.2012.11.007

Poblete-Castro I, Escapa IF, Jaeger C, Puchalka J, Lam CMC, Schomburg D, Prieto MA, dos Santos VAPM (2012) The metabolic response of *P. putida* KT2442 producing high levels of polyhydroxyalkanoate under single- and multiple-nutrient-limited growth: Highlights from a multi-level omics approach. Microb Cell Fact 11:34. doi:10.1186/1475-2859-11-34

Rehm BHA (2003) Polyester synthases: natural catalysts for plastics. Biochem J 376:15–33

Shozui F, Matsumoto K, Motohashi R, Sun JA, Satoh T, Kakuchi T, Taguchi S (2011) Biosynthesis of a lactate (LA)-based polyester with a 96 mol% LA fraction and its application to stereocomplex formation. Polym Degrad Stab 96(4):499–504. doi:10.1016/j.polymdegradstab.2011.01.007

Song YY, Matsumoto K, Yamada M, Gohda A, Brigham CJ, Sinskey AJ, Taguchi S (2012) Engineered *Corynebacterium glutamicum* as an endotoxin-free platform strain for lactate-based polyester production. Appl Microbiol Biotechnol 93(5):1917–1925. doi:10.1007/s00253-011-3718-0

Taguchi S, Doi Y (2004) Evolution of polyhydroxyalkanoate (PHA) production system by "enzyme evolution": Successful case studies of directed evolution. Macromol Biosci 4(3):145–156

Taguchi S, Maehara A, Takase K, Nakahara M, Nakamura H, Doi Y (2001) Analysis of mutational effects of a polyhydroxybutyrate (PHB) polymerase on bacterial PHB accumulation using an in vivo assay system. FEMS Microbiol Lett 198(1):65–71

Taguchi S, Yamada M, Matsumoto K, Tajima K, Satoh Y, Munekata M, Ohno K, Kohda A, Shimamura T, Kambe H, Obata S (2008) A microbial factory for lactate-based polyesters using a lactate-polymerizing enzyme. Proc Natl Acad Sci USA 105(45):17323–17327

Tajima K, Satoh Y, Satoh T, Itoh R, Han XR, Taguchi S, Kakuchi T, Munekata M (2009) Chemo-enzymatic synthesis of poly(lactate-*co*-(3-hydroxybutyrate)) by a lactate-polymerizing enzyme. Macromolecules 42(6):1985–1989

Takase K, Matsumoto K, Taguchi S, Doi Y (2004) Alteration of substrate chain-length specificity of type II synthase for polyhydroxyalkanoate biosynthesis by in vitro evolution: in vivo and in vitro enzyme assays. Biomacromolecules 5(2):480–485

Yamada M, Matsumoto K, Nakai T, Taguchi S (2009) Microbial production of lactate-enriched poly[(R)-lactate-*co*-(R)-3-hydroxybutyrate] with novel thermal properties. Biomacromolecules 10(4):677–681

Zhou B, Xiao JF, Tuli L, Ressom HW (2012) LC-MS-based metabolomics. Mol Biosyst 8(2):470–481. doi:10.1039/C1mb05350g

9

Effect of high pressure on hydrocarbon-degrading bacteria

Martina Schedler[1], Robert Hiessl[1], Ana Gabriela Valladares Juárez[1], Giselher Gust[2] and Rudolf Müller[1*]

Abstract

The blowout of the Deepwater Horizon in the Gulf of Mexico in 2010 occurred at a depth of 1500 m, corresponding to a hydrostatic pressure of 15 MPa. Up to now, knowledge about the impact of high pressure on oil-degrading bacteria has been scarce. To investigate how the biodegradation of crude oil and its components is influenced by high pressures, like those in deep-sea environments, hydrocarbon degradation and growth of two model strains were studied in high-pressure reactors. The alkane-degrading strain *Rhodococcus qingshengii* TUHH-12 grew well on n-hexadecane at 15 MPa at a rate of 0.16 h^{-1}, although slightly slower than at ambient pressure (0.36 h^{-1}). In contrast, the growth of the aromatic hydrocarbon degrading strain *Sphingobium yanoikuyae* B1 was highly affected by elevated pressures. Pressures of up to 8.8 MPa had little effect on growth of this strain. However, above this pressure growth decreased and at 12 MPa or more no more growth was observed. Nevertheless, *S. yanoikuyae* continued to convert naphthalene at pressure >12 MPa, although at a lower rate than at 0.1 MPa. This suggests that certain metabolic functions of this bacterium were inhibited by pressure to a greater extent than the enzymes responsible for naphthalene degradation. These results show that high pressure has a strong influence on the biodegradation of crude oil components and that, contrary to previous assumptions, the role of pressure cannot be discounted when estimating the biodegradation and ultimate fate of deep-sea oil releases such as the Deepwater Horizon event.

Keywords: Biodegradation; High pressure; Hydrocarbons; Naphthalene; n-Hexadecane

Introduction

From April to July 2010, 779 million litres of oil were released into the Gulf of Mexico when the Deepwater Horizon (DWH) drilling rig platform exploded (Atlas and Hazen 2011). This event was the largest marine oil spill in history. However, there are other anthropogenic and natural sources of oil released into the oceans. The National Research Council (2003) estimated an overall input of about 1.3 Mt oil per year into the marine environment from all sources. Approximately 47% originates from natural seeps and the remaining 53% comes from activities related to the extraction, transportation and consumption of crude oil or refined products (National Research Council 2003).

In case of the Deepwater Horizon accident, it is estimated that a substantial proportion of the hydrocarbons entering deep plumes was converted to biomass (about $0.8–2 \cdot 10^{10}$ mol carbon) (Shiller and Joung 2012). Substantial bacterial blooms were observed in deep waters in the months following the blowout, indicating that indigenous hydrocarbon-degrading bacteria were enriched by the released crude oil and methane (Bælum *et al.* 2012; Hazen *et al.* 2010; Kessler *et al.* 2011; Redmond and Valentine 2012; Valentine *et al.* 2010, 2012). Oil-degrading bacteria have evolved over millions of years and are ubiquitous in the marine environment. Up to now, more than 200 bacterial, algal and fungal genera, representing over 500 species, are described as capable of hydrocarbon degradation (Yakimov *et al.* 2007). Therefore, natural bacterial activity is an important mechanism for environmental remediation of oil spills. Much research has been done on crude oil biodegradation in the marine environment (e.g. Colwell *et al.* 1977; Head *et al.* 2006; Leahy and Colwell 1990; Powell *et al.* 2004; Yakimov *et al.* 2007), especially in the context of the DWH blowout (Hazen

---

* Correspondence: ru.mueller@tu-harburg.de
[1]Institute of Technical Biocatalysis, Hamburg University of Technology, Hamburg 21073, Germany
Full list of author information is available at the end of the article

et al. 2010; Kessler *et al.* 2011; Redmond and Valentine 2012; Valentine *et al.* 2012).

The DWH drilling rig well, from which oil and gas flowed out uncontrollably for three months, was located 1,500 m below the sea surface. Such deep-sea environments are characterized by extreme conditions. These include low temperatures of 3°C (±1°C) (Jannasch and Taylor 1984) and high hydrostatic pressures up to 110 MPa in the Mariana Trench, the deepest site existing in the ocean at 10,994 m (±40 m) (Abe and Horikoshi 2001; Gardner and Armstrong 2011). However, only a limited number of studies regarding the capabilities of bacteria to degrade oil and hydrocarbons have been conducted under high pressure (Grossi *et al.* 2010). Despite the detection of pressure-induced differences in growth and hydrocarbon utilisation (Schwarz *et al.* 1974, 1975), most reports have investigated oil biodegradation only at surface pressure (0.1 MPa) (Cui *et al.* 2008; Tapilatu *et al.* 2010; Wang *et al.* 2008), and corresponding results may not be applicable to the deep sea. Thus, oil biodegradation processes under extreme deep-water conditions are not well understood. Although the rate and extent of hydrocarbon degradation at elevated pressures has been understudied, an understanding of the impact of elevated pressures on biodegradation is increasingly critical in the wake of expanding drilling in deep waters.

This study aims to improve our understanding of microbial degradation processes of crude oil at in situ deep-sea conditions. In particular, we consider the effects of high pressure on hydrocarbon biodegradation using high-pressure lab technology. Oil is one of the most complex mixtures of organic compounds known, containing more than 17,000 distinct components (Hassanshahian and Cappello 2013; Head *et al.* 2006). To simplify our approach, we investigated the biodegradation of two representatives of the main fractions of oil by two different bacterial model strains: *Rhodococcus qingshengii* TUHH-12, a degrader of the alkane n-hexadecane; and *Sphingobium yanoikuyae* B1, a bacterium capable of utilizing naphthalene, a polycyclic aromatic hydrocarbon (PAH) (Gibson *et al.* 1973). *R. qingshengii* and *S. yanoikuyae,* as well as other species of the genera *Rhodococcus* and *Sphingobium* have been isolated from deep-sea sediments (Colquhoun *et al.* 1998a; Cui *et al.* 2008; Heald *et al.* 2001; Peng *et al.* 2008; Tapilatu *et al.* 2010; European Nucleotide Archive 2014; Wang and Gu 2006). Recently, *Rhodococcus* sp. and *Sphingobium* sp. were found to be present in sediment samples collected in May 2011, about 2 and 6 km away from the wellhead of the DWH (Liu and Liu 2013). In our experiments, both strains *R. qingshengii* TUHH-12 and *S. yanoikuyae* B1 were incubated at 0.1 MPa and at increasing pressures to determine the influence of pressure on the growth and hydrocarbon-degradation abilities of these strains.

Materials and methods

Microorganisms

The alkane-degrading bacterium *R. qingshengii* TUHH-12 [DSM No.: 46766] was isolated from a seawater sample located directly beneath an ice cap swimming on the water during an expedition to the island of Spitzbergen, Norway, by Prof. Hauke Trinks (Hamburg University of Technology). The genome of this strain was recently sequenced (Lincoln SA, Penn State University, unpublished data).

S. yanoikuyae B1, purchased from DSMZ [DSM No.: 6900], was originally isolated from a polluted stream by Gibson *et al.* 1973. This bacterium was preliminary identified as *Beijerinckia* species, but has later been reclassified as *Sphingobium yanoikuyae* B1 (Khan *et al.* 1996). This strain is known to degrade different aromatic and polycyclic aromatic hydrocarbons (Gibson *et al.* 1973; Lang 1996).

Cultivation conditions

R. qingshengii TUHH-12 was cultivated on minimal-mineral medium with n-hexadecane as the sole carbon source. The medium consisted of 2.6 g Na_2HPO_4, 1.33 g KH_2PO_4, 1 g $(NH_4)_2SO_4$ and 0.20 g $MgSO_4 \cdot 7$ H_2O dissolved in 1000 mL of demineralized water. The medium was adjusted to pH 7. After sterilisation, 5 mL of trace element solution and 1 mL of vitamin solution were added. Both solutions were prepared as described in DSMZ methanogenium medium 141 and autoclaved or sterile filtered separately (DSMZ 2012). Cultures were incubated at room temperature and mixed at 200 rpm. The strain was kept on agar plates containing the same medium with 15 g/L agar added for solidification.

S. yanoikuyae B1 was cultured in Brunner mineral medium or on agar plates according to the DSMZ medium 457 (DSMZ 2012b) at 30°C or at room temperature and 200 rpm. Naphthalene was used as sole carbon source.

Biodegradation experiments at high pressure

Ten high-pressure reactors consisting of stainless steel cylinders capped with bronze lids were used to simulate and to investigate biodegradation under elevated pressures as they occur in deep-sea environments. Additionally, ten aluminium reactors with the same geometry were used, serving as controls to monitor biodegradation under atmospheric pressure in simultaneous experiments (Figure 1). Both reactor types had a volume of 160 mL. For experiments with *R. qingshengii* TUHH-12, 20 mL mineral medium was filled into sterilized glass vials and supplemented with 1 mM n-hexadecane. For cultivation of *S. yanoikuyae* B1, 20 mL Brunner medium and 1.77 mM naphthalene were used. The amount of carbon in 1.77 mM naphthalene is equal to the amount of carbon in 1 mM n-hexadecane. The media were inoculated with a grown preculture of the respective bacterial

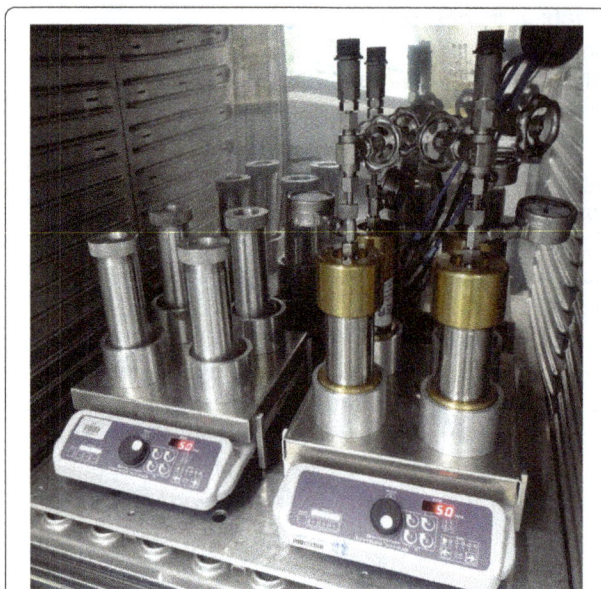

Figure 1 High-pressure reactors and control reactors. High-pressure reactors (right, made from stainless steel and bronze, max. pressure 40 MPa, pressurizing with N_2 gas, 160 mL volume) and aluminum control reactors (left, max. 0.1 MPa, 160 mL volume) were used for cultivation of hydrocarbon degraders at high and ambient pressure. The cultures were mixed with magnetic stirrers.

strain, constituting 10% of the total volume. The vials were placed inside the reactors. The high-pressure reactors were pressurized with nitrogen gas up to 15 MPa (equivalent to 1,500 m DWH well depth). The cultures were incubated at room temperature. Since the oil components used in these experiments are nearly insoluble in water, stirring rates affect biodegradation rates; therefore, efficient mixing of the cultures was ensured by magnetic stirring at 200 rpm. Due to the immiscible two-phase system of oil and water and the impracticality of subsampling at high pressure, no representative samples could be taken from the reactors to monitor oil concentrations. Thus, for each point in time in a diagram the content of one reactor was processed. Before opening a reactor containing n-hexadecane, it was cooled for 5 h at 4°C to minimise evaporation of n-hexadecane. Bacterial growth was measured and the hydrocarbon concentrations were analysed to quantify the degree of biodegradation. In several repetitions of the experiments the effects of pressure were the same. Only slightly different growth and degradation rates were observed due to different sampling times and slightly differing inoculation cell numbers. Thus, the diagrams presented represent the typical course of growth and hydrocarbon degradation.

Determination of growth of *S. yanoikuyae* B1 with glucose at ambient and high pressure

In order to determine, whether the growth of *S. yanoikuyae* B1 was inhibited when growing on a non-toxic substrate

at high pressure, we incubated the bacteria in Brunner medium with 1% glucose (w/v) at room temperature. After incubation for 44.5 hours at 0.1 MPa and at 15 MPa, the reactors were opened and cell numbers were determined.

Bacterial growth

Colony forming units (CFUs) of *R. qingshengii* TUHH-12 were determined by spreading 5 μL of the culture on Luria-Bertani (LB) agar plates in triplicate. The colonies were counted after 3 to 4 days of incubation at room temperature.

For *S. yanoikuyae* B1, plate counts were performed with R2A agar medium (DSMZ medium 830 (DSMZ 2012c)). The colonies were counted after 2 to 3 days of incubation at room temperature.

Quantification of the biodegradation of hydrocarbons by gas chromatography–mass spectrometry

After growth of the cultures, the remaining n-hexadecane or naphthalene was extracted from the complete culture medium of each reactor with 5 mL of n-hexane. Dodecane or n-hexadecane were added before extraction as internal standard respectively. An aliquot of 1 μL of the apolar phase, containing the hydrocarbon, was injected with a split ratio of 28:1 into a Hewlett-Packard 5890 Series II gas chromatograph (GC) coupled to a Hewlett-Packard 5971A mass selective detector. The GC was equipped with an Agilent HP-5MS column (30 m length, 0.25 mm internal diameter) and helium was used as carrier gas. The injector temperature for both n-hexadecane and naphthalene was increased from 80°C to 200°C at a rate of 0.5°C/s. The oven temperature program was as follows: an initial temperature of 80°C was increased to a final temperature of 200°C at a rate of 15°C/min, with a final 1 min hold at 200°C. The mass spectrometer was operated in full scan mode over 50–650 amu. The MS transfer line temperature was held at 320°C and the ion source temperature at 180°C.

Detection of hydroxylated intermediates in naphthalene conversion

For detection of hydroxylated intermediates in naphthalene conversion, the colourimetric method described by Arnow (1937) was used. After 3 min centrifugation of 1 mL of a grown culture at 13,000 rpm, 200 μL of the supernatant were supplemented by 200 μL of the following reagents, in the order given, mixing well after each addition: 0.5 N HCl, nitrite-molybdate reagent, 1 N NaOH. If catechol or 1,2-dihydroxynaphthalene were present, a yellow colour resulted after addition of HCl and nitrite-molybdate reagent, and a red colour appeared after addition of NaOH. In the case of monohydroxylated compounds like salicylate or monohydroxynaphthalene the solution remained yellow.

Results

Degradation of n-hexadecane by *R. qingshengii* TUHH-12 at ambient and high pressure

R. qingshengii TUHH-12 was cultivated on n-hexadecane as the sole source of carbon and energy. *R. qingshengii* TUHH-12 was found to grow well and to mineralize this hydrocarbon at atmospheric pressure (0.1 MPa) as well as at high pressure (15 MPa). At 15 MPa the degradation and growth behaviour was slightly different from that at atmospheric pressure (Figure 2). In both cases, a lag phase of 16 to 17 h was followed by an exponential growth phase and a stationary phase starting after 43 to 44 h of incubation. However, the growth rate of *R. qingshengii* TUHH-12 in the exponential phase was 0.36 h^{-1} at ambient pressure, from 17 h to 43 h, compared to 0.16 h^{-1} at high pressure, from 16 h to 44 h. In the stationary phase a higher cell density was reached at 0.1 MPa than at 15 MPa. The rate of degradation of n-hexadecane was 0.035 mM/h at ambient pressure from 17 to 43 h, and 0.019 mM/h at high pressure, from 16 to 44 h. In control experiments without bacteria at 15 MPa the n-hexadecane also slowly decreased although with a much slower rate of 0.007 mM/h.

Degradation of naphthalene by *S. yanoikuyae* B1 at ambient and high pressure

S. yanoikuyae B1 was incubated on naphthalene at high pressure (13.9 MPa) and at atmospheric pressure. The growth of *S. yanoikuyae* B1 on this PAH was strongly inhibited by high pressure. Bacteria grew at 0.1 MPa with a lag phase of 15 h, an exponential phase with a

growth rate of 0.33 h^{-1} (from 15 to 28 h) and reached stationary phase at 28 h of incubation (Figure 3). At 13.9 MPa, however, CFUs of *S. yanoikuyae* B1 decreased after 15 h cultivation time until no CFUs could be counted after 66 h.

In 0.1 MPa experiments, the analysis of remaining naphthalene in the medium showed that the substrate was degraded completely after 19 h. The degradation rate of naphthalene, from 7 h to 19 h, was 0.064 mM/h. Because the bacteria did not grow at 13.9 MPa, we expected that at this pressure no naphthalene would be degraded at all. However, we also observed a decrease in substrate concentration at this pressure. With a conversion rate of 0.054 mM/h (from 7 h to 25 h), 96.6% of the naphthalene was converted after 75 h of incubation. After 66 h of incubation at elevated pressure, the initially colourless culture medium turned brown, while at ambient pressure the culture showed no change of colour.

In control experiments without bacteria about 20.8% of the initial naphthalene was found to be lost after 19 days of incubation at 14.2 MPa by evaporation of the highly volatile naphthalene. With a loss of about 25.3% at 0.1 MPa there was no significant difference to the incubation at high pressure.

Since *S. yanoikuyae* B1 did not grow at 13.9 MPa, we conducted additional experiments to determine the maximum pressure at which growth was possible. *S. yanoikuyae* B1 was incubated for 70 h on naphthalene at different pressures in the range between 0.1 MPa and 13 MPa (Figure 4). In the range of 0.1 to 8.8 MPa, the CFU counts remained relatively constant, but decreased when 8.8 MPa was exceeded. In incubations at and above 12 MPa no growth

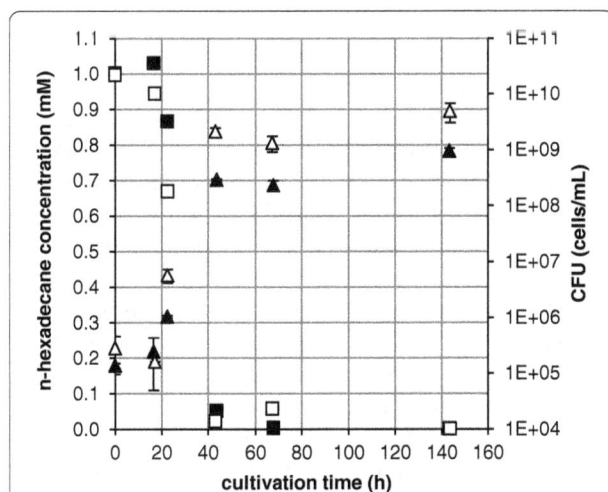

Figure 2 Degradation of n-hexadecane at 0.1 MPa vs. 15 MPa by *R. qingshengii* TUHH-12. The CFUs were determined by plate counting and n-hexadecane concentrations were measured by gas chromatography–mass spectrometry. CFUs were determined in triplicate and standard deviations are shown. △ CFU at 0.1 MPa, ▲ CFU at 15 MPa, □ n-hexadecane concentration at 0.1 MPa, ■ n-hexadecane concentration at 15 MPa.

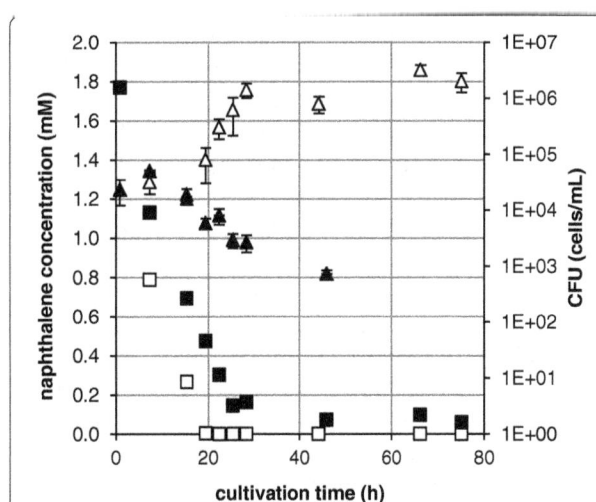

Figure 3 Growth of *S. yanoikuyae* B1 on naphthalene at 0.1 MPa vs. 13.9 MPa. *S. yanoikuyae* B1 was cultivated at room temperature. CFUs were determined in triplicate and standard deviations are shown. △ CFU at 0.1 MPa, ▲ CFU at 13.9 MPa, □ concentration of naphthalene at 0.1 MPa, ■ concentration of naphthalene at 13.9 MPa.

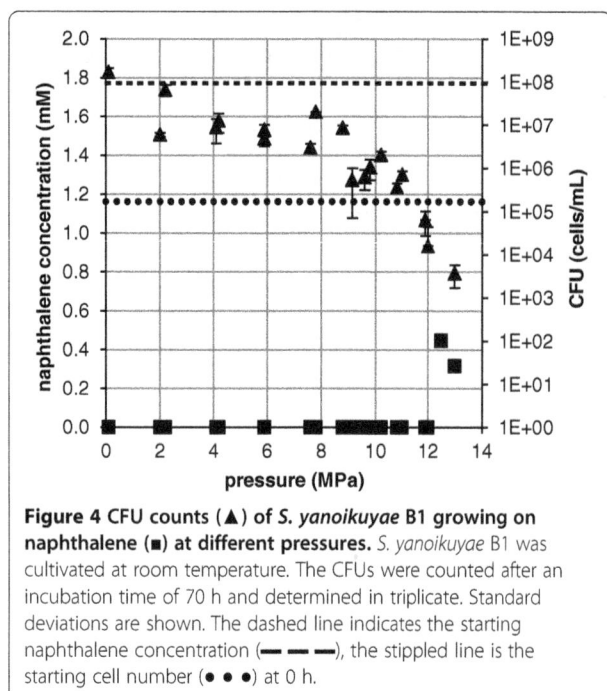

Figure 4 CFU counts (▲) of *S. yanoikuyae* B1 growing on naphthalene (■) at different pressures. *S. yanoikuyae* B1 was cultivated at room temperature. The CFUs were counted after an incubation time of 70 h and determined in triplicate. Standard deviations are shown. The dashed line indicates the starting naphthalene concentration (▬ ▬ ▬), the stippled line is the starting cell number (● ● ●) at 0 h.

occurred after 70 h cultivation time and viable cell counts were lower than at the start of the incubation.

The naphthalene concentration decreased to below the limit of detection under both ambient pressure and pressures up to 12 MPa. At 12.5 MPa and 13 MPa, 25.2% and 17.9% of the original naphthalene remained, respectively, although CFUs did not increase. While at 0.1 MPa no change of colour could be observed, at 12.5 MPa, after 70 h of incubation, the medium had turned brown. In the test with the colour reagent of Arnow (1937), we found no colour, indicating that neither mono- nor dihydroxylated compounds like 1,2-dihydroxynaphthalene, catechol or salicylate were formed. We therefore assume that the formation of the brown colour was due to the polymerization of either quinones or aromatic ring cleavage products.

Growth of *S. yanoikuyae* B1 with glucose at different pressures

While *S. yanoikuyae* grew well on glucose at 0.1 MPa, at 15 MPa no growth at all was observed. After 44.5 hours the cell number had decreased from $5.98 \cdot 10^5$ to $1.7 \cdot 10^5$ cells per mL at 15 MPa. These results indicate that at high pressure it is not the conversion of naphthalene or its metabolites but rather another central function in *S. yanoikuyae* B1, which is inhibited.

Discussion

Degradation of n-hexadecane by *R. qingshengii* TUHH-12 at ambient and high pressure

R. qingshengii TUHH-12 degraded the alkanes n-hexadecane, decane and tetracosane. *R. qingshengii* has been found to assimilate and mineralize different hydrocarbons including benzene, toluene, xylenes, naphthalene, n-dodecane (Benedek *et al.* 2013). However, until now, nothing was known about degradation capabilities of *R. qingshengii* at other than atmospheric pressure. Our experiments showed that the growth rate of *R. qingshengii* TUHH-12 at ambient pressure (0.36 h^{-1}) was slightly higher than at 15 MPa (0.16 h^{-1}). This leads to the conclusion that a pressure of 15 MPa has a slightly negative effect on the growth of this bacterium, suggesting it can be classified as a piezotolerant organism. These findings are confirmed by the work of Colquhoun *et al.* (1998b) and Heald *et al.* (2001), who showed that certain *Rhodococcus* strains were able to grow at even higher pressures of 40 MPa and 60 MPa on glucose yeast extract medium.

Effects of elevated pressure on the naphthalene degradation by *S. yanoikuyae* B1

The model strain *S. yanoikuyae* B1, used for PAH degradation in our high-pressure experiments, is capable of utilising a variety of aromatic compounds including biphenyl, anthracene, phenanthrene and naphthalene (Gibson *et al.* 1973), as well as toluene, cyclohexane and 1,3,5-trimethylbenzene (Lang 1996) as carbon sources for growth. We found significant differences in growth of *S. yanoikuyae* B1 with naphthalene at different pressures. These effects occurred at pressures lower than those typically assumed to be the threshold for pressure effects. First significant effects of high pressure on cellular components and processes of bacteria were found to start at 20 MPa, affecting the RNA transcription (Yayanos and Pollard 1969). Modifications of membrane fluidity were shown to occur at pressures above 100 MPa (Hauben *et al.* 1997) and protein denaturation was observed at more than 400 MPa (Aertsen *et al.* 2009).

In our experiments at 0.1 MPa *S. yanoikuyae* B1 was able to grow with naphthalene, whereas at 13.5 MPa bacteria did not grow at all and after 66 h of incubation, cells were no longer viable. A similar behaviour was found with glucose as carbon source. Thus, we conclude that *S. yanoikuyae* B1 is a piezosensitive strain that grew best and utilized naphthalene at an optimal rate at ambient pressure. Nevertheless, *S. yanoikuyae* B1 still metabolized naphthalene at 13.9 MPa, although slower than at 0.1 MPa, so that after 75 h of incubation 96.6% of the substrate was converted. We need to emphasize that the naphthalene-degradation capability of *S. yanoikuyae* B1 is much less sensitive to high pressure than growth is. This indicates a new type of piezosensitivity, which, to our knowledge, was not described in literature before.

We can only speculate about the reasons for the observed changes in growth and degradation ability of *S. yanoikuyae* B1 at high pressure. Even a combination of several pressure-induced effects is possible. The investigation of the specific reasons for the strong inhibition of growth and rather slight inhibition of naphthalene conversion of *S. yanoikuyae* B1 by high pressure emerges from our study as a new research topic.

So far in predicting the behavior and degradation of oil spills in deep-sea environments all models use data obtained at ambient pressure for calculating degradation rates. The results presented here show, that the effect of pressure cannot be neglected. When our data obtained under high pressure were included in a model describing the fate of the oil in the case of the Deepwater Horizon, the model predicted the observed changes in oil concentrations much better (Lindo-Atichati *et al.* 2014).

Our experiments show that pressure affects both bacterial growth and hydrocarbon-degrading activity at pressures much lower than previously reported in the literature as they occur at modern deep-sea drilling sites. Consequently, pressure effects need to be considered as a crucial factor in predictions of oil biodegradation in deep waters.

Competing interests

The authors declare that they have no competing interests.

Authors' contributions

MS participated in the design and performance of the experiments and drafted the manuscript. RH carried out the experiments and analysed the data. AGVJ participated in the design and coordination of the experiment, isolated the strain *R. qingshengii* TUHH-12, and contributed in the drafting of the manuscript. GG participated in the experimental design and contributed in the drafting of the manuscript. RM conceived of the study, participated in its design and contributed in the drafting of the manuscript. All authors read and approved the final manuscript.

Acknowledgments

This research was made possible by a grant from British Petroleum (BP)/ The Gulf of Mexico Research Initiative. This publication was supported by the German Research Foundation (DFG) and the Hamburg University of Technology (TUHH) in the funding programme "Open Access Publishing". The authors thank the company Technik Service Andreas Meyer for providing the high-pressure reactors and for technical advice. We are grateful to Eva Mong Su for her support in the lab and we thank Sara Lincoln from Penn State University for critical reading of the manuscript.

Author details

[1]Institute of Technical Biocatalysis, Hamburg University of Technology, Hamburg 21073, Germany. [2]Institute for Product Development and Mechanical Engineering Design, Hamburg University of Technology, Hamburg 21073, Germany.

References

Abe F, Horikoshi K (2001) The biotechnological potential of piezophiles. Trends Biotechnol 19:102–108. doi:10.1016/S0167-7799(00)01539-0

Aertsen A, Meersman F, Hendrickx MEG, Vogel RF, Michiels CW (2009) Biotechnology under high pressure: applications and implications. Trends Biotechnol 27:434–441. doi:10.1016/j.tibtech.2009.04.001

Arnow LE (1937) Colorimetric determination of the components of 3,4-dihydroxyphenylalaninetyrosine mixtures. J Biol Chem 118:531–537

Atlas RM, Hazen TC (2011) Oil biodegradation and bioremediation: a tale of the two worst spills in U.S. history. Environ Sci Technol 45:6709–6715. doi:10.1021/es2013227

Bælum J, Borglin S, Chakraborty R, Fortney JL, Lamendella R, Mason OU, Auer M, Zemla M, Bill M, Conrad ME, Malfatti SA, Tringe SG, Holman H-Y, Hazen TC, Jansson JK (2012) Deep-sea bacteria enriched by oil and dispersant from the Deepwater Horizon spill. Environ Microbiol 14:2405–2416. doi:10.1111/j.1462-2920.2012.02780.x

Benedek T, Vajna B, Táncsics A, Márialigeti K, Lányi S, Máthé I (2013) Remarkable impact of PAHs and TPHs on the richness and diversity of bacterial species in surface soils exposed to long-term hydrocarbon pollution. World J Microbiol Biotechnol 29:1989–2002. doi:10.1007/s11274-013-1362-9

Colquhoun JA, Mexson J, Goodfellow M, Ward AC, Horikoshi K, Bull AT (1998a) Novel rhodococci and other mycolate actinomycetes from the deep sea. Antonie Van Leeuwenhoek 74:27–40. doi:10.1023/A:1001743625912

Colquhoun JA, Heald SC, Li L, Tamaoka J, Kato C, Horikoshi K, Bull AT (1998b) Taxonomy and biotransformation activities of some deep-sea actinomycetes. Extremophiles 2:269–277. doi:10.1007/s007920050069

Colwell RR, Walker JD, Cooney JJ (1977) Ecological aspects of microbial degradation of petroleum in the marine environment. Crit Rev Microbiol 5:423–445. doi:10.3109/10408417709102813

Cui Z, Lai Q, Dong C, Shao Z (2008) Biodiversity of polycyclic aromatic hydrocarbon-degrading bacteria from deep sea sediments of the Middle Atlantic Ridge. Environ Microbiol 10:2138–2149. doi:10.1111/j.1462-2920.2008.01637.x

DSMZ GmbH (2012a) 141 METHANOGENIUM MEDIUM., http://www.dsmz.de/microorganisms/medium/pdf/DSMZ_Medium141.pdf. Accessed 12 May 2014

DSMZ GmbH (2012b) 457 MINERAL MEDIUM (BRUNNER)., http://www.dsmz.de/microorganisms/medium/pdf/DSMZ_Medium457.pdf. Accessed 12 May 2014

DSMZ GmbH (2012c) 830 R2A MEDIUM., http://www.dsmz.de/microorganisms/medium/pdf/DSMZ_Medium830.pdf. Accessed 12 May 2014

European Nucleotide Archive (2014) The EMBL. European Bioinformatics Institute, Hinxton, Cambridge, http://www.ebi.ac.uk/ena/data/view/KC894023. Accessed 29. May 2014

Gardner JV, Armstrong AA (2011) The Mariana Trench: A new view based on multibeam echosounding, vol 1, AGU Fall Meeting Abstracts., p 1517

Gibson DT, Roberts RL, Wells MC, Kobal VM (1973) Oxidation of biphenyl by a Beijerinckia species. Biochem Biophys Res Co 50:211–219. doi:10.1016/0006-291X(73)90828-0

Grossi V, Yakimov MM, Al Ali B, Tapilatu Y, Cuny P, Goutx M, La Cono V, Giuliano L, Tamburini C (2010) Hydrostatic pressure affects membrane and storage lipid compositions of the piezotolerant hydrocarbon-degrading *Marinobacter hydrocarbonoclasticus* strain #5. Environ Microbiol 12:2020–2033. doi:10.1111/j.1462-2920.2010.02213.x

Hassanshahian M, Cappello S (2013) Crude Oil Biodegradation in the Marine Environments. In: Rosenkranz F (ed) Chamy R. Biodegradation - Engineering and Technology, InTech, pp 101–135

Hauben KJ, Bartlett DH, Soontjens CC, Cornelis K, Wuytack EY, Michiels CW (1997) *Escherichia coli* mutants resistant to inactivation by high hydrostatic pressure. Appl Environ Microbiol 63:945–950

Hazen TC, Dubinsky EA, DeSantis TZ, Andersen GL, Piceno YM, Singh N, Jansson JK, Probst A, Borglin SE, Fortney JL, Stringfellow WT, Bill M, Conrad ME, Tom LM, Chavarria KL, Alusi TR, Lamendella R, Joyner DC, Spier C, Baelum J, Auer M, Zemla ML, Chakraborty R, Sonnenthal EL, D'haeseleer P, Holman H-YN, Osman S, Lu Z, Nostrand JDV, Deng Y et al (2010) Deep-sea oil plume enriches indigenous oil-degrading bacteria. Science 330:204–208. doi:10.1126/science.1195979

Head IM, Jones DM, Röling WFM (2006) Marine microorganisms make a meal of oil. Nat Rev Micro 4:173–182. doi:10.1038/nrmicro1348

Heald SC, Brandão PFB, Hardicre R, Bull AT (2001) Physiology, biochemistry and taxonomy of deep-sea nitrile metabolising *Rhodococcus* strains. Antonie Van Leeuwenhoek 80:169–183. doi:10.1023/A:1012227302373

Jannasch HW, Taylor CD (1984) Deep-sea microbiology. Annu Rev Microbiol 38:487–514. doi:10.1146/annurev.mi.38.100184.002415

Kessler JD, Valentine DL, Redmond MC, Du M, Chan EW, Mendes SD, Quiroz EW, Villanueva CJ, Shusta SS, Werra LM, Yvon-Lewis SA, Weber TC (2011) A persistent oxygen anomaly reveals the fate of spilled methane in the Deep Gulf of Mexico. Science 331:312–315. doi:10.1126/science.1199697

Khan AA, Wang R-F, Cao W-W, Franklin W, Cerniglia CE (1996) Reclassification of a polycyclic aromatic hydrocarbon-metabolizing bacterium, *beijerinckia* sp. strain B1, as *sphingomonas yanoikuyae* by fatty acid analysis, protein pattern analysis, DNA-DNA hybridization, and 16S ribosomal DNA sequencing. Int J Syst Bacteriol 46:466–469. doi:10.1099/00207713-46-2-466

Lang E (1996) Diversity of bacterial capabilities in utilizing alkylated benzenes and other aromatic compounds. Lett Appl Microbiol 23:257–260. doi:10.1111/j.1472-765X.1996.tb00078.x

Leahy JG, Colwell RR (1990) Microbial degradation of hydrocarbons in the environment. Microbiol Rev 54:305–315

Lindo-Atichati D, Paris CB, Le Hénaff M, Schedler M, Valladares Juárez AG, Müller R (2014) Simulating the effects of droplet size, high-pressure biodegradation, and variable flow rate on the subsea evolution of deep plumes from the Macondo blowout. Deep Sea Res Part II: Topical Stud Oceanogr, 10.1016/j.dsr2.2014.01.011 (in press)

Liu Z, Liu J (2013) Evaluating bacterial community structures in oil collected from the sea surface and sediment in the northern Gulf of Mexico after the Deepwater Horizon oil spill. Microbiol Open 2:492–504. doi:10.1002/mbo3.89

National Research Council (2003) Oil in the Sea III: Inputs, Fates, and Effects. The National Academies Press, Washington, DC

Peng F, Wang Y, Sun F, Liu Z, Lai Q, Shao Z (2008) A novel lipopeptide produced by a Pacific Ocean deep-sea bacterium, *Rhodococcus* sp. TW53. J Appl Microbiol 105:698–705. doi:10.1111/j.1365-2672.2008.03816.x

Powell SM, Bowman JP, Snape I (2004) Degradation of nonane by bacteria from Antarctic marine sediment. Polar Biol 27:573–578. doi:10.1007/s00300-004-0639-8

Redmond MC, Valentine DL (2012) Natural gas and temperature structured a microbial community response to the *Deepwater Horizon* oil spill. Proc Natl Acad Sci U S A 109:20292–20297. doi:10.1073/pnas.1108756108

Schwarz JR, Walker JD, Colwell RR (1974) Deep-sea bacteria: growth and utilization of hydrocarbons at ambient and in situ pressure. Appl Microbiol 28:982–986

Schwarz JR, Walker JD, Colwell RR (1975) Deep-sea bacteria: growth and utilization of *n*-hexadecane at *in situ* temperature and pressure. Can J Microbiol 21:682–687. doi:10.1139/m75-098

Shiller AM, Joung D (2012) Nutrient depletion as a proxy for microbial growth in Deepwater Horizon subsurface oil/gas plumes. Environ Res Lett 7:045301. doi:10.1088/1748-9326/7/4/045301

Tapilatu Y, Acquaviva M, Guigue C, Miralles G, Bertrand J-C, Cuny P (2010) Isolation of alkane-degrading bacteria from deep-sea Mediterranean sediments. Lett Appl Microbiol 50:234–236. doi:10.1111/j.1472-765X.2009.02766.x

Valentine DL, Kessler JD, Redmond MC, Mendes SD, Heintz MB, Farwell C, Hu L, Kinnaman FS, Yvon-Lewis S, Du M, Chan EW, Tigreros FG, Villanueva CJ (2010) Propane respiration jump-starts microbial response to a deep oil spill. Science 330:208–211. doi:10.1126/science.1196830

Valentine DL, Mezić I, Maćešić S, Črnjarić-Žic N, Ivić S, Hogan PJ, Fonoberov VA, Loire S (2012) Dynamic autoinoculation and the microbial ecology of a deep water hydrocarbon irruption. Proc Natl Acad Sci U S A 109:20286–20291. doi:10.1073/pnas.1108820109

Wang YP, Gu J-D (2006) Degradability of dimethyl terephthalate by *Variovorax paradoxus* T4 and *Sphingomonas yanoikuyae* DOS01 isolated from deep-ocean sediments. Ecotoxicology 15:549–557. doi:10.1007/s10646-006-0093-1

Wang B, Lai Q, Cui Z, Tan T, Shao Z (2008) A pyrene-degrading consortium from deep-sea sediment of the West Pacific and its key member Cycloclasticus sp. P1. Environ Microbiol 10:1948–1963. doi:10.1111/j.1462-2920.2008.01611.x

Yakimov MM, Timmis KN, Golyshin PN (2007) Obligate oil-degrading marine bacteria. Curr Opin Biotechnol 18:257–266. doi:10.1016/j.copbio.2007.04.006

Yayanos AA, Pollard EC (1969) A study of the effects of hydrostatic pressure on macromolecular synthesis in *escherichia coli*. Biophys J 9:1464–1482. doi:10.1016/S0006-3495(69)86466-0

Co-production of bioethanol and probiotic yeast biomass from agricultural feedstock: application of the rural biorefinery concept

Claire M Hull[1], E. Joel Loveridge[1], Iain S Donnison[2], Diane E Kelly[1] and Steven L Kelly[1]*

Abstract

Microbial biotechnology and biotransformations promise to diversify the scope of the biorefinery approach for the production of high-value products and biofuels from industrial, rural and municipal waste feedstocks. In addition to bio-based chemicals and metabolites, microbial biomass itself constitutes an obvious but overlooked by-product of existing biofermentation systems which warrants fuller attention. The probiotic yeast *Saccharomyces boulardii* is used to treat gastrointestinal disorders and marketed as a human health supplement. Despite its relatedness to *S. cerevisiae* that is employed widely in biotechnology, food and biofuel industries, the alternative applications of *S. boulardii* are not well studied. Using a biorefinery approach, we compared the bioethanol and biomass yields attainable from agriculturally-sourced grass juice using probiotic *S. boulardii* (strain MYA-769) and a commercial *S. cerevisiae* brewing strain (Turbo yeast). Maximum product yields for MYA-769 (39.18 [±2.42] mg ethanol mL^{-1} and 4.96 [±0.15] g dry weight L^{-1}) compared closely to those of Turbo (37.43 [±1.99] mg mL^{-1} and 4.78 [±0.10] g L^{-1}, respectively). Co-production, marketing and/or on-site utilisation of probiotic yeast biomass as a direct-fed microbial to improve livestock health represents a novel and viable prospect for rural biorefineries. Given emergent evidence to suggest that dietary yeast supplementations might also mitigate ruminant enteric methane emissions, the administration of probiotic yeast biomass could also offer an economically feasible way of reducing atmospheric CH_4.

Keywords: Bioethanol; Biomass; Biorefinery; Cholesterol; Probiotic; *Saccharomyces boulardii*

Introduction

There is global impetus towards development of biorefineries that utilise industrial, rural and municipal waste for the production of bioenergy and marketable bio-based compounds. The biorefinery concept has been identified as a significant opportunity for rural economic development (Charlton et al. 2009; Leistritz and Hodur 2008; Sharma et al. 2011) and perennial ryegrass (*Lolium perenne* L.) is currently under investigation as a non-food crop that could be processed as feedstock in a rural biorefinery setting (Farrar et al. 2012). Production of biogas from ryegrass pulp (Kyazze et al. 2008) and bioethanol from grass juice (Martel et al. 2010; Martel et al. 2011) has already been achieved. Microbial biotechnology and metabolic engineering promises to

diversify the application of the biorefinery approach for production of novel products and several 'designer yeast strains' capable of using polyfructose have been reported (Martel et al. 2011; Wang et al. 2011; Zhang et al. 2010). Nonetheless, yeast species that already have GRAS (generally regarded as safe) status remain readily applicable to biorefinery processes and novel uses for yeast biomass warrant consideration. In the present study, we investigate the co-production of bioethanol and probiotic yeast biomass from enzyme-pretreated grass juice (Martel et al. 2010).

In addition to its use in fermentation, food and biofuel industries, the brewing yeast *Saccharomyces cerevisiae* has several health applications. It is used as a protein supplement, immune enhancer and is employed as a vehicle for the introduction of dietary compounds as a commercialised health product (Moyad 2008). The yeast *S. boulardii* is administered for the treatment of gastrointestinal disorders (Buts 2009; Vandenplas et al. 2009;

* Correspondence: S.L.Kelly@swansea.ac.uk
[1]Institute of Life Science, College of Medicine, Swansea University, Swansea SA2 8PP, Wales, UK
Full list of author information is available at the end of the article

Zanello et al. 2009) and is currently the only commercially available probiotic yeast. The ability of *S. boulardii* to ferment ethanol has been documented (Gurgu et al. 2011) as have certain physiological and growth characteristics (Edwards-Ingram et al. 2007) including evidence that it can assimilate cholesterol (Chen et al. 2010; Psomas et al. 2003). However, despite its genetic relatedness to *S. cerevisiae* (Edwards-Ingram et al. 2004) and use as a human probiotic for over 50 years, the alternative applications of *S. boulardii* are not well studied. Given growing interest in the biotherapeutic properties of different yeasts (Foligne et al. 2010) there is now a clear incentive to develop and apply research knowledge about food grade yeasts.

The purpose of this study was to investigate if the co-production of bioethanol and probiotic yeast biomass is a feasible strategy for enhancing the productivity and value of rural biorefineries of the future. We sought to determine the potential bioethanol and yeast biomass yields attainable from agriculturally-sourced grass juice using *S. boulardii* (MYA-769) and a commercial *S. cerevisiae* brewing strain (Turbo yeast). Both strains of yeast are safe and the methodology reported in the present study (from feedstock extraction to product utilisation) compatible with land availability, rural land use patterns, current legislation and the existing technology base in the United Kingdom (Charlton et al. 2009; Farrar et al. 2012; Martel et al. 2010). The applications of yeast biomass as a feed additive and/or probiotic for livestock in the rural biorefinery setting are discussed.

Materials and methods
Yeast strains and growth media
Bioethanol and biomass co-production studies were undertaken using a commercial brewing strain of *Saccharomyces cerevisiae* (Turbo yeast; Gert Strand AB) and a probiotic strain of *Saccharomyces boulardii* (MYA-769; ATCC). Both were maintained at 30°C on yeast-peptone-dextrose (YPD) medium containing (w/v): 2% glucose, 2% bacto peptone and 1% yeast extract (±2% agar). All media components were supplied by Difco^{TM} (BBL/Difco Laboratories). All other chemicals were supplied by Sigma (Sigma-Aldrich Ltd) unless otherwise stated.

Grass juice (GJ) feedstock was extracted from ryegrass *Lolium perenne* supplied by the Institute of Biological, Environmental and Rural Sciences (IBERS, UK) (Martel et al. 2010). GJ was screened to remove large particulates, autoclaved (121°C, 30 min) and frozen (-80°C). When required for use as a growth and fermentation substrate, particle-free GJ was thawed and component fructans enzymatically hydrolysed using truncated *L. paracasei* β fructosidase ($_t$fosEp) as

previously described (Martel et al. 2010). The concentration of free monosaccharides in untreated and enzyme pre-treated GJ + $_t$fosEp was determined and the latter chosen for use as feedstock for all experimental work undertaken in the present study.

For sterol assimilation studies, cholesterol-supplemented glucose yeast minimal media ($_{glc}$YM^{+chol}) containing 1.34% yeast nitrogen base without amino acids, 2% glucose and 10 μg mL^{-1} cholesterol (final concentration) was prepared. Cholesterol was dissolved in 1:1 Tween 80:ethanol to give a 2 mg mL^{-1} stock and filter-sterilised prior to use.

Sugar assay
Sugar analyses were performed on 2500-fold diluted GJ and GJ + $_t$fosEp in 100 mM potassium phosphate, pH 7.0, containing 10 mM MgSO$_4$, 1 mM NAD$^+$, 1.5 mM ATP and 20 U mL^{-1} *Leuconostoc mesenteroides* glucose-6-phosphate dehydrogenase (Worthington Biochemical Corporation). Concentrations of glucose, fructose, sucrose and fructan were determined from the changes in absorbance at 340 nm following sequential addition of 20 U mL^{-1} *S. cerevisiae* hexokinase (Worthington Biochemical Corporation), 20 U mL^{-1} *E. coli* phosphoglucose isomerase (Megazyme International Ireland Ltd), 1.5 U mL^{-1} *S. cerevisiae* sucrase/maltase (Megazyme International Ireland Ltd) and 150 μg $_t$fosEp (purified as previously described Martel et al. 2010) respectively. Standards of glucose, fructose, sucrose and chicory inulin were used to calibrate the assay.

Growth and fermentation studies
Growth and fermentation experiments were performed in 100-well honeycomb microplates using a Bioscreen C (Oy Growth Curves Ab Ltd, Finland). Uniform starting (t_0h) culture densities were achieved by resuspending individual yeast colonies in GJ and diluting to obtain 5×10^5 cells mL^{-1} in GJ. Starting cultures were vortexed and aliquotted into Bioscreen wells (300 μL volumes). All experiments were incubated at 20°C for optimal bioethanol production as previously described (Martel et al. 2011) and optical density readings (at 600 nm) taken every 45 min. Data was exported from the Bioscreen in ASCII format prior to analysis using Excel (Microsoft Office 2003).

Growth parameters were derived using standard methodology. Briefly, ΔOD values describe maximum OD – minimum OD; the lag phase is defined as the length of time a culture spends at < 10% of maximum OD; $T_{1/2}$Max values are equivalent to the time taken to reach half the maximum increase in growth of a culture (ΔOD × 0.5). Fastest doubling times (DT) were estimated by dividing the natural logarithm of 2 by the fastest culture growth rates (μ), where μ is the gradient of the linear trend line fitted to log-transformed OD data.

Screening of experimental cultures for bacterial contamination and observations of the cell morphology of both yeast strains were made using a Nikon Eclipse E600 microscope.

Bioethanol and biomass

At specific time intervals (t_0h, t_{24}h, t_{48}h, t_{72}h, t_{96}h t_{100}h and t_{124}h) Bioscreen measurements were suspended and a 10 µL volume of culture supernatant removed from representative experimental wells. These 10 µL volumes were immediately diluted (10-, 100- and 1000-fold) with distilled water and frozen for subsequent ethanol analysis. Ethanol determinations were made using a spectrophotometric assay kit (K-ETOH 11/06; Megazyme Ltd) according to manufacturer's instructions. For biomass yield estimations (g dry weight L^{-1}), the contents of 10 unsampled Bioscreen wells were pooled at t_{124}h and dried to constant mass using a centrifugal evaporator (Heto Maxi Dry Plus).

Cholesterol assimilation experiments

Individual colonies from Turbo yeast (*S. cerevisiae*) and MYA-796 (*S. boulardii*) agar plates were used to inoculate 10 mL volumes of $_{glc}YM^{+chol}$ media. These starter cultures (3 replicates per yeast species) were maintained at 37°C for 48 h in static (no agitation) 30 mL sterilin vials to attain low-oxygen conditions. At t_{48}h, cell pellets were harvested by centrifugation and washed three times with sterile water prior to sterol extraction, derivatisation and analysis using gas chromatography-mass spectrometry (GC-MS).

GC-MS sterol analysis

Washed cell pellets from cultures grown using YPD, GJ + $_t$fosEp and $_{glc}YM^{+chol}$ were resuspended in 7:3 methanol:water containing 18% (w/v) potassium hydroxide and 0.1% (w/v) pyrogallol and heated at 90°C for 2 h.

Non-saponifiable sterols were extracted into glass HPLC vials using 3×2 mL volumes of hexane. Extracts were evaporated to dryness using a centrifugal evaporator (Heto Maxi Dry Plus) and derivatised using 100 µL N, O-bis(trimethylsilyl)trifluoroacetamide and trimethyl-chlorosilane (BSTFA-TMCS [99:1]) and 50 µL anhydrous pyridine at 70°C for 2 h.

Trimethylsilyl (TMS)-derivatised sterols were analyzed using a 7890A GC-MS system (Agilent Technologies) with a DB-5MS fused silica column (30 m × 0.25 mm × 0.25 µm film thickness; J&W Scientific). The oven temperature was initially held at 70°C for 4 min, then increased at 25°C min^{-1} to a final temperature of 280°C, which was held for a further 25 min. Samples were analyzed in splitless mode (1 µL injection volume) using helium carrier gas, electron impact ionization (ion source temperature of 150°C) and scanning from m/z 40 to 850. GC-MS data files were analysed using MSD Enhanced ChemStation software (Agilent Technologies) to determine sterol profiles for all isolates and for derivation of integrated peak areas. Sterols were identified by reference to retention times and mass fragmentation patterns for known standards.

Results

Results from the present study demonstrate the potential to co-produce bioethanol and probiotic yeast biomass from grass juice feedstock and identify avenues for process development and application in rural birefinery settings.

Pre-treatment of grass juice (GJ) feedstock with the soluble, truncated core domain of *Lactobacillus paracasei* β-fructosidase ($_t$fosEp) purified from recombinant *Escherichia coli* (Martel et al. 2010) resulted in the complete hydrolysis of non-fermentable fructan moieties (Figure 1). The total monosaccharide (glucose and fructose) content of GJ + $_t$fosEp (73.31 [±0.67] mg mL^{-1}) was over two-fold higher than that of untreated GJ (30.39

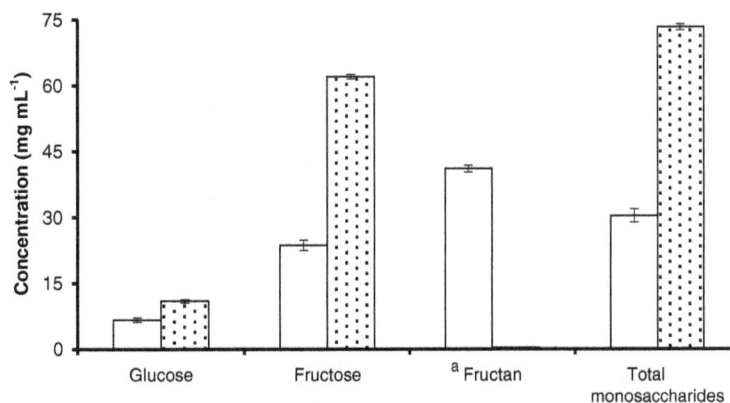

Figure 1 Mean [±S.D] glucose, fructose and fructan concentrations in untreated GJ (open bars) and GJ + $_t$fosEp (filled bars).
ᵃ Fructan = polyfructose.

Table 1 Mean [±S.D] growth parameters for Turbo yeast and MYA-796 grown on enzyme-pretreated grass juice (GJ + $_t$fosEp)

	Growth parameters				Maximum product yield	
	ΔOD_{600}	Lag (h)	$T_{1/2}$Max (h)	Max DT (h)	Bioethanol (mg mL^{-1})	Biomass (g L^{-1})
Turbo	1.70 [0.02]	11.25 [0.5]	17.25 [0.5]	3.5 [0.5]	37.43 [1.99]	4.78 [0.10]
MYA-796	1.74 [0.02]	12.00 [0.5]	20.25 [0.5]	4.5 [0.5]	39.18 [2.42]	4.96 [0.15]

ΔOD_{600} = maximum – minimum optical density reading (at 600 nm); Lag phase = length of time population remains at < 10% of maximum OD; $T_{1/2}$Max = time taken to achieve half maximal population growth (maximum OD – minimum OD × 0.5); Max DT = maximum observed doubling time.

[±1.51] mg mL^{-1}); the sucrose content of both was negligible (0.60 [±0.06] mg mL^{-1}). Grass juice contains smaller amounts of other sugars (e.g., galactooligosaccharides and maltosaccharides) in addition to proteins which can also be used for fermentation and growth; non-fermentable carbohydrates (i.e., lignin, cellulose and hemicellulose) are found in grass pulp and the fibrous biomass fraction (Charlton et al. 2009). GJ + $_t$fosEp was found to support optimal yeast growth, bioethanol and biomass production and was consequently employed as the feedstock for all experimental work reported in the present study.

The growth parameters determined for *S. cerevisiae* (Turbo yeast) and *S. boulardii* (MYA-769) were closely comparable (Table 1 and Figure 2) indicating that the composition of GJ + $_t$fosEp did not inhibit the growth of either yeast. The maximum bioethanol and biomass product yields for Turbo (37.43 [±1.99] mg ethanol mL^{-1} and 4.78 [±0.10] g dry weight L^{-1}) and MYA-769 (39.18 [±2.42] mg ethanol mL^{-1} and 4.96 [±0.15] g dry weight L^{-1}) were also very similar.

Microscope observations revealed that the two yeasts exhibited different growth morphologies (Figure 3). Turbo cultures were characterised by round solitary blastoconidia and normal cellular budding (Figure 3A) while MYA-769 cultures comprised a mixture of yeast-like, elongated and psuedohyphal growth forms (Figure 3B).

Results from cholesterol uptake experiments (Table 2 and Figures 4 and 5) indicate that, under oxygen-limited conditions and at a growth temperature compatible with that of the human body (37°C) MYA-769 assimilated more cholesterol than Turbo (Table 2 and Figure 5).

Discussion

Bioethanol and biomass

S. cerevisiae has traditionally been used in food production, biotechnology, brewing and biofuel industries; however, the bioethanol and biomass production observed in the present study highlights the potential to utilise *S. boulardii* for industrial ethanol fermentation processes (Gurgu et al. 2011). That GJ + $_t$fosEp is a suitable feedstock for yeast biofermentations is evidenced by the sterol composition of both yeast species following growth on standard YPD media and GJ + $_t$fosEp (Table 2 and Figure 4). Neither yeast was affected by perturbations

in ergosterol biosynthesis; ergosterol is an essential yeast membrane sterol needed to maintain membrane viability and healthy growth (Daum et al. 1998). Morphological differences were observed when Turbo and MYA-769 were grown using both GJ + $_t$fosEp and standard YPD media suggesting that the psuedohyphal growth of MYA-769 (Figure 3) is a typical strain characteristic and not a response to nutrient limitation; nutrient limitation is understood to be a prerequisite for filamentous growth in wild-type *S. cerevisiae* (Gimeno et al. 1992).

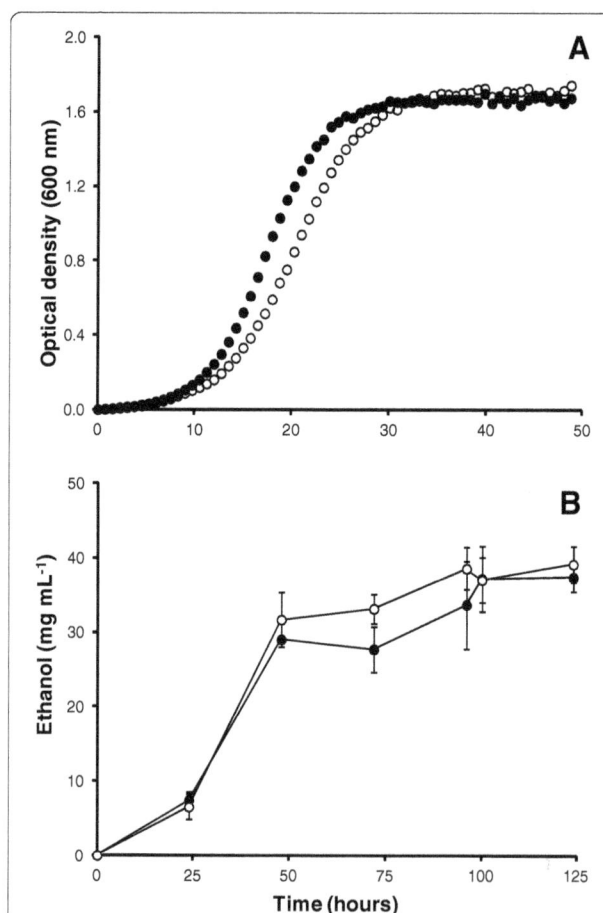

Figure 2 Growth (A) and bioethanol fermentation (B; mean values [±S.D]) of turbo yeast (●) and MYA-796 (○) grown using GJ + $_t$fosEp. Note that ethanol concentrations were sampled after yeast cultures had reached stationary phase (typically t$_{48}$h); ethanol concentrations decreased after t$_{125}$h.

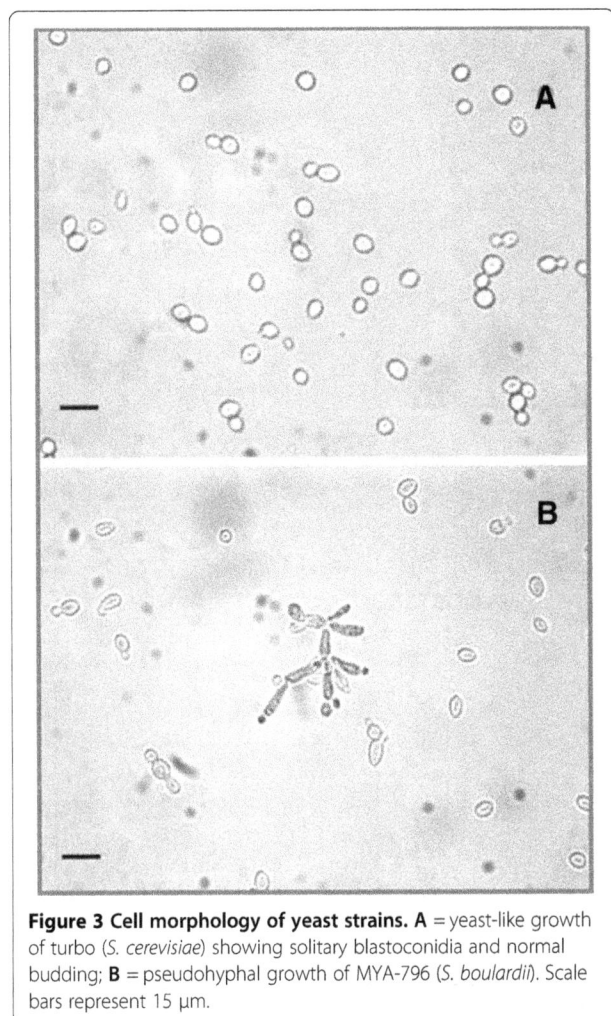

Figure 3 Cell morphology of yeast strains. A = yeast-like growth of turbo (*S. cerevisiae*) showing solitary blastoconidia and normal budding; **B** = pseudohyphal growth of MYA-796 (*S. boulardii*). Scale bars represent 15 μm.

S. boulardii is reported to possess an enhanced ability for pseudohyphal switching and is understood to survive better at low pH than other strains of *S. cerevisiae* (Edwards-Ingram et al. 2007).

Given these morphological and physiological characteristics, *S. boulardii* presents itself as an attractive microorganism for biotechnology and industrial applications where flocculent and sedimenting strains of *S. cerevisiae* are particularly valuable (Kida et al. 1990;

Seong et al. 2006). A flocculation mutant of *Candida glabrata*, another species showing genetic relatedness to *S. cerevisiae* (Roetzer et al. 2011), has been identified as a potentially useful strain for bioethanol production because of its growth at higher temperatures (Watanabe et al. 2009).

Cholesterol assimilation

Research and commercial interest surrounds the biotherapeutic properties of different yeasts (Foligne et al. 2010) and those with cholesterol-lowering activity have attracted specific attention (Chen et al. 2010; Psomas et al. 2003). *S. cerevisiae* is already known to sequester extracellular cholesterol under anaerobic conditions (Lorenz et al. 1986); however, results from the present study indicate cholesterol uptake by *S. boulardii* (strain MYA-769) is superior. In view of the impetus towards lowering cholesterol through dietary modifications and speculation that probiotic yeast could provide a means to lower serum cholesterol (Chen et al. 2010; Krasowska et al. 2007) work to characterise cholesterol uptake in the host environment using a wider number of strains is now required. Phytosterols were not detectable in the sterol chromatograms for Turbo yeast or MYA-796 grown using GJ + ₜfosEp (Table 2 and Figure 4), and indeed the sterol content of GJ was negligible; however, we did find residual ryegrass pulp, a by-product of the grass juice extraction process, to be rich in plant sterols (data not shown). The potential to extract phytosterols from ryegrass and alternative plant biomass feedstock requires consideration not least because diet supplementations containing plant sterols could offer protection against a variety of chronic ailments including cardiovascular diseases, obesity, diabetes, and cancer (Bradford and Awad 2007).

Applications for yeast biomass

Given legislation restricting the addition of antibiotics to animal feed (Seo et al., 2010) the potential to market food-grade yeast biomass and/or utilise it in a rural biorefinery setting is important. The use of *S. boulardii* for treatment of human gastrointestinal disorders is well documented (Buts 2009; Vandenplas et al. 2009;

Table 2 Mean [±S.D] cellular sterol composition (%) of turbo yeast and MYA-796

	Sterol composition (%)									
	Zymosterol		Ergosterol		Lanosterol		[a] Intermediates		Cholesterol	
	Turbo	MYA-769	Turbo	MYA-769	Turbo	MYA-769	Turbo	MYA-769	Turbo	MYA-769
YPD	17.92 [0.38]	17.79 [0.49]	**61.71 [3.29]**	**61.91 [1.39]**	5.93 [1.65]	8.06 [0.59]	14.45 [2.02]	12.25 [1.49]	—	—
GJ + ₜfosEp	15.53 [2.80]	12.31 [0.10]	**67.31 [0.49]**	**61.86 [1.10]**	5.90 [0.09]	9.31 [0.76]	11.26 [3.20]	16.52 [0.23]	—	—
glcYM[+chol]	5.7 [0.78]	1.6 [0.41]	26.9 [1.27]	12.2 [2.12]	**35.7 [0.74]**	**38.2 [3.64]**	27.10 [1.79]	27.30 [2.05]	4.6 [0.48][b]	20.7 [0.96][b]

The most abundant sterol in each experiment is emboldened. [a] = sum of all minor sterol intermediates (each comprising < 5% of total cellular sterol fraction); [b] = exogenously supplied cholesterol; Strikethrough = not detected.

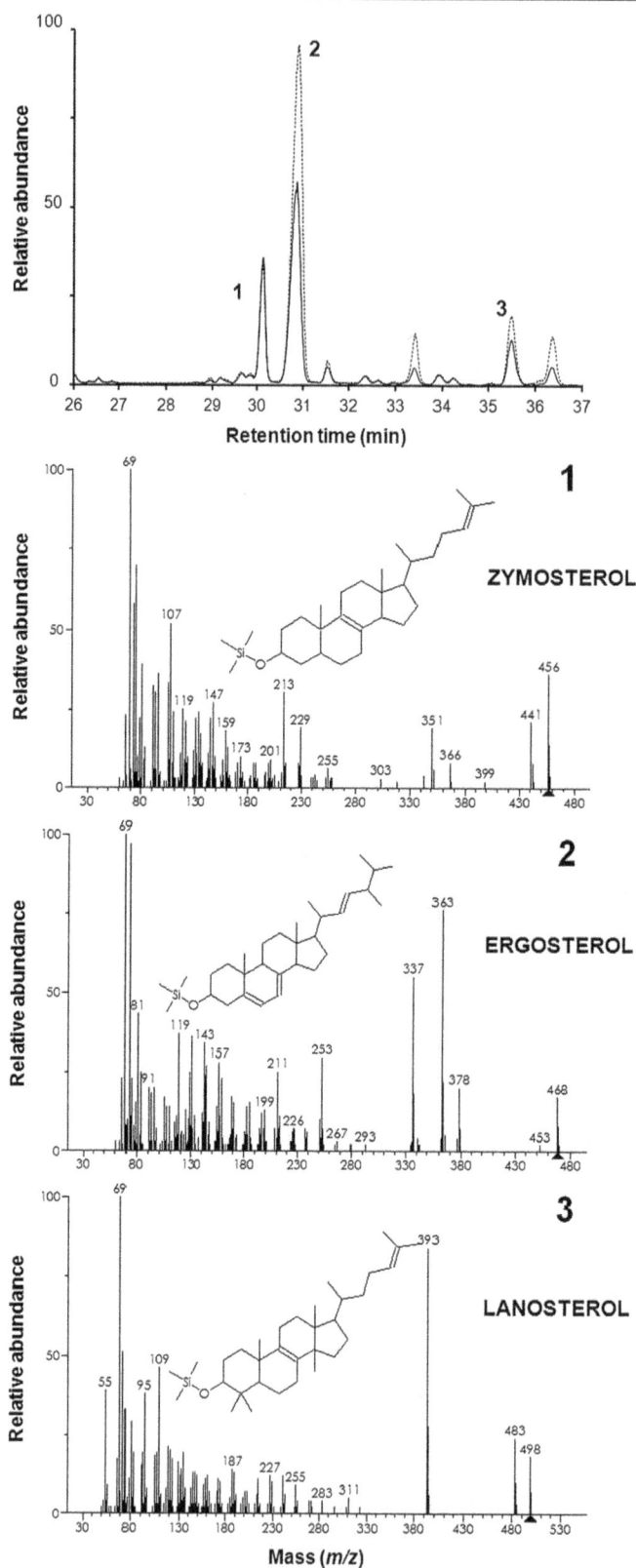

Figure 4 Sterol composition. Overlay of GC-MS sterol chromatograms for turbo yeast (unbroken trace) and MYA-796 (broken trace) grown on GJ + ₜfosEp. Diagnostic fragmentation spectra for **1**) zymosterol, **2**) ergosterol and **3**) lanosterol are shown; note the presence of minor sterol intermediates (retention times 31.5-34.5 min and 36.5 min).

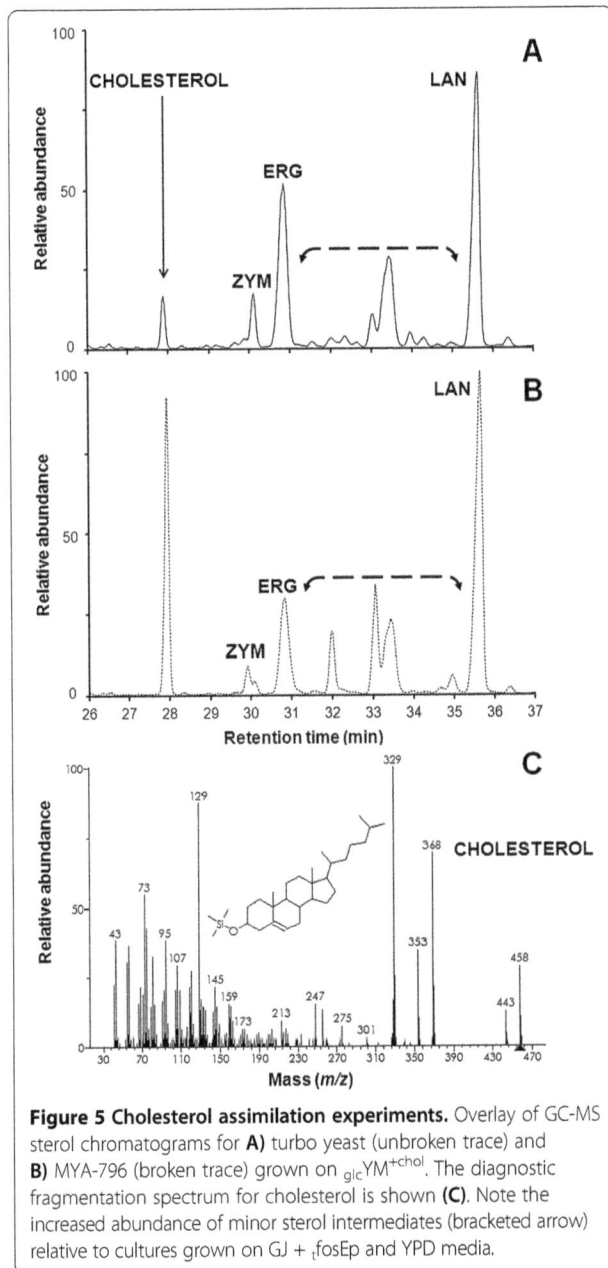

Figure 5 Cholesterol assimilation experiments. Overlay of GC-MS sterol chromatograms for **A)** turbo yeast (unbroken trace) and **B)** MYA-796 (broken trace) grown on $_{glc}$YM^{+chol}. The diagnostic fragmentation spectrum for cholesterol is shown **(C)**. Note the increased abundance of minor sterol intermediates (bracketed arrow) relative to cultures grown on GJ + $_t$fosEp and YPD media.

In conclusion, rural biorefineries have been identified as a potential means to facilitate social and economic regeneration in regions where a low GDP affects communities (Charlton et al. 2009). Here we demonstrate the possibility of generating bioethanol and probiotic yeast biomass from agriculturally-sourced grass juice. Breeding for bio-ethanol production using *L. perenne* L. with higher water-soluble carbohydrate content is already underway in the United Kingdom (Farrar et al. 2012) and results from the present study highlight further opportunities for integrated microbial biotechnology and large scale biorefining using high sugar perennial ryegrasses (Charlton et al. 2009; Martel et al. 2010; Martel et al. 2011). The experiments described can be a base towards extraction of maximum value from grass as proteins and chlorophyll are some of the other products that can be envisaged. The grass cellulosic fibre could also be fermented after appropriate enzymological and process treatments. Increased content of water soluble carbohydrates in breeding processes and engineering solutions to concentrate the juice to generate higher concentrations of bioethanol for distillation are also avenues that need to be explored. Finally it is evident *S. boulardii* could be used in other bioethanol processes if a use for biomass on that scale was desirable as in reducing methane emissions.

Competing interests
The authors declare that they have no competing interests.

Authors' contributions
CMH designed and undertook the growth studies and conceived and drafted the manuscript. EJL carried out the sugar assays. ISD and DEK participated in the design of the study and written work. SLK conceived the study, and participated in its design and coordination and helped to draft the manuscript. All authors read and approved the final manuscript.

Authors' information
EJL undertook experimental work at Swansea University, his current contact address is now: School of Chemistry, Cardiff University, Main Building, Park Place, Cardiff, CF10 3AT, Wales, UK.

Acknowledgements
We thank Marcus Hull for technical support. Funding for this study was provided by the Biotechnology and Biological Sciences Research Council (2/E13805) and the BEACON Convergence project supported by Welsh Government and the European Regional Development Fund (ERDF) of the European Union. Analytical facilities were provided by the EPSRC National Mass Spectrometry Service Centre (Swansea University, UK).

Author details
[1]Institute of Life Science, College of Medicine, Swansea University, Swansea SA2 8PP, Wales, UK. [2]Institute of Biological, Environmental & Rural Sciences, Aberystwyth University, Gogerddan, Aberystwyth SY23 3EE, Wales, UK.

Zanello et al. 2009) and studies indicate that probiotic yeast could also be administered as a direct-fed microbial to improve livestock quality (Collier et al. 2010; Keyser et al. 2007). The effects of dietary yeast (*S. cerevisiae*) autolysate on poultry health have also been documented (Yalçin et al. 2012). Finally, there is emerging evidence to show that dietary yeast supplementations could mitigate ruminant enteric methane emissions (Cottle et al. 2011; Lila et al. 2004). Methane is the second most important greenhouse gas and considering the importance of ruminant livestock, probiotic administration could offer an economically feasible way of reducing ruminant CH_4 production while improving productivity (Shibata and Terada 2010).

References
Bradford PG, Awad AB (2007) Phytosterols as anticancer compounds. Mol Nutr Food Res 51(2):161–170
Buts J-P (2009) The probiotic *Saccharomyces boulardii* upgrades intestinal digestive functions by several mechanisms. Acta Gastroenterol Belg 72(2):274–276

Charlton A, Elias R, Fish S, Fowler P, Gallagher J (2009) The biorefining opportunities in Wales: Understanding the scope for building a sustainable, biorenewable economy using plant biomass. Chem Eng Res Des 87(9A):1147–1161

Chen L-S, Ma Y, Maubois JL, He SH, Chen LJ, Li HM (2010) Screening for the potential probiotic yeast strains from raw milk to assimilate cholesterol. Dairy Sci Technol 90(5):537–548

Collier CT, Carroll JA, Ballou MA, Starkey JD, Sparks JC (2010) Oral administration of Saccharomyces cerevisiae boulardii reduces mortality associated with immune and cortisol responses to Escherichia coli endotoxin in pigs. J Anim Sci 89(1):52–58

Cottle DJ, Nolan JV, Wiedemann SG (2011) Ruminant enteric methane mitigation: a review. Anim Prod Sci 51(6):491–514

Daum G, Lees ND, Bard M, Dickson R (1998) Biochemistry, cell biology and molecular biology of lipids of Saccharomyces cerevisiae. Yeast 14(16):1471–1510

Edwards-Ingram LC, Gent ME, Hoyle DC, Hayes A, Stateva LI, Oliver SG (2004) Comparative genomic hybridization provides new insights into the molecular taxonomy of the Saccharomyces sensu stricto complex. Genome Res 14(6):1043–1051

Edwards-Ingram L, Gitsham P, Burton N, Warhurst G, Clarke I, Hoyle D, Oliver SG, Stateva L (2007) Genotypic and physiological characterization of Saccharomyces boulardii, the probiotic strain of Saccharomyces cerevisiae. Appl Environ Microb 73(8):2458–2467

Farrar K, Bryant DN, Turner L, Gallagher JA, Thomas A, Farrell M, Humphreys MO, Donnison IS (2012) Breeding for bio-ethanol production in Lolium perenne L.: Association of allelic variation with high water-soluble carbohydrate content. abbreviation. Bioenerg Res 5(1):149–157

Foligne B, Dewulf J, Vandekerckove P, Pignede G, Pot B (2010) Probiotic yeasts: Anti-inflammatory potential of various non-pathogenic strains in experimental colitis in mice. World J Gastroenterol 16(17):2134–2145

Gimeno CJ, Ljungdahl PO, Styles CA, Fink GR (1992) Unipolar cell divisions in the yeast Saccharomyces cerevisiae lead to filamentous growth - Regulation by starvation and Ras. Cell 68(6):1077–1090

Gurgu L, Lafraya A, Polaina J, Marin-Navarro J (2011) Fermentation of cellobiose to ethanol by industrial Saccharomyces strains carrying the beta-glucosidase gene (BGL1) from Saccharomycopsis fibuligera. Bioresour Technol 102(8):5229–5236

Keyser SA, McMeniman JP, Smith DR, MacDonald JC, Galyean ML (2007) Effects of Saccharomyces cerevisiae subspecies boulardii CNCM I-1079 on feed intake by healthy beef cattle treated with florfenicol and on health and performance of newly received beef heifers. J Anim Sci 85(5):1264–1273

Kida K, Asano SI, Yamadaki M, Iwasaki K, Yamaguchi T, Sonoda Y (1990) Ethanol-production by flocculating yeast.4. Continuous high-ethanol fermentation from cane molasses by a flocculating yeast. J Ferment Bioeng 69(1):39–45

Krasowska A, Kubik A, Prescha A, Lukaszewicz M (2007) Assimilation of omega 3 and omega 6 fatty acids and removing of cholesterol from environment by Saccharomyces cerevisiae and Saccharomyces boulardii strains. J Biotechnol 131(2):S63–S64

Kyazze G, Dinsdale R, Hawkes FR, Guwy AJ, Premier GC, Donnison IS (2008) Direct fermentation of fodder maize, chicory fructans and perennial ryegrass to hydrogen using mixed microflora. Bioresour Technol 99(18):8833–8839

Leistritz FL, Hodur NM (2008) Biofuels: a major rural economic development opportunity. Biofuel Bioprod Bior Biofpr 2(6):501–504

Lila ZA, Mohammed N, Yasui T, Kurokawa Y, Kanda S, Itabashi H (2004) Effects of a twin strain of Saccharomyces cerevisiae live cells on mixed ruminal microorganism fermentation in vitro. J Anim Sci 82(6):1847–1854

Lorenz RT, Rodriguez RJ, Lewis TA, Parks LW (1986) Characteristics of sterol uptake in Saccharomyces cerevisiae. J Bacteriol 167(3):981–985

Martel CM, Warrilow AGS, Jackson CJ, Mullins JGL, Togawa RC, Parker JE, Morris MS, Donnison IS, Kelly DE, Kelly SL (2010) Expression, purification and use of the soluble domain of Lactobacillus paracasei beta-fructosidase to optimise production of bioethanol from grass fructans. Bioresour Technol 101(12):4395–4402

Martel CM, Parker JE, Jackson CJ, Warrilow AGS, Rolley N, Greig C, Morris SM, Donnison IS, Kelly DE, Kelly SL (2011) Expression of bacterial levanase in yeast enables simultaneous saccharification and fermentation of grass juice to bioethanol. Bioresour Technol 102(2):1503–1508

Moyad MA (2008) Brewer's/baker's yeast (Saccharomyces cerevisiae) and preventive medicine: Part II. Urol Nurs 28(1):73–75

Psomas EI, Fletouris DJ, Litopoulou-Tzanetaki E, Tzanetakis N (2003) Assimilation of cholesterol by yeast strains isolated from infant feces and Feta cheese. J Dairy Sci 86(11):3416–3422

Roetzer A, Gabaldon T, Schueller C (2011) From Saccharomyces cerevisiae to Candida glabrata in a few easy steps: important adaptations for an opportunistic pathogen. FEMS Microbiol Lett 314(1):1–9

Seo JK, Kim S-W, Kim MH, Upadhaya SD, Kam DK, Ha JK (2010) Direct-fed microbials for ruminant animals. Asian Aust J Anim Sci 23(12):1657–1667

Seong KT, Katakura Y, Ninomiya K, Bito Y, Katahira S, Kondo A, Ueda M, Shioya S (2006) Effect of flocculation on performance of arming yeast in direct ethanol fermentation. Appl Microbiol Biotechnol 73(1):60–66

Sharma HSS, Lyons G, McRoberts C (2011) Biorefining of perennial grasses: A potential sustainable option for Northern Ireland grassland production. Chem Eng Res Des 89(11A):2309–2321

Shibata M, Terada F (2010) Factors affecting methane production and mitigation in ruminants. Anim Sci J 81(1):2–10

Vandenplas Y, Brunser O, Szajewska H (2009) Saccharomyces boulardii in childhood. J Pediatr 168(3):253–265

Wang J-M, Zhang T, Chi Z, Liu G-L, Chi Z-M (2011) 18S rDNA integration of the exo-inulinase gene into chromosomes of the high ethanol producing yeast Saccharomyces sp W0 for direct conversion of inulin to bioethanol. Biomass Bioenerg 35(7):3032–3039

Watanabe I, Nakamura T, Shima J (2009) Characterization of a spontaneous flocculation mutant derived from Candida glabrata: A useful strain for bioethanol production. J Biosci Bioeng 107(4):379–382

Yalçin S, Uzunoglu K, Duyum HM, Eltan O (2012) Effects of dietary yeast autolysate (Saccharomyces cerevisiae) and black cumin seed (Nigella sativa L.) on performance, egg traits, some blood characteristics and antibody production of laying hens. Livest Sci 145(1–3):13–20

Zanello G, Meurens F, Berri M, Salmon H (2009) Saccharomyces boulardii effects on gastrointestinal diseases. Curr Issues Mol Biol 11:47–58

Zhang T, Chi Z, Chi Z, Parrou J-L, Gong F (2010) Expression of the inulinase gene from the marine-derived Pichia guilliermondii in Saccharomyces sp W0 and ethanol production from inulin. Microbial Biotech 3(5):576–582

11

Exoproteome analysis of *Clostridium cellulovorans* in natural soft-biomass degradation

Kohei Esaka[1], Shunsuke Aburaya[1], Hironobu Morisaka[1,2], Kouichi Kuroda[1] and Mitsuyoshi Ueda[1,2*]

Abstract

Clostridium cellulovorans is an anaerobic, cellulolytic bacterium, capable of effectively degrading various types of soft biomass. Its excellent capacity for degradation results from optimization of the composition of the protein complex (cellulosome) and production of non-cellulosomal proteins according to the type of substrates. In this study, we performed a quantitative proteome analysis to determine changes in the extracellular proteins produced by *C. cellulovorans* for degradation of several types of natural soft biomass. *C. cellulovorans* was cultured in media containing bagasse, corn germ, rice straw (natural soft biomass), or cellobiose (control). Using an isobaric tag method and a liquid chromatograph equipped with a long monolithic silica capillary column/mass spectrometer, we identified 372 proteins in the culture supernatant. Of these, we focused on 77 saccharification-related proteins of both cellulosomal and non-cellulosomal origins. Statistical analysis showed that 18 of the proteins were specifically produced during degradation of types of natural soft biomass. Interestingly, the protein Clocel_3197 was found and commonly involved in the degradation of every natural soft biomass studied. This protein may perform functions, in addition to its known metabolic functions, that contribute to effective degradation of natural soft biomass.

Keywords: *Clostridium cellulovorans*; Cellulosome; Soft-biomass degradation; Proteome analysis; Monolithic column

Introduction

Cellulosic and herbaceous types of biomass (soft biomass) such as rice straw, switchgrass, and bagasse show promise as substrates for the production of chemical products and fuels. However, it is difficult to degrade soft biomass (Lynd et al. 1999). Cellulose is comprised of a glucose-linked structure that is resistant to degradation due to the number of hydrogen bonds in its crystalline structure (Mansfield et al. 1999). Chemical procedures, including processing with strong acids, high pressures, or high temperatures, are generally employed to degrade cellulose to glucose; however, these methods impose an environmental burden. In addition, degradation strategies must be optimized according to the type of soft biomass, based on individual structures and components.

The artificial and commercial cellulase cocktails currently available are expensive; however, several naturally occurring microbes present an attractive alternative. We have focused on the cellulolytic bacterium *Clostridium cellulovorans*. *C. cellulovorans* is a mesophilic, anaerobic bacterium that can degrade various components of plant cell walls, including not only cellulose, but also hemicelluloses and pectin (Sleat et al. 1984). Previously, we performed genome analysis of *C. cellulovorans* and demonstrated that it produced a "cellulosome" (Tamaru et al. 2010), a multi-enzyme complex that is known to be produced by several types of cellulolytic and anaerobic bacteria (Bae et al. 2013; Bayer et al. 2004; Doi and Kosugi 2004) such as *C. thermocellum* (Lamed et al. 1983b; Bayer et al. 1983; Lamed et al. 1983a) and *C. cellulolyticum* (Desvaux 2005; Gal et al. 1997). *C. cellulovorans* has high cellulolytic activity due to the presence of numerous polysaccharide degradation-related proteins that show synergistic effects (Fierobe et al. 2002). Genomic analysis of *C. cellulovorans* indicated the presence of 57 cellulosome-related genes, including four scaffold and 53 cellulosomal protein-encoding genes (Tamaru et al. 2010). The major scaffold protein, CbpA, is composed of nine cohesin domains that bind to various cellulosomal proteins (Tamaru 2001). Using proteome analysis, we reported that *C. cellulovorans* optimized the composition of its cellulosomal protein according to the

* Correspondence: miueda@kais.kyoto-u.ac.jp
[1]Division of Applied Life Sciences, Graduate School of Agriculture, Kyoto University, Sakyo-ku, Kyoto, Japan
[2]Kyoto Integrated Science and Technology Bio-Analysis Center, Shimogyo-ku, Kyoto, Japan

type of basal substrates (cellobiose, avicel, and xylan) (Morisaka et al. 2012), and that this ability played a major role in polysaccharide degradation (Matsui et al. 2013). However, compared to the genomes of other cellulosome-producing clostridial species, the genome of *C. cellulovorans* contains a very large number (190) of non-cellulosomal protein-encoding genes (Tamaru et al. 2011). Non-cellulosomal proteins do not form a complex (cellulosome) and function as free saccharification-related enzymes. We reported that non-cellulosomal proteins also played a key role in effective degradation of basal biomass (Matsui et al. 2013). *C. cellulovorans* could effectively degrade various types of natural soft biomass via the cooperative activity of cellulosomal and non-cellulosomal proteins. This unique and cooperative feature offers the potential to enhance the efficiency of soft-biomass degradation. However, few comprehensive and molecular studies of the degradation of natural soft biomass have been reported. To improve the efficiency of soft-biomass utilization, it will be useful to study the changes in the *C. cellulovorans* protein profile in response to various types of natural soft biomass.

In this study, we performed a quantitative analysis of the cellulosomal and non-cellulosomal proteins produced by *C. cellulovorans* during the degradation of several types of natural soft biomass. We used bagasse (the byproduct of sugar cane processing), corn germ (corn embryos), and rice straw as carbon sources. Proteins in the culture supernatant (exoproteome) were analyzed using a LC-MS/MS system equipped with a long monolithic silica capillary column (470 cm), as described previously (Matsui et al. 2013; Morisaka et al. 2012). We identified the individual protein profiles of the exoproteomes, including both cellulosomal and non-cellulosomal proteins. Additionally, integrated proteome and genome analysis indicated that *C. cellulovorans* produced proteins that showed promise for improving the efficiency of degradation of natural soft biomass.

Methods
Cell culture and medium
C. cellulovorans 743B (ATCC35296) was grown anaerobically as previously described (Sleat et al. 1984), differing only in carbon source, which was replaced by 0.3% (w/v) cellobiose and 0.3% (w/v) soft biomass.

Growth substrates
Cellobiose (Sigma, St Louis, MO, USA) and cellulosic soft biomass were used in the growth experiments. Bagasse, corn germ, and rice straw were used as soft biomass. Bagasse, containing 39.6% cellulose, 20.2% hemicellulose, 25.9% lignin, 14.3% other components, was provided by H. Nonaka, Mie University (Nonaka et al. 2013; Ren and Funaoka 2009); corn germ, containing 10.9% cellulose,

23.3% hemicellulose, 0.6% lignin, 65.2% other components, was provided by Tsuji Oil Mill Co. Ltd (Furuya et al. 2010); and rice straw (Nakanishi et al. 2012), containing 39.2% cellulose, 27.4% hemicellulose, 4.4% lignin, 29.0% other components, was provided by H. Miyake, Mie University. These were crushed for 1 min by using a Hi-Power Blender MX1100XTS (Waring Commercial, Torrington, CT, USA), and the resulting soft biomass (diameter < 250 μm) was used for *C. cellulovorans* culture.

Preparation of extracellular proteins (exoproteome) for quantitative proteome analysis
Samples from *C. cellulovorans* cultures were prepared for proteome analysis as previously described (Matsui et al. 2013). Each stationary-phase culture (50 mL) was centrifuged ($6,000 \times g$, 25°C), and the supernatant was subjected to ultrafiltration using an Amicon Ultra-15 Centrifugal Filter Unit (MWCO 10 kDa, Millipore, Darmstadt, Hessen, Germany) to obtain the extracellular proteins. The concentrated samples were independently dissolved in 100 μL of triethylammonium bicarbonate buffer (200 mM), to which 5 μL of Tris(2-carboxyethyl) phosphine (200 mM) was added, and the reaction was allowed to proceed for 60 min at 55°C. To this mixture, 5 μL of iodoacetamide (375 mM) was added, and the reaction continued for 30 min, protected from light, at room temperature. Sequencing grade modified trypsin (1 μg/μL; Promega, Madison, WI, USA) was added (2 μL), and the proteins were digested overnight at 37°C. The four proteome samples (cellobiose, bagasse, corn germ, and rice straw) were labeled using a tandem mass tag (TMT) 6-plex labeling kit (Thermo Fisher Scientific, Waltham, MA, USA) with reporters at m/z = 128, 129, 130, and 131, respectively, in 41 μL acetonitrile. After 60 min of reaction at room temperature, 8 μL of 5% (w/v) hydroxylamine was added to each tube and mixed for 15 min. In addition, a mixture of tryptic fragments from all substrates was combined with TMT-126 (reporter at m/z = 126) as an internal standard for quantification. The aliquots were then pooled and evaporated under vacuum and dissolved in 100 μL of trifluoroacetic acid (0.1%) and used for LC-MS/MS analysis.

Exoproteome analysis
Proteome analysis was performed using an LC (Ultimate 3000°; Thermo Fisher Scientific)-MS/MS (LTQ Orbitrap Velos Mass Spectrometer°; Thermo Fisher Scientific) system equipped with a long monolithic column, as previously described (Matsui et al. 2013; Morisaka et al. 2012). Tryptic digests were separated by reversed-phase chromatography using a monolithic silica capillary column (470 cm long, 0.1 mm ID), at a flow rate of 500 nL/min. The gradient was provided by changing the mixing ratio of the two eluents: A, 0.1% (v/v) formic acid and B, 80% acetonitrile containing 0.1% (v/v) formic acid. The

gradient was started with 5% B, increased to 45% B for 600 min, further increased to 95% B to wash the column, returned to the initial condition, and held for re-equilibration. The separated analytes were detected using a mass spectrometer with a full scan range of 350–1,500 m/z (resolution 60,000), followed by 10 data-dependent higher-energy c-trap dissociation (HCD) MS/MS scans acquired for TMT reporter ions, using 40% normalized collision energy in HCD with 0.1 ms activation time and an electrospray ionization (ESI) voltage of 2.3 kV. The ion transfer tube temperature was set to 280°C. Triplicate analyses were performed for each sample in three independent experiments, and the collected data were reviewed for protein identification and quantification.

Data analysis was performed using Proteome Discoverer software (Thermo Fisher Scientific). Protein identification was performed using the Mascot algorithm against the *C. cellulovorans* protein database (4,254 sequences) from NCBI (National Center for Biotechnology Information, http://www.ncbi.nlm.nih.gov/), with a precursor mass tolerance of 20 ppm and a fragment ion mass tolerance of 50 mmu. Carbamidomethylation of cysteine and a TMT 6-plex at the N-terminus were set as fixed modifications. Protein quantification was performed using the Reporter Ions Quantifier with the TMT 6-plex method. The data were then filtered with a cut-off criteria of q-value ≤ 0.05, corresponding to a 5% false discovery rate (FDR) on a spectral level. The values for the exponentially modified protein abundance index

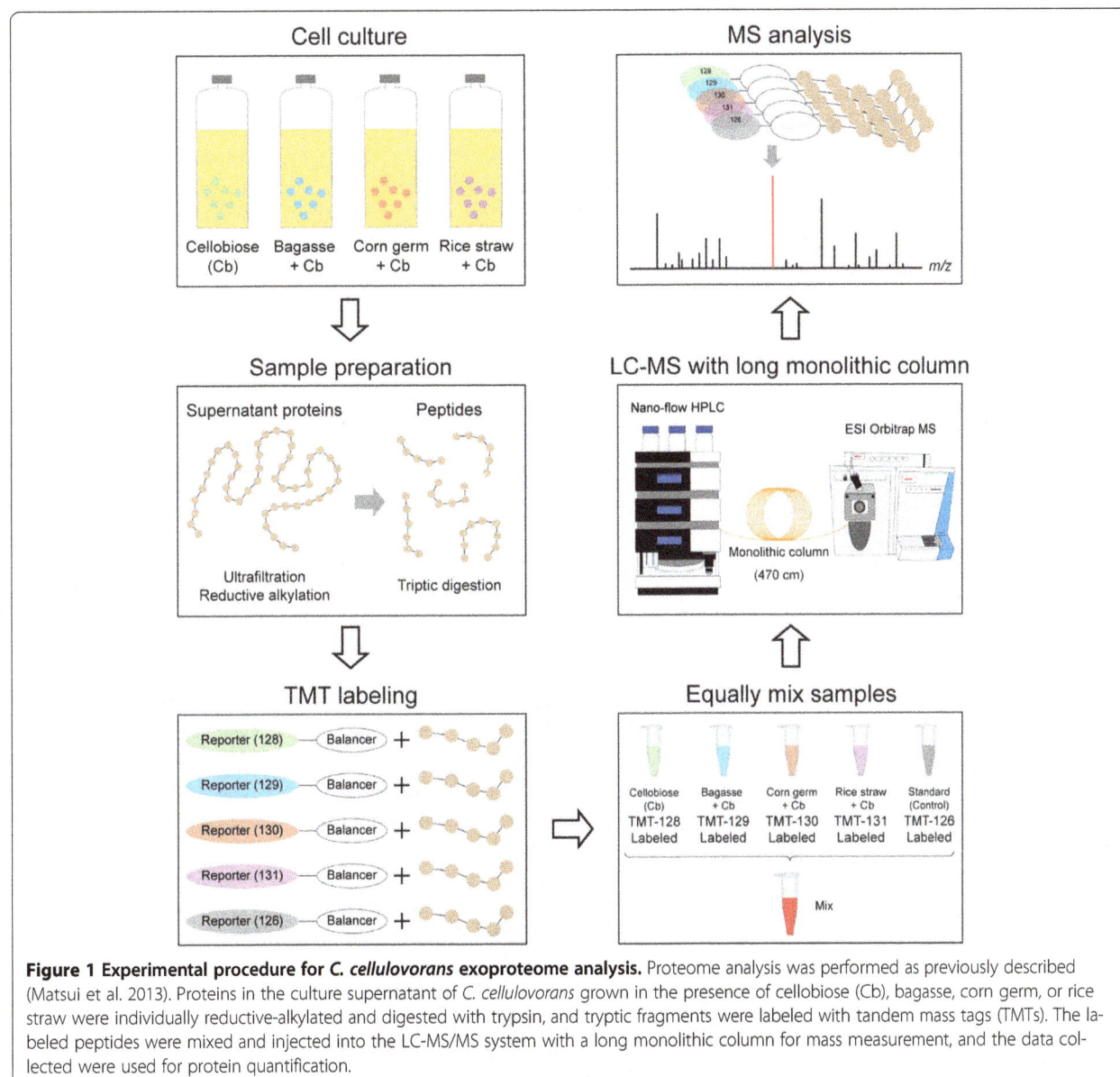

Figure 1 Experimental procedure for *C. cellulovorans* exoproteome analysis. Proteome analysis was performed as previously described (Matsui et al. 2013). Proteins in the culture supernatant of *C. cellulovorans* grown in the presence of cellobiose (Cb), bagasse, corn germ, or rice straw were individually reductive-alkylated and digested with trypsin, and tryptic fragments were labeled with tandem mass tags (TMTs). The labeled peptides were mixed and injected into the LC-MS/MS system with a long monolithic column for mass measurement, and the data collected were used for protein quantification.

(emPAI) (Ishihama et al. 2005) were used to estimate the abundance of cellulosomal and non-cellulosomal proteins. Proteins with no missing values in three replicates were accepted in the protein quantification analysis. Global median normalization was performed to normalize the quantity of each tryptic digest injected into the mass spectrometer.

Results

Analysis of proteins in *C. cellulovorans* culture supernatants

To investigate the degradation of natural soft biomass, proteins were isolated from the supernatant of stationary-phase *C. cellulovorans* cultures grown on bagasse, corn germ, or rice straw, and subjected to LC-MS/MS analysis (Figure 1) (Matsui et al. 2013). The mass spectrometry data collected were used for exoproteome analysis, as shown in Figure 2.

To normalize the amount of each tryptic digest injected into the mass spectrometer, the identified proteins were standardized with the median. To determine which proteins were differentially produced for each type of soft biomass, we focused on 77 cellulosomal and non-

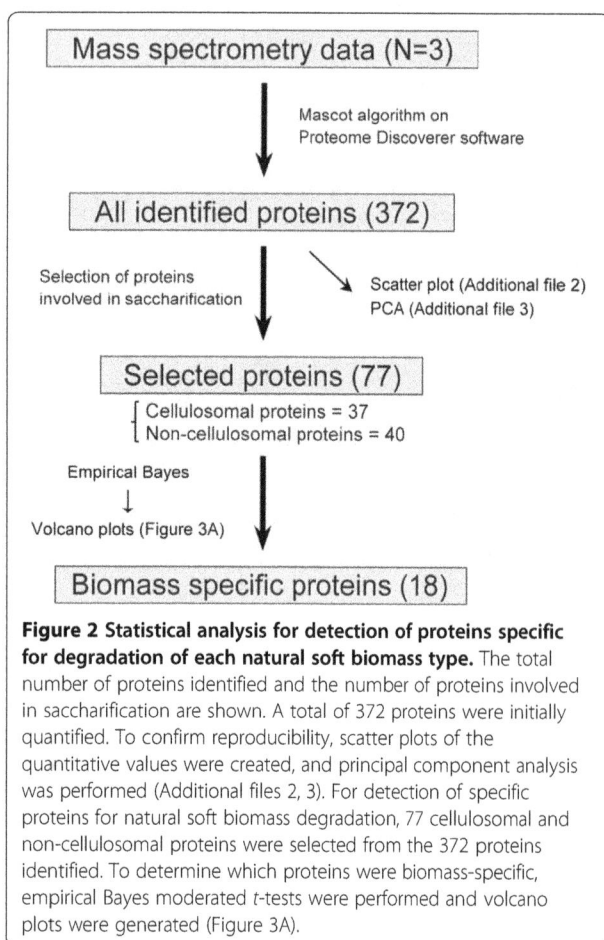

Figure 2 Statistical analysis for detection of proteins specific for degradation of each natural soft biomass type. The total number of proteins identified and the number of proteins involved in saccharification are shown. A total of 372 proteins were initially quantified. To confirm reproducibility, scatter plots of the quantitative values were created, and principal component analysis was performed (Additional files 2, 3). For detection of specific proteins for natural soft biomass degradation, 77 cellulosomal and non-cellulosomal proteins were selected from the 372 proteins identified. To determine which proteins were biomass-specific, empirical Bayes moderated *t*-tests were performed and volcano plots were generated (Figure 3A).

cellulosomal saccharification-related enzymes, chosen from among the 372 proteins identified (Additional file 1). Next, we performed empirical Bayes moderated *t*-tests and created volcano plots (Figure 3A) using the quantitative values for each of the 77 proteins (Matsui et al. 2013); proteins produced in culture with cellobiose were used as controls. *P*-values were adjusted using the Benjamini-Hochberg method to avoid the problem of multiple testing. Proteins which met the criteria (FDR-adjusted *P*-value < 0.05 and fold-change of protein ratio > 2 as compared to cellobiose) were defined as individual "biomass-specific proteins." Four bagasse-, 11 corn germ-, and six rice straw-specific proteins were identified (Figure 3B and Table 1).

Bagasse-specific proteins

Four bagasse-specific proteins were identified, including one cellulosomal and three non-cellulosomal proteins. The cellulosomal protein was Clocel_2820 (HbpA), a scaffolding protein with no enzymatic activity, similar to CbpA, which is the main scaffolding protein of the cellulosome. HbpA has been reported to enhance cellulosomal cellulase (Clocel_1150 (EngB) or Clocel_2819 (EngL)) activity on solid substrates such as avicel or corn fiber, but not on soluble carboxymethyl cellulose (Matsuoka et al. 2007). Thus, *C. cellulovorans* might increase the production of HbpA to accelerate cellulase activity on solid biomass. Among the non-cellulosomal proteins, Clocel_0873, which is classified as a member of the PL9 family by the Carbohydrate-Active enZymes (CAZy) database (Lombard et al. 2014), is considered to have pectate lyase activity. Clocel_1430, classified as a member of the GH31 family, is considered to have α-xylosidase activity. Clocel_0873 is a pectin-specific protein, and Clocel_1430 is a xylan-specific protein (Matsui et al. 2013). *C. cellulovorans* might recognize the pectin contained in bagasse and produce these proteins in response. Clocel_3197, a member of the GH130 family, was also detected.

Corn germ-specific proteins

One cellulosomal protein, Clocel_3650, and 10 non-cellulosomal proteins were identified. Clocel_3650, a member of the GH44 family, is considered to have endoglucanase activity. Among the non-cellulosomal proteins, Clocel_2606, classified as a member of the GH5 and CBM46 families, is considered to have cellulose-binding as well as cellulose-degrading activities (Aspeborg et al. 2012); it is a phosphoric acid swollen cellulose (PASC)-specific protein (Matsui et al. 2013). Clocel_3650 and Clocel_2606 are considered to be involved in acceleration of cellulose degradation. Clocel_0034, a member of the GH31 family, is considered to have α-xylosidase activity, while Clocel_2535, classified as a member of the GH43 family, is considered to have β-xylosidase

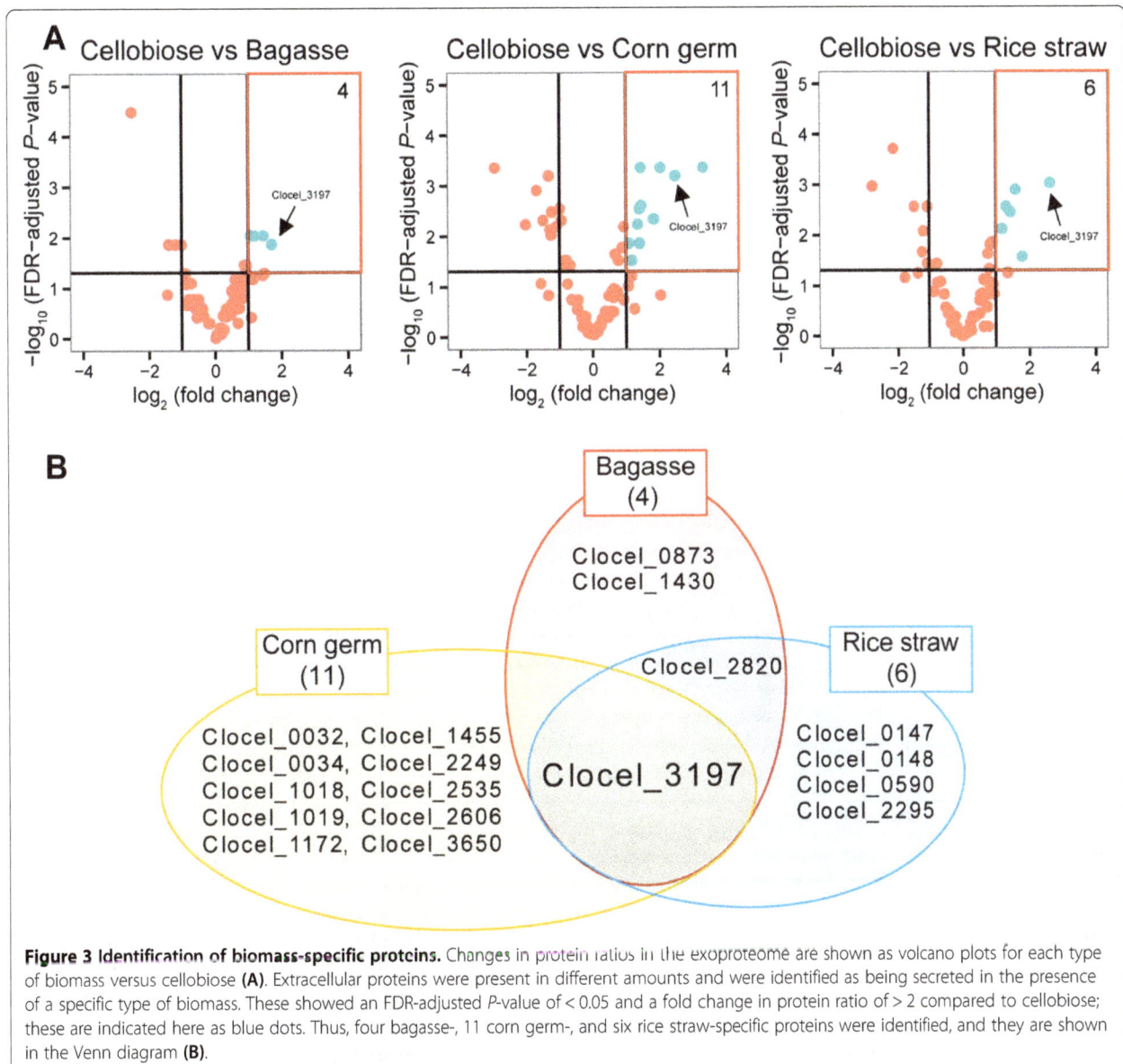

Figure 3 Identification of biomass-specific proteins. Changes in protein ratios in the exoproteome are shown as volcano plots for each type of biomass versus cellobiose **(A)**. Extracellular proteins were present in different amounts and were identified as being secreted in the presence of a specific type of biomass. These showed an FDR-adjusted *P*-value of < 0.05 and a fold change in protein ratio of > 2 compared to cellobiose; these are indicated here as blue dots. Thus, four bagasse-, 11 corn germ-, and six rice straw-specific proteins were identified, and they are shown in the Venn diagram **(B)**.

activity. These xylan degradation-related proteins are thought to contribute to the acceleration of xylan degradation. Clocel_1455, a member of the GH53 and CBM61 families, is considered to have endo-1, 4-β-galactosidase activity and 1, 4-β-galactan binding ability (Cid et al. 2010). Clocel_0032 is also a PASC-specific protein (Matsui et al. 2013). It is a member of the GH94 family and is considered to have cellodextrin phosphorylase activity. Clocel_1172 is a pectin-specific protein (Matsui et al. 2013). It is a member of the PL1 family and is thought to have pectate lyase activity. Clocel_2249 is considered to be a glucuronate isomerase, and it also contributes to the acceleration of pectin degradation. Clocel_1018 and Clocel_1019 are members of the GH1 family; enzymes in this family are thought to have various enzymatic activities, such as β-glucosidase

and β-xylosidase activities. Clocel_3197, which was identified among the bagasse-specific proteins, was also detected.

Rice straw-specific proteins

The rice straw-specific proteins consisted of four cellulosomal and two non-cellulosomal proteins. Among the cellulosomal proteins, Clocel_2295 (XynA), a member of the GH11 and CE4 families, has endoxylanase and acetyl-xylan esterase activities (Kosugi et al. 2002) and is a xylan-specific protein (Matsui et al. 2013). Since xylan (xylose) is abundant in the hemicellulose of rice straw (Yoswathana et al. 2010), *C. cellulovorans* is considered to recognize xylose and increase the production of XynA for effective degradation of rice straw. Clocel_0147 is a cysteine protease inhibitor (Meguro et al. 2011), which is

Table 1 The list of biomass-specific proteins

Biomass specifically	Protein type	Locus	Name[a]	CAZy family(ies)[b]	vs Cellobiose		emPAI[d]
					log$_2$-fold change	FDR-adjusted P-value[c]	
Bagasse (4)	Cellulosomal	Clocel_2820	HbpA	NA	1.45	9.13E-03	2.49
	Non-cellulosomal	Clocel_1430		GH31	1.19	9.13E-03	0.55
		Clocel_3197		GH130	1.71	1.37E-02	0.79
		Clocel_0873		PL9	1.07	9.13E-03	1.19
Corn germ (11)	Cellulosomal	Clocel_3650		GH44	1.45	4.39E-04	0.53
	Non-cellulosomal	Clocel_1018		GH1	1.41	2.85E-03	0.77
		Clocel_1019		GH1	3.29	4.39E-04	0.34
		Clocel_2606		GH5,CBM46	1.36	5.78E-03	2.44
		Clocel_0034		GH31	1.46	2.48E-03	0.51
		Clocel_2535		GH43	1.09	1.38E-02	0.20
		Clocel_1455		GH53,CBM61,CBM61,CBM61	2.03	4.39E-04	0.13
		Clocel_0032		GH94	1.83	4.50E-03	0.36
		Clocel_3197		GH130	2.46	6.34E-04	0.79
		Clocel_1172		PL1	1.41	1.38E-02	0.50
		Clocel_2249		NA	1.18	3.00E-02	1.01
Rice straw (6)	Cellulosomal	Clocel_2295	XynA	GH11,CE4	1.79	2.64E-02	0.69
		Clocel_0148		NA	1.59	1.24E-03	0.65
		Clocel_0147	CpiA	NA	1.3	2.72E-03	9.10
		Clocel_2820	HbpA	NA	1.43	3.46E-03	2.49
	Non-cellulosomal	Clocel_3197		GH130	2.63	9.16E-04	0.79
		Clocel_0590		NA	1.19	7.51E-03	5.27

[a]Name: Names of only the reported proteins were shown.
[b]See http://www.cazy.org/. NA, not annotated (not included in CAZy database).
[c]P-values were adjusted for multiple testing with the Benjamini-Hochberg method.
[d]emPAI: the values of exponentially modified protein abundance index.

considered to protect microbe and their cellulosomes from plant protease attack. Our results suggest that *C. cellulovorans* recognizes rice as its substrate and produces large amounts of cyspin as a defense mechanism. Clocel_0148, a protein of unknown function, was identified as a rice straw-specific protein. Clocel_2820 (HbpA), found among the bagasse-specific proteins, was also detected. Clocel_0590, detected among the non-cellulosomal proteins, is thought to be a xylose isomerase (Ota et al. 2013) and to be involved in degradation of the xylan contained in rice straw. Clocel_3197, found in both the bagasse- and corn germ-specific proteins, was also detected.

Discussion

We identified several proteins involved in degradation of various types of biomass; analysis of replicates showed that the results were reproducible. Scatter plots of normalized quantitative values for all combinations showed high correlation factors (Additional file 2). Principal component analysis (PCA) was performed for all 372 identified proteins to confirm that the proteome profile was similar between three biological replicates. The PCA

score plots (Additional file 3) showed a degree of high similarity between biological replicates, and the plots for each substrate formed individual groups. These results indicate that the quantitative proteome analysis showed a high degree of reproducibility and reliability.

Of the total 372 proteins identified, 77 proteins were determined to be involved in saccharification. Of these, 37 were cellulosomal proteins and 40 were non-cellulosomal proteins (Additional file 1). *C. cellulovorans* possesses 57 cellulosomal protein-encoding genes and 190 non-cellulosomal protein-encoding genes. Therefore, *C. cellulovorans* produced 64.9% (37 of 57) of its cellulosomal proteins and 21.1% (40 of 190) of its non-cellulosomal proteins for degradation of the types of natural soft biomass examined here.

From the statistical analysis, Clocel_3197 was commonly identified as all biomass-specific proteins in *C. cellulovorans* exoproteome. Interestingly, SignalP analysis did not detect a signal peptide-encoding sequence for Clocel_3197. Clocel_3197 is likely to be localized to the exterior of cells, based on the emPAI of Clocel_3197, which was nearly equal to that of Clocel_2295 (XynA)

(Table 1), which has a signal peptide-encoding sequence. This result also indicated that sample preparation did not lyse cells. Clocel_3197, a member of the GH130 family, has been annotated as a D-fructose 4-o-β-D-mannosyl-D-glucose phosphorylase. Some species, such as *Bacteroides fragilis*, have the homologue of Clocel_3197 containing operon and it plays an important role in mannan catabolic pathway (Senoura et al. 2011). However, all enzymes related to mannan degradation or metabolism were not detected (Additional file 1). Thus, this protein may also possess an extracellular function. Previous reports have indicated the presence of several proteins lacking signal peptides that are secreted by unconventional pathways (López-Villar et al. 2006; Kinseth et al. 2007). For example, several metabolites produced by Clocel_3197, which have different functions like those observed in moon-lighting proteins (Kinseth et al. 2007; López-Villar et al. 2006) may play an important role during substrate recognition and natural soft-biomass degradation by *C. cellulovorans*. The mechanisms of substrate degradation and recognition remain unknown, and this protein may be useful in future investigations of the substrate degradation and recognition mechanisms of *C. cellulovorans*.

In conclusion, we quantified the cellulosomal and non-cellulosomal protein profiles produced by *C. cellulovorans* cultured on various types of soft biomass. A total of 77 cellulosomal and non-cellulosomal proteins were identified from the *C. cellulovorans* culture supernatant by using an LC-MS/MS system equipped with a long monolithic silica capillary column. Empirical Bayes moderated *t*-tests and volcano plots identified four bagasse-, 11 corn germ-, and six rice straw-specific proteins. Clocel_3197 was identified from the supernatant of cultures grown on all three types of biomass, and may perform as-yet-unknown functions that contribute to effective degradation of natural soft biomass.

Additional files

Additional file 1: The 372 proteins identified. Proteome analytes were injected to LC-MS/MS system. Collected data were used for protein identification by Proteome Discoverer software. Three independent biological experiments were performed, and proteins identified in every replicates with a number of used peptides per protein (≥3) were accepted. As a result, 372 proteins were successfully identified.

Additional file 2: Scatter plots of the three biological replicates for each substrate. The fold-change values of identified 372 proteins (Additional file 1) by using the Reporter Ions Quantifier with the TMT 6-plex method were normalized using global median. Scatter plots of normalized values were depicted using the data derived from three biological replicates of each culture (cellobiose, bagasse, corn germ, and rice straw). The values of Pearson's correlation were successfully high in each combination.

Additional file 3: Principal component analysis of the data from the three biological replicates. Principal component analysis was performed using normalized fold-change values of identified 372 proteins (Additional file 1) for investigation of the similarity of protein production profile between each biological replicate. Proteome data from each substrate clustered in close proximity. The cumulative contribution rate for principal component (PC) PC1 to PC3 was 87.5% (green: cellobiose; red: bagasse; purple: corn germ; blue: rice straw).

Abbreviations

ATCC: American type culture collection CAZy, carbohydrate-active enZymes; Cb: Cellobiose; CBM: Carbohydrate-binding modules; CE: Carbohydrate esterases; Cyspin: Cystein protease inhibitor; emPAI: Exponentially modified protein abundance index; ESI: Electrospray ionization; FDR: False discovery rate; GH: Glycoside hydrolases; GT: Glycosyl transferases; HCD: Higher-energy c-trap dissociation; HPLC: High performance liquid chromatography; ID: Internal diameter; LC-MS: Liquid chromatography-mass spectrometry; NCBI: National Center for Biotechnology Information; PCA: Principal component analysis; PL: Polysaccharide lyases; PMI: Phosphomannose isomerase; PASC: Phosphoric acid swollen cellulose; SDS: Sodium sodecyl sulfate; TMT: Tandem mass tag.

Competing interests

The authors declare that they have no competing interest.

Authors' contributions

KE and SA generated the strains used. KE, SA and HM performed most of the mass measurement. KE, SA and HM performed most of the statistical analyses. KE, HM, KK and MU designed the study and drafted the manuscript. All authors read and approved the final manuscript.

Acknowledgements

We thank Dr. Hiroshi Nonaka, Dr. Hideo Miyake (Mie University), and Tsuji Oil Mill Co. Ltd. for kindly providing soft biomass. We thank Mr. Daisuke Higo (Thermo Fisher Scientific) for valuable discussion for mass spectrometry. This research was supported by JST, CREST. This work was also partially supported by the Program for Promotion of Basic and Applied Researches for Innovations in Bio-oriented Industry (BRAIN).

References

Aspeborg H, Coutinho PM, Wang Y, Brumer H, Henrissat B (2012) Evolution, substrate specificity and subfamily classification of glycoside hydrolase family 5 (GH5). BMC Evol Biol 12(1):186

Bae J, Morisaka H, Kuroda K, Ueda M (2013) Cellulosome complexes: natural biocatalysts as arming microcompartments of enzymes. J Mol Microbiol Biotechnol 23(4–5):370–378

Bayer EA, Kenig R, Lamed R (1983) Adherence of *Clostridium thermocellum* to cellulose. J Bacteriol 156(2):818–827

Bayer EA, Belaich J-P, Shoham Y, Lamed R (2004) The cellulosomes: multienzyme machines for degradation of plant cell wall polysaccharides. Annu Rev Microbiol 58:521–554

Cid M, Pedersen HL, Kaneko S, Coutinho PM, Henrissat B, Willats WG, Boraston AB (2010) Recognition of the helical structure of β-1, 4-galactan by a new family of carbohydrate-binding modules. J Biol Chem 285(46):35999–36009

Desvaux M (2005) The cellulosome of *Clostridium cellulolyticum*. Enzyme Microb Technol 37(4):373–385

Doi RH, Kosugi A (2004) Cellulosomes: plant-cell-wall-degrading enzyme complexes. Nat Rev Microbiol 2(7):541–551

Fierobe HP, Bayer EA, Tardif C, Czjzek M, Mechaly A, Belaich A, Lamed R, Shoham Y, Belaich JP (2002) Degradation of cellulose substrates by cellulosome chimeras substrate targeting versus proximity of enzyme components. J Biol Chem 277(51):49621–49630

Furuya H, Ide Y, Hamamoto M, Asanuma N, Hino T (2010) Isolation of a novel bacterium, *Blautia glucerasei* sp. nov., hydrolyzing plant glucosylceramide to ceramide. Arch Microbiol 192(5):365–372

Gal L, Pages S, Gaudin C, Belaich A, Reverbel-Leroy C, Tardif C, Belaich J-P (1997) Characterization of the cellulolytic complex (cellulosome) produced by *Clostridium cellulolyticum*. Appl Environ Microbiol 63(3):903–909

Ishihama Y, Oda Y, Tabata T, Sato T, Nagasu T, Rappsilber J, Mann M (2005) Exponentially modified protein abundance index (emPAI) for estimation of absolute protein amount in proteomics by the number of sequenced peptides per protein. Mol Cell Proteomics 4(9):1265–1272

Kinseth MA, Anjard C, Fuller D, Guizzunti G, Loomis WF, Malhotra V (2007) The Golgi-associated protein GRASP is required for unconventional protein secretion during development. Cell 130(3):524–534

Kosugi A, Murashima K, Doi RH (2002) Xylanase and acetyl xylan esterase activities of XynA, a key subunit of the *Clostridium cellulovorans* cellulosome for xylan degradation. Appl Environ Microbiol 68(12):6399–6402

Lamed R, Setter E, Bayer E (1983a) Characterization of a cellulose-binding, cellulase-containing complex in *Clostridium thermocellum*. J Bacteriol 156(2):828–836

Lamed R, Setter E, Kenig R, Bayer E (1983b) The cellulosome: a discrete cell surface organelle of *Clostridium thermocellum* which exhibits separate antigenic, cellulose-binding and various cellulolytic activites. Biotechnol Bioeng Symp 13:163–181

Lombard V, Ramulu HG, Drula E, Coutinho PM, Henrissat B (2014) The carbohydrate-active enzymes database (CAZy) in 2013. Nucleic Acids Res 42(D1):D490–D495

López-Villar E, Monteoliva L, Larsen MR, Sachon E, Shabaz M, Pardo M, Pla J, Gil C, Roepstorff P, Nombela C (2006) Genetic and proteomic evidences support the localization of yeast enolase in the cell surface. Proteomics 6(S1):S107–S118

Lynd LR, Wyman CE, Gerngross TU (1999) Biocommodity engineering. Biotechnol Prog 15(5):777–793

Mansfield SD, Mooney C, Saddler JN (1999) Substrate and enzyme characteristics that limit cellulose hydrolysis. Biotechnol Prog 15(5):804–816

Matsui K, Bae J, Esaka K, Morisaka H, Kuroda K, Ueda M (2013) Exoproteome profiles of *Clostridium cellulovorans* grown on various carbon sources. Appl Environ Microbiol 79(21):6576–6584

Matsuoka S, Yukawa H, Inui M, Doi RH (2007) Synergistic interaction of *Clostridium cellulovorans* cellulosomal cellulases and HbpA. J Bacteriol 189(20):7190–7194

Meguro H, Morisaka H, Kuroda K, Miyake H, Tamaru Y, Ueda M (2011) Putative role of cellulosomal protease inhibitors in *Clostridium cellulovorans* based on gene expression and measurement of activities. J Bacteriol 193(19):5527–5530

Morisaka H, Matsui K, Tatsukami Y, Kuroda K, Miyake H, Tamaru Y, Ueda M (2012) Profile of native cellulosomal proteins of *Clostridium cellulovorans* adapted to various carbon sources. AMB Express 2(1):1–5

Nakanishi A, Bae J, Fukai K, Tokumoto N, Kuroda K, Ogawa J, Nakatani M, Shimizu S, Ueda M (2012) Effect of pretreatment of hydrothermally processed rice straw with laccase-displaying yeast on ethanol fermentation. Appl Microbiol Biotechnol 94(4):939–948

Nonaka H, Kobayashi A, Funaoka M (2013) Lignin isolated from steam-exploded eucalyptus wood chips by phase separation and its affinity to *Trichoderma reesei* cellulase. Bioresour Technol 140:431–434

Ota M, Sakuragi H, Morisaka H, Kuroda K, Miyake H, Tamaru Y, Ueda M (2013) Display of *Clostridium cellulovorans* xylose isomerase on the cell surface of *Saccharomyces cerevisiae* and its direct application to xylose fermentation. Biotechnol Prog 29(2):346–351

Ren H, Funaoka M (2009) Potential of herbaceous lignocellulosics as industrial raw materials. Trans Materials Res Soc Japan 34(4):727–730

Senoura T, Ito S, Taguchi H, Higa M, Hamada S, Matsui H, Ozawa T, Jin S, Watanabe J, Wasaki J (2011) New microbial mannan catabolic pathway that involves a novel mannosylglucose phosphorylase. Biochem Biophys Res Commun 408(4):701–706

Sleat R, Mah RA, Robinson R (1984) Isolation and characterization of an anaerobic, cellulolytic bacterium, *Clostridium cellulovorans* sp. nov. Appl Environ Microbiol 48(1):88–93

Tamaru Y (2001) The *Clostridium cellulovorans* cellulosome: an enzyme complex with plant cell wall degrading activity. Chem Rec 1(1):24–32

Tamaru Y, Miyake H, Kuroda K, Nakanishi A, Kawade Y, Yamamoto K, Uemura M, Fujita Y, Doi RH, Ueda M (2010) Genome sequence of the cellulosome-producing mesophilic organism *Clostridium cellulovorans* 743B. J Bacteriol 192(3):901–902

Tamaru Y, Miyake H, Kuroda K, Nakanishi A, Matsushima C, Ueda M (2011) Comparison of the mesophilic cellulosome-producing *Clostridium cellulovorans* genome with other cellulosome-related clostridial genomes. Microb Biotechnol 4(1):64–73

Yoswathana N, Phuriphipat P, Treyawutthiwat P, Eshtiaghi MN (2010) Bioethanol production from rice straw. Energy Res J 1(1):26

12

Feasibility study of an alkaline-based chemical treatment for the purification of polyhydroxybutyrate produced by a mixed enriched culture

Yang Jiang[1], Gizela Mikova[2], Robbert Kleerebezem[1], Luuk AM van der Wielen[1] and Maria C Cuellar[1*]

Abstract

This study focused on investigating the feasibility of purifying polyhydroxybutyrate (PHB) from mixed culture biomass by alkaline-based chemical treatment. The PHB-containing biomass was enriched on acetate under non-sterile conditions. Alkaline treatment (0.2 M NaOH) together with surfactant SDS (0.2 w/v% SDS) could reach 99% purity, with more than 90% recovery. The lost PHB could be mostly attributed to PHB hydrolysis during the alkaline treatment. PHB hydrolysis could be moderated by increasing the crystallinity of the PHB granules, for example, by biomass pretreatment (e.g. freezing or lyophilization) or by effective cell lysis (e.g. adjusting alkali concentration). The suitability of the purified PHB by alkaline treatment for polymer applications was evaluated by molecular weight and thermal stability. A solvent based purification method was also performed for comparison purposes. As result, PHB produced by mixed enriched cultures was found suitable for thermoplastic applications when purified by the solvent method. While the alkaline method resulted in purity, recovery yield and molecular weight comparable to values reported in literature for PHB produced by pure cultures, it was found unsuitable for thermoplastic applications. Given the potential low cost and favorable environmental impact of this method, it is expected that PHB purified by alkaline method may be suitable for other non-thermal polymer applications, and as a platform chemical.

Keywords: Polyhydroxybutyrate; Alkaline treatment; Crystallinity; Thermal stability; Mixed cultures

Introduction

Polyhydroxyalkanoates (PHAs) have received much attention as bio-based plastics that may contribute to future replacement of petroleum based plastics. Their performance ranges from stiff and brittle to soft and tough (Sudesh et al. 2000 and Laycock et al. 2013). The most common PHA is polyhydroxybutyrate (PHB), which has similar thermal and some mechanical properties (e.g. tensile strength) compared to isotactic polypropylene (Sudesh et al. 2000). In contrast to petroleum based plastics, PHA's biodegradability in various natural environments makes them suitable as disposables for packaging, agricultural or medical applications (Williams and Martin 2002, Bucci et al. 2005, Markets and Markets, 2013). The fact that more and more varieties of PHAs have been discovered and/or synthesized suggests that PHAs are not limited to thermoplastic applications. Moreover, PHA derivatives such as hydroxy fatty acid monomers may serve as chiral building blocks for the production of biochemicals and the methyl esters of their monomers could be used as a biofuel (Chen, 2009).

Chen (2009) summarized the current status of commercial PHA production. Many types of commercial PHAs are available on the market. For example, Polyhydroxybutyrate-co-hydroxyvalerate (PHBV) can be synthesized by pure culture of either *Ralstonia eutropha* or recombinant *E. coli* from glucose and propionic acid. Middle chain length PHAs, such as polyhydroxyhydroxyhexanoate (PHHx), can be produced by pure culture of *Pseudomonas putida*. Despite of the above mentioned advantages of PHAs compared to conventional petroleum based plastics their large scale application is still constrained by their high price in

* Correspondence: m.c.cuellar@tudelft.nl
[1]Department of Biotechnology, Delft University of Technology, Julianalaan 67, 2628 BC Delft, The Netherlands
Full list of author information is available at the end of the article

the market. Economic evaluations of the PHA production process identified the following cost drivers (Choi and Lee, 1997; van Wegen et al. 1998): (a) raw materials (fermentation feedstock), (b) downstream processes for product recovery and purification, and (c) costs associated to maintaining a pure culture during the fermentation (e.g. fermentor costs and energy required for sterilization). Several studies have integrated the PHA production process with wastewater treatment with a dynamic feast-famine enrichment system, aiming at intracellular PHB content up to 90% (Johnson et al. 2009), in order to reduce the cost from raw material and energy consumption aspects (reviewed by Dias et al. 2006). Recent results showed that such process is capable of producing PHAs as good as the current pure-culture process in terms of intracellular PHAs content and biomass specific PHAs production rates (Jiang et al. 2012). However, the challenge in terms of cost reduction in downstream process still remains.

PHAs are present in microorganisms as hydrophobic and water insoluble inclusion bodies which need to be separated from cell material. Plenty of techniques for PHA recovery and purification from pure cultures have been evaluated in literature and reviewed by Jacquel et al. (2008) and by Kunasundari and Sudesh (2011). The conventional organic solvent based purification method is still the best in terms of final product purity and recovery yield, although organic solvents may generate environmental issues (Ramsay et al. 1994; de Koning and Witholt, 1997). Several less toxic organic solvents have been reported for PHAs extraction (summarized in Jacquel et al. 2008; Kunasundari and Sudesh 2011; Riedel et al. 2013). Most of those solvents are specific for middle chain length PHAs purification, instead of short chain length PHAs (e.g. PHB, PHV) (Jiang et al. 2006; Elbahloul and Steinbüchel, 2009; Terada and Marchessault 1999). Nevertheless, short chain length PHAs are usually the main products when wastewater is used as feedstock (Dionisi et al. 2005; Bengtsson et al. 2008; Albuquerque et al. 2010; Jiang et al. 2012). Moreover, solvents such as 1, 2-proplene bicarbonate, require high temperature (>140°C) during the purification process, which typically leads to high energy consumption (Fiorese et al. 2009; Riedel et al. 2013).

Removal of cell materials by alkaline treatment was considered more economically feasible by Choi and Lee (1997, 1999) as compared to an organic solvent based PHA purification process. The alkaline treatment method has been widely reported in literature for pure cultures, resulting in purity and recovery yield as high as 98% and 97%, respectively (Choi and Lee 1999, Mohammadi et al. 2012a, b). An open culture process is based on the enrichment of a mixture of different microorganisms; it is unclear whether alkaline treatment can equally remove cell materials from microorganisms from enriched mixed

cultures. Furthermore, the chemicals used in the treatment could degrade the PHA granules, as well as negatively influence the thermal stability of PHAs during processing as thermoplastics (Kim et al. 2006).

The fate of PHAs during alkaline treatment and the thermal stability of the chemically treated PHA have hardly been reported. Moreover, only few studies have been published on recovery and purification of PHAs from mixed cultures (Serafim et al. 2008). In this study, the feasibility of the alkaline method for recovery and purification of PHB obtained from mixed cultures was evaluated. This study focused on the PHA degradation during the chemical treatment and on product properties such as molecular weight and thermal stability. PHB recovery and purification by extraction with dichloromethane was used for comparison purposes.

Material and methods
Biomass preparation and PHB recovery
The biomass used in this study was obtained from a 2 L sequencing batch reactor (SBR) fed with acetate under feast-famine condition. The composition of the working medium was: 125 mM $NaAc \cdot 3H_2O$, 3.93 mM NH_4Cl, 1.87 mM KH_2PO_4, 0.42 mM $MgSO_4 \cdot 7H_2O$, 0.54 mM KCl, 1.13 ml/L trace elements solution according to Vishniac and Santer (1957) and 3.71 mg/L allythiourea (to prevent nitrification). The operational conditions of the bioreactor were 30°C, pH 7, 1 day sludge retention time (SRT) and hydraulic retention time (HRT) and 18 h cycle length. The length of the feast phase was about 2.5 h during the steady state. The PHB was the sole storage polymer produced due to the fact that acetate was the sole carbon source. The biomass was collected at the end of the feast phase, when the cellular PHB content was between 62 wt% and 72 wt%. The dominant bacterial species in the SBR operated under such condition was P. acidivorans, a gram-negative bacterium (Jiang et al. 2011).

Fresh biomass from bioreactor was collected by centrifugation (Heraeus, Germany) at 10000 g for 10 min at room temperature. The supernatant was removed and the pellet was resuspended with Milli-Q water to reach a final biomass concentration of approximately 20 g/L. 10 mL of this biomass suspension was used for PHB recovery. Two types of chemicals were applied either solely or together to remove the cell materials: (1) alkalis (NaOH at concentrations varying between 0.02 M and 1 M, or 0.2 M NH_4OH), and (2) surfactant (SDS at concentrations varying between 0.025% and 0.2%). The biomass suspension with the added chemicals was incubated in 50 mL tubes at 200 rpm and 30°C for 1 hour unless otherwise stated. The suspension was subsequently centrifuged at 10000 g for 10 min at 4°C. The supernatant was separated from the pellet and collected for soluble polymer or monomer measurements. The pellet was washed twice with Milli-Q water and dried at 60°C overnight.

Besides fresh biomass, pre-treated biomass was also evaluated in this study. The fresh biomass pellet collected after centrifugation was subjected to either freezing at −20°C or lyophilization. The same chemical treatment procedures as for the fresh biomass were applied to the pre-treated biomass in order to study the influence of pre-treatment on the PHB recovery. Lyophilized biomass was additionally used for solvent extraction for comparison purposes. The PHB was firstly purified by dichloromethane, following the procedure described in Ramsay et al. (1994). Further purification was achieved by dissolving 1 wt% of PHB in chloroform at 60°C for 50 min. The chloroform sample was subsequently slowly poured into cold ethanol (10 times volume to chloroform) while stirring rigorously. The precipitate was filtered of the solution, washed with ethanol and vacuum dried at 50°C.

The setup of all the experiments in this study is summarized in the Table 1. All experiments were performed in at least duplicate.

Analytical methods

In order to evaluate the PHB mass balance of all experiments, the PHB quantity in fresh biomass, in final products and in the supernatant were determined. The PHB content in the biomass and in the final products was determined by gas chromatography (GC) according to the method described in Johnson et al. (2009). Commercial PHB (SigmaAldrich, the Netherlands) was used as standard. Based on the PHB mass present in the biomass ($PHA_{initial}$) and the dried pellet (PHA_{end}), the recovery yield was calculated by equation 1:

$$Recovery = \frac{PHA_{end}}{PHA_{initial}} \cdot 100\% \qquad [g/g] \qquad (1)$$

The PHB losses in the supernatant after chemical treatment ($PHB_{supernatant}$) was analyzed by gas chromatography (GC) with a modified procedure: 0.5 ml of the supernatant from chemical treatment was used for PHB concentration analysis. Commercial PHB mixed with 0.5 ml of chemical solution for PHB purification was used as standard. The remaining procedures were the

Table 1 Summary of all experiments conducted in this study

Chemical	Concentration	Biomass state	Time	Initial PHB content	Purity	Recovery	Mass balance	HB/PHB[c]
[–]	[M;w/v%]	[–]	[h]	[wt%]	[%]	[%]	[%]	[%]
NaOH	0.02	Fresh	1	71.8 ± 5.7	77.3 ± 4.0	92.2 ± 4.2	−1.0 ± 3.7	97.9 ± 5.3
NaOH	0.05	Fresh	1	72.7 ± 7.1	84.0 ± 1.4	94.7 ± 3.2	−3.9 ± 2.9	96.7 ± 4.2
NaOH	0.10	Fresh	1	65.3 ± 3.1	83.8 ± 4.4	98.0 ± 1.4	−1.0 ± 1.4	92.1 ± 8.5
NaOH	0.20	Fresh	1	69.4 ± 1.1	86.6 ± 3.0	96.7 ± 1.9	−0.7 ± 2.6	97.5 ± 16.7
NaOH	0.20	Fresh	0.3	68.6 ± 0.7	87.3 ± 2.2	96.4 ± 2.6	−2.9 ± 2.4	85.9 ± 14.8
NaOH	0.20	Fresh	0.5	68.6 ± 0.7	88.8 ± 0.8	98.5 ± 1.8	−0.7 ± 1.8	90.9 ± 18.9
NaOH	0.20	Fresh	3	68.6 ± 0.7	92.1 ± 0.8	93.5 ± 2.4	−1.5 ± 0.6	92.0 ± 3.2
NaOH	0.40	Fresh	1	65.3 ± 3.1	87.9 ± 5.4	95.2 ± 3.7	0.7 ± 3.1	89.9 ± 4.8
NaOH	0.70	Fresh	1	65.3 ± 3.1	89.7 ± 5.8	90.9 ± 5.0	0.4 ± 4.4	89.4 ± 7.1
NaOH	1.00	Fresh	1	65.3 ± 3.1	90.6 ± 4.7	85.6 ± 2.3	−0.3 ± 2.1	89.1 ± 4.0
NH₄OH	0.20	Fresh	1	68.6 ± 0.7	62.6 ± 2.8	63.3 ± 16.4	−3.9 ± 0.9	36.3 ± 10.9
SDS	0.20	Fresh	1	68.0 ± 0.0	79.0 ± 1.4	63.5 ± 0.7	3.6 ± 0.6	14.0 ± 1.4
NaOH + SDS	0.20 + 0.025	Fresh	1	66.1 ± 2.2	94.9 ± 2.6	92.6 ± 6.9	−2.7 ± 3.8	94.2 ± 6.4
NaOH + SDS	0.20 + 0.050	Fresh	1	66.1 ± 2.2	96.9 ± 1.3	93.5 ± 4.8	−2.4 ± 2.4	92.4 ± 3.1
NaOH + SDS	0.20 + 0.100	Fresh	1	66.1 ± 2.2	98.3 ± 0.5	91.5 ± 5.9	−3.9 ± 4.8	96.3 ± 4.8
NaOH + SDS	0.20 + 0.200	Fresh	1	66.1 ± 2.2	99.1 ± 0.5	91.0 ± 4.9	−3.1 ± 1.9	92.5 ± 5.0
NaOH	0.20	Freezing	1	65.9 ± 2.4	94.1 ± 3.5	95.6 ± 2.5	−2.9 ± 2.1	94.3 ± 5.4
NaOH	0.20	Freeze dried	1	69.9 ± 2.2	95.9 ± 3.7	95.5 ± 0.6	−3.2 ± 0.8	98.8 ± 0.9
NH₄OH	0.20	Freeze dried	1	69.9 ± 2.2	87.4 ± 2.1	95.0 ± 1.8	−3.9 ± 0.9	87.1 ± 12.9
SDS	0.20	Freeze dried	1	69.9 ± 2.2	93.5 ± 4.1	93.7 ± 2.2	−3.1 ± 1.6	91.3 ± 8.7
CH₂Cl₂	30[a]	Freeze dried	o/n[b]	72.2 ± 0.4	97.6	55.9	ND	ND

[a]30 times of TSS.
[b]Overnight.
[c]Fraction of hydrolyzed monomer in total polymer in the supernatant.

same as described in Johnson et al. (2009). The potential by-products of chemical treatment (e.g. hydrobutyric acid, HB and crotonic acid, CA) (Yu et al. 2005) were analyzed by high-performance liquid chromatography (HPLC) with a BioRad Animex HPX-87H column and a UV detector (Waters 484, 210 nm). The mobile phase, 1.5 mM H_3PO_4 in Milli-Q water, had a flow rate of 0.6 mL/min and a temperature of 59°C.

The overall mass balance was calculated by equation 2:

$$MassBalance = \frac{\left(PHA_{end} + PHB_{supernatant} + CA_{supernatant} - PHA_{initial}\right)}{PHA_{initial}}$$
$$\times 100\% \quad [g/g]$$
$$\tag{2}$$

where, $PHB_{supernatant}$ means the total PHB loss within the supernatant measured by GC and $CA_{supernatant}$ indicates the identified crotonic acid in the supernatant by HPLC. As a consequence, a closer value to 0% indicates a better mass balance. In this study, most of the experiments had mass balance errors smaller than 5% (see Table 1).

A degree of PHA degradation was defined as the fraction of HB or CA concentration over total initial PHB mass in the biomass (equation 3 or 4).

$$HB/PHB_{initial} = \frac{HB_{supernatant}}{PHA_{initial}} \cdot 100\% \quad [g/g] \tag{3}$$

or,

$$CA/PHB_{initial} = \frac{CA_{supernatant}}{PHA_{initial}} \cdot 100\% \quad [g/g] \tag{4}$$

Chemical PHB degradation may occur either randomly in the middle of the polymer chain or from the end of the polymer chain. The GC method measured the overall lost PHB in the supernatant in terms of both soluble PHB oligomers and HB monomer, while HPLC method only quantified the HB monomers. A ratio between soluble HB monomer and overall PHB in the supernatant was used to assess the chemical PHB degradation mechanism (equation 5). A higher value (close to 1) indicates that HB is sole product of PHB degradation, suggesting PHB is degraded from the end of the polymer chain. Otherwise, PHB is more likely hydrolyzed by chemicals randomly from the middle of the chain, generating oligomers as products.

$$HB/PHB_{supernatant} = \frac{HB_{supernatant}}{PHA_{supernatant}} \cdot 100\% \quad [g/g]$$
$$\tag{5}$$

Fourier transform infrared spectroscopy (FTIR)
The composition and the crystallinity of dry pellets were examined using a spectrum 100 FT-IR spectrometer

(PerkinElmer). The solid powders were pressed on a germanium crystal window of a microhorizontal ATR for measurement of single reflection and absorption of infrared by the specimens.

Thermal stability
Around 100 mg of an untreated biomass, PHB isolated from biomass by a chemical or an organic solvent treatment and/or a commercial PHB (Tianan, China) were isothermally treated in a compression molding machine (Dr Collins) at 170°C for a certain period of time (1, 3, 5, 10 and 15 min). The molecular weight of PHB before and after the thermal treatment was determined by a size exclusion chromatography (SEC). For SEC analysis, around 3 mg of a sample was dissolved in 1 ml hexafluoroisopropanol (HFIP) at room temperature overnight. The sample was subsequently filtered using 0.2 μm filter. Molar mass distribution was determined using a Waters model 510 pump and a Waters 712 WISP chromatograph with PL-gel mix D columns (300 × 7.5 mm, Polymer Laboratories). HFIP was used as an eluent with a flow rate of 1 ml/min. The system was calibrated with PMMA standards.

The thermal degradation rate can be expressed by the equation 6 (Grassie et al. 1984a, b):

$$\left(\frac{1}{Pn,t} - \frac{1}{Pn,0}\right) = k_D\, t \quad [1/s] \tag{6}$$

where, P_n,t and $P_{n,0}$ are number average degrees of polymerization at time t and time 0 s, respectively. The rate constant (k_D) was determined from the slope of the equation 6 function. P_n,t and $P_{n,0}$ were calculated using number average of molecular weight (M_n) in time t and time 0 s according to equations 7a and 7b.

$$P_{n,t} = \frac{M_{n,t}}{M_m} \quad [(g/mol)/(g/mol)] \tag{7a}$$

$$P_{n,0} = \frac{M_{n,0}}{M_m} \quad [(g/mol)/(g/mol)] \tag{7b}$$

M_m is the molecular weight of a PHB monomer unit, i.e. 86.09 g/mol.

Results
PHB recovery and purification
Alkalis and surfactant were two chemicals used in this study in order to purify and recover PHB from fresh biomass. Initially sole NaOH treatments with different concentration and treatment time were conducted (see Table 1). The final product purity increased by increasing NaOH concentration or by the prolonged treatment time, but the recovery yield was negatively influenced by those two parameters. On the basis of the final product purity and recovery yield, the treatment with 0.2 M NaOH for 1 h was chosen as the standard condition (see Table 1).

Under this standard condition, the final product purity and the recovery yield can reach 87% and 97%, respectively. In order to improve the purity from the standard condition, and to favor the sustainability of the process, different chemicals combinations were tested. With the purpose of improving the purity, surfactant was added to the standard condition to remove the cell materials further. With additional dosage of SDS to our standard condition, the purity can be improved up to 99% with a slight decrease in recovery yield (91%). NH_4OH was aimed at replacing NaOH, because it is potentially easier to be recycled than NaOH (van Hee et al. 2005). However, significant decrease was observed in both purity (to 63%) and recovery yield (63%) when treating fresh biomass with 0.2 M NH_4OH.

Besides recovering PHB from fresh biomass, the effect of pre-treatment such as lyophilization or freezing, was also studied. These pre-treatments led to a higher purity in all cases and an improved recovery yield in sole SDS and NH_4OH treatment (see Table 1). For comparison purposes, recovery and purification by solvent extraction was also conducted in this study. Extraction with dichloromethane reached 98% purity from lyophilized biomass. However, the recovery yield was very low (55%) in this study.

Thermal stability of purified PHB

In order to utilize PHAs as thermoplastics, thermal stability is a crucial parameter. Thermoplastic polymers are usually processed at temperatures at least 10°C above their melting point and typical residential time in an extruder does not exceed one minute. The processing temperature of PHB is usually between 170 and 180°C. Therefore, the thermal stability of the samples was studied in terms of PHB degradation during the first minute at 170°C.

Number average of molecular weight (M_n) of PHB as a function of time during the thermal treatment is shown in Table 2. PHB isolated from biomass by a solvent method and the commercial PHB showed the highest thermal stability with less than 7% M_n drop within the first minute of the treatment ($\Delta M_{n,1}$). The resulting molecular weight after the processing was still acceptable for a plastic application ($M_n > 169$ kg/mol). The sample purified by 0.2 M NaOH or by 0.2 M NaOH and 0.2% SDS showed much more pronounced molecular weight decrease ($\Delta M_{n,1} > 70\%$). The consequent molecular weights were below 45 kg/mol. As compared to the chemically purified PHB, the degradation of the polymer in the untreated biomass was less detrimental ($\Delta M_{n,1} = 62\%$). The rate of the polymer chain scission, i.e. the degradation rate constant (k_D), was calculated from the slope of the kinetic function shown in Figure 1. Thermal stability results at 170°C are summarized in Table 3, in terms of a ratio between k_D of a specific sample and k_D of the commercial PHB reference ($k_{D,ref}$). It can be observed that both, the untreated biomass and the chemically purified PHB showed significant deterioration in terms of a faster degradation rate. On the other side, the solvent isolated PHB performed even better than the commercial sample.

PHB degradation by alkalis

The thermal instability of PHB purified by alkalis based method could be due to PHB hydrolysis. As it has been reported in Yu et al. (2005), abiotic hydrolysis of PHB by alkalis was observed in this study as well. Both HB monomer and CA were found as PHB hydrolysis products. Our data showed that the PHB degradation by NaOH in the fresh biomass was dependent on the treatment time and NaOH concentration. The hydrolysis products concentration showed linear relation with NaOH treatment time (Figure 2), while the relation between the NaOH concentration and the hydrolyzed products concentration is non-linear (Figure 3). In the tested NaOH concentration range, the HB monomer decreased with the increasing NaOH concentration before 0.1 M NaOH and then increased with NaOH concentration. For an initial PHB content of 68%, at the standard condition in this study (i.e. 0.2 M NaOH treatment for 1 h with fresh biomass), about 1.3% of initial PHB was hydrolyzed into HB monomer and about 0.6% of initial PHB was converted to CA.

The pre-treatment step also showed some influence on the PHB hydrolysis. Much less HB or CA was produced after lyophilization or freezing pre-treatment (Figure 4).

The spectrum of hydrolysis products in the supernatant can be used as an indication of the chemical PHB degradation mechanism (equation 5). When the biomass with or without pre-treatment was treated by NaOH, $HB/PHB_{supernatant}$ ratio was always close to 100% (see Table 1). The closed mass balance in this study suggested that no other forms of soluble PHB oligomers were formed during NaOH treatment.

FTIR spectra

The effect of NaOH concentration and pre-treatment on PHB hydrolysis was investigated further by evaluating the crystallinity state of several samples through FTIR analysis (Xu et al. 2002; Yu and Chen 2006). An intensity ratio of the absorbance at 1230 cm^{-1} to that at 1453 cm^{-1} was used to calculate the polymer crystallinity index (CI, Xu et al. 2002). Larger CI value corresponds to higher crystallinity whilst smaller values reflect lower crystalline portion. As can be seen from Table 4, both chemical treatment and pre-treatment process show influence on PHB CI value. NH_4OH treated sample showed the lowest crystallinity compared to the rest of the samples.

Table 2 Molecular weight (number average M_n and weight average M_w) and molecular weight change $\left(\left(\frac{M_{n,0}-M_{n,t}}{M_{n,0}}\right) \times 100\right)$ of various PHB samples as a function of thermal treatment at 170°C.

Sample	Chemical treatment	PHB purity [wt.%]	Time of thermal treatment at 170°C [min]	M_n [kg/mol]	M_w [kg/mol]	$\left(\frac{M_{n,0}-M_{n,t}}{M_{n,0}}\right) \times 100$ [%]
Commercial PHB	-	99	0	182	647	0
			1	169	583	7
			3	175	541	4
			5	119	391	35
			10	150	435	18
			15	135	373	26
PHB from biomass	-	67	0	135	224	0
			1	51	111	62
			3	33	62	76
			5	30	42	78
			10	25	34	81
			15	19	25	86
	Solvent	99	0	915	1755	0
			1	883	1731	3
			3	824	1573	10
			5	771	1562	15
			10	516	1144	44
			15	560	1255	39
	0.2 M NaOH	85	0	119	315	0
			1	19	39	84
			3	8	13	93
			5	6	8	95
			10	2	3	98
			15	1.8	2.2	98
	0.2 M NaOH + 0.2% SDS	95	0	163	484	0
			1	45	73	72
			3	14	23	91
			5	11	20	93
			10	4	8	98
			15	3	4	98

Water content in the samples was in between 0.01 and 0.02 wt.%.

FTIR can also be used to qualitatively detect both PHB and proteins in the final products (Yu and Chen 2006). Therefore, all purified products were analyzed by FTIR, together with commercial PHB as control of PHB absorbance, and lyophilized biomass as a control of both PHB and protein absorbance. Figure 5a shows the spectrum of PHB from fresh biomass purified by different chemicals in comparison with commercial PHB and lyophilized biomass. The absorption at 1720 cm^{-1} and 1278 cm^{-1} respectively indicates C = O stretch and C-O stretch of the ester bonds. They both represent the presence of PHB. The absorption peaks at 1650 cm^{-1} and

1540 cm^{-1} represent amide I and amide II band in proteins. As can be seen in Figure 5, the commercial PHB and the PHB purified by NaOH-SDS mixture show highly similar spectra. In contrast, proteins were detected in all other samples.

Discussion
PHB recovery and purification
In this study, a high PHB purity was obtained from fresh biomass by treatment with alkali and surfactant. In principle, both alkali and surfactant can react with lipid and proteins, solubilizing the cell wall material

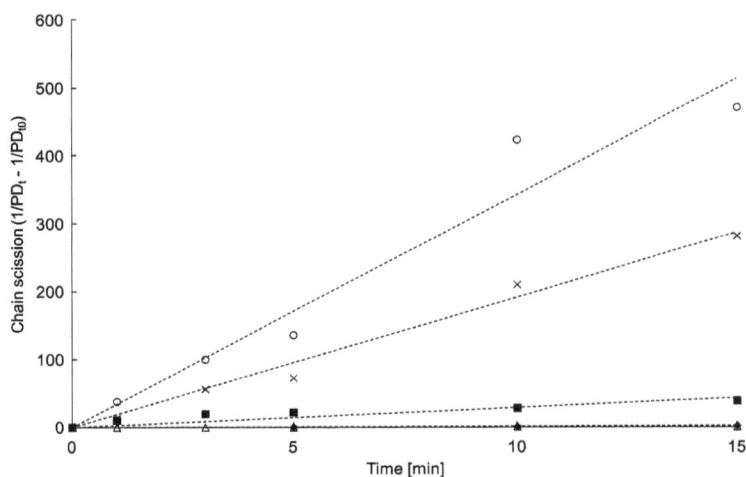

Figure 1 The effect of chemical treatment on thermal stability of commercial PHB (Tianan) and PHB isolated from biomass represented here by polymer chain scission ($1/P_{n,t} - 1/P_{n,0}$) as a function of time at 170°C. Water content in the samples was in between 0.01 and 0.02 wt%. The numbers in brackets represent PHB purity. PHB purified by 0.2 M NaOH (empty circle, 85% pure); PHB purified by 0.2 M NaOH and 0.2% SDS (cross, 95% pure); Unpurified PHB within biomass (solid square, 67% pure); PHB purified by solvent (empty triangle, 99% pure); Commercial PHB (solid diamond, 99% pure).

and releasing the intracellular contents. Our results suggest that sole NaOH treatment can lyse cells but it is insufficient to remove all cell materials. Under our standard condition (0.2 M NaOH, for 1 hour at 30°C), still about 13.4% of cell material impurities remained in the final pellets. Those remaining impurities are likely water insoluble proteins and lipids. We observed that those hydrophobic impurities can be effectively removed by combined NaOH and SDS treatment. Higher SDS concentrations resulted in a higher final purity, likely due to micelle formation by SDS. Once the SDS concentration approached its critical micelle concentration (CMC), which is between 0.17-0.23 w/v%, more proteins and lipids were removed. However, SDS micelles might also solubilize PHB granules. Indeed, our data showed that PHB recovery yield decreased at high SDS concentration (Table 1).

The hydroxide ion concentration was also observed to have influence on cell materials removal and PHB recovery. In the case of NH_4OH and low NaOH concentration, for example, both purity and recovery were observed to be lower than at the standard condition. Since NH_4OH is a weak base, at the same solution concentration the amount of dissociated hydroxide ion from NH_4OH is much lower than for NaOH (about 100 times less). In fact, samples treated by 0.2 M NH_4OH and 0.02 M NaOH displayed the lowest purity in this study (respectively 63% and 77%, see Table 1). Next to a decreased removal of cell materials, treatment at 0.2 M NH_4OH showed more severe PHB degradation, which resulted in a lower recovery yield. This may be related to the PHB granules crystallinity status, which is discussed in the next section.

Table 3 Thermal degradation rate constants (k_D) of various PHB samples at 170°C and thermal degradation rate constants relative to the commercial PHB reference ($k_{D,ref}$) as a function of chemical treatment, purification method and purity

Sample	Chemical treatment	PHB purity [wt.%]	k_D 10^{-6} [1/s]	$k_D/k_{D,ref}$ 10^{-6} [1/s]
Commercial PHB	-	99	0.18 ± 0.02*	1.0
	0.2 M NaOH	99	1.40 ± 0.10	8.0
	0.2% SDS	99	0.80 ± 0.10	4.0
PHB from biomass	-	67	5.40 ± 0.80	30.0
	Solvent	99	0.08 ± 0.01	0.4
	0.2 M NaOH	85	54.00 ± 5.00	300.0
	0.2 M NaOH + 0.2% SDS	95	29.00 ± 2.00	161.0

*k_D of dried commercial PHB was used as a reference ($k_{D,ref}$).
Water content in all samples was in between 0.01 – 0.02 wt%.

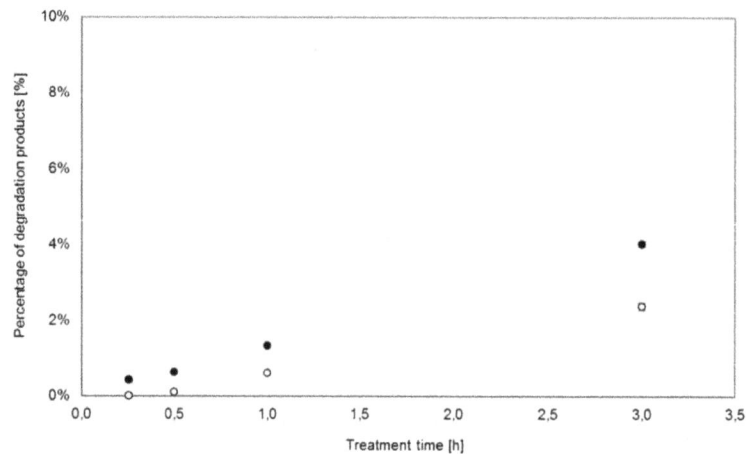

Figure 2 The relation between monomers production from PHB and NaOH treatment time. The fraction of two monomer products, hydroxybutyric acid (HB) and crotonic acid (CA) over total initial PHB (equations 3 and 4) are indicated by solid circle and empty diamond, respectively. The experiment was performed with fresh biomass at 0.2 M NaOH and 30°C in duplicate. Initial PHB content was 68%.

PHA recovery by chemical treatment has been widely reported in literature, but to our knowledge, on pure cultures only. The results are very diverse (Table 5). Considering the variability across studies in terms of microorganism, cell pre-treatment, temperature, initial PHB content and chemical concentration, among others, and their lack of PHB mass balance data, it is difficult to compare these results directly to our observations. Here we focus on the studies performed on fresh biomass, because at production scale it might be preferred to avoid any pre-treatment step.

Choi and Lee (1999) reported that direct treatment of fresh recombinant *E.coli* by 0.2 M NaOH can result in 97% purity and 91% recovery. This is the best result described for sole NaOH treatment method. The major difference between their research and our study is that pure culture of recombinant bacteria were used in their research in contrast to mixed culture in our study. It is possible that some microorganism species in the mixed culture biomass are not efficiently treated by NaOH. Anis et al. (2013), for example, treated wet biomass of recombinant *C. necator* by 0.1 M NaOH, resulting in final purity (84%) and recovery yield (91%) more similar to our observations.

Regarding studies with sole surfactant treatment, Kim et al. (2003) applied SDS to *Ralstonia eutropha* cells, but additional heating at 121°C and washing steps were required to remove proteins and achieve a final purity of 97%. Interestingly, their PHB recovery (>92%) was remarkably higher than our results (63%, see Table 1). This

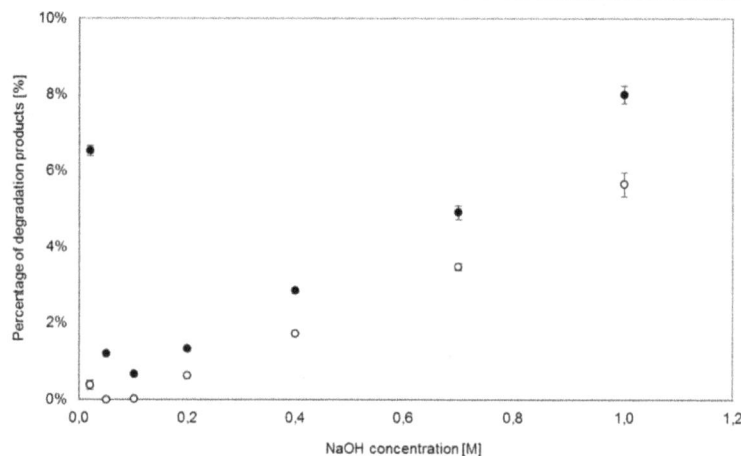

Figure 3 Fraction of degradation products, HB (solid round) and CA (empty diamond), over total initial PHB (equations 3 and 4). The experiment was performed with fresh biomass at 30°C for 1 hour in duplicate. Initial PHB content was 68%.

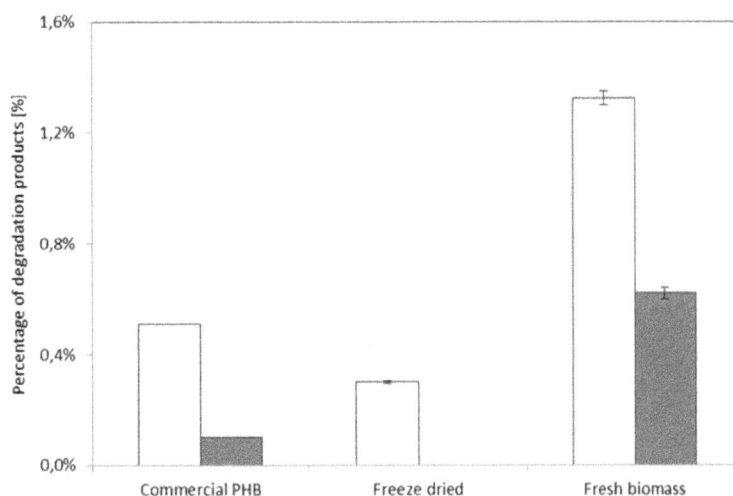

Figure 4 The influence of lyophilization on the PHB degradation by NaOH, expressed as fraction of degradation products over total initial PHB (equations 3 and 4). White color indicates HB and gray color represents CA. Samples were treated with 0.2 M NaOH for 1 hour at 30°C in duplicate. Initial PHB content was 68%.

suggests that temperature plays a significant role in the interaction between SDS and PHB – for example, due to altered critical micelle concentration (Bayrak 2003) – resulting in less PHB loss with the supernatant.

The synergistic effect of alkalis and surfactants on PHB recovery and purification has not been well studied yet. Peng et al. (2013) combined SDS and NaOH for PHB purification of dried cells, resulting in lower purity (87%) but comparable recovery yield (96%) as in our study (99% and 95%, respectively).

PHB degradation by alkalis

We observed that a weak alkaline condition, 0.2 M NH_4OH and NaOH at concentration lower than 0.1 M, resulted in a larger degree of PHB hydrolysis. On the other hand, cell pre-treatment by lyophilization improved the recovery yield (Table 1) and resulted in less

Table 4 Crystallinity index (CI = A_{1230}/A_{1453})

Biomass status	Chemicals	CI
Commercial PHB	-	4.7
Lyophilized	CH_2Cl_2	5.7
Lyophilized	SDS	4.7
Lyophilized	NaOH	4.5
Fresh biomass	NaOH + SDS	4.4
Lyophilized	NH_4OH	4.2
Fresh biomass	NaOH	3.8
Lyophilized	-	2.9
Fresh biomass	NH_4OH	2.1

Larger value means that PHB is at a more crystallinity status and smaller value means that PHB is at a more amorphous status.

HB and CA monomers formed when compared to fresh cells (Figure 4). This effect may be related to the crystalline state of PHB granules during treatment. In the microbial cell, PHB granules are present as hydrophobic amorphous inclusions containing 5–10% of water (Yu and Chen 2006). PHB granules at amorphous status are fragile to chemical hydrolysis. In fact, Yu and Chen (2006) and Valappil et al. (2007) suggested that PHB crystallization can increase PHB resistance to chemical treatment. PHB crystallization can be induced either by complete removal of water or by damaging the cell membrane (de Koning and Lemstra 1992), the crystallization extent being dependent on the damage level of the membrane (Kawaguchi and Doi 1990; Harrison et al. 1992). Our results confirm their observations. At weak alkaline condition and without pre-treatment, PHB in the biomass seems to maintain its amorphous status (Table 4).

PHB hydrolysis decreases the molecular weight of final products, the rate and extent of decrease being dependent on the degradation mechanism. In this study, most of the lost PHB in the supernatant could be traced back in terms of HB monomer. Furthermore, a linear relation between HB concentration and treatment time also suggested that PHB degradation occurs at the end of the polymer chain (Figure 2). This is in agreement with the observations from Yu et al. (2005) on PHB from pure cultures.

Thermal stability

Several studies have reported molecular weight and thermal properties as an indication of PHA quality for polymer applications, for PHAs obtained from pure cultures

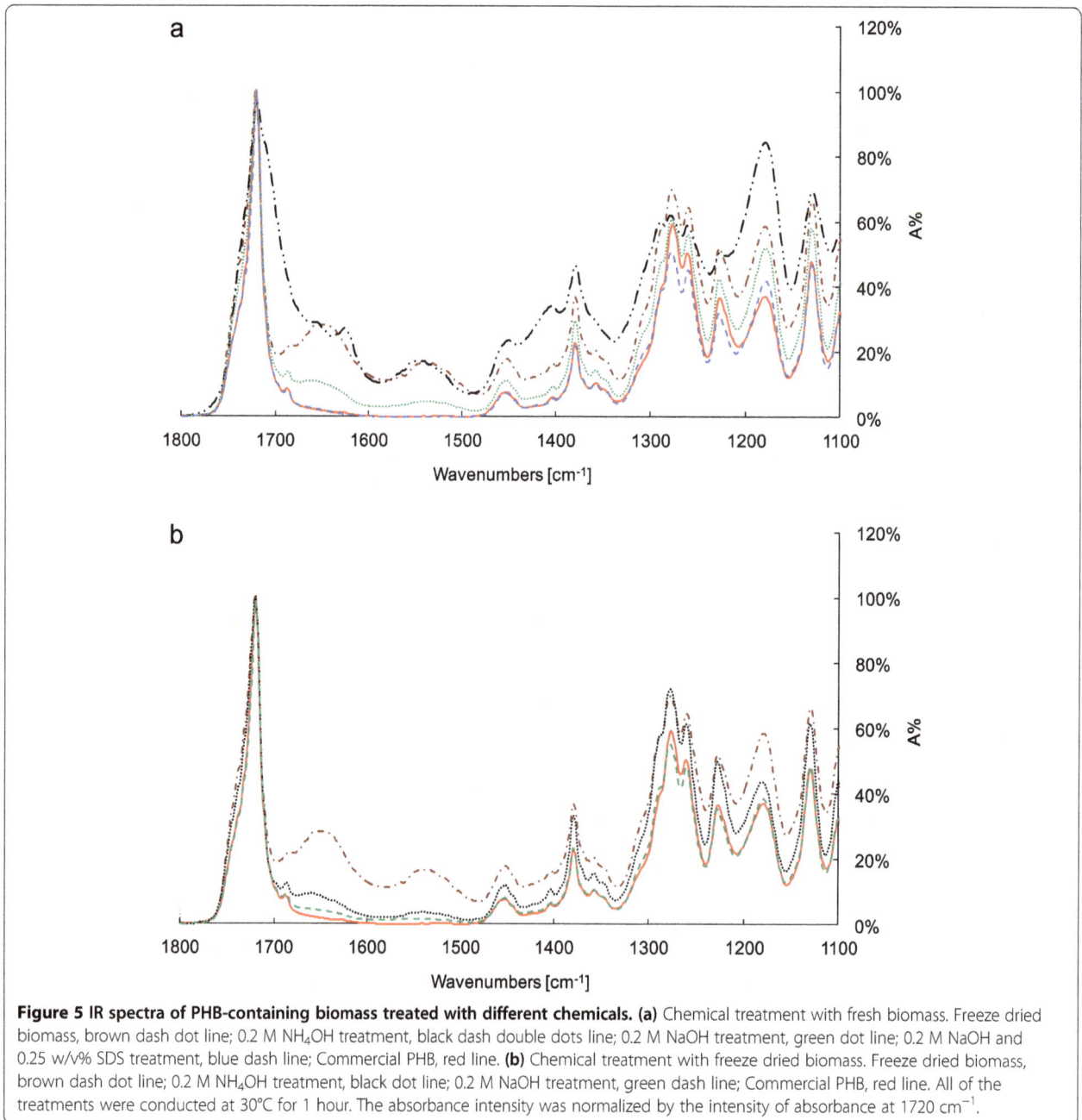

Figure 5 IR spectra of PHB-containing biomass treated with different chemicals. **(a)** Chemical treatment with fresh biomass. Freeze dried biomass, brown dash dot line; 0.2 M NH$_4$OH treatment, black dash double dots line; 0.2 M NaOH treatment, green dot line; 0.2 M NaOH and 0.25 w/v% SDS treatment, blue dash line; Commercial PHB, red line. **(b)** Chemical treatment with freeze dried biomass. Freeze dried biomass, brown dash dot line; 0.2 M NH$_4$OH treatment, black dot line; 0.2 M NaOH treatment, green dash line; Commercial PHB, red line. All of the treatments were conducted at 30°C for 1 hour. The absorbance intensity was normalized by the intensity of absorbance at 1720 cm^{-1}.

(e.g. Kim et al. 2003, Fiorese et al. 2009, Anis et al. 2012) and from mixed cultures (summarized by Laycock et al. 2013), and for several PHA recovery and purification methods. For thermoplastic applications, thermal stability is an important parameter. An instable polymer degrades during melt processing resulting in lower molecular weight material. At a certain critical molecular weight, mechanical properties substantially deteriorate. Kanesawa and Doi (1990) studied the effect of molecular weight on mechanical properties of PHBV copolymer. They reported that the tensile strength

started to deteriorate at M_n of 50 kg/mol and at around 20 kg/mol the sample had no strength anymore. Hablot et al. (2008) studied the effect of fermentation residues, surfactants and processing conditions on both the thermal properties and thermal degradation of PHB obtained from pure cultures by a solvent method. To our knowledge, our study provides the first data on thermal stability of PHB obtained from mixed cultures.

The sample purified by solvent showed very similar thermal stability as compared to the commercial PHB, suggesting that the quality of PHB produced by the

Table 5 Literature comparison

Bacteria species	Biomass status	Chemical	Concentration	Initial PHA content	Purity	Recovery	Reference
E. coli(rec)	Frozen	NaOH	0.1	77%	91%	90%	Choi and Lee (1999)
E. coli	Frozen	SDS	0.5%	77%	98%	87%	Choi and Lee (1999)
E. coli	Frozen	NH$_4$OH	0.1	77%	85%	95%	Choi and Lee (1999)
E. coli	Frozen	H$_2$SO$_4$	0.1	77%	79%	87%	Choi and Lee (1999)
C.necator	Lyophilized	NaOH	0.1	38%	97%	97%	(Mohammadi et al. 2012a, b)
C.necator	Lyophilized	NaOH	0.1	60%	80%	90%	Anis et al. (2012)
C.necator	Lyophilized	NH$_4$OH	0.1	60%	60%	62%	Anis et al. (2012)
Comamonas	Lyophilized	NaOH	0.05	34%	89%	97%	(Mohammadi et al. 2012a, b)
R.eutropha	Lyophilized	NaOH	N.D.	70%	78%	45%	Yang et al. (2011)
R.eutropha	Lyophilized	SDS	5%	70%	90%	81%	Yang et al. (2011)
R.eutropha	Lyophilized	SDS	1%	50%	87%	N.D.	Ramsay et al. (1990)
P.putida	Lyophilized	NaOH	0.1	20%	40%	95%	Jiang et al. (2006)
E. coli	Oven dried	NaOH + SDS	0.1 + 10%	60%	87%	96%	Peng et al. (2013)
R.eutropha	Fresh	SDS	0.5%-20%	75%	97%	92%	Kim et al. (2003)
R.eutropha	Fresh	H$_2$SO$_4$	1	60%	76%	94%	Yu and Chen (2006)
A.vinelandii	Fresh	NH$_3$	1	84%	94%	N.D.	Page and Cornish (1993)
C.nector	Fresh	NaOH	0.1	68%	84%	91%	Anis et al. (2013)
E. coli	Fresh	NaOH	0.2	79%	97%	91%	Choi and Lee (1999)
P.acidivorans*	Fresh	NaOH	0.2	68%	89%	97%	This study
P.acidivorans*	Fresh	NH$_4$OH	0.2	68%	65%	78%	This study
P.acidivorans*	Lyophilized	NH$_4$OH	0.2	68%	87%	96%	This study
P.acidivorans*	Fresh	NaOH + SDS	0.2 + 0.2%	68%	99%	95%	This study

*Dominant bacterial species in the mixed culture at the cultivation conditions of this study.

mixed microbial culture is comparable to PHB from pure cultures. On the other hand, PHB purified by chemical treatment showed severe thermal stability deterioration. By comparing the thermal degradation rate constants of several samples relative to the commercial PHB (Table 3), this effect could be attributed to 1) residues from the chemical treatment and 2) remaining biomass impurities. The inorganics used in the treatment could either attach to the polymer chain or stay as free molecules in the polymer. In both cases, they could catalyze a polymer chain scission either via β-elimination (Kim et al. 2006) or hydrolysis mechanism (Yu and Marchessault 2000, Yu et al. 2005). These results clearly show that the choice of recovery and purification method has a large impact on material properties.

In summary, this work studied the feasibility of purifying PHB from mixed culture biomass by alkaline-based chemical treatment. The purity and recovery obtained were comparable to those reported for pure cultures. PHB losses could be attributed to hydrolysis during the chemical treatment with HB monomer as main product, also in line with what has been observed for material from pure cultures. The extent of hydrolysis

can be moderated by increasing the crystallinity of the PHB granules; in this study, by either adjusting the alkali concentration, or by cell pretreatment.

The recovery and purification method had a large influence on the quality of the product for thermoplastic applications. PHB purified by solvent displayed thermal stability comparable to commercial PHB. However, PHB obtained by alkaline treatment resulted in significant thermal stability deterioration, despite of the high purity and recovery yield obtained. The quality of the product for thermoplastic applications might be improved by further optimizing the alkaline treatment process, targeting residual inorganics and biomass components. Given the potential advantages of the alkaline treatment in terms of process economics and environmental impact, it is expected that this method can be of interest for other PHB applications.

Competing interests
The authors declare that they have no competing interests.

Authors' contributions
YJ, GM and MC participated in the design of the experiments and analysis of results. YJ performed the experiments. GM performed the thermal stability

testing. YJ and GM wrote the manuscript. RK, LW, MC edited the manuscript. All authors read and approved the final manuscript.

Acknowledgements

These investigations were financially supported by the Technology Foundation STW (W2R, project nr. 11605). We thank Leonie Marang for kindly biomass supplies and Judith van Gorp and Lijing Xue for lab assistance.

Author details

[1]Department of Biotechnology, Delft University of Technology, Julianalaan 67, 2628 BC Delft, The Netherlands. [2]Polymer Technology Group Eindhoven BV, De Lismortel 31, 5612 AR Eindhoven, The Netherlands.

References

Albuquerque MGE, Torres CAV, Reis MAM (2010) Polyhydroxyalkanoate (PHA) production by a mixed microbial culture using sugar molasses: effect of the influent substrate concentration on culture selection. Water Res 44(11):3419–3433

Anis SNS, Iqbal NM, Kumar S, Amirul A-A (2013) Effect of different recovery strategies of P(3HB-co-3HHx) copolymer from Cupriavidus necator recombinant harboring the PHA synthase of Chromobacterium sp USM2. SepPurif Technol 102:111–117

Anis SNS, Nurhezreen MI, Sudesh K, Amirul AA (2012) Enhanced recovery and purification of P(3HB-co-3HHx) from recombinant Cupriavidus necator using alkaline digestion method. Appl Biochem Biotech 167(3):524–535

Bayrak Y (2003) Micelle formation in sodium dodecyl sulfate and dodecyltrimethylammonium bromide at different temperatures. Turk J Chem 27(4):487–492

Bengtsson S, Werker A, Christensson M, Welander T (2008) Production of polyhydroxyalkanoates by activated sludge treating a paper mill wastewater. Bioresource Technol 99(3):509–516

Bucci DZ, Tavares LBB, Sell I (2005) PHB packaging for the storage of food products. Polym Test 24(5):564–571

Chen G-Q (2009) A microbial polyhydroxyalkanoates (PHA) based bio- and materials industry. Chem Soc Rev 38(8):2434–2446

Choi JI, Lee SY (1997) Process analysis and economic evaluation for poly(3-hydroxybutyrate) production by fermentation. Bioprocess Eng 17(6):335–342

Choi JI, Lee SY (1999) Efficient and economical recovery of poly(3-hydroxybutyrate) from recombinant Escherichia coli by simple digestion with chemicals. Biotechnol Bioeng 62(5):546–553

de Koning GJM, Lemstra PJ (1992) The amorphous state of bacterial Poly[(R)-3-Hydroxyalkanoate] Invivo. Polymer 33(15):3292–3294

de Koning GJM, Witholt B (1997) A process for the recovery of poly(hydroxyalkanoates) from Pseudomonads Part 1: solubilization. Bioprocess Eng 17(1):7–13

Dias JML, Lemos PC, Serafim LS, Oliveira C, Eiroa M, Albuquerque MGE, Ramos AM, Oliveira R, Reis MAM (2006) Recent advances in polyhydroxyalkanoate production by mixed aerobic cultures: from the substrate to the final product. Macromol Biosci 6(11):885–906

Dionisi D, Carucci G, Papini MP, Riccardi C, Majone M, Carrasco F (2005) Olive oil mill effluents as a feedstock for production of biodegradable polymers. Water Res 39(10):2076–2084

Elbahloul Y, Steinbuechel A (2009) Large-scale production of Poly(3-Hydroxyoctanoic Acid) by Pseudomonas putida GPo1 and a simplified downstream process. Environ Microb 75(3):643–651

Fiorese ML, Freitas F, Pais J, Ramos AM, de Aragao GMF, Reis MAM (2009) Recovery of polyhydroxybutyrate (PHB) from Cupriavidus necator biomass by solvent extraction with 1,2-propylene carbonate. Eng Life Sci 9(6):454–461

Grassie N, Murray EJ, Holmes PA (1984a) The thermal degradation of poly(−(D)-β-hydroxybutyric acid): Part 1—identification and quantitative analysis of products. Polym Degrad Stabil 6(1):47–61

Grassie N, Murray EJ, Holmes PA (1984b) The thermal degradation of poly(−(D)-β-hydroxybutyric acid): Part 2—changes in molecular weight. Polym Degrad Stabil 6(2):95–103

Hablot E, Bordes P, Pollet E, Avérous L (2008) Thermal and thermo-mechanical degradation of poly(3-hydroxybutyrate)-based multiphase systems. Polym Degrad Stabil 93(2):413–421

Harrison STL, Chase HA, Amor SR, Bonthrone KM, Sanders JKM (1992) Plasticization of Poly(Hydroxybutyrate) Invivo. Int J Biol Macromol 14(1):50–56

Jacquel N, Lo C-W, Wei Y-H, Wu H-S, Wang SS (2008) Isolation and purification of bacterial poly (3-hydroxyalkanoates). Biochem Eng J 39(1):15–27

Jiang X, Ramsay JA, Ramsay BA (2006) Acetone extraction of mcl-PHA from Pseudomonas putida KT2440. J Microbiol Meth 67(2):212–219

Jiang Y, Marang L, Tamis J, van Loosdrecht MCM, Dijkman H, Kleerebezem R (2012) Waste to resource: converting paper mill wastewater to bioplastic. Water Res 46(17):5517–5530

Jiang Y, Sorokin DY, Kleerebezem R, Muyzer G, van Loosdrecht M (2011) Plasticicumulans acidivorans gen. nov., sp nov., a polyhydroxyalkanoate-accumulating gammaproteobacterium from a sequencing-batch bioreactor. Int J Syst Evol Micr 61:2314–2319

Johnson K, Jiang Y, Kleerebezem R, Muyzer G, Van Loosdrecht MCM (2009) Enrichment of a mixed bacterial culture with a high polyhydroxyalkanoate storage capacity. Biomacromolecules 10(4):670–676

Kanesawa Y, Doi Y (1990) Hydrolytic degradation of microbial poly(3-hydroxybutyrate-co-3-hydroxyvalerate) fibers. Macromol Chem Rapid Commun 11(12):679–682

Kawaguchi Y, Doi Y (1990) Structure of native Poly(3-Hydroxybutyrate) granules characterized by X-ray-diffraction. FEMS Microbiol Lett 70(2):151–156

Kim KJ, Doi Y, Abe H (2006) Effects of residual metal compounds and chain-end structure on thermal degradation of poly(3-hydroxybutyric acid). Polym Degrad Stabil 91(4):769–777

Kim M, Cho KS, Ryu HW, Lee EG, Chang YK (2003) Recovery of poly(3-hydroxybutyrate) from high cell density culture of Ralstonia eutropha by direct addition of sodium dodecyl sulfate. Biotechnol Lett 25:55–59

Kunasundari B, Sudesh K (2011) Isolation and recovery of microbial polyhydroxyalkanoates. Express Polym Lett 5(7):620–634

Laycock B, Halley P, Pratt S, Werker A, Lant P (2013) The chemomechanical properties of microbial polyhydroxyalkanoates. Prog Polym Sci 38(3–4):536–583

Markets and Markets (2013) Polyhydroxyalkanoate (PHA) Market, By Application (Packaging, Food Services, Bio-medical, Agriculture) & Raw Material — Global Trends & Forecasts to 2018. Report code:CH1610

Mohammadi M, Hassan MA, Phang L-Y, Shirai Y, Man HC, Ariffin H, Amirul AA, Syairah SN (2012a) Efficient Polyhydroxyalkanoate recovery from recombinant Cupriavidus necator by using low concentration of NaOH. Environ Eng Sci 29(8):783–789

Mohammadi M, Hassan MA, Shirai Y, Man HC, Ariffin H, Yee L-N, Mumtaz T, Chong M-L, Phang L-Y (2012b) Separation and purification of Polyhydroxyalkanoates from newly isolated Comamonas sp EB172 by simple digestion with sodium hydroxide. Separ Sci Technol 47(3):534–541

Page WJ, Cornish A (1993) Growth of Azotobacter-Vinelandii Uwd in fish peptone medium and simplified extraction of poly-beta-hydroxybutyrate. Appl Environl Microb 59(12):4236–4244

Peng Y-C, Lo C-W, Wu H-S (2013) The isolation of poly(3-hydroxybutyrate) from recombinant Escherichia coli XL1-blue using the digestion method. Can J Chem Eng 91(1):77–83

Ramsay JA, Berger E, Ramsay BA, Chavarie C (1990) Recovery of poly-3-hydroxyalkanoic acid granules by a surfactant-hypochlorite treatment. Biotechnol Tech 4(4):221–226

Ramsay JA, Berger E, Voyer R, Chavarie C, Ramsay BA (1994) Extraction of poly-3-hydroxybutyrate using chlorinated solvents. Biotechnol Tech 8(8):589–594

Riedel SL, Brigham CJ, Budde CF, Bader J, Rha C, Stahl U, Sinskey AJ (2013) Recovery of poly(3-hydroxybutyrate-co-3-hydroxyhexanoate) from Ralstonia eutropha cultures with non-halogenated solvents. Biotechnol Bioeng 110(2):461–470

Serafim LS, Lemos PC, Torres C, Reis MAM, Ramos AM (2008) The influence of process parameters on the characteristics of polyhydroxyalkanoates produced by mixed cultures. Macromol Biosci 8(4):355–366

Sudesh K, Abe H, Doi Y (2000) Synthesis, structure and properties of polyhydroxyalkanoates: biological polyesters. Prog Polym Sci 25(10):1503–1555

Terada M, Marchessault RH (1999) Determination of solubility parameters for poly (3-hydroxyalkanoates). Int J Biol Marcromol 25:207–215

Valappil SP, Misra SK, Boccaccini AR, Keshavarz I, Bucke C, Roy I (2007) Large-scale production and efficient recovery of PHB with desirable material properties, from the newly characterised Bacillus cereus SPV. J Biotechnol 132(3):251–258

Van Hee P, van der Wielen LAM, van der Lans RGJM (2005) Method for the Production of a Fermentation Product from an Organism, Patent application US2005158817 (A1)

van Wegen RJ, Ling Y, Middelberg APJ (1998) Industrial production of polyhydroxyalkanoates using Escherichia coli: an economic analysis. Chem Eng Res Des 76(A3):417–426

Vishniac W, Santer M (1957) Thiobacilli. Bacteriol Rev 21(3):195–213

Williams SF, Martin DP (2002) Applications of PHAs in Medicine and Pharmacy. A Chapter in Biopolymers, vol 4, Polyesters III - Applications and Commercial Products, John Wiley and Sons., pp 91–127

Xu J, Guo BH, Yang R, Wu Q, Chen GQ, Zhang ZM (2002) In situ FTIR study on melting and crystallization of polyhydroxyalkanoates. Polymer 43(25):6893–6899

Yang Y-H, Brigham C, Willis L, Rha C, Sinskey A (2011) Improved detergent-based recovery of polyhydroxyalkanoates (PHAs). Biotechnol Lett 33(5):937–942

Yu J, Chen LXL (2006) Cost-effective recovery and purification of polyhydroxyalkanoates by selective dissolution of cell mass. Biotechnol Progr 22(2):547–553

Yu J, Plackett D, Chen LXL (2005) Kinetics and mechanism of the monomeric products from abiotic hydrolysis of poly[(R)-3-hydroxybutyrate] under acidic and alkaline conditions. Polym Degrad Stabil 89(2):289–299

Yu G, Marchessault RH (2000) Characterization of low molecular weight poly(β-hydroxybutyrate)s from alkaline and acid hydrolysis. Polymer 41(3):1087–1098

13

Characterization of *Aspergillus aculeatus* β-glucosidase 1 accelerating cellulose hydrolysis with *Trichoderma* cellulase system

Yutaro Baba, Jun-ichi Sumitani[*], Shuji Tani and Takashi Kawaguchi

Abstract

Aspergillus aculeatus β-glucosidase 1 (AaBGL1), which promotes cellulose hydrolysis by *Trichoderma* cellulase system, was characterized and compared some properties to a commercially supplied orthologue in *A. niger* (AnBGL) to elucidate advantages of recombinant AaBGL1 (rAaBGL1) for synergistic effect on *Trichoderma* enzymes. Steady–state kinetic studies revealed that rAaBGL1 showed high catalytic efficiency towards β-linked glucooligosaccharides. Up to a degree of polymerization (DP) 3, rAaBGL1 prefered to hydrolyze β-1,3 linked glucooligosaccharides, but longer than DP 3, preferred β-1,4 glucooligosaccharides (up to DP 5). This result suggested that there were different formation for subsites in the catalytic cleft of AaBGL1 between β-1,3 and β-1,4 glucooligosaccharides, therefore rAaBGL1 preferred short chain of laminarioligosaccharides and long chain of cellooligosaccharides on hydrolysis. rAaBGL1 was more insensitive to glucose inhibition and more efficient to hydrolyze the one of major transglycosylation product, gentiobiose than AnBGL, resulting that rAaBGL1 completely hydrolyzed 5% cellobiose to glucose faster than AnBGL. These data indicate that AaBGL1 is valuable for the use of cellulosic biomass conversion.

Keywords: *Aspergillus aculeatus*; β-glucosidase; Biomass conversion; *Trichoderma reesei*; Cellulase

Introduction

Cellulosic biomass has been most abundant biomass that widely distributed on earth. Cellulose is degraded into monomeric sugar by cellulases, and is appropriate source of biofuels and biochemicals production. However, high crystallinity and insolubility of cellulose makes difficult to degrade to soluble sugar, such as glucose. Cellulose degradation is accomplished by the synergistic action among endoglucanases (E.C. 3.2.1.4), cellobiohydrolases (E.C. 3.2.1.91), and β-glucosidases (BGLs, E.C. 3.2.1.21, Woodward 1991). However it has been known that cellobiohydrolases and BGLs are significantly inhibited by the hydrolysis products such as cellobiose and glucose, and that the product inhibition reduces the overall rate of cellulose hydrolysis (Andric et al. 2010; Xiao et al. 2004). BGLs hydrolyze β-glucosidic bonds to release glucose units from the non-reducing end of β-glucoologosaccharides or glucosides. BGLs are classified in the glycoside hydrolase (GH) family 1, 3, 5, 9, 30, and 116 in the CAZy database (Henrissat 1991; Henrissat and Bairoch 1993,1996; URL: http://www.cazy.org/). In the fungal cellulase system, BGL mainly hydrolyze cellooligosaccharides to glucose on the final step of cellulose degradation. Thus the BGL having high hydrolytic activity is required to forestall the product inhibition by cellobiose against cellobiohydrolases (Du et al. 2010). It has been known that the cellulase mixture secreted by the filamentous fungus *Trichoderma reesei*, which is used for industrial application, has very low activity of BGL, and this problem has been tried to solve by addition of exogenous BGL, for example from *A. niger* (Berlin et al. 2007; Chauve et al. 2010; Dekker 1986; Singhania et al. 2013). Since the BGL is important for cellulase system, the BGL having more powerful activity for hydrolysis and accelerating cellulase system of *T. reesei* is required to promote saccharification of the cellulosic biomass.

Aspergillus aculeatus no. F-50 [NBRC 108796] was isolated in soil whose cellulose- and hemicellulose-degrading enzymes effectively hydrolyzed pulp in combination with

* Correspondence: monger@biochem.osakafu-u.ac.jp

Graduate School of Life and Environmental Sciences, Osaka Prefecture University, 1-1 Gakuen-cho, Naka-ku, Sakai, Osaka 599-8531, Japan

T. reesei (Murao et al. 1979). AaBGL1, a dominant BGL in the culture supernatant of *A. aculeatus* no F-50, was purified and characterized (Sakamoto et al. 1985a,b), and its cDNA was cloned and sequenced (Kawaguchi et al. 1996). AaBGL1 has unique features for hydrolysis of cellooligosaccharides in terms of not only showing high specific activity for cellooligosaccharides with increasing degree of polymerization (DP) up to 5, but also being detected no transglycosylation products on hydrolysis of cellobiose and insoluble cellooligosaccharides by paper chromatography (Sakamoto et al. 1985b; unpublished data). Moreover, Nakazawa et al. reported that *T. reesei* expressing AaBGL1 gene under the control of *xyn3* promoter (X3AB1 strain) exhibited 63- and 25-fold higher BGL activity than that of PC-3-7 strain and X3TB1 strain which expressed *T. reesei* BGL I gene under the control of *xyn3* promoter respectively (Nakazawa et al. 2011). In addition, JN11, which is the crude cellulase preparation from X3AB1, released more glucose than commercially available cellulases, Accellerase 1500 and Cellic CTec from various pretreated biomasses if those have rich hemicelluloses (e.g. NaOH pretreatment; Kawai et al. 2012). Kawai et al. mentioned for these results that JN11 has the best balance of BGL and hemicellulase activities for the degradation of cellulosic biomasses. As described above, AaBGL1 is a useful BGL for biomass conversion. However, AaBGL1 has not been investigated the enzymatic analysis in detail.

There are many fungal BGLs which share high similarity of amino acid sequence with AaBGL1 in the GH family 3. Nevertheless, the reason why AaBGL1 was selected for use in cellulosic biomass conversion, was unclear due to partial investigation of the detailed enzymatic properties. Here we demonstrated the availability of rAaBGL1 by characterization of rAaBGL1, especially substrate specificity and transglycosylation products, and comparing capability to hydrolyze cellobiose with BGL from *A. niger* (AnBGL) which shares high amino acid sequence similarity with AaBGL1.

Materials and methods

Strains and medium

Escherichia coli DH5αF' was used as a host for construction of recombinant plasmids. *A. oryzae* niaD300 strain was used as a host for AaBGL1 gene expression.

A. oryzae transfomant was cultivated at 30°C with shaking (160 rpm) in Erlenmeyer flasks in the minimal medium (MM) with 5% glucose, 1.5% NaNO3, 0.00008% Mo7O24•2H2O, 0.00111% H3BO3, 0.00016% CoCl•6H2O, 0.00016% CuSO4•5H2O, 0.005% EDTA•2Na, 0.0005% FeSO4•7H2O, 0.0005% MnCl2•4H2O, 0.0022% ZnSO4•7H2O, 0.00013% KCl, 0.00013% MgSO4•7H2O, 0.00038% KH2PO4 (pH 6.5) for appropriate days.

Expression of AaBGL1 gene in *A. oryzae* and purification of recombinant AaBGL1

The AaBGL1 gene was amplified by PCR using *A. aculeatus* genomic DNA as a template, with 5'- aactgcaggcggc cgcatcatgaagctcagttggcttg-3' as a sense primer and 5'-aagc atgctcattgcaccttcgggagc-3' as an antisense primer. PCR condition is the following thermal settings: 30 cycles of 10 s initial denaturation step at 98°C, followed by 5 s annealing step at 55°C, and 3 min of extension step at 72°C using PrimeSTAR HS DNA polymerase (TaKaRa, Japan). The PCR product was digested by *Not* I and *Sph* I, and inserted into the same sites of *Aspergillus* expression vector, pNAN8142 (Minetoki et al. 2003). The resultant plasmid was named as pNPN-AaBGL1. *A. oryzae* niaD300 strain was transformed with pNAN-AaBGL1 by the protoplast-PEG method (Gomi et al. 1987; Kanamasa et al. 2001). Several transformants was isolated and cultivated to confirm the production of rAaBGL1. The methods of the expression and the purification of rAaBGL1 is previously described (Suzuki et al. 2013). Briefly describing below; *A. oryzae* transformant overexpressing AaBGL1 gene was cultivated in 2.4 L MM liquid medium for 3 days, and the mycelia were harvested and washed with 5 volume 20 mM sodium acetate buffer (pH 5.0). To release rAaBGL1 from cell surface, the mycelia were incubated at 30°C for 2 days in 2.4 L releasing buffer (10 mg/ml cycloheximide, 1 mM benzylsulfonyl fluoride, 0.02% sodium azide in 20 mM sodium acetate buffer (pH 5.0)) with shaking. After releasing AaBGL1 from cell surface, supernatant was obtained by filtration as a crude enzyme. The crude enzyme was applied to a DEAE-TOYOPEARL® 650 M column equilibrated with 20 mM sodium acetate buffer (pH 5.0). rAaBGL1 was eluted with a linear gradient of NaCl (0–0.3 M). The active fractions were collected, added to ammonium sulfate at 30% saturation, and subjected to a Butyl-TOYOPEARL® 650 M column equilibrated with 30% saturation of ammonium sulfate in same buffer. The rAaBGL1 was eluted with a reverse linear gradient of ammonium sulfate (30–0% saturation). After collecting rAaBGL1 containing fractions, the enzyme was precipitated with ammonium sulfate at 80% saturation, dissolved in 20 mM sodium acetate buffer (pH 5.0), and dialyzed in the same buffer. Homogeneity of rAaBGL1 was confirmed by sodium dodecyl sulfate-polyacrylamide gel electrophoresis (SDS-PAGE).

Purification of BGL from *A. niger*

Glucosidase from *A. niger* (SIGMA-ALDRICH, Co.) was dissolved in 5 ml of 20 mM sodium acetate buffer (pH 5.0). After dialyzing in the same buffer, the same steps of purification as those of rAaBGL1 were performed (Suzuki et al. 2013), followed by gel filtration on HiLaod™ 16/60 Superdex™ 200 pg column (GE healthcare) equilibrated with 20 mM sodium acetate buffer (pH 5.0), and hydrophobic interaction chromatography on Hiprep™ 16/10 Phenyl FF column (low sub; GE healthcare) with reverse linear gradient of ammonium sulfate (30–0% saturation).

Protein assay

Protein concentration was determined from the absorbance at 280 nm using extinction coefficient (ε) as $162,000$ $M^{-1} \cdot cm^{-1}$ with the exception of the comparison of saccharification between rAaBGL1 and AnBGL.

According to the comparison of BGL ability from *A. acuelatus* and *A. niger*, the protein concentration was determined with the Bio-Rad Protein Assay, based on the method of Bradford (Bio-Rad Laboratories). Bovine γ-globulin was used as a standard.

Enzyme assays

Enzymatic reaction was performed by incubating 100 μl enzyme with 100 μl of 3 mM *p*-nitrophenyl (pNP)-monosaccharides in 100 mM sodium acetate buffer (pH 5.0). Reaction was stopped by adding 2 ml of 1 M Na_2CO_3. Released *p*-nitrophenol was quantified by measuring the absorbance at 405 nm using ε as 18.5 $mM^{-1} \cdot cm^{-1}$. One unit of BGL activity was defined as the amount of enzyme required for the release of 1 μmol of *p*-nitrophenol per minute from the substrate.

Effect of temperature and pH on the activity and stability of rAaBGL1

The optimum temperature was determined by incubating rAaBGL1 (6.13 nM) with 1.5 mM *p*-nitrophenyl-β-D-glucopyranoside (pNP-Glc) in 20 mM sodium acetate buffer (pH 5.0) at various temperature (30–70°C) for 10 min. The optimum pH was determined by incubating rAaBGL1 (6.13 nM) with 1.5 mM pNP-Glc in various pH range of buffer (3.3–6.3, sodium citrate buffer; 6.5–7.3, sodium phosphate buffer; 7.0–8.9, Tris–HCl buffer; 8.8–10.7, glycine-NaOH buffer) at 37°C for 10 min. The thermal stability was determined by incubating rAaBGL1 (6.13 nM) at various temperature (30–70°C) for 30 min. After incubation, sample was on ice for 5 min, followed by incubating with 1.5 mM pNP-Glc in 20 mM sodium acetate buffer (pH 5.0) at 37°C for 10 min. The pH stability was determined by incubating rAaBGL1 (61.3 nM) in 100 mM various pH range of buffer (1.8–3.2, glycine-HCl buffer; 3.4–6.0, sodium acetate buffer; 6.5–7.3, sodium phosphate buffer; 7.2–9.1, Tris–HCl buffer; 8.8–10.5, glycine-NaOH buffer) at room temperature for 1 h. After incubation, sample was diluted 10 fold by adding 200 mM sodium acetate buffer (pH 5.0), followed by incubation with 1.5 mM pNP-Glc in 20 mM sodium acetate buffer (pH 5.0) at 37°C for 10 min.

Inhibition by glucose

For determination of inhibition constant for glucose on AaBGL1, enzymatic reaction was performed by incubation of rAaBGL1 with 0.1, 0.2, 0.3 and 0.4 mM pNP-Glc in the presence of 1.0, 2.5, 5.0, 10.0, 20.0 and 40.0 mM glucose in 20 mM sodium acetate buffer (pH 5.0) at 37°C. Initial rate of released *p*-nitrophenol were measured, and then, K_i value for glucose on AaBGL1 was calculated with Dixon plot.

For the effect of inhibition by glucose for rAaBGL1 and AnBGL, enzymatic reaction was performed by incubating each enzyme with 1.5 mM pNP-Glc in the presence of 0.05, 0.25, 0.5, 1.0, 2.0, and 4.0% glucose in 100 mM sodium acetate buffer (pH 5.0) at 37°C for 10 min.

Detection of reaction products by HPAEC-PAD

For detection of enzymatic reaction products, high-performance anion exchange column chromatography (HPAEC) with a pulsed amperometoric detector (PAD) equipped with a CarboPac PA10 guard column (4 × 50 mm) and a CarboPac PA10 analytical column (4 × 250 mm; Dionex Co.) was used. Enzymatic reaction was performed by incubation with equivalent volume of rAaBGL1 (20.0 nM) and each substrate in 20 mM sodium acetate buffer (pH 5.0) at 37°C. Reaction mixture was sampled at appropriate time, and added into equal volume of 0.2 M NaOH. Resultant mixtures were subjected to HPAEC-PAD using mobile phase of 100 mM NaOH with 10 mM sodium acetate. Glucose, cellobiose (Wako Pure Chemical Industries, Ltd.), cellotriose, cellotetraose, cellopentaose, laminaribiose, laminaritriose, laminaritetraose, laminaripentase (Megazyme), gentiobiose, sophorose (SIGMA-ALDRICH, Co.) were used as standards.

Kinetic analysis

For the kinetic analysis, cellobiose, cellotriose, cellotetraose, cellopentaose, lamianaribiose, laminaritriose, laminaritetraose, laminaripentaose and gentiobiose were used as substrates. Appropriate concentrations of each substrate were mixed with equivalent volume of enzyme in 20 mM sodium acetate buffer (pH 5.0). Every 1 or 2 min, reaction was stopped by adding 50 μl of 1 N HCl, and after 5 min, neutralized with adding 50 μl of neutralizton solution (0.4 N NaOH and 0.8 M Tris). The amount of released glucose was determined by using Glucose CII-Test Wako (Wako Pure Chemical Industries, Ltd.). Kinetic constants were determined using Hanes-Woolf plot according to Michaelis-Menten equation. In the case of disaccharides, k_{cat} value was calculated by half of glucose production velocity because one glucodisaccharide molecule composed 2 glucose molecules. Equivalent molar of glucose (G_1) and G_{n-1} production from G_n (n = 3–5) was confirmed at the end point of the reaction by HPAEC-PAD.

For the detection of reaction products by rAaBGL1, laminaribiose, cellobiose, and gentiobiose (25 mM) were reacted with rAaBGL1 (10.0 nM) in 10 mM sodium acetate buffer (pH 5.0) at 37°C. Reaction was stopped by addition of equivalent volume of 0.2 N NaOH and reaction mixtures were analyzed for HPEAC-PAD as described above.

Results

Purification and characterization of rAaBGL1

To investigate the biochemical characterization of rAaBGL1, we purified rAaBGL1 as described in the Materials and methods, and confirmed the homogeneity by SDS-PAGE (Figure 1). The molecular mass calculated from the amino acid sequence was 91.3 kDa, however purified rAaBGL1 was approximately 130 kDa.

Enzymatic properties of rAaBGL1 were determined using pNP-Glc as a substrate. The enzyme was stable between 40–50°C, and in a pH range of 3.0–10.0 with over 80% of its maxmum activity. The optimum temperature was 65°C, and optimum pH was 5.5 (Figure 2).

Specific activity for several pNP-monosaccharides was determined (Table 1). rAaBGL1 was shown highest activity toward pNP-Glc, and slight activity toward pNP-β-D-fucopyranoside, pNP-α-L-arabinofuranoside, pNP-β-D-xylopyranoside, pNP-β-D-galactopyranoside. No activity was detected for pNP-β-D-mannopyranoside, pNP-N-acetyl-β-D-glucosaminide.

Substrate specificity and kinetic parameters for natural β-glucooligosaccharides was determined (Table 2). For disaccharide hydrolysis, rAaBGL1 was shown the highest k_{cat}/K_m value toward laminaribiose among three disaccharides because of the lowest K_m value. Cellobiose was not preferable substrate for rAaBGL1 because of the highest K_m value and lowest k_{cat} value. The k_{cat}/K_m value for cellooligosaccharides and laminarioligosaccharides were increased up to tetra- and trisaccharide, respectively. AaBGL1 exhibited stationary high k_{cat} value for cellopentaose, whereas displayed the lower affinity and turnover number for laminaritetraose and laminaripentaose than laminaritriose.

Detection of transglycosylation product

To identify the transglycosylation products by rAaBGL1, the time course of the reaction products using cellobiose, gentiobiose, and laminaribiose as a substrate were analyzed by HPAEC-PAD (Figure 3). In the early stage of the reaction (0–1 h), the reaction product of rAaBGL1 with each substrate was glucose. In the middle stage of the reaction (2–4 h), the reaction products of rAaBGL1 with cellobiose and laminaribiose were glucose and gentiobiose from transglycosylation (Figure 3A,C). In the reaction with gentiobiose, it is expected that gentiobiose was produced as a transglycosylation product as in the

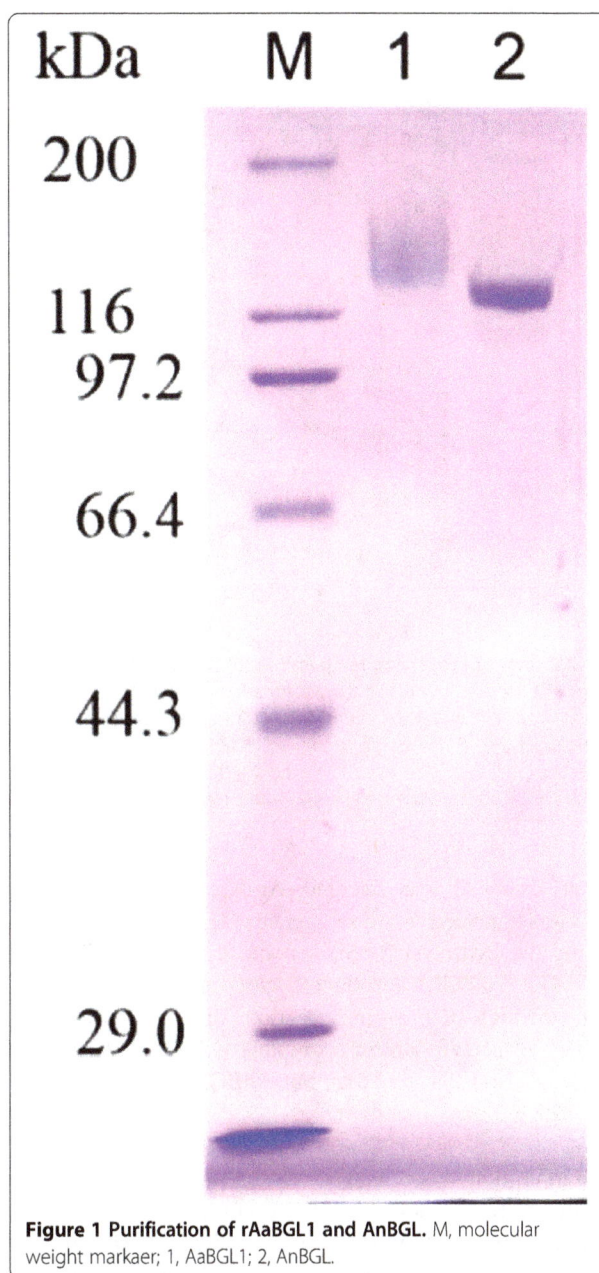

Figure 1 Purification of rAaBGL1 and AnBGL. M, molecular weight markaer; 1, AaBGL1; 2, AnBGL.

case with cellobiose and laminaribiose, because any oligosaccharides other than gentiobiose were not detected (Figure 3B). Thus, these results indicated that gentiobiose was only or main product of transglycosylation by rAaBGL1 under the condition used in this study. In the final stage, the reaction product of rAaBGL1 with each substrate was glucose because of the high hydrolytic activity toward a transglycosylation product, gentiobiose (Table 2).

Comparison of the saccharification ability between rAaBGL1 and AnBGL on the hydrolysis of 5% cellobiose

To evaluate the performance of rAaBGL1, AnBGL which shared 82.4% identity with AaBGL1 (Dan et al. 2000;

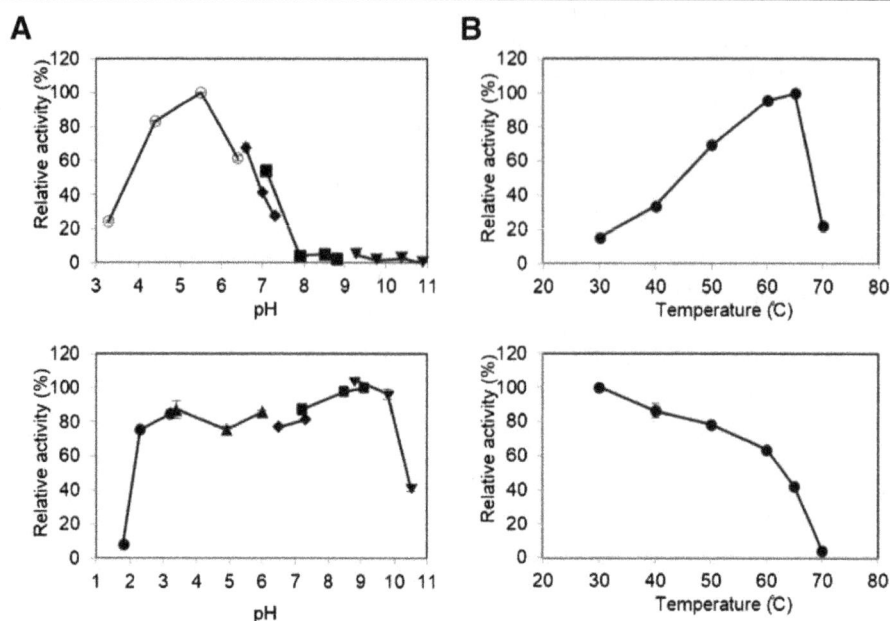

Figure 2 Effects of pH (A) and temperature (B) on the activity (upper panels) and the stability (lower panels) of purified rAaBGL1.
To determine the effect of pH on the activity and the stability (A), enzyme was incubated with 1.5 mM pNP-Glc for 10 min in 100 mM following buffers: glycine-HCl, pH 1.9–2.8 (closed circle); sodium acetate, pH 3.4–5.9 (closed triangle); sodium citrate, pH 3.3–6.3 (open circle); sodium phosphate, pH 6.4–7.3 (closed diamond); Tris–HCl pH 6.8–8.9 (closed square); Glycine-NaOH pH 9.5–11.0 (closed inveted triangle). To determine the effect of temperature on the activity and the stability (B), enzyme was incubated with 1.5 mM pNP-Glc at 30–70°C in 100 mM sodium acetate buffer (pH 5.0). Data are expressed at the mean ± the standard deviation of three independent experiments.

Seidle et al. 2004) was selected. AnBGL from Sigma-aldrich was purified to homogeneity (Figure 1), and compared the hydrolysis of 5% cellobiose with rAaBGL1 (Figure 4). rAaBGL1 produced glucose faster than AnBGL through the entire reaction time measured. rAaBGL hydrolyzed almost completely 5% cellobiose after 8 h reaction (94.1 ± 0.8%), but AnBGL was not sufficient (82.8 ± 0.4%).

Glucose inhibition

Generally, BGL is inhibited by the reaction product, glucose. Therefore we compared the sensitivity to various concentration of glucose on the hydrolysis of pNP-Glc between rAaBGL1 and AnBGL (Figure 5). As a result,

rAaBGL1 was lower sensitivity to glucose at all the concentration tested than AnBGL. The K_i value for gluose on rAaBGL1 used pNP-Glc as a substrate was 9.99 ± 0.94 mM (data not shown).

Discussion

In this study, we performed detailed investigation of enzymatic properties of rAaBGL1 that have already demonstrated synergistic effects by adding to the cellulase system of *T. reesei*.

**Table 1 Specific activity of rAaBGL1 for various
p-nitrophenyl-β-D-glycopyranosides**

Substrate	Specific activity (U/mg)
p-nitrophenyl-α-L-arabinofuranoside	0.057
p-nitrophenyl-β-D-fucopyranoside	0.017
p-nitrophenyl-β-D-xylopyranoside	0.428
p-nitrophenyl-β-D-glucopyranoside	128
p-nitrophenyl-β-D-galactopyranoside	0.006
p-nitrophenyl-β-D-mannopyranoside	ND
p-nitrophenyl-N-acetyl-β-D-glucosaminide	ND

ND:Not Detected.

**Table 2 Kinetic parameters of rAaBGL1 for various
natural substrates**

Substrate	K_m (mM)	k_{cat} (s^{-1})	k_{cat}/K_m (s^{-1}·mM^{-1})
gentiobiose	0.52 ± 0.02	457 ± 4	873 ± 29
laminaribiose	0.41 ± 0.02	444 ± 17	1080 ± 20
laminaritriose	0.22 ± 0.01	337 ± 3	1550 ± 80
laminaritetraose	0.72 ± 0.03	304 ± 4	423 ± 11
laminaripentaose	1.13 ± 0.01	285 ± 5	251 ± 2
cellobiose	2.06 ± 0.07	354 ± 10	172 ± 3
cellotriose	0.45 ± 0.02	477 ± 3	1060 ± 50
cellotetraose	0.32 ± 0.01	433 ± 4	1340 ± 40
cellopentaose	0.41 ± 0.02	433 ± 9	1070 ± 30

Each value is the mean of triplicate experiments.

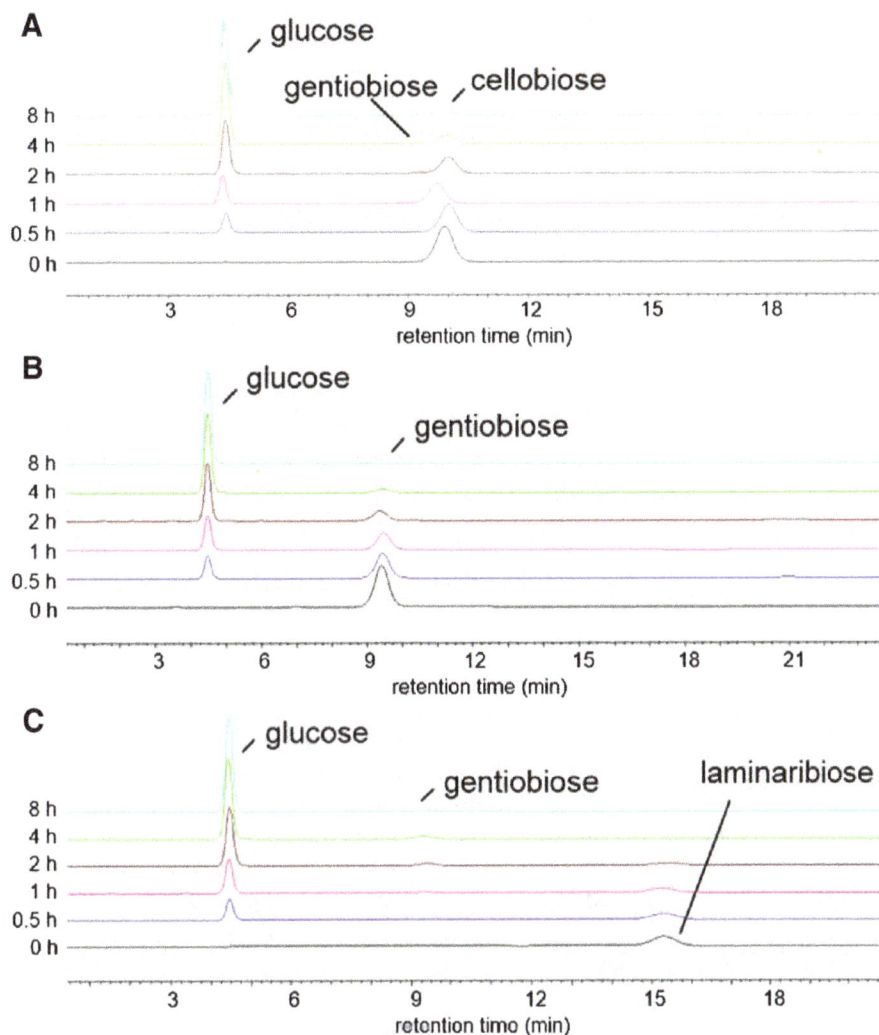

Figure 3 Hydrolysis of disaccharides by rAaBGL1. The hydrolysis of 25 mM cellobiose (**A**), gentiobiose (**B**), and laminaribiose (**C**) was performed by incubation with 10.0 nM AaBGL1 at 37°C for 8 h. The hydrolysis products of indicating times were analyzed by HPAEC-PAD, and identified by comparison of retention time of each peak with those of standards.

Mature AaBGL1 was consisted 841 amino acids, and confirmed secretion in culture supernatant due to possessing signal peptide. The molecular mass calculated from the amino acid sequence was 91.3 kDa, however, purified rAaBGL1 showed approximately 130 kDa by SDS-PAGE analysis (Figure 1). This molecular mass was similar to native AaBGL1 in culture supernatant from *A. aculeatus* (Sakamoto et al. 1985a). Recently, crystalline structure of rAaBGL1, which treated with the endoglycosidase H at undenaturing condition was solved at a 1.80 Å resolution (Suzuki et al. 2013). AaBGL1 has 9 *N*-glycans out of 16 potential *N*-glycosylation sites in the monomer, and *O*-glycosylation was not observed. BGLs from *A. kawachii*, *A. niger* and *A. oryzae* that shared high similarity of amino acid sequence with AaBGL1 have more than 10 potential *N*-glycosylation sites in their amino acid sequence and

occur several *N*-glycosylations, in consequence, these BGLs indicated the similar molecular mass with AaBGL1 (Iwashita et al. 1998, 1999; Langston et al. 2006; Seidle et al. 2004).

In previous study, enzymatic properties of authentic AaBGL1 were investigated (Sakamoto et al. 1985b). Thus, we compared enzymatic properties of AaBGL1 between recombinant protein from *A. oryzae* and authentic one. Thermal and pH profiles were almost similar between authentic and recombinant AaBGL1 with exception that rAaBGL1 had higher optimum temperature and thermal stability, and wider range of pH stability than authentic AaBGL1, although assay conditions were different. BGL2, which is probably isoform of AaBGL1 generated by different glycosylation, had higher stability than BGL1. The slight differences of thermal and pH

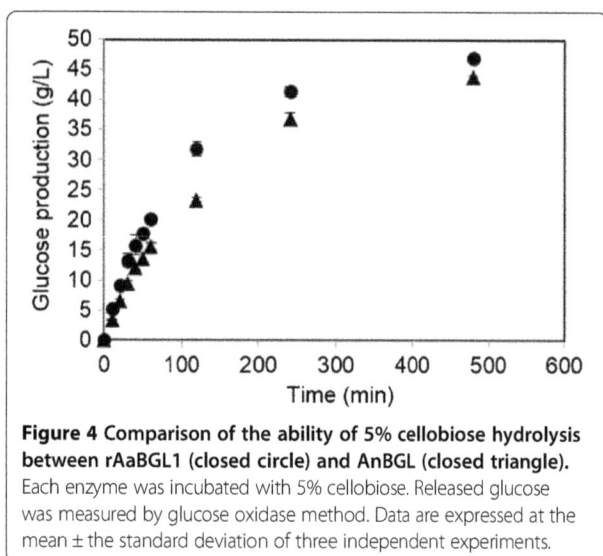

Figure 4 Comparison of the ability of 5% cellobiose hydrolysis between rAaBGL1 (closed circle) and AnBGL (closed triangle). Each enzyme was incubated with 5% cellobiose. Released glucose was measured by glucose oxidase method. Data are expressed at the mean ± the standard deviation of three independent experiments.

profiles between authentic and recombinant AaBGL1 might be developed from the difference of modification by glycosylation between *A. aculeatus* and *A. oryzae*.

It is known that GH family 3 is composed of β-glucosidase (EC 3.2.1.21), glucan 1,3-β-glucosidase (EC 3.2.1.58), glucan 1,4-β-glucosidase (EC 3.2.1.74), exo-1,3-1,4-glucanase (EC 3.2.1.-), xylan 1,4-β-xylosidase (EC 3.2.1.37), β-*N*-acetylhexosaminidase (EC 3.2.1.52), α-L-arabinofuranosidase (EC 3.2.1.55). In addition, BGLs are divided into three groups on the basis of their substrate specificity, such as aryl-β-glucosidase, aryl- and alkyl-β-glucosidase, and enzymes with broad substrate specificity (Takahashi et al. 2011). Moreover, Harvey et al. classified GH family 3 BGLs into 6 distinct

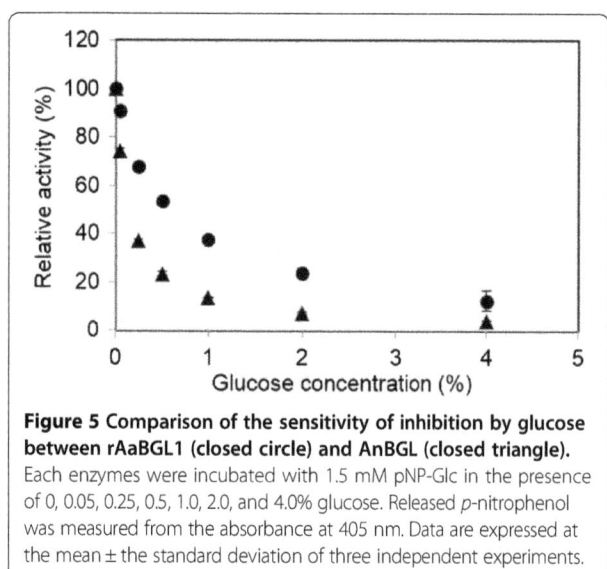

Figure 5 Comparison of the sensitivity of inhibition by glucose between rAaBGL1 (closed circle) and AnBGL (closed triangle). Each enzymes were incubated with 1.5 mM pNP-Glc in the presence of 0, 0.05, 0.25, 0.5, 1.0, 2.0, and 4.0% glucose. Released *p*-nitrophenol was measured from the absorbance at 405 nm. Data are expressed at the mean ± the standard deviation of three independent experiments.

phylogenetic branches by their amino acid sequence similarities (Harvey et al. 2000), and Suzuki et al. mentioned that diverse structures of the subsite +1 makes different substrate specificity in GH family 3 BGLs (Suzuki et al. 2013). However, substrate specificity of AaBGL1 has been investigated only for β-1,4 linked glucan and oligosaccharides (Sakamoto et al. 1985b). Therefore, in this study, the detailed substrate specificity of AaBGL1 was investigated. As a result, rAaBGL1 exhibited high specificity toward pNP-Glc, that is β-glucosidic linkage (Table 1). In kinetic analysis for natural disaccharides, catalytic efficiency increased in the order of cellobiose < gentiobiose < laminaribiose among three β-disaccharides. In all substrates used in this study, rAaBGL1 exhibited highest catalytic efficiency (k_{cat}/K_m) for laminaritriose due to lowest K_m value, but longer than laminaritriose, k_{cat}/K_m value was decreased because of lowering the affinity and the turnover number. Eukaryotic GH family 3 BGLs have been characterized that there is a tendency for laminaribiose to be the best substrate among the other β-linked disaccharides (Hrmova et al. 2002; Igarashi et al. 2003; Langston et al. 2006; Nakajima et al. 2012; Seidle et al. 2004; Takahashi et al. 2011). On the other hand, for cellooligosaccharides, cellotetraose was the best substrate for rAaBGL1 owing to the stationary low K_m value and high k_{cat} value. In addition, several kinetic analyses of GH 3 BGLs imply that there are three subsites, −1, +1, +2, in the active site, and that subsite +1 is the most important subsite for the substrate binding (Hrmova et al. 2002; Kawai et al. 2004; Nakatani et al. 2010; Yazaki et al. 1997). Subsite map determined by kinetic analysis of rAaBGL1 for cellooligosaccharides hydrolysis, using the method for subsite analysis of exo-acting enzyme (Hiromi et al. 1973), revealed that AaBGL1 had the subsite +2 like other GH family 3 BGLs, but subsite map for laminarioligosaccharides, showed weak affinity at subsite +2 (Figure 6). Previously, it was demonstrated that

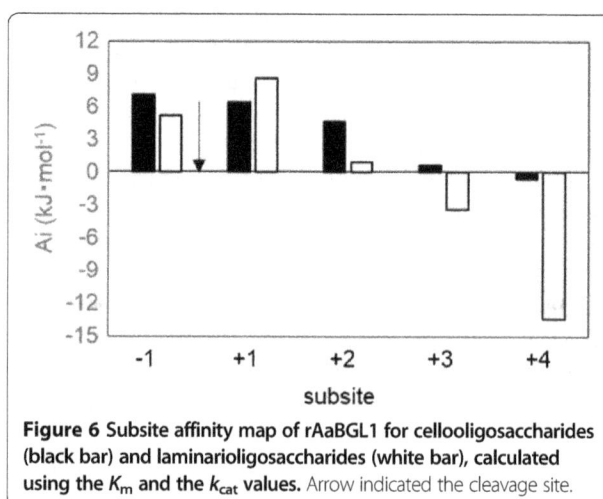

Figure 6 Subsite affinity map of rAaBGL1 for cellooligosaccharides (black bar) and laminarioligosaccharides (white bar), calculated using the K_m and the k_{cat} values. Arrow indicated the cleavage site.

AaBGL1 had long, aromatic residue-rich cleft in the active site (Suzuki et al. 2013), and these results suggested that this structure contributed to the activity for long chain (DP 4–6) of cellooligosaccharides, but not of laminariolignosaccharides. Thus, rAaBGL1 preferred short chain (DP 2–3) of laminarioligosaccharides and long chain (DP 4–6) of cellooligosaccharides on hydrolysis.

It has not been clear whether rAaBGL1 has transglycosylation activity because of its potent saccharifying activity. During hydrolysis of gentiobiose, we could not detect any oligosaccharide except for substrate (Figure 3B). On the other hand, gentiobiose was detected in the middle stage of hydrolysis for cellobiose and laminaribiose as a substrate (Figure 3A,C). These results indicated that rAaBGL1 has transglycosylation activity, and can produce gentiobiose at least. This result is consistent with the report the glucose molecule at subsite +1 points its C6 hydroxyl to C1 hydroxyl of the glucose molecule at subsite −1 in AaBGL1-glucose complex (Suzuki et al. 2013). In fact, it has been reported that BGL from *A. niger* also produces gentiobiose as one of transglycosylation products (Seidle et al. 2004). Moreover, it has been reported that some BGLs transferred glucose from non-reducing end of substrate to O6 position of non-reducing end of β-linked disaccharides (Kawai et al. 2004; Kono et al. 1999; Seidle et al. 2004; Watanabe et al. 1992). However, we could not clearly identify to produce trisaccharide as a transglycosylation product by rAaBGL1 under the condition used in this study.

In cellulosic biomass hydrolysis using *Trichoderma* cellulase system, cellobiose is accumulated by the dominant enzyme, cellobiohydrolase I that plays an important role for the hydrolysis of crystalline cellulose and by low activity of BGLs. The accumulation of cellobiose causes the lowering of overall activity of cellulase system via the inhibition of cellobiohydrolase activity. Therefore, we attempted to compare the saccharifying activity toward cellobiose between rAaBGL1 and commercially available AnBGL, an orthologue of AaBGL1 (82.4% identity). The optimum pH of AaBGL1 and AnBGL was around 4 (Sakamoto et al. 1985b, Seidle et al. 2004). In this study, we compared the cellobiose hydrolytic activity at pH 5.0, because enzymatic saccharification of cellulosic biomasses was performed at pH between 4.8–5.0 in the many studies (Berlin et al. 2007; Chen et al. 2013; Dashtban and Qin 2012; Harrison et al. 2013; Kawai et al. 2012). As a result, rAaBGL1 hydrolyzed 5% cellobiose faster than AnBGL (Figure 4), because of the difference of the sensitivity for the product inhibition, especially by glucose. In fact, rAaBGL1 was more insensitive to inhibition by glucose than AnBGL at all the glucose concentration tested (Figure 5). K_i value for glucose of AaBGL1 is 9.99 ± 0.94 mM, on the other hand, that of AnBGL was 3 mM (Seidle et al. 2004). Moreover,

the specific activity for a major transglycosylation product, gentiobiose of rAaBGL1 and AnBGL was 170 ± 4 U/mg and 57 ± 1 U/mg, respectively, under the condition of 25 mM gentiobiose in 10 mM sodium acetate buffer (pH 5.0) at 37°C (data not shown). AnBGL exhibited higher K_m value (1.3 mM) toward gentiobiose than AaBGL1 (0.52 mM), although K_m value toward cellooligosaccharides of AnBGL was almost similar to AaBGL1 at each optimum pH condition (Seidle et al. 2004). Thus, we concluded that AaBGL1 has high saccharifying activity toward cellobiose compared with AnBGL, because AaBGL1 was insensitive to competitive inhibition by glucose, and has high hydrolytic activity toward gentiobiose, which is transglycosylation product during cellobiose hydrolysis. Recently, the saccahrifying activity of JN11, a crude enzyme preparation from *T. reesei* X3AB1 strain expressing AaBGL1 gene was compared with those of commercially available cellulase, Accellerase 1500 and Cellic CTec using various types of pretreated biomass as substrates (Kawai et al. 2012). Using JN11, especially glucose yield was improved, suggesting that the advantage of JN11 for saccharification of cellulosic biomass resulted in the increase of cellobiase activity by AaBGL1.

In this study, we investigated the biochemical characterization of rAaBGL1 from *Aspergillus acuelatus* no. F-50 strain, and compared the cellobiose-saccharifying activity between rAaBGL1 and AnBGL, an orthologue of AaBGL1. Here we demonstrated the potent cellobiase activity of rAaBGL1 for cellulosic biomass degradation, is combined with the high hydrolyzing efficiency toward gentiobiose derived from transglycosylation and low sensitivity to the product inhibition by glucose.

Competing interests
The authors declare that they have no competing interests.

Authors' contributions
YB: Performed the experiments for purification and characterization of the enzymes, analyzed data, and drafted manuscript. JS: Conceived the study, designed the experiments, critically analyzed data, and revised manuscript. ST: Performed experiments for DNA manipulation and gene expression. TK: Designed the experiments, fritically analyzed data, and revised manuscript. All authors read and approved the final manuscript.

Acknowledgements
This work was financially supported by The New Energy and Industrial Technology Deveropment Organization and by Grant-in-Aid for Scientific Reserch (22580090) from the Japan Society for the Promotion of Science.

References
Andric P, Meyer AS, Jensen PA, Johansen KD (2010) Reactor design for minimizing product inhibition during enzymatic lignocellulose hydrolysis: I. Significance and mechanism of cellobiose and glucose inhibition on cellulolytic enzymes. Biotechnol Adv 28:308–324
Berlin A, Maximenko V, Gilkes N, Saddler J (2007) Optimization of enzyme complexes for lignocellulose hydrolysis. Biotechnol Bioeng 97:287–296
Chauve M, Mathis H, Huc D, Casanave D, Monot F, Ferreira NL (2010) Comparative kinetic analysis of two fungal β-glucosidases. Biotechnol Biofuels 3:3

Chen Y, Stevens MA, Zhu Y, Holmes J, Xu H (2013) Understanding of alkaline pretreatment parameters for corn stover enzymatic saccharification. Biotechnol Biofuels 6:8

Dan S, Marton I, Dekel M, Bravdo BA, He S, Withers SG, Shoseyov O (2000) Cloning, expression, characterization, and nucleophile identification of family 3, Aspergillus niger β-glucosidase. J Biol Chem 275:4973–4980

Dashtban M, Qin W (2012) Overexpression of an exotic thermotolerant β-glucosidase in trichoderma reesei and its significant increase in cellulolytic activity and saccharification of barley straw. Microb Cell Fact 11:63

Dekker RF (1986) Kinetic, inhibition, and stability properties of a commercial β-D-glucosidase (cellobiase) preparation from Aspergillus niger and its suitability in the hydrolysis of lignocellulose. Biotechnol Bioeng 28:1438–1442

Du F, Wolger E, Wallace L, Liu A, Kaper T, Kelemen B (2010) Determination of product inhibition of CBH1, CBH2, and EG1 using a novel cellulase activity assay. Appl Biochem Biotechnol 161:313–317

Gomi K, Iimura Y, Hara S (1987) Integrative transformation of Aspergillus oryzae with a plasmid containing the Aspergillus nidulans argB gene. Agric Biol Chem 51:2549–2555

Harrison MD, Zhang Z, Shand K, O'Hara IM, Doherty WOS, Dale JL (2013) Effect of pretreatment on saccharification of sugar cane bagasse by complex and simple enzyme mixtures. Bioresour Technol 148:105–113

Harvey AJ, Hrmova M, De Gori R, Varghese JN, Fincher GB (2000) Comparative modeling of the three-dimensional structures of family 3 glycoside hydrolases. Proteins 41:257–269

Henrissat B (1991) A classification of glycosyl hydrolases based on amino-acid sequence similarities. Biochem J 280:309–316

Henrissat B, Bairoch A (1993) New families in the classification of glycosyl hydrolases based on amino- acid sequence similarities. Biochem J 293:781–788

Henrissat B, Bairoch A (1996) Updating the sequence-based classification of glycosyl hydrolases. Biochem J 316:695–696

Hiromi K, Niita Y, Numata C, Ono S (1973) Subsite affinities of glucoamylase: examination of the validity of the substie theory. Biochim Biophys Acta 302:362–375

Hrmova M, Gori RD, Smith BJ, Fairweather JK, Driguez H, Varghese JN, Fincher GB (2002) Structural basis for broad substrate specificity in higher plant β-D-glucan glucohydrolases. Plant Cell 14:1033–1052

Igarashi K, Tani T, Kawai R, Samejima M (2003) Family 3 β-glucosidase from cellulose-degrading culture of the white-rot fungus Phanerochaete chrysosporium is a glucan 1,3-β-glucosidase. J Biosci Bioeng 95:572–576

Iwashita K, Todoroki K, Kimura H, Shimoi H, Ito K (1998) Purification and characterization of extracellular and cell wall bound β-glucosidases from Aspergillus kawachii. Biosci Biotechnol Biochem 62:1938–1946

Iwashita K, Todoroki K, Kimura H, Shimoi H, Ito K (1999) The bglA gene of Aspergillus kawachii encodes both extracellular and cell wall-bound β-glucosidases. Appl Environ Micorbiol 65:5546–5553

Kanamasa S, Takada G, Kawaguchi T, Sumitani J, Arai M (2001) Overexpression and purification of Aspergillus aculeatus β-mannosidase and analysis of the integrated gene in Aspergillus oryzae. J Biosci Bioeng 92:131–137

Kawaguchi T, Enoki T, Tsurumaki S, Sumitani J, Ueda M, Ooi T, Arai M (1996) Cloning and sequencing of the cDNA encoding beta-glucosidase 1 from Aspergillus aculeatus. Gene 173:287–288

Kawai R, Igarashi K, Kitaoka M, Ishii T, Samejima M (2004) Kinetics of substrate transglycosylation by glycoside hydrolase family 3 glucan (1→3)-β-glucosidase from the white-rot fungus Phanerochaete chrysosporium. Carbohydr Res 339:2851–2857

Kawai T, Nakazawa H, Ida N, Okada H, Tani S, Sumitani J, Kawaguchi T, Ogasawara W, Morikawa Y, Kobayashi Y (2012) Analysis of the saccharification capability of high-functional cellulase JN11 for various pretreated biomasses through a comparison with commercially available counterparts. J Ind Microbiol Biotechnol 39:1741–1749

Kono H, Kawano S, Tajima K, Erata T, Takai M (1999) Structural analyses of new tri- and tetrasaccharides produced from disaccharides by transglycosylation of purified Trichoderma viride β-glucosidase. Glycoconj J 16:415–423

Langston J, Sheehy N, Xu F (2006) Substrate specificity of Aspergillus oryzae family 3 β-glucosidase. Biochim Biophys Acta 1764:972–978

Minetoki T, Tsuboi H, Koda A, Ozeki K (2003) Development of high expression system with the improved promoter using the cis-acting element in Aspergillus species. J Biol Macromol 3:89–96

Murao S, Kanamoto J, Arai M (1979) Isolation and identification of a cellulolytic enzyme producing microorganism. J Ferment Technol 57:151–156

Nakajima M, Yamashita T, Takahashi M, Nakano Y, Takeda T (2012) Identification, cloning, and characterization of β-glucosidase from Ustilago esculenta. Appl Microbiol Biotechnol 93:1989–1998

Nakatani Y, Lamont IL, Cutfield JF (2010) Discovery and characterization of a distinctive Exo-1,3/1,4-β-glucanase from the marine bacterium Pseudoalteromonas sp. strain BB1. Appl Enviorn Micorbiol 76:6760–6768

Nakazawa H, Kawai T, Ida N, Shida Y, Kobayashi Y, Okada H, Tani S, Sumitani J, Kawaguchi T, Morikawa Y, Ogasawara W (2011) Construction of a recombinant Trichoderma reesei strain expressing Aspergillus aculeatus β-glucosidase 1 for efficient biomass conversion. Biotechnol Bioeng 109:92–99

Sakamoto R, Arai M, Murao S (1985a) Purification and Physicochemical Properties of Three β-Glucosidases from Aspergillus aculeatus No. F-50. Agric Biol Chem 49:1275–1281

Sakamoto R, Arai M, Murao S (1985b) Enzymic properties of three β-glucosidases from Aspergillus aculeatus No F-50. Agric Biol Chem 49:1283–1290

Seidle HF, Marten I, Shoseyov O, Huber RE (2004) Physical and kinetic properties of the family 3 β-glucosidase from Aspergillus niger which is important for cellulose breakdown. Protein J 23:11–23

Singhania RR, Patel AK, Sukumaran RK, LarrocheC PA (2013) Role and significance of β-glucosidases in the hydrolysis of cellulose for bioethanol production. Bioresour Technol 127:500–507

Suzuki K, Sumitani J, Nam Y, Toru N, Tani S, Wakagi T, Kawaguchi T, Fushinobu S (2013) Crystal structures of glycoside hydrolase family 3 β-glucosidase 1 from Aspergillus aculeatus. Biochem J 452:211–221

Takahashi M, Konishi T, Takeda T (2011) Biochemical characterization of Magnaporthe oryzae β-glucosidases for efficient β-glucan hydrolysis. Appl Microbiol Biotechnol 91:1073–1082

Watanabe T, Sato T, Yoshioka S, Koshijima T, Kuwahara M (1992) Purification and properties of Aspergillus niger β-glucosidase. Eur J Biochem 209:651–659

Woodward J (1991) Synergism in cellulase systems. Bioresource Technol 36:67–75

Xiao Z, Zhang X, Gregg DJ, Saddler JN (2004) Effects of sugar inhibition on cellulases and β-glucosidase during enzymatic hydrolysis of softwood substrates. Appl Biochem Biotechnol 113–116:1115–1126

Yazaki T, Ohnishi M, Rokushika S, Okada G (1997) Subsite structure of the β-glucosidase from Aspergillus niger, evaluated by steady-state kinetics with cello-oligosaccahrides as substrates. Carbohydr Res 298:51–57

14

Characterization of the *Kluyveromyces marxianus* strain DMB1 *YGL157w* gene product as a broad specificity NADPH-dependent aldehyde reductase

Hironaga Akita[1*], Masahiro Watanabe[1], Toshihiro Suzuki[1], Nobutaka Nakashima[2,3] and Tamotsu Hoshino[1,2]

Abstract

The open reading frame *YGL157w* in the genome of the yeast *Kluyveromyces marxianus* strain DMB1 encodes a putative uncharacterized oxidoreductase. However, this protein shows 46% identity with the *Saccharomyces cerevisiae* S288c NADPH-dependent methylglyoxal reductase, which exhibits broad substrate specificity for aldehydes. In the present study, the YGL157w gene product (KmGRE2) was purified to homogeneity from overexpressing *Escherichia coli* cells and found to be a monomer. The enzyme was strictly specific for NADPH and was active with a wide variety of substrates, including aliphatic (branched-chain and linear) and aromatic aldehydes. The optimal pH for methylglyoxal reduction was 5.5. With methylglyoxal as a substrate, the optimal temperature for enzyme activity at pH 5.5 was 45°C. The enzyme retained more than 70% of its activity after incubation for 30 min at temperatures below 35°C or at pHs between 5.5 and 9.0. In addition, the KmGRE2-overexpressing *E. coli* showed improved growth when cultivated in cedar hydrolysate, as compared to cells not expressing the enzyme. Taken together, these results indicate that KmGRE2 is potentially useful as an inhibit decomposer in *E. coli* cells.

Keywords: Aldehyde inhibitor; BICES; GRE2; *Kluyveromyces marxianus*; Lignocellulosic biomass; Reductase

Introduction

The NADPH-dependent methylglyoxal reductase (EC 1.1.1.283) in *Saccharomyces cerevisiae* is termed GRE2. Using NADPH as a coenzyme, GRE2 catalyzes the stereoselective reduction of a broad range of substrates, including aldehydes and diketones, as well as aliphatic and aromatic ketones (Chen et al. 2003; Murata et al. 1985). In *S. cerevisiae*, this enzyme functions within the high osmolarity glycerol pathway (Garay-Arroyo and Covarrubias 1999), and its expression is induced by environmental conditions, including ionic, osmotic, oxidative, heat shock and heavy metal-related stresses (Garay-Arroyo and Covarrubias 1999; Krantz et al. 2004; Liu et al. 2008; Rep et al. 2001; Rutherford and Bird 2004). GRE2 also shows isovaleraldehyde reductase activity and so acts as a suppressor of filamentation (Chen et al. 2003; Hauser et al. 2007). To date, GRE2

and homologues have been purified to homogeneity from *S. cerevisiae* (Chen et al. 2003; Murata et al. 1985), *Aspergillus niger* (Inoue et al. 1988) and goat liver (Ray and Ray 1984), and their enzymatic properties have been characterized. In addition, the three-dimensional structures of the *S. cerevisiae* S288c GRE2 apo enzyme and the enzyme-NADP⁺ complex expressed in *Escherichia coli* have been solved (Guo et al. 2014). Based on its structural features, GRE2 is classified as a member of the extended short-chain-dehydrogenase/reductase superfamily (Müller et al. 2010).

S. cerevisiae GRE2 is currently being used as a versatile biocatalyst for the stereoselective synthesis of hydroxy compounds, which serve as building blocks in the production of pharmaceuticals and other fine chemicals (Choi et al. 2010; Ema et al. 2008; Müller et al. 2010; Park et al. 2010). Another advantageous feature of GRE2 is a decomposer in bacteria. For example, GRE2 is used for glycolaldehyde degradation during bioethanol production in *S. cerevisiae* (Jayakody et al. 2013). In addition, a *S. cerevisiae* strain overexpressing a GRE2 with site-directed mutagenesis exhibited enhanced

* Correspondence: h-akita@aist.go.jp
[1]Biomass Refinery Research Center, National Institute of Advanced Industrial Sciences and Technology (AIST), 3-11-32 Kagamiyama, Higashi-Hiroshima, Hiroshima 739-0046, Japan
Full list of author information is available at the end of the article

furfural and 5-hydroxymethylfurfural (HMF) detoxification (Moon and Liu 2012). Conversely, in a *S. cerevisiae* GRE2 knockout strain growth was suppressed by environmental stress (Warringer and Blomberg 2006), and filament formation was increased in the presence of isoamyl alcohol (Hauser et al. 2007). Hence, GRE2 is regarded as a key enzyme necessary for inhibitor and stress tolerance in *S. cerevisiae.*

We recently isolated *Kluyveromyces marxianus* strain DMB1, a thermotolerant yeast, from sugarcane bagasse hydrolysate and determined its genomic sequence (Suzuki et al. 2014). Within the sequence, we identified open reading frame *YGL157w*, which shows 46% identity with *S. cerevisiae* S288c GRE2 (Figure 1). In the hope of identifying a more stable GRE2 homologue, in the present study, we purified and characterized the enzyme from *K. marxianus* strain DMB1 after its overexpression in *E. coli* cells. In addition, we examined its ability to improve growth of cells cultured in cedar hydrolysate.

Materials and Methods

Construction of expression vectors

The plasmid pET-16b/YGL157w was constructed for production of *K. marxianus* YGL157w protein with a N-terminal hexahistidine tag. After preparation of genomic DNA from *K. marxianus* strain DMB1 (strain number: HUT7412), the *YGL157w* gene (accession number: LC016711) was amplified using PCR with KOD -plus- DNA polymerase (Toyobo, Osaka, Japan) and the primers 5′-**CAT*ATG***ACGTACGTTGTGGTTACTGGT GC-3′ (the *Nde*I site is in bold and the initiation codon is in italics) and 5′-**GGATCC***TTA*GTTGTT AGCCTTTAGTATTTG-3′ (the *Bam*HI site is in bold and termination codon is in italics). The PCR product was cloned into pTA2 (Toyobo, Osaka, Japan) and sequenced to check for PCR errors. The *YGL157w* gene was then excised from the resulting plasmid using *Nde*I and *Bam*HI and subcloned into pET-16b (Novagen, Hessen, Germany) to give pET-16b/YGL157w.

Expression of proteins

YGL157w protein was expressed in *E. coli* BL21 (DE3) cells harboring pET-16b/YGL157w and then purified to homogeneity. The cells were grown at 37°C for 3 h in Luria-Bertani (LB) medium (1 L) containing 100 mg/L ampicillin. After inducing expression by addition of isopropyl β-D-1-thiogalactopyranoside (IPTG) to a final concentration of 1.0 mM, the culture was incubated for an additional 3 h. The cells were then harvested, suspended in 20 mM Tris–HCl buffer (pH 7.9) containing 500 mM NaCl (buffer A) and 5 mM imidazole, and disrupted by ultrasonication. The resultant lysate was clarified by centrifugation (27,500 × *g* for 15 min at 4°C), after which the supernatant was applied to a Chelating Sepharose Fast Flow column (20 mL; GE Healthcare, Buckinghamshire, UK) charged with Ni^{2+} and equilibrated with buffer A containing 5 mM imidazole. After washing the column with buffer A containing 5 mM imidazole (40 mL) and then 60 mM imidazole (60 mL), the recombinant YGL157w protein was eluted with buffer A containing 500 mM imidazole. The active fractions were pooled, concentrated using a Vivaspin 20 concentrator (10,000 MWCO, Sartorius AG, Goettingen,

```
KmGRE2   1 MTYVVVTGASGYLAQHVIKQLLERNYKVIGTVRNQQKAKDIAKLFQND-NLTLELVPDLL 59
  GRE2   1 -MSVFVSGANGFIAQHIVDLLLKEDYKVIGSARSQEKAENLTEAFGNNPKFSMEVVPDIS 59
              *.*:**.*::***::. **:.:*****:.*.*:**:::: * *: ::::*.***:

KmGRE2  60 QADVFDDLFLKYTGQIKHVIHTASPCRFDTTEYENEMLLPAINGTKRVLESIKKYASETV 119
  GRE2  60 KLDAFDHVFQKHGKDIKIVLHTASPFCFDITDSERDLLIPAVNGVKGILHSIKKYAADSV 119
           : *.**.:* *:  :** *:*****  ** *: *.::*:**:**.* :*.******:::*

KmGRE2 120 ETVVYTSSVSALANPAGILDSNLTLTEESWNPDSFEDGKKDVFSAYYVSKTFAERTAWDF 179
  GRE2 120 ERVVLTSSYAAVFDMAKENDKSLTFNEESWNPATWESCQSDPVNAYCGSKKFAEKAAWEF 179
           * ** *** :*: : *   *..**:.****** ::*. :.* ..**  **.***::**:*

KmGRE2 180 WKENKDQVKFQLTTICPSYIFGPQAFEENAKGTLNFSTEIVNKILHSKPGDELDKNFAGA 239
  GRE2 180 LEENRDSVKFELTAVNPVYVFGPQMFDKDVKKHLNTSCELVNSLMHLSPEDKIPELFGG- 238
            :**:*.***:**:: * *:**** *:::.*  ** * *:**.::* .* *:: : *.*

KmGRE2 240 FVDVRDVARAHVLALEKPELKSKRLVLLNDVYALQDVADYINKHFPELRGKIATGVPGEG 299
  GRE2 239 YIDVRDVAKAHLVAFQKRETIGQRLIVSEARFTMQDVLDILNEDFPVLKGNIPVGKPGSG 298
           ::******:**::*::* *   .:**:: :  :::*** * :*:.** *:*:*..* **.*

KmGRE2 300 KEIVKYIAHYDNSKTKKLLGFEFISFGQAITDTVAQILKANN-- 341
  GRE2 299 ATHNTLGATLDNKKSKKLLGFKFRNLKETIDDTASQILKFEGRI 342
               . * **.*:******:* .: ::* **.:**** :.
```

Figure 1 Multiple sequence alignment of KmGRE2 and GRE2. The sequence alignment was prepared using ClustalW (Thompson et al. 1994). Residues involved in the substrate-binding site in GRE2 are boxed.

Germany) and loaded onto a HiLoad 26/60 Superdex 200 pg column (GE Healthcare) equilibrated with 20 mM Tris–HCl buffer (pH 8.0) containing 50 mM NaCl. The active fractions were pooled and dialyzed against 20 mM Tris–HCl buffer (pH 7.2). Finally, the dialysate was concentrated and the resultant solution was used for biochemical experiments.

Protein concentrations were determined using the Bradford method with bovine serum albumin (BSA) serving as the standard (Bradford 1976).

Molecular mass determination
SDS-PAGE was carried out on a 10% polyacrylamide gel using the method of Laemmli (1970). EzStandard PrestainBlue (ATTO, Tokyo, Japan) was used as the molecular mass standards. The protein sample was boiled for 5 min in EzApply (ATTO). Protein bands were visualized by staining with EzStainAqua (ATTO).

The molecular mass of the native enzyme was determined by gel filtration column chromatography using a Superdex 200 Increase 10/300 GL column. Conalbumin (75 kDa), ovalbumin (43 kDa), carbonic anhydrase (29 kDa), ribonuclease A (13.7 kDa) and aprotinin (6.5 kDa) served as molecular standards (GE Healthcare).

Assay of enzyme activity
KmGRE2 activity was measured by monitoring the decreases in the absorbance at 340 nm caused by the reduction of aldehyde, or the increases in the absorbance caused by the oxidation of alcohol. The mixture (1 mL) used for the reductive reaction contained 100 mM acetate buffer (pH 5.5), 5 mM aldehyde, 0.2 mM NADPH and YGL157w protein. The mixture (1 mL) used for the oxidative reaction contained 100 mM bicarbonate-NaOH (pH 10.0), 5 mM alcohol, 1.25 mM NADP$^+$ and YGL157w protein. The reaction was started by the addition of coenzymes, and the absorbance at 340 nm was monitored at 25°C using a Shimadzu UV-2450 (Kyoto, Japan). The extinction coefficient of NADPH was 6.22 mM^{-1} cm^{-1}. One unit of enzyme was defined as the amount of enzyme producing 1 μmol of NADPH per min at 25°C in the reductive reaction of methylglyoxal.

Effects of pH and temperature on enzyme activity
The pH dependence of the reduction catalyzed by YGL157w protein was determined at 25°C using 100 mM concentrations of acetate (pH 4.0–5.5) and citrate (pH 5.5–6.5). The temperature dependence was evaluated by measuring the reductive reaction at temperatures ranging from 25 to 45°C.

Effects of pH and temperature on enzyme stability
The effect of pH on enzyme stability was evaluated by incubating 100 nM YGL157w protein for 30 min at 35°C with 50 mM concentrations of acetate (pH 5.0–5.5), citrate (pH 5.5–6.5), phosphate (pH 6.5–8.0), borate-NaOH (pH 8.0–9.0) and bicarbonate-NaOH (pH 9.0–11.0). The enzyme solution was then rapidly cooled on ice, and the remaining activity was determined using the standard reduction assay. The thermal stability was determined by incubating YGL157w protein in 20 mM Tris–HCl buffer (pH 7.2) for 30 min at temperatures ranging from 25–45°C. The enzyme solution was then rapidly cooled on ice, and the remaining activity was determined using the standard reduction assay.

Determination of kinetic parameters
The initial velocity of the reductive reaction was analyzed using the standard assay conditions. To determine the kinetic constants for methylglyoxal and NADPH, several concentrations of methylglyoxal (0.05–15 mM) or NADPH (0.01–0.15 mM) were used. The initial velocity was then plotted against the substrate concentration, and the K_m and k_{cat} values were determined by curve fitting using Igor Pro ver. 3.14 (WaveMetrics, Tigard, OR, USA).

Preparation of hydrolysate
Lignocellulosic biomass material (Japanese cedar) was milled using a cutter mill (MKCM-3; Masuko Sangyo, Saitama, Japan), after which the resulting particles were used as the initial raw material. According to Lee et al. (2010), mechanochemical and hydrothermal pretreatment was carried out. The resulting sample was hydrolyzed using 20 FPU/g of Acremonium cellulase (Meiji Seika Pharma, Nagoya, Japan) and 40 μL/g of Optimash BG (Genencor International, Rochester, NY, USA) in 50 mM citrate buffer (pH 5.0) at 50°C and 150 rpm. After incubation for 48 h, the reaction mixture was harvested by centrifugation, and the supernatant was filtered through a 0.2 μm filter (Merck Millipore, Billerica, MA, USA). The pH of the mixture was then adjusted to 6.5, the mixture was diluted, and the resulting solution was used as the hydrolysate. Further details of the procedure are provided elsewhere (Akita et al. 2015).

Effect of KmGRE2 expression on cell growth
The effect of KmGRE2 expression was evaluated by cultivation in a test tube using 3 mL of hydrolysate containing 0.5 mM IPTG, which was incubated at 37°C and 180 rpm. E. coli BL21 (DE3) cells harboring pET-16b/YGL157w or pET-16b were pregrown overnight and then diluted 1:100 with fresh hydrolysate. Cultures were monitored for cell growth at OD$_{600}$ using an Eppendorf BioSpectrometer (Eppendorf, Hamburg, Germany).

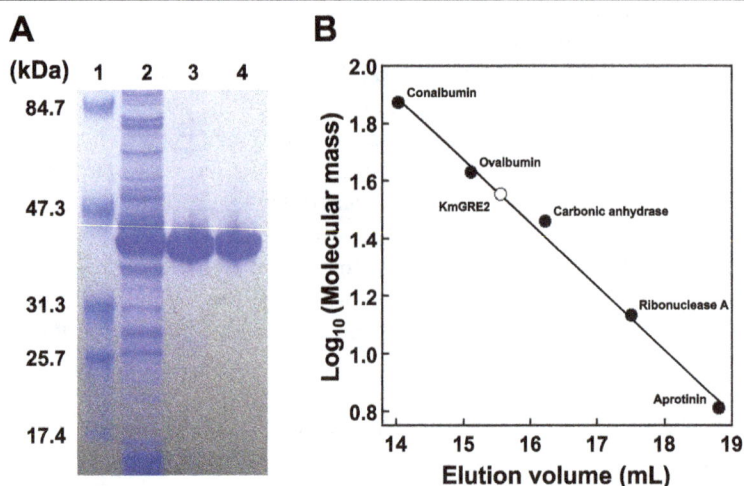

Figure 2 Purification of YGL157w protein. (A) Purification steps followed by SDS-PAGE. Proteins were separated by SDS-PAGE and visualized by EzStainAqua staining: lane 1, protein molecular size markers; lane 2, crude extract; lane 3, Chelating Sepharose Fast Flow column chromatography pool; lane 4, HiLoad 26/60 Superdex 200 pg column pool. **(B)** Determination of molecular mass using gel filtration chromatography.

Quantification of sugars and aldehydes

After clarifying the culture by centrifugation and filtration, the supernatant was subjected to high performance liquid chromatography (HPLC). Quantification was performed using an Aminex HPX-87H cationic exchange column connected to an Aminex 85H Micro-Guard Column (Bio-Rad Labs, Richmond, CA, USA). The chromatographic conditions for sugars and aldehydes were as follows: mobile phase, 4.5 mM H_2SO_4 or 8 mM H_2SO_4; flow rate, 0.6 mL·min^{-1}; and the column oven temperature, 65°C or 35°C. Sugars and aldehydes were detected using a Jasco RI-2031 Plus Intelligent Refractive Index Detector (Jasco, Tokyo, Japan) or a Jasco UV-2070 Plus Intelligent UV/VIS Detector at 278 nm (Jasco).

Results

Purification and molecular mass determination of KmGRE2

After expression in 4.86 g (wet weight) of *E. coli* cells harboring pET-16b/YGL157w, YGL157w protein was purified using two successive purification steps: Chelating Sepharose Fast Flow column chromatography and gel filtration chromatography (Figure 2A). Ultimately, a pure protein was obtained with an overall yield of 82.4% (Table 1).

The apparent molecular mass of the YGL157w protein was determined to be about 36 kDa using Superdex 200

Increase 10/300 GL column gel filtration chromatography (Figure 2B). SDS-PAGE of the enzyme showed one major band of 40 kDa (Figure 2A), suggesting the native protein exists as a monomer.

Substrate specificity and kinetic properties of KmGRE2

When assessed the enzymatic activity of recombinant YGL157w protein, we found that it catalyzed the reduction of linear, branched-chain and aromatic aldehydes using NADPH as the coenzyme (Table 2). High levels of activity were observed with isovaleraldehyde (C5), methylglyoxal (C3) and valeraldehyde (C5), while the lower levels were observed with octanal (C8), benzaldehyde (C7) and HMF (C6). YGL157w protein showed no activity toward *p*-anisaldehyde (C8), *p*-hydroxy benzaldehyde (C7), D-alanine, L-alanine, D-lactate, L-lactate and pyruvate. Only NADPH was utilized as the cofactor for reduction of methylglyoxal by the enzyme; NADH was not inactive. In addition, using $NADP^+$ as the coenzyme, no activity was observed for oxidative reactions toward alcohol-analogs such as isoamyl alcohol, isobutanol, 2-propanol, 1-hexanol, 1-heptanol and 1-octanol under the described conditions. These results demonstrate that *YGL157w* gene encodes a NADPH-dependent GRE2, which we are calling KmGRE2.

After measuring the initial rates at various methylglyoxal or NADPH concentrations, regression analyses

Table 1 Purification of KmGRE2 from *E. coli* BL21 (DE3)

Purification step	Total protein (mg)	Total activity (U)	Specific activity (U/mg)	Yield (%)
Crude extract	215	410	1.91	100
Chelating sepharose fast flow column	79.1	371	4.70	90.5
HiLoad 26/60 superdex 200 pg column	40.0	338	8.45	82.4

Table 2 Substrate specificity

Substrate	Relative activity (%)[a]
Isovaleraldehyde	244 ± 1.9
Methylglyoxal	100
Valeraldehyde	95.6 ± 1.6
Hexanal	81.4 ± 1.4
Heptanal	80.8 ± 2.5
Furfural	60.1 ± 1.7
Propionaldehyde	49.3 ± 0.7
Octanal	22.3 ± 1.5
Benzaldehyde	14.3 ± 2.0
HMF	1.0$\geqq$
Cinnamaldehyde	N/A[b]
Vanillin	N/A[b]

[a]Reductive activities were measured in 100 mM acetate buffer (pH 5.5) containing 3 mM substrate, 0.1 mM NADPH and 100 nM enzyme.
[b]N/A means no measurable activity. Due to the high absorbance of this substrate at 340 nm, activity was not determined under the assay conditions.

were used to fit the data to the Michaelis-Menten equation (data not shown). The K_m and k_{cat} values for methylglyoxal were calculated as 0.30 ± 0.018 mM and $1.3 \times 10^3 \pm 15$ min^{-1}, respectively. The kinetic parameters for NADPH were 0.028 ± 0.0012 mM and $1.4 \times 10^3 \pm 22$ min^{-1} mM^{-1}, respectively. In addition, the k_{cat}/K_m for methylglyoxal and NADPH were 4.4×10^3 and 5.1×10^4 min^{-1} mM^{-1}, respectively. These results are similar to those of *S. cerevisiae* (Murata et al. 1985).

Effects of pH and temperature on enzyme activity and stability

The effect of pH on the reduction of methylglyoxal was determined by assessing the enzyme activity at several pHs. At a temperature of 25°C, the optimum pH was about 5.5 (Figure 3A). When the temperature dependence at pH 5.5 was examined, maximum activity was observed at around 45°C (Figure 3B). Moreover, when KmGRE2 was incubated for 30 min at various temperatures in 20 mM Tris–HCl buffer (pH 7.2), KmGRE2 retained more than 80% of its activity at temperatures below 35°C (Figure 3C). On the other hand, there was a complete loss of activity when the enzyme was incubated

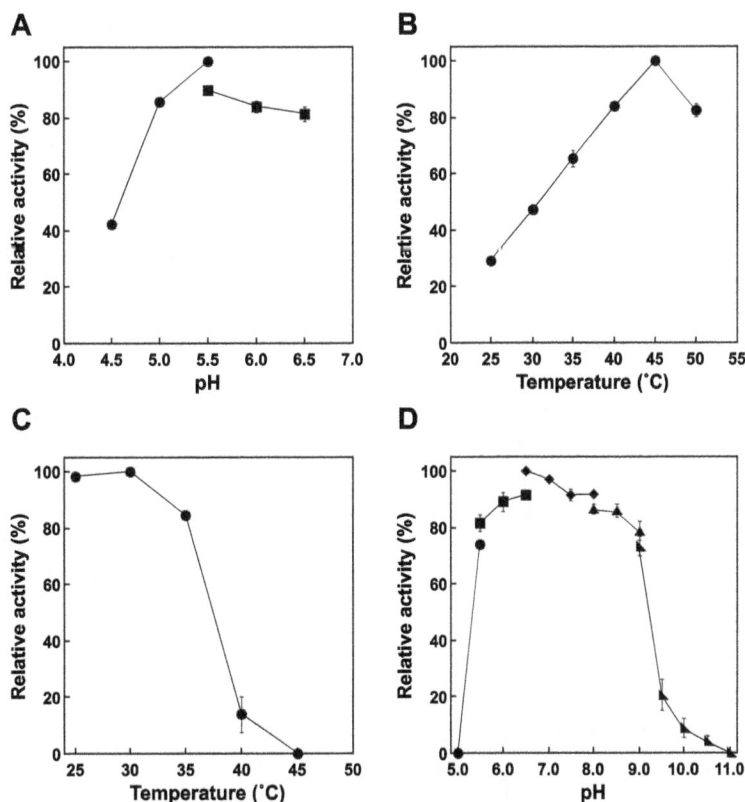

Figure 3 Effects of pH and temperature on KmGRE2 activity and stability. The markers of buffer were indicated following: circles, acetate; squares, citrate; diamonds, phosphate, isosceles triangles, borate-NaOH and right triangles, bicarbonate-NaOH. **(A)** Effect of pH on KmGRE2 activity. **(B)** Effect of temperature on KmGRE2 activity. **(C)** KmGRE2 activity after incubation for 30 min at various temperatures in the 20 mM Tris–HCl buffer (pH 7.2). **(D)** KmGRE2 activity after incubation for 30 min at 35°C in buffer solutions of various pHs. Error bars indicate SE (*n* = 3).

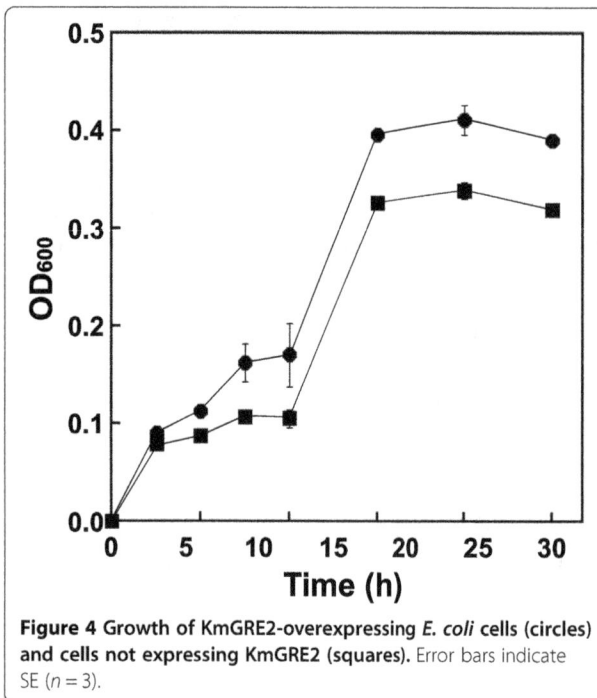

Figure 4 Growth of KmGRE2-overexpressing *E. coli* cells (circles) and cells not expressing KmGRE2 (squares). Error bars indicate SE (*n* = 3).

at temperatures above 45°C. When the effect of pH on the stability of the enzyme was evaluated based on the activity remaining after incubation at 35°C for 30 min, we found that KmGRE2 retained more than 70% of its activity at pHs between 5.5 and 9.0 (Figure 3D). Somewhat disappointingly, however, KmGRE2 showed nearly the same temperature and pH stability as *S. cerevisiae* GRE2 (Park et al. 2010).

Cell growth in hydrolysate from cedar

To assess the ability of KmGRE2 as decomposer, the effect of KmGRE2 expression on the growth of cell in cedar hydrolysate was determined by monitoring the *E. coli* growth. When cedar hydrolysate was prepared, glucose and xylose were mainly included as the sugars, whereas aldehyde inhibitors such as furfural and HMF were mainly generated. Thus, these sugars and aldehydes were detected by HPLC. The pH of hydrolysate was decided based on the optimal pH for KmGRE2 activity and the possible growth pH of *E. coli*. When cultivated in the hydrolysate under these conditions, KmGRE2-overexpressing *E. coli* showed more rapid growth than *E.*

coli not expressing the enzyme (Figure 4), with enhanced furfural degradation (Table 3).

Discussion

In the present study, we succeeded in expressing the *YGL157w* gene from *K. marxianus* strain DMB1 in *E. coli* cells and purifying the product. Characterization of the purified enzyme showed that KmGRE2 harbored strong NADPH-dependent reductive activities toward at least 10 aldehyde substrates (Tables 2). The higher activities were observed on C3 branched-chain and C3 to C7 linear aldehydes, whereas lower or no activities were detected for C8 linear aldehyde and C6 to C8 aromatic aldehydes. Conversely, *S. cerevisiae* GRE2 showed the highest activity for phenyglyoxal (C8) in the presence of NADPH (Murata et al. 1985). When we compared the amino acid sequences of KmGRE2 and *S. cerevisiae* S288c GRE2, we found that Ser127, Tyr165 and Lys169 in GRE2 were completely conserved in KmGRE2 as Ser127, Tyr165 and Lys169 (Figure 1). The three aforementioned residues in GRE2 are considered the crucial roles for the substrate dehydrogenation: Ser127 stabilizes the substrate, Tyr165 acts on a catalytic base and Lys169 facilitates the catalysis at neutral pH (Guo et al. 2014). However, two residues responsible for the substrate binding differ between the two enzymes: Phe85 and Tyr128 in GRE2 are respectively replaced by Cys85 and Val128 in KmGRE2 (Guo et al. 2014) (Figure 1). These substitutions may reduce the hydrophobic interactions for aromatic aldehydes in KmGRE2, which suggests that the molecular mechanism for substrate recognition differs between KmGRE2 and GRE2. To assess the molecular mechanism, we are now trying to obtain crystals of cofactor and/or substrate-bound KmGRE2.

The utilization of biofuel from lignocellulosic biomass holds promise as a means of abating global warming. This has prompted the development of a number of bioconversion methods for biofuel production (Akita et al. 2015; Lan and Liao 2013; Nakashima et al. 2014; da Silva et al. 2014). But while those methods produced several kinds of biofuels from hydrolysate derived of lignocellulosic biomass, the productivities and yields were often low (Akita et al. 2015; Lan and Liao 2013; Nakashima et al. 2014; da Silva et al. 2014). One of the mentioned causes of the low productivity is microbial growth inhibition by aldehyde inhibitors (Mills et al. 2009). Because aldehyde inhibitors such as furfural,

Table 3 Sugar and aldehyde components in cedar hydrolysate

Incubation time (h)	KmGRE2 expression	Glucose (mM)	Xylose (mM)	Furfural (mM)	HMF (mM)
0	Not overexpressing	278.8	197.8	34.6	21.1
30	Overexpressing	262.5 ± 4.4	183.4 ± 3.4	20.5 ± 2.0	16.8 ± 0.6
	Not overexpressing	264.3 ± 1.9	184.4 ± 4.7	23.1 ± 0.2	15.9 ± 0.2

HMF, glycolaldehyde, methylglyoxal and vanillin are generated mainly during the biomass hybridization process (Jarboe and Chi 2013; Jayakody et al. 2011), they are able to inhibit microbial growth and interfere with subsequent fermentation (Jayakody et al. 2011; Liu et al. 2008; Mills et al. 2009; Moon and Liu 2012). Consequently, we proposed that KmGRE2 utilizes as inhibitor decomposer. To confirm the ability of KmGRE2 to play decomposer, we assessed the effect of KmGRE2 expression on cell growth in cedar hydrolysate, production of which led to the formation of both furfural and HMF. As anticipated, the KmGRE2-overexpressing *E. coli* showed substantial growth improvement (Figure 4). We think that the growth improvement was achieved by enhanced furfural degradation, which provided for the preferable culture conditions at early culture phase. In fact, the OD_{600} of KmGRE2-overexpressing *E. coli* at 6 to 12 h were 1.3–1.6-fold higher than these of not expressing *E. coli*. On the other hand, the less activity toward HMF in KmGRE2-overexpressing *E. coli* remained unclear. The omics analysis on the metabolic response of KmGRE2-overexpressing *E. coli* may reveal this phenomenon. Recently, we developed a simple and efficient method involving biomass-inducible chromosome-based expression system (BICES) for expressing foreign genes without the use of plasmids or expensive inducers (Akita et al. 2015; Nakashima et al. 2014). This method can also be used to produce biofuels, but the productivity and yield were markedly diminished when hydrolysate from Japanese cedar as the carbon source for isobutanol production (Akita et al. 2015). We are now planning to integrate KmGRE2 gene into the genome of the *E. coli* strain involving BICES. We anticipate that this will improve growth rates, thereby increasing the productivity and yield.

Competing interests

The authors declare that they have no competing interests.

Authors' contribution

HA designed this study, performed experiments, participated in the interpretation of the results and drafted the manuscript. MW and TS participated in the design and coordination of this study and helped to revise the manuscript. NN and TH conceived and designed this study, coordinated the experiments, interpreted the results and revised the manuscript for important intellectual content. All authors read and approved the final manuscript.

Acknowledgment

We are grateful to all members of the Bio-conversion Research Team at our Institute [Biomass Refinery Research Center, National Institute of Advanced Industrial Sciences and Technology (AIST)] for their technical assistance and valuable discussion.

Author details

[1]Biomass Refinery Research Center, National Institute of Advanced Industrial Sciences and Technology (AIST), 3-11-32 Kagamiyama, Higashi-Hiroshima, Hiroshima 739-0046, Japan. [2]Bioproduction Research Institute, National Institute of Advanced Industrial Sciences and Technology (AIST), 2-17-2-1 Tsukisamu-Higashi, Toyohira-ku, Sapporo 062-8517, Japan. [3]Department of Biological Information, Graduate School of Bioscience and Biotechnology, Tokyo Institute of Technology, 2-12-1-M6-5 Ookayama, Meguro-ku, Tokyo 152-8550, Japan.

References

Akita H, Nakashima N, Hoshino T (2015) Bacterial production of isobutanol without expensive reagents. Appl Microbiol Biotechnol 99:991–999

Bradford MM (1976) A rapid and sensitive method for the quantitation of microgram quantities of protein utilizing the principle of protein-dye binding. Anal Biochem 72:248–254

Chen CN, Porubleva L, Shearer G, Svrakic M, Holden LG, Dover JL, Johnston M, Chitnis PR, Kohl DH (2003) Associating protein activities with their genes: rapid identification of a gene encoding a methylglyoxal reductase in the yeast Saccharomyces cerevisiae. Yeast 20:545–554

Choi YH, Choi HJ, Kim D, Uhm KN, Kim HK (2010) Asymmetric synthesis of (S)-3-chloro-1-phenyl-1-propanol using Saccharomyces cerevisiae reductase with high enantioselectivity. Appl Microbiol Biotechnol 87:185–193

da Silva TL, Gouveia L, Reis A (2014) Integrated microbial processes for biofuels and high value-added products: the way to improve the cost effectiveness of biofuel production. Appl Microbiol Biotechnol 98:1043–1053

Ema T, Ide S, Okita N, Sakai T (2008) Highly efficient chemoenzymatic synthesis of methyl (R)-o-chloromandelate, a key intermediate for clopidogrel, via asymmetric reduction with recombinant Escherichia coli. Adv Synth Catal 350:2039–2044

Garay-Arroyo A, Covarrubias AA (1999) Three genes whose expression is induced by stress in Saccharomyces cerevisiae. Yeast 22:879–892

Guo PC, Bao ZZ, Ma XX, Xia Q, Li WF (2014) Structural insights into the cofactor-assisted substrate recognition of yeast methylglyoxal/isovaleraldehyde reductase Gre2. Biochim Biophys Acta 1844:1486–1492

Hauser M, Horn P, Tournu H, Hauser NC, Hoheisel JD, Brown AJ, Dickinson JR (2007) A transcriptome analysis of isoamyl alcohol-induced filamentation in yeast reveals a novel role for Gre2p as isovaleraldehyde reductase. FEMS Yeast Res 7:84–92

Inoue Y, Rhee H, Watanabe K, Murata K, Kimura A (1988) Metabolism of 2-oxoaldehyde in mold. Purification and characterization of two methylglyoxal reductases from Aspergillus niger. Eur J Biochem 171:213–218

Jarboe LR, Chi Z (2013) Inhibition of microbial biocatalysts by biomass-derived aldehydes and methods for engineering tolerance. In: Torrioni L, Pescasseroli E (eds) New Developments in Aldehydes Research. Nova Science Publishers, New York

Jayakody LN, Hayashi N, Kitagaki H (2011) Identification of glycolaldehyde as the key inhibitor of bioethanol fermentation by yeast and genome-wide analysis of its toxicity. Biotechnol Lett 33:285–292

Jayakody LN, Horie K, Hayashi N, Kitagaki H (2013) Engineering redox cofactor utilization for detoxification of glycolaldehyde, a key inhibitor of bioethanol production, in yeast Saccharomyces cerevisiae. Appl Microbiol Biotechnol 97:6589–6600

Krantz M, Nordlander B, Valadi H, Johansson M, Gustafsson L, Hohmann S (2004) Anaerobicity prepares Saccharomyces cerevisiae cells for faster adaptation to osmotic shock. Eukaryot Cell 3:1381–1390

Laemmli UK (1970) Cleavage of structural proteins during the assembly of the head of bacteriophage T4. Nature 227:680–685

Lan EI, Liao JC (2013) Microbial synthesis of n-butanol, isobutanol, and other higher alcohols from diverse resources. Bioresour Technol 135:339–349

Lee SH, Chang F, Inoue S, Endo T (2010) Increase in enzyme accessibility by generation of nanospace in cell wall supramolecular structure. Bioresour Technol 101:7218–7223

Liu ZL, Moon J, Andersh BJ, Slininger PJ, Weber S (2008) Multiple gene-mediated NAD(P)H-dependent aldehyde reduction is a mechanism of in situ detoxification of furfural and 5-hydroxymethylfurfural by Saccharomyces cerevisiae. Appl Microbiol Biotechnol 81:743–753

Mills TY, Sandoval NR, Gill RT (2009) Cellulosic hydrolysate toxicity and tolerance mechanisms in Escherichia coli. Biotechnol Biofuels 2:26

Moon J, Liu ZL (2012) Engineered NADH-dependent GRE2 from Saccharomyces cerevisiae by directed enzyme evolution enhances HMF reduction using additional cofactor NADPH. Enzyme Microb Technol 50:115–120

Müller M, Katzberg M, Bertau M, Hummel W (2010) Highly efficient and stereoselective biosynthesis of (2S,5S)-hexanediol with a dehydrogenase from Saccharomyces cerevisiae. Org Biomol Chem 8:1540–1550

Murata K, Fukuda Y, Simosaka M, Watanabe K, Saikusa T, Kimura A (1985) Metabolism of 2-oxoaldehyde in yeasts. Purification and characterization of

NADPH-dependent methylglyoxal-reducing enzyme from *Saccharomyces cerevisiae*. Eur J Biochem 151:631–636

Nakashima N, Akita H, Hoshino T (2014) Establishment of a novel gene expression method, BICES (biomass-inducible chromosome-based expression system), and its application for production of 2,3-butanediol and acetoin. Metab Eng 25:204–214

Park HJ, Jung J, Choi H, Uhm KN, Kim HK (2010) Enantioselective bioconversion using *Escherichia coli* cells expressing *Saccharomyces cerevisiae* reductase and *Bacillus subtilis* glucose dehydrogenase. J Microbiol Biotechnol 20:1300–1306

Ray M, Ray S (1984) Purification and partial characterization of a methylglyoxal reductase from goat liver. Biochim Biophys Acta 802:119–127

Rep M, Proft M, Remize F, Tamás M, Serrano R, Thevelein JM, Hohmann S (2001) The *Saccharomyces cerevisiae* Sko1p transcription factor mediates HOG pathway-dependent osmotic regulation of a set of genes encoding enzymes implicated in protection from oxidative damage. Mol Microbiol 40:1067–1083

Rutherford JC, Bird AJ (2004) Metal-responsive transcription factors that regulate iron, zinc, and copper homeostasis in eukaryotic cells. Eukaryot Cell 3:1–13

Suzuki T, Hoshino T, Matsushika A (2014) Draft genome sequence of *Kluyveromyces marxianus* strain DMB1, isolated from sugarcane bagasse hydrolysate. Genome Announc 2:e00733-14

Thompson JD, Higgins DG, Gibson TJ (1994) CLUSTAL W: improving the sensitivity of progressive multiple sequence alignment through sequence weighting, position-specific gap penalties and weight matrix choice. Nucleic Acids Res 22:4673–4680

Warringer J, Blomberg A (2006) Involvement of yeast YOL151W/GRE2 in ergosterol metabolism. Yeast 23:389–398

15

Enhanced xylose fermentation and ethanol production by engineered *Saccharomyces cerevisiae* strain

Leonardo de Figueiredo Vilela[1], Verônica Parente Gomes de Araujo[1], Raquel de Sousa Paredes[1], Elba Pinto da Silva Bon[1], Fernando Araripe Gonçalves Torres[2], Bianca Cruz Neves[1] and Elis Cristina Araújo Eleutherio[1*]

Abstract

We have recently demonstrated that heterologous expression of a bacterial xylose isomerase gene (*xylA*) of *Burkholderia cenocepacia* enabled a laboratorial *Saccharomyces cerevisiae* strain to ferment xylose anaerobically, without xylitol accumulation. However, the recombinant yeast fermented xylose slowly. In this study, an evolutionary engineering strategy was applied to improve xylose fermentation by the *xylA*-expressing yeast strain, which involved sequential batch cultivation on xylose. The resulting yeast strain co-fermented glucose and xylose rapidly and almost simultaneously, exhibiting improved ethanol production and productivity. It was also observed that when cells were grown in a medium containing higher glucose concentrations before being transferred to fermentation medium, higher rates of xylose consumption and ethanol production were obtained, demonstrating that xylose utilization was not regulated by catabolic repression. Results obtained by qPCR demonstrate that the efficiency in xylose fermentation showed by the evolved strain is associated, to the increase in the expression of genes *HXT2* and *TAL1*, which code for a low-affinity hexose transporter and transaldolase, respectively. The ethanol productivity obtained after the introduction of only one genetic modification and the submission to a one-stage process of evolutionary engineering was equivalent to those of strains submitted to extensive metabolic and evolutionary engineering, providing solid basis for future applications of this strategy in industrial strains.

Keywords: Evolutionary engineering; Xylose isomerase; Saccharomyces cerevisiae; TAL1; Xylose; Ethanol

Introduction

The ethanol production from lignocellulosic biomass is a promising alternative energy source. *Saccharomyces cerevisiae* is the preferred microorganism used to produce ethanol due to its excellent ability to ferment glucose in addition to its high tolerance to ethanol and inhibitors presented in lignocellulosic hydrolysates (Stambuk et al. 2008). However, it is not capable to ferment xylose, present in significant amounts in biomass hydrolysates. In *S. cerevisiae*, xylose is converted into xylulose via two enzymes that use different cofactors, leading to a redox imbalance and, consequently, prevents xylose fermentation. Two main strategies have been commonly applied to solve this problem:

the cloning of a xylose reductase and a xylitol dehydrogenase which are linked to the same coenzyme or the cloning of a xylose isomerase which converts xylose directly into its isomer xylulose. Nevertheless, yeasts engineered through that strategy to ferment xylose still do it slowly and accumulate xylitol (Kim et al. 2012). Thus, additional genetic modifications have been carried out as an attempt to increase the specific consumption of xylose as well as the rate and yield of production of ethanol: i) overexpression of the enzymes necessary for the conversion of xylulose into glycolysis intermediates; ii) deletion of the endogenous aldose reductase which converts xylose into xylitol; iii) overexpression of heterologous xylose transporters (Cai et al. 2012). Besides the metabolic engineering approach, evolutionary engineering has been employed to improve the cell performance for ethanol production and to increase the stability of the recombinant strains (Cai et al. 2012). Evolutionary

* Correspondence: eliscael@iq.ufrj.br
[1]Department of Biochemistry, Institute of Chemistry, Federal University of Rio de Janeiro, Rio de Janeiro, Brazil
Full list of author information is available at the end of the article

engineering strategies are in fact complementary to metabolic engineering in the search for desired phenotypes through the imposition of one or more selective pressures. Although a great diversity of metabolic engineering and adaptation approaches have been tested to boost xylose fermentation in biomass hydrolysates, yield and productivity of ethanol by genetically engineered *S. cerevisiae* strains are still much lower than those of glucose fermentation.

The simultaneous conversion of xylose and glucose is another bottleneck to the economic ethanol production from biomass hydrolysates (Ha et al. 2011; Eiteman et al. 2008). Yeasts engineered to ferment xylose are not able to consume xylose until glucose is completely exhausted. One possible explanation for this phenomenon is that glucose represses the expression of genes necessary to the catabolism of xylose through Mig1 an important and essential transcription factor for the process of catabolic repression. In the presence of high levels of glucose, Mig1 rapidly moves from the cytoplasm into the nucleus and binds to the promoters of glucose-repressible genes. When the cells are deprived of glucose, Mig1 is transported back to the cytoplasm, releasing glucose repression (Rolland et al. 2002). In an attempt to overcome the inhibitory effect of glucose over the use of xylose, a recent report described an engineered yeast strain, designed to perform intracellular hydrolysis of cellobiose, allowing co-consumption of cellobiose and xylose (Ha et al. 2011). Noteworthy, the sequential use of xylose after glucose depletion could also be attributed to the competition between xylose and glucose during uptake. In *S. cerevisiae*, this pentose is transported by facilitated transport mediated by hexose permeases which transport xylose with very low affinity compared with uptake of glucose (Subtil and Boles 2012). Discovery or engineering of a xylose-specific transporter which is not inhibited by glucose and which shows high affinity and capacity of transport might improve cellular performance to ferment xylose in biomass hydrolysates (Weber et al. 2010).

Recently, we described the functional expression of *Burkholderia cenocepacia* xylose isomerase and its effect on the ability of *S. cerevisiae* to ferment xylose-glucose blends (De Figueiredo Vilela et al. 2013). A major breakthrough, the ethanol yields obtained by the sole heterologous expression of the bacterial enzyme was equivalent to those of strains submitted to extensive metabolic and evolutionary engineering. The recombinant strain did not accumulate xylitol, but it still consumed xylose very slowly compared to glucose, resulting in relatively low ethanol productivities. In the present work, we report an evolutionary engineering approach which promoted an increase on the consumption rate of xylose and ethanol production by the *S. cerevisiae* expressing xylose isomerase from *B. cenocepacia*. Aiming to unravel the changes between the metabolically engineered and the subsequently evolved strain, the expression profile of several genes involved in xylose fermentation was analyzed.

Methods
Strain and adaptive evolution experiment
Saccharomyces cerevisiae BY4741 (*MATa, his3, leu2, met15, ura3*) expressing xylose isomerase from *Burkholderia cenocepacia* (de Figueiredo Vilela et al. 2013) was used in this study. The adaptive evolution experiments were performed in 500 mL flasks at 28°C/160 rpm filled with 100 mL of YNB (0.7% yeast nitrogen base without amino acids; 0.01% histidine; 0.01% leucine; 0.01% metionine) containing 2% xylose. Initial cell concentration was 0.5 mg/mL ($8,3 \times 10^6$ UFC/mL). A serial transfer of 50 mg of cells (dry weight) into fresh medium was performed every 24 h. After 40 passages, cells were plated on solid YNB medium. An isolated colony was used in the experiments (Subtil and Boles 2012). Both original (un-evolved) and evolved strains were grown in YNB medium containing different proportions of glucose-xylose (D4%, D3% X1%, D2%X2%, D1%X3%, X4%) until exponential growth phase. Next, they were collected by centrifugation and transferred to fermentation media.

Fermentation conditions
Fermentation was carried out at 30°C and pH 5.0, in flasks of 50 mL filled with 25 mL of fermentation medium (4.0% Total sugar; 0.4% $(NH_4)_2SO_4$ and 0.4% KH_2PO_4, pH 5) (Zhou et al. 2012). Initial cell concentration was 1.5 mg/mL ($2,5 \times 10^7$ UFC/mL). Over the time, samples were collected, centrifuged and the cell-free supernatants were used for the determination of glucose and xylose consumption as well as xylitol and ethanol production. The concentrations of glucose, xylose and xylitol were determined by HPLC (Shimadzu) equipped with refractive index detector. The column used for separation was a LiChrospher NH_2 (Merck). The HPLC apparatus was operated with a mobile phase of 80% acetronitrile and 20% H_2O at a flow rate of 1.0 mL/min (Ferreira et al. 1997). The ethanol concentration was assessed by dichromate oxidation method with supernatants after distillation (Seo et al. 2009).

Expression analysis
The un-evolved and evolved strains were cultivated until exponential growth phase in YNB medium containing glucose as carbon source. For enzyme assays, cells were harvested and disrupted with glass beads (diameter 0.45 mm, Sigma Aldrich, U.S.A.). Protein concentration was determined with Stickland assays using bovine serum albumin as a standard (Stickland 1951). Xylose isomerase activity was measured as previously described (de Figueiredo Vilela et al. 2013). Quantitative qPCR was used to measure the expression of *XKS1, RPE1, RKI1, TAL1, TKL1, HXT1,*

HXT2 and *HXT7. TAF10* (RNA pol II transcription factor activity/transcription initiation and chromatin modification) was used as an endogenous control (Teste et al. 2009). Gene sequences were analyzed using File Builder® 3.1 v2.0 (Applied Biosystems, U.S.A.). TaqMan probes were synthesized by Life Technologies (Applied Biosystems, USA). QuickPrep mRNA Purification Kit (GE Healthcare Life Sciences) was used to extract mRNA from cells harvested in the mid-log phase. Ready to Go RT-PCR Beads (GE Healthcare Life Sciences) was used to for cDNA synthesis. All real time PCR reactions were performed on a Stepone System Real Time PCR cycler (Applied Biosystems, U.S.A.) according to manufacturer's instructions. The $2^{-\Delta\Delta CT}$ method was used to analyze the relative changes in gene expression in evolved and un-evolved strains (Livak and Schmittgen 2001).

Results

Evolutionary engineering promotes high levels of xylose consumption in the presence of glucose

The adaptive evolution experiments were performed with 40 passages in YNB medium and isolated clones were screened. A single clone was than selected for further analyses outstanding results. To investigate the effect of this engineering approach on the capacity of the evolved clone to consume xylose in the presence of glucose, cells were pre-grown on in YNB medium containing different proportions of glucose-xylose until exponential growth phase and then transferred to a fermentation medium containing 2% xylose and 2% glucose. Interestingly, the evolved strain showed a substantially higher xylose consumption rate than the un-evolved strain (around 10-fold higher), as shown in Table 1. Furthermore, the rate of xylose consumption during fermentation shown by cells pre-grown under higher glucose concentrations was more than 80% of the rate of glucose consumption.

Evolutionary engineering improves xylose-specific consumption and ethanol productivity

To investigate and compare the fermentative performance of the evolved and un-evolved strains, cells were pre-grown in YNB medium supplemented with 4% glucose until

exponential growth phase. Eventually, strains were transferred to a fermentation medium containing glucose and xylose at the same concentration and proportion found in sugar cane hydrolyzates (3.0% glucose plus 1.0% xylose).

The evolutionary engineering strategy significantly improved the specific xylose consumption rate, and provided efficient ethanol production from this sugar cane-like xylose-glucose mixture. The ethanol yield and productivity showed by the evolved strain were 13% (0.51 × 0.45 g ethanol/g sugar) and 120% (0.42 × 0.19 g ethanol/g cell/h) higher, respectively, than that of the un-evolved strain (Figure 1A, B). The evolved strain did not show xylitol accumulation, unlike the un-evolved strain (results not shown).

Analyses of gene expression profiles demonstrate up-regulation of genes for transaldolase and a hexose transporter

The un-evolved and evolved strains were grown to exponential growth phase in YNB medium containing glucose as the sole carbon source. Quantitative qRT-PCR was used to measure the expression of genes known to be involved in xylose fermentation (Additional file 1: Table S1). The $2^{-\Delta\Delta CT}$ method was used to analyze the relative changes in gene expression in evolved and un-evolved strains. Xylose isomerase activity was measured but no significant differences in xylose isomerase activity between both strains were observed (data not shown). Results obtained by qRT-PCR (Figure 2) suggest that the efficient xylose fermentation showed by the evolved strain can be attributed, at least in part, to the elevated expression of *TAL1*, which codes for transaldolase. The evolved strain presented a 3.0 fold higher *TAL1* expression than the un-evolved strain. Expression of *HXT2*, that codes for an hexose-transporter was also measured, whilst no important change was observed within the expression of the other target genes.

Discussion

In this work, an evolutionary engineering approach was applied to select a spontaneous mutant with higher specific xylose consumption rate. For this, the recombinat

Table 1 Specific glucose and xylose consumption rates

Growth medium containing YNB	Specific Consumption (g/L/h) in the fermentation medium (D2%X2%)									
	D4%		D3%X1%		D2%X2%		D1%X3%		X4%	
Strains	Glucose	Xylose	Glucose	Xylose	Glucose	Xylose	Glucose	Xylose	Glucose	Xylose
Non-adapted	0.48 ± 0.07	0.08 ± 0.04	0.47 ± 0.02	0.08 ± 0.02	0.47 ± 0.02	0.06 ± 0.04	0.45 ± 0.02	0.06 ± 0.03	0.41 ± 0.01	0.05 ± 0.02
Adapted	0.73 ± 0.03	0.60 ± 0.02	0.73 ± 0.07	0.59 ± 0.09	0.73 ± 0.05	0.55 ± 0.03	0.73 ± 0.06	0.55 ± 0.02	0.70 ± 0.09	0.49 ± 0.02

Initially, both original (un-evolved) and evolved on xylose strains were grown in YNB medium containing different proportions of glucose-xylose until mid-log phase of growth and, then, shifted to the fermentation medium containing 2% glucose and 2% xylose. The aliquots were harvested in time of 24 hours, centrifuged and the supernatants were used to determine the concentration of glucose and xylose, which were used to calculate the rate of sugar consumption. The results represent the mean ± standard deviation of at least three independent experiments.

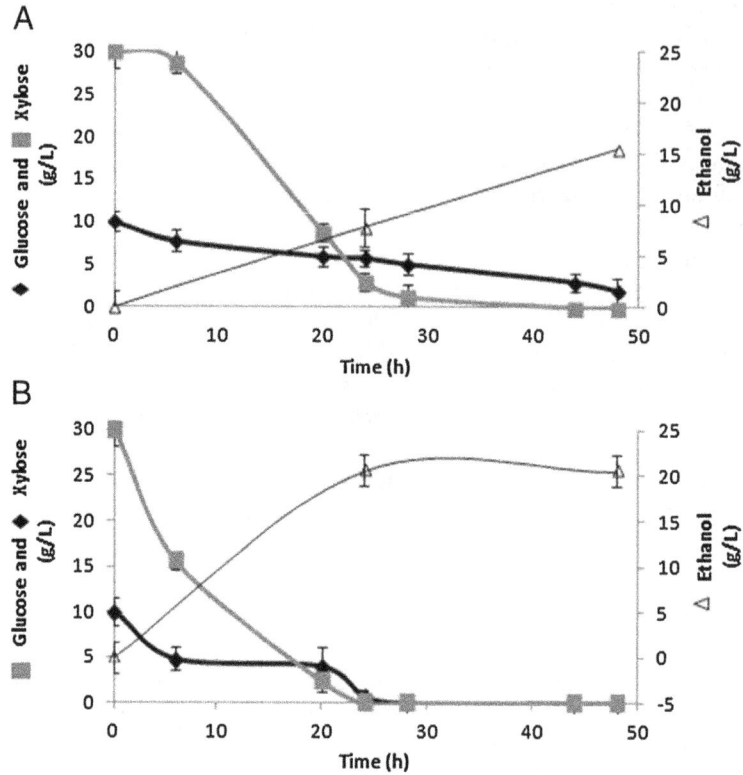

Figure 1 Sugar consumption and ethanol production by *S. cerevisiae*. Cells of un-evolved **(A)** and evolved strains **(B)** were grown in YNB-medium supplemented with 4% glucose until mid-log phase, collected by centrifugation, washed with distilled water and transferred to fermentation medium containing 3% glucose and 1% xylose. The samples were collected in times of 0, 6, 20, 24, 28, 44, 48 hours and the supernatants were used to determine the concentration of glucose and xylose, in times of 24 and 48 hours the supernatants were used to determine the ethanol concentration. The results represent the mean ± standard deviation of at least three independent experiments.

xylose-fermenting strain expressing xylose isomerase from *B. cenocepacia* was submitted to sequential batch cultivations in YNB-medium supplemented with 2% xylose. According to some authors (Ha et al. 2011), the delay in the xylose consumption caused by glucose is related to the catabolic repression phenomenon. To investigate the effect of this engineering approach on the capacity of cells to consume xylose in the presence of glucose, cells were pregrown on YNB-media containing different glucose-xylose mixtures and subsequently transferred to a fermentation

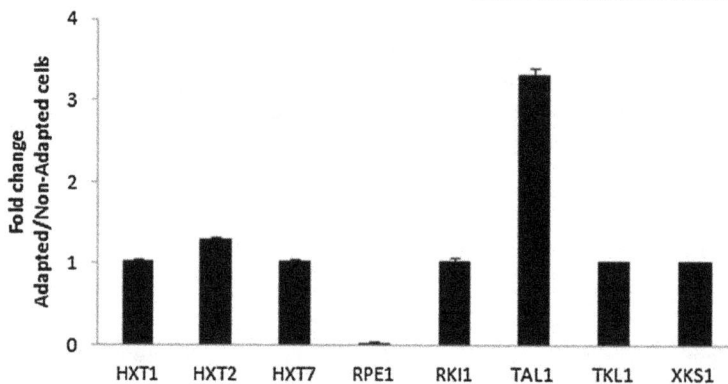

Figure 2 Gene expression profiles of genes involved with xylose utilization by *S. cerevisiae*. The real-time quantitative (qRT-PCR) was used to detect the mRNA expression level of genes involved with xylose catabolism. Gene expression, calculated as fold change compared to the endogenous control gene *TAF10*, was determined by qRT-PCR in cells harvested at the middle of log-phase of growth. Fold change between un-evolved and evolved strains was evaluated by the $2^{-\Delta\Delta CT}$ method. All the results were expressed as the mean ± standard deviation of at least three independent experiments.

medium containing 2% xylose and 2% glucose. Similarly to a previous report (Stambuk et al. 2008), our fermentation processes used high yeast cell densities and media poor in nutrients to increase ethanol yield, in detriment of cell growth rate. According to Table 1, the evolved strain showed a substantially higher xylose consumption rate when compared to the un-evolved strain (around 10-fold higher). Furthermore, the rate of xylose consumption by cells pre-grown under higher glucose concentrations was more than 80% of the rate of glucose consumption. Interestingly, the beneficial effect of glucose on xylose consumption was observed in both strains, strongly indicating that xylose metabolism is not subjected to catabolic repression. These results lead to the conclusion that the success of evolutionary engineering should not relate to the selection of unrepressed derivatives, which no longer respond to the presence of glucose, as previously suggested in the literature (Ha et al. 2011).

According to some authors, glucose repression is a significant barrier to successful utilization of mixed sugars in cellulosic hydrolyzates (Ha et al. 2011). In contrast, the present work demonstrated that the xylose specific consumption during fermentation of a glucose-xylose blend was improved when cells were previously grown on glucose, as opposed to xylose alone.

In xylose-grown cells, the glycolytic flux to piruvate is 25 to 30 times lower than in glucose-grown cells (Klimacek et al. 2010). Probably, it is a consequence of the increased expression of genes repressed via Mig1 in the xylose-grown cells (Salusjärvi et al. 2008) as Mig1 is dephosphorylated (a form in which it represses its target genes) only at high glycolytic rates (Elbing et al. 2004b). Confirming this hypothesis, it has been reported that the expression of genes encoding respiratory, tricarboxylic acid (TCA) cycle, and gluconeogenic enzymes is higher during growth on xylose than in glucose repressed cells (Elbing et al. 2004b). Thus, when growing in the presence of xylose, yeast cells switch the mode of metabolism from fermentation to respiratory, leading to a reduction in the glycolytic flux. By reducing the concentration of intracellular piruvate, respiration is favored in detriment to the ethanol production since the Km for the mitochondrial piruvate dehydrogenase alpha is 0.65 mM (Complex 1981) versus 2.29 mM for the piruvate decarboxylase 1, the key enzyme in alcoholic fermentation (Sergienko and Jordan 2001). On the other hand, xylose seems to be sensed by yeast as a non-fermentable carbon source. The presence of this pentose reduces significantly the activity of hexokinase 2 (Hxk2), a crucial regulator of the glucose repression signal in S. cerevisiae (Bergdahl et al. 2013). Hxk2 is dephosphorylated on fermentable carbon sources, whereas both phosphorylated and dephosphorylated forms exist on poorly fermentable carbon sources (Randez-Gil et al.

1998). Two of the three phosphorylated forms of Hxk2 are present in cells grown on xylose (Salusjärvi et al. 2008). Therefore, increasing the intracellular concentration of glycolytic intermediates would affect, in turn, the transcription of glycolytic and ethanologenic enzymes. This may provide the metabolic basis to explain why xylose was utilized at a higher rate in fermentation media when cells were pre-grown in the presence of increased concentrations of glucose.

Next, we compared the fermentative performance of the evolved stain to the un-evolved strain. Cells were pre-grown in YNB medium supplemented with 4% glucose until exponential growth phase and then transferred to a fermentation medium containing glucose and xylose at the same concentration and proportion found in sugar cane hydrolyzates (3.0% glucose plus 1.0% xylose). According to Figure 1A and B, the evolutionary engineering strategy improved significantly the xylose specific consumption rate besides the efficiency of ethanol production from this xylose-glucose mixture. The ethanol yield and productivity showed by the evolved strain were 13% (0.51 × 0.45 g ethanol/g sugar) and 120% (0.42 × 0.19 g ethanol/g cell/h) higher, respectively, than that of the un-evolved strain. Furthermore the evolved strain did not show xylitol accumulation, such as occurred in un-evolved strain (results not shown). The ethanol yield and productivity obtained by the strategy used in this work was equivalent, or even better, to those of strains submitted to extensive metabolic and evolutionary engineering. Recent studies described the construction of genetically modified and evolved strains of S. cerevisiae, which were able to ferment xylose as a sole carbon with productivities between 0.2 and 0.8 g ethanol/g cells/h (Zhou et al. 2012; Kim et al. 2013; Demeke et al. 2013; Hector et al. 2013). According to Cai and collaborators (Cai et al. 2012), who compared the performance of 30 engineered S. cerevisiae strains, the best result was achieved with the recombinant and evolved strain RWB218, carrying the xylose isomerase of the fungus Piromyces sp., besides several additional genetic modifications (deletion of GRE3 and over expression of XKS1, TAL1, TKL1, RKI1 and RPE1). That strain showed ethanol yield of 0.43 g ethanol/g sugar and productivity of 0.2 g ethanol/g cell/h. A similar fermentation performance was achieved by the engineered S. cerevisiae strain overexpressing Piromyces sp. XYLA, Pichia stipitis XYL3 and all genes of the non-oxidative pentose phosphate pathway, besides being submitted to a three-stage process of xylose adaptation (Zhou et al. 2012). Although most of the recombinant S. cerevisiae strains carrying heterologous xylose reductase-xylose dehydrogenase genes produced considerable amounts of xylitol at a high yield, the strain F106KR, expressing a xylose reductase, which preferred NADPH to NADH and containing several other genetic modifications, showed a high capacity to produce ethanol from high xylose concentrations (Xiong et al. 2011). The yields of F106KR from 100 g/L glucose and 100 g/L xylose in 72 h were 0.42 g ethanol/g and 0.07 g xylitol/g.

Subsequently, the expression of genes known to be involved in xylose fermentation (Additional file 1: Table S1) was analyzed. No significant differences in xylose isomerase activity between both strains were observed (data not shown). Results obtained by qRT-PCR (Figure 2) suggest that the efficient xylose fermentation showed by the evolved strain can be attributed, at least in part, to the elevated expression of TAL1, which codes for transaldolase. The evolved strain presented a 3.0 fold-higher TAL1 expression than the un-evolved strain. According to Klimacek and collaborators (Klimacek et al. 2010) the utilization of xylose instead of glucose has several effects on the yeast metabolome that are specific to anaerobic consumption of xylose. For example, the reaction catalyzed by Tal1 is strongly shifted away from its equilibrium, indicating that this reaction is a rate-limiting step for the conversion of xylose into ethanol. Confirming our study that transaldolase has a great influence on xylose utilization, it has been previously demonstrated that the overexpression of the transaldolase from Pichia stipitis in Fusarium oxysporum significantly improves ethanol production from xylose (Fan et al. 2011).

According to Figure 2, in addition to TAL1, HXT2 (encoding a low-affinity hexose transporter) was up-regulated significantly in the evolved strain. In the evolved strain, the expression of HXT2 was at least 30% higher when compared with un-evolved strain. S. cerevisiae take up xylose through the family of hexose transporters, which have a much higher affinity for glucose than xylose (Cai et al. 2012). The transporters Hxt1 to Hxt4 plus Hxt6 and Hxt7 are the most important for the uptake of xylose; they display distinct transport capacities and affinities for this pentose (Cai et al. 2012). Among them, Hxt2 has the second highest transport capacity, being able to take up xylose at a rate of 8.74 g/h/g dry weight of cell at high sugar concentrations. Therefore, it should be expected a positive effect on xylose utilization under increased HXT2 expression, such as occurred in the evolved strain. The expression of the genes which code for hexose transporters depends on the glucose concentration in the medium (Elbing et al. 2004a). In our experiments, cells were pre-grown in YNB-medium containing 4% glucose, a condition which activates HXT2 expression, which might have greatly improved the xylose utilization.

Zha et al. (2014) conducted a comparative analysis of gene expression of XDH in S. cerevisiae, which is $NADP^+$ dependent, before and after the evolutionary engineering, Which lead to an increase in the expression of RPE1. This report suggested that the increase in RPE1 expression is due to the greater need for the formation of ribose-5-phosphate, since a sufficient amount of precursor is required when cells are grown on xylose, to maintain the balance between glycolysis and aromatic amino acids and nucleic acids biosynthesis pathways. In conclusion, the results showed in our work confirmed that the reduction on RPE1 expression was observed and could be related to the low need of ribose-5-phosphate by S. cerevisiae in fermentation process, with no problem on biomass yield. Therefore maintaining the balance between glycolysis and biosynthetic pathways of aromatic amino acids and nucleic acids was not necessary.

The efficiency of the evolutionary engineering strategy used herein is promising and the fermentation performance of the xylose-evolved recombinant strain was remarkable when compared to previous reports in the literature (Cai et al. 2012; Zhou et al. 2012). However, genetic alterations that occurred during the evolution process remains to be further assessed. Additional studies are necessary to gain insight into the possible mutations which are related to the observed physiology phenotype.

Finally, the evolutionary engineering of our recombinant yeast strain fermented glucose and xylose rapidly and almost simultaneously, showing a substantial improvement in ethanol production and productivity. It was also observed that when cells were grown in a medium containing higher glucose concentration, before being transferred to fermentation medium, higher xylose consumption rates were obtained, demonstrating that xylose utilization was not regulated by catabolic repression. In addition, results obtained by qRT-PCR suggested that the efficiency in xylose fermentation should be attributed to, at least in part, the increasing on HXT2 and TAL1 expression.

Additional file

Additional file 1: Table S1. The target genes used in the gene expression studies.

Competing interests

The authors declare that they have are no competing interests and certify that there is no conflict of interest with any financial or non-financial organization regarding the contents of the manuscript.

Authors' contributions

LV carried out the experimental work, participated in its design, coordination and writing the manuscript, VA participated in the acquisition of data, RP helped with the acquisition of data, EB participated in the project's conception and coordination, FT participated in its design and coordination, BN participated in its design, coordination and critical review of the manuscript, EE participated in its conception, design, coordination and writing of the manuscript. All authors have read and approved the final manuscript.

Acknowledgements

This work was supported by FINEP, CNPq and CAPES.

Author details

[1]Department of Biochemistry, Institute of Chemistry, Federal University of Rio de Janeiro, Rio de Janeiro, Brazil. [2]Department of Cellular Biology, Institute of Biology, University of Brasília, Brasília, DF, Brazil.

References

Bergdahl B, Sandström AG, Borgström C, Boonyawan T, van Niel EWJ, Gorwa-Grauslund MF (2013) Engineering yeast hexokinase 2 for improved tolerance toward xylose-induced inactivation. PLoS One 8:e75055

Cai Z, Zhang B, Li Y (2012) Engineering *Saccharomyces cerevisiae* for efficient anaerobic xylose fermentation: reflections and perspectives. Biotechnol J 7:34–46

Complex PD (1981) Pyruvate Dehydrogenase Complex from Baker's Yeast 1: Purification and Some Kinetic and Regulatory Properties. Eur J Biochem 579:573–9

De Figueiredo VL, de Mello VM, Reis VCB, Bon EPDS, Gonçalves Torres FA, Neves BC, Eleutherio ECA (2013) Functional expression of *Burkholderia cenocepacia* xylose isomerase in yeast increases ethanol production from a glucose-xylose blend. Bioresour Technol 128:792–6

Demeke MM, Dietz H, Li Y, Foulquié-Moreno MR, Mutturi S, Deprez S, Den Abt T, Bonini BM, Liden G, Dumortier F, Verplaetse A, Boles E, Thevelein JM (2013) Development of a D-xylose fermenting and inhibitor tolerant industrial *Saccharomyces cerevisiae* strain with high performance in lignocellulose hydrolysates using metabolic and evolutionary engineering. Biotechnol Biofuels 6:89

Eiteman MA, Lee SA, Altman E (2008) A co-fermentation strategy to consume sugar mixtures effectively. J Biol Eng 2:3

Elbing K, Larsson C, Bill RM, Albers E, Snoep JL, Boles E, Hohmann S, Gustafsson L, Goethe-universita W (2004a) Role of Hexose Transport in Control of Glycolytic Flux in *Saccharomyces cerevisiae* Role of Hexose Transport in Control of Glycolytic Flux in Saccharomyces cerevisiae. Appl. Environ, Microbiol

Elbing K, Ståhlberg A, Hohmann S, Gustafsson L (2004b) Transcriptional responses to glucose at different glycolytic rates in *Saccharomyces cerevisiae*. Eur J Biochem 271:4855–64

Fan JX, Yang Q, Liu ZH, Huang XM, Song JZ, Chen ZX, Sun Q, Wang S (2011) The characterization of transaldolase gene tal from *Pichia stipitis* and its heterologous expression in *Fusarium oxysporum*. Mol Biol Rep 38(3):1831–40

Ferreira JC, Thevelein JM, Hohmann S, Paschoalin VMF, Trugo LC, Panek AD (1997) Trehalose accumulation in mutants of *Saccharomyces cereÕisiae* deleted in the UDPG-dependent trehalose synthase-phosphatase complex. Biochim. Biophys. Acta - Gen. Subj. 40–50

Ha S-J, Galazka JM, Kim SR, Choi J-H, Yang X, Seo J-H, Glass NL, Cate JHD, Jin Y-S (2011) Engineered *Saccharomyces cerevisiae* capable of simultaneous cellobiose and xylose fermentation. Proc Natl Acad Sci U S A 108:504–9

Hector RE, Dien BS, Cotta M, Mertens J (2013) Growth and fermentation of D-xylose by *Saccharomyces cerevisiae* expressing a novel D-xylose isomerase originating from the bacterium *Prevotella ruminicola* TC2-24. Biotechnol biofuels 6:84

Kim SR, Ha S-J, Wei N, Oh EJ, Jin Y-S (2012) Simultaneous co-fermentation of mixed sugars: a promising strategy for producing cellulosic ethanol. Trends Biotechnol 30:274–82

Kim SR, Park Y-C, Jin Y-S, Seo J-H (2013) Strain engineering of *Saccharomyces cerevisiae* for enhanced xylose metabolism. Biotechnol Adv 31:851–61

Klimacek M, Krahulec S, Sauer U, Nidetzky B (2010) Limitations in xylose-fermenting *Saccharomyces cerevisiae*, made evident through comprehensive metabolite profiling and thermodynamic analysis. Appl Environ Microbiol 76:7566–74

Livak KJ, Schmittgen TD (2001) Analysis of relative gene expression data using real-time quantitative PCR and the 2(−Delta Delta C(T)) Method. Methods (San Diego, Calif) 25:402–8

Randez-Gil F, Sanz P, Entian KD, Prieto J a (1998) Carbon source-dependent phosphorylation of hexokinase PII and its role in the glucose-signaling response in yeast. Mol Cell Biol 18:2940–8

Rolland F, Winderickx J, Thevelein JM (2002) Glucose-sensing and -signalling mechanisms in yeast. FEMS Yeast Res 2:183–201

Salusjärvi L, Kankainen M, Soliymani R, Pitkänen J-P, Penttilä M, Ruohonen L (2008) Regulation of xylose metabolism recombinant *Saccharomyces cerevisiae*. Microb Cell Fact 7:18

Seo H-B, Kim H-J, Lee O-K, Ha J-H, Lee H-Y, Jung K-H (2009) Measurement of ethanol concentration using solvent extraction and dichromate oxidation and its application to bioethanol production process. J Ind Microbiol Biotechnol 36:285–92

Sergienko E, Jordan F (2001) Catalytic acid–base groups in yeast pyruvate decarboxylase. 2. Insights into the specific roles of D28 and E477 from the rates and stereospecificity of formation of carboligase side products. Biochemistry 40:7369–81

Stambuk BU, Eleutherio ECA, Marina L, Maria FA, Bon EPS (2008) Brazilian potential for biomass ethanol : Challenge of using hexose and pentose co- fermenting yeast strains. J Sci Ind Res 67:918–26

Stickland H (1951) of of by. The Determination of Small Quantities of Bacteria by means of the Biuret Reaction. J Gen Microbiol 5:698–703

Subtil T, Boles E (2012) Competition between pentoses and glucose during uptake and catabolism in recombinant *Saccharomyces cerevisiae*. Biotechnol biofuels 5:14

Teste M-A, Duquenne M, François JM, Parrou J-L (2009) Validation of reference genes for quantitative expression analysis by real-time RT-PCR in *Saccharomyces cerevisiae*. BMC Mol Biol 10:99

Weber C, Farwick A, Benisch F, Brat D, Dietz H, Subtil T, Boles E (2010) Trends and challenges in the microbial production of lignocellulosic bioalcohol fuels. Appl Microbiol Biotechnol 87:1303–15

Xiong M, Chen G, Barford J (2011) Alteration of xylose reductase coenzyme preference to improve ethanol production by *Saccharomyces cerevisiae* from high xylose concentrations. Bioresour Technol 102:9206–15

Zha J, Shen M, Hu M, Song H, Yuan Y (2014) Enhanced expression involved in initial xylose metabolism and the oxidative pentose phosphate pathway in the improved xylose-utilizing Saccharomyces cerevisiae through evolutionary engineering. J Ind Microbiol Biotechnol 41:27–39

Zhou H, Cheng J-S, Wang BL, Fink GR, Stephanopoulos G (2012) Xylose isomerase overexpression along with engineering of the pentose phosphate pathway and evolutionary engineering enable rapid xylose utilization and ethanol production by *Saccharomyces cerevisiae*. Metab Eng 14:611–22

16

Combining the effects of process design and pH for improved xylose conversion in high solid ethanol production from *Arundo donax*

Benny Palmqvist[*] and Gunnar Lidén

Abstract

The impact of pH coupled to process design for the conversion of the energy crop *Arundo donax* to ethanol was assessed in the present study under industrially relevant solids loadings. Two main process strategies were investigated, i.e. the traditional simultaneous saccharification and co-fermentation (SSCF) and a HYBRID design, where a long high temperature enzymatic hydrolysis step was carried out prior to continued low temperature SSCF, keeping the same total reaction time. Since acetic acid was identified as the major inhibitor in the slurry, the scenarios were investigated under different fermentation pH in order to alleviate the inhibitory effect on, in particular, xylose conversion. The results show that, regardless of fermentation pH, a higher glucan conversion could be achieved with the HYBRID approach compared to SSCF. Furthermore, it was found that increasing the pH from 5.0 to 5.5 for the fermentation phase had a large positive effect on xylose consumption for both process designs, although the SSCF design was more favored. With the high sugar concentrations available at the start of fermentation during the HYBRID design, the ethanol yield was reduced in favor of cell growth and glycerol production. This finding was confirmed in shake flask fermentations where an increase in pH enhanced both glucose and xylose consumption, but also cell growth and cell yield with the overall effect being a reduced ethanol yield. In conclusion this resulted in similar overall ethanol yields at the different pH values for the HYBRID design, despite the improved xylose uptake, whereas a significant increase in overall ethanol yield was found with the SSCF design.

Keywords: Bioethanol; High solids loading; Xylose fermentation; SSCF; Enzymatic hydrolysis

Introduction

The lignocellulosic ethanol industry is moving from pilot scale to demonstration/full scale operation, which is evidenced by the construction of several production facilities worldwide (Balan et al. 2013; Janssen et al. 2013). Driven by this transition, research has been intensified towards problems related to high solid operation, an important factor for process economy (Humbird et al. 2010; Macrelli et al. 2012). By increasing the content of water insoluble solids (WIS) throughout the process a number of benefits can be gained, e.g. reduced water usage with lowered distillation and waste water treatment costs as a result. Benefits can also be gained by reductions in investment and production costs since equipment size and energy consumption can be reduced (Galbe et al. 2007; Wingren et al. 2003).

Looking at some of the larger plants in operation today, various agricultural residues (or in some cases dedicated energy crops) are used as raw material to a large extent. One example is Beta Renewables' full scale plant in Crescentino, Italy, which will convert wheat straw and *Arundo donax* (is a perennial cane used as a dedicated energy crop) to bioethanol, with biogas and bioelectricity as the main co-products. A common feature of these straw based raw materials is their relatively high content of the C5 sugar xylose (Wiselogel et al. 1996), which makes good xylose conversion another high priority target in order to reach an economically feasible process (Sassner et al. 2008). Today a number of *Saccharomyces cerevisiae* strains (the common workhorse in the bioethanol industry) have been genetically engineered in order to convert also xylose to ethanol. The

---
* Correspondence: benny.palmqvist@chemeng.lth.se
Department of Chemical Engineering, Lund University, Box 124, SE-221 00 Lund, Sweden

main genetic modifications made are the insertion of either a bacterial xylose isomerase, or a fungal xylose reductase and xylitol dehydrogenase, together with over-expression of several genes in the pentose phosphate pathway (PPP) in order to convert xylose to xylulose and further on to ethanol through the PPP (Almeida et al. 2011; Van Vleet and Jeffries 2009; Hahn-Hägerdal et al. 2007). Although impressive achievements have been accomplished with genetic and evolutionary engineering, glucose is still the preferred substrate over xylose. It has, however, previously been shown that xylose conversion can be increased with clever process design (Olofsson et al. 2010a; Olofsson et al. 2008; Olofsson et al. 2010b).

As previously mentioned, working at high WIS content potentially improves process economy. However, high solid operation has also been shown to generally decrease the yields of both enzymatic hydrolysis and simultaneous saccharification and fermentation (SSF) (Kristensen et al. 2009b). Two of the main issues when increasing the WIS loading are the dramatically increased viscosities as a result of the fibrous nature of the biomass (Knutsen and Liberatore 2009; Roche et al. 2009; Viamajala et al. 2009; Wiman et al. 2011) and the increased concentrations of biomass degradation products, e.g. hydroxymethylfurfural (HMF), furfural and acetic acid, which potentially inhibit the fermenting microorganism (Almeida et al. 2011). The formation of inhibitory degradations products, for example HMF and furfural, can be avoided by designing a mild pretreatment step. Acetic acid, in contrast, is inherent in the biomass material itself where acetyl groups are present on the xylan backbone. During pretreatment (and possibly enzymatic hydrolysis) the acetyl groups are released from the hemicellulose, hence forming acetic acid, which may inhibit the fermentation. It has been shown that the inhibitory effects can be decreased by operating at a higher pH-value, since it is the undissociated form of the acid that causes inhibition. The pKa value of acetic acid is 4.76, so large effects can be anticipated around a pH-value of 5. The positive effect has been shown to be particularly strong for xylose fermentation (Bellissimi et al. 2009; Casey et al. 2010).

The increased viscosities of the high solid slurries can create mixing problems in the reactors (Viamajala et al. 2010) as well as problems in pumping of the slurry. Well mixed hydrolysis, and fermentation, processes are important in order to avoid temperature, pH and concentration gradients, since these could cause yield losses. One way to quickly reduce viscosity, and ease mixing, is the introduction of a high temperature hydrolysis step, commonly referred to as viscosity reduction (VR), or liquefaction (Jorgensen et al. 2007), prior to the more traditional SSF concept. This HYBRID process design,

where a high temperature hydrolysis is followed by a low temperature fermentation (and continued hydrolysis) of the whole remaining slurry is potentially a good process option. It allows (partly) independent optimization of the hydrolysis and fermentation steps, which is beneficial since the optimal temperature for the cellulose mixture typically is about 50°C whereas the yeast grows optimally at temperatures around 30–35°C. Furthermore, it may be beneficial to increase the fermentation pH above 5.0 (the typical optimum for enzymatic hydrolysis) in order to decrease the toxic effects of the hydrolyzate if high amounts of acetic acid are present. The improvement of commercially available enzyme mixtures (McMillan et al. 2011) works towards favoring of this HYBRID process concept due to improved temperature stability and decreased end-product inhibition, as argued by Cannella and Jorgensen (2013). In the work by Cannella and Jorgensen, however, the added complexity of co-fermenting xylose and glucose to ethanol was not addressed. It is well-known that high glucose concentrations inhibit xylose uptake by the yeasts (Lee et al. 2002; Saloheimo et al. 2007). Therefore, in the case of co-fermentation of xylose, the SSCF process (with "C" indicating co-fermentation) has the advantage over a HYBRID process that you can keep a low, but non-zero, concentration of glucose in the media, which has been proven very beneficial for xylose consumption (Bertilsson et al. 2008; Meinander et al. 1999).

In this work, we assess how ethanol yields and xylose conversion are affected by the choice of process design at different pH levels. The raw material used in the study is pretreated *Arundo donax* at industrially relevant solids concentrations. Importantly, by analyzing fiber composition after each experiment the effects on the enzymatic hydrolysis, and the fermentation of both xylose and glucose could be assessed separately for the different scenarios. The effects of an increased pH at high acetic acid and sugar concentration were furthermore assessed separately in shake flask fermentations of fiber-free *Arundo donax* hydrolyzate.

Materials and methods
Raw materials
Steam pretreated *Arundo donax* slurry was kindly provided by Biochemtex S.p.A. Italia (Rivalta, Italy). The material was kept frozen until used. The fiber composition of the pretreated *Arundo donax* as well as the soluble sugars (Table 1) was determined according to NREL (National renewable energy laboratory) procedures (Sluiter et al. 2008a; Sluiter et al. 2008b). It can be noted from Table 1 that the major part of the soluble sugars still remains in oligomeric form after pretreatment and that the major inhibitor present is acetic acid. The WIS content of the material was determined to 22.5% by washing repeatedly with deionized water over filter paper (Whatman No.1).

Table 1 Composition of the pretreated *Arundo donax* slurry

Solid composition (% of WIS)

	Average	Std dev
Glucan	48.2	0.2
Xylan	3.8	0.0
Galactan	n.d[a]	-
Arabinan	n.d	-
Mannan	n.d	-
Lignin	41.7	0.0

Soluble components (g L^{-1})

Sugars	Monomers	Total sugars (including monomers)
Glucose	2.5	14.2
Xylose	4.0	18.4
Galactose	n.d	n.d
Arabinose	n.d	n.d
Mannose	n.d	n.d

Inhibitors and degradation products

Acetic Acid	5.6
HMF	n.d
Furfural	0.2

[a]n.d. = not detected, i.e. below detection limit.
The solid composition is based on wt-% of the WIS content and the soluble components are reported in g L^{-1} liquid. The WIS content of the pretreated material was measured to 22.5 wt-%.

Cell cultivation

The recombinant xylose-fermenting strain *S. cerevisiae* TMB3400 (Wahlbom et al. 2003) was used in all experiments. The yeast was produced by an initial pre-culture in shake flask, followed by an aerobic batch cultivation on glucose, and finally an aerobic fed-batch cultivation on *Arundo donax* hydrolyzate liquid, in order to improve inhibitor tolerance by adaptation as previously shown by Alkasrawi *et al.* (Alkasrawi et al. 2006).

The yeast was inoculated (from agar plate) in 300 ml shake flasks (liquid volume of 100 ml) containing 16.5 g L^{-1} glucose, 7.5 g L^{-1} (NH$_4$)$_2$SO$_4$, 3.5 g L^{-1} KH$_2$PO$_4$, 0.74 g L^{-1} MgSO$_2$·7H$_2$O, trace metals and vitamins. The cells were grown for 24 h at 30°C and a starting pH of 5.2 in a rotary shaker at 180 rpm. Subsequently, aerobic batch cultivation was performed in a 2.5 L bioreactor (Biostat A, B. Braun Biotech International, Melsungen, Germany) at 30°C. The working volume was 0.7 L and the medium contained 20.0 g L^{-1} glucose, 20.0 g L^{-1} (NH$_4$)$_2$SO$_4$, 10.0 g L^{-1} KH$_2$PO$_4$, 2.0 g L^{-1} MgSO$_4$, 27.0 mL L^{-1} trace metal solution and 2.7 mL L^{-1} vitamin solution. The cultivation was initiated by adding 20.0 mL of the pre-culture to the bioreactor. The pH was maintained at 5.0 throughout the cultivation,

by automatic addition of 3 M NaOH. The trace metal and vitamin solutions were prepared according to Taherzadeh *et al.* (Taherzadeh et al. 1996). Aeration was maintained at 1.2 L min^{-1}, and the stirrer speed was kept at 800 rpm. When the ethanol produced in the batch phase was depleted, the feeding of clarified hydrolyzate liquid (obtained by pressing the Arundo donax slurry) was initiated. 1.0 L of liquid (supplemented with 35 g/L glucose to ensure a sufficiently high final cell concentration) was fed to the reactor with an initial feed rate of 0.04 L h^{-1} which was increased linearly to 0.10 L h^{-1} during 16 h of cultivation. The aeration during the fed-batch phase was maintained at 1.5 L min^{-1}, and the stirrer speed was kept at 800 rpm.

After cultivation, the cells were harvested by centrifugation in 700 mL flasks using a HERMLE Z 513 K centrifuge (HERMLE Labortechnik, Wehingen, Germany). The pellets were resuspended in 9 g L^{-1} NaCl-solution in order to obtain a cell suspension with a cell mass concentration of 80 g dry weight L^{-1}. The time between cell harvest and initiation of the following SSCF/shake flask fermentation was no longer than 3 h.

Hybrid SSCF experiments

All experiments were carried out under anaerobic conditions using 2.5 L bioreactors (Biostat A, B. Braun Biotech International, Melsungen, Germany) with an initial WIS content of 21% and a final working broth weight of 1.0 kg. The experiments were run for a total time of 96 h. SSCF experiments were compared to HYBRID experiments were a high-temperature hydrolysis step had been conducted for the first 48 hours (keeping a total time of 96 hours). During the enzymatic hydrolysis a temperature of 45°C was maintained and when yeast was added (i.e. during the SSCF phase) the temperature was lowered and kept at 34°C. The enzyme solution used was Cellic CTec 3 provided by Novozymes (Novozymes A/S, Bagsvaerd, Denmark) at a dose of 0.075 g enzyme solution g^{-1} glucan. The pH was maintained at either 5.0 or 5.5 throughout the fermentations by automatic addition of 4 M NaOH. The *Arundo donax* slurry was supplemented with 0.5 g L^{-1} NH$_4$H$_2$PO$_4$, 0.025 g L^{-1} MgSO$_4$·7H$_2$O and 1.0 g L^{-1} yeast extract at the start of the fermentation phase. An initial yeast concentration of 3 g dry weight L^{-1} of the cultivated cells was used. In addition, a hydrolysis experiment was performed at 34°C in order to see how well the material hydrolyzed at the lower SSCF temperature. All experiments were carried out in duplicates.

Shake flask fermentations

100 mL shake flask fermentations were carried out in duplicates under anaerobic conditions in 300 mL Erlenmeyer flask. The medium used for the fermentations was the clarified liquid fraction obtained by separating the

remaining solids (with vacuum filtration) after enzymatic hydrolysis of steam pretreated *Arundo donax*. The enzymatically hydrolyzed steam pretreated material was provided by Biochemtex and the separation of solids was carried out at Lund University. Two different acetic acid levels were investigate by first diluting the hydrolyzate slightly to achieve an acetic acid concentration of approximately 4 g L^{-1} and then supplementing one part of it with acetic acid to reach 8 g L^{-1}. pH was set initially to 5.0, 5.5 and 6.0, respectively, for both acetate levels. Prior to fermentation the medium was autoclaved, to ensure sterile conditions, and supplemented with 0.5 g L^{-1} $NH_4H_2PO_4$, 0.025 g L^{-1} $MgSO_4 \cdot 7H_2O$ and 1.0 g L^{-1} yeast extract (same supplements as in the hybrid SSCF experiments). The temperature was kept constant at 34°C. An initial yeast concentration of 3 g dry weight L^{-1} of cultivated yeast was added to start the experiment. 2.5 mL liquid samples were taken repeatedly during 48 hours and analyzed for sugars and metabolites as well as pH and optical density.

Analysis

HPLC was used for analysis of sugars and metabolites. Samples from the hydrolysis liquid were centrifuged (16,000 × g) in 2 mL eppendorf tubes at 14,000 rpm for 5 min. (Z 160 M, HERMLE Labortechnik, Wehingen, Germany). The supernatant was filtered through 0.2 μm filters, and stored at −20°C. The concentration of sugars, glycerol and xylitol were determined using a polymer column (Aminex HPX-87P, Bio-Rad Laboratories, München, Germany) at 85°C. MilliQ-water was used as eluent, with a flow rate of 0.6 mL min^{-1}. Organic acids, furans and ethanol concentrations were determined with an Aminex HPX-87H column (Bio-Rad Laboratories, München, Germany) at 60°C. 50 mM H_2SO_4 was used as eluent, with a flow rate of 0.6 mL min^{-1}. The sugars, acetate, glycerol, xylitol and ethanol were detected with a refractive index detector (Waters 2410; Waters, Milford, MA, USA) and HMF and furfural with a UV detector at a wavelength of 210 nm (Waters 2487; Waters, Milford, MA, USA). A larger sample volume was taken for the final (96 h) and liquid densities were determined (solids separated by centrifugation and filtration, as described above) by pipeting 0.500 ml on to an analytical scale to record the weight. Triplicates were performed with a standard deviation of less than 1%.

Optical density (OD) was measured with a spectrophotometer at 600 nm (Helios Gamma; Spectronic Camspec Ltd, Leeds, UK) in order to follow cell growth during the shake flask fermentations. The OD values were then correlated to cell dry-weight through a calibration curve obtained from dry-weight measurements and dilutions of the concentrated yeast suspension obtained after the cultivation procedure.

A one-way analysis of variance (ANOVA) was performed in MATLAB R2011b (Mathworks, Natick, USA) to established significant differences in fermentation yields and xylose uptake for both the reactor and shake flask experiments.

Calculation of yields and carbon recovery

The standard assumptions when calculating glucose and ethanol yields are generally those of a constant volume and liquid density throughout the reaction. However, significant errors in yield calculations will result from using these assumptions during high solids operation (Kristensen et al. 2009a; Zhu et al. 2011). Therefore, in order to accurately estimate the hydrolysis yield in this study, we measured the WIS content and analyzed the solid composition according to standard NREL procedure (Sluiter et al. 2008b) after each fermentation. Hence the hydrolysis yields (Y_{Hyd}) could be calculated as the difference in total amount of glucan between start and end samples, according to Equation 1. The degree of xylan hydrolysis was calculated analogously.

In all following equations, m_{reac} denotes the total mass of the slurry, WIS the fraction of water insoluble solids in the total slurry (g/g), x_i the mass fraction of the respective polymer in the insoluble solids (g/g) and ϕ_i the molecular ratio of the polymer to its corresponding monomer. V_{Liq} denotes the volume of hydrolysate liquid (L) and ρ_{Liq} the corresponding density of the liquid (g/L). [i] denotes the concentration (g/L) of compound i, measured by HPLC, and M_i the c-mole mass of the compound. The short notations for the compounds are Glu (glucose), Xyl (xylose), Xyli (xylitol), Gly (glycerol), Cel (cellobiose), X (biomass) and EtOH (ethanol). The subscript tot, indicates the concentration determined by total sugar analysis for the respective sugars and subscript 0 and end indicates values at 0 and 96 hour respectively. The initial and final mass, m_{reac}, will be slightly different due to the loss of carbon dioxide in the fermentation. The loss will be in the order of a few percent, and was in the calculations made here not compensated for.

$$Y_{Hyd} = \frac{m_{reac\,0}\,WIS_0 X_{Glu\,0} - m_{reac\,end}\,WIS_{end} X_{Glu\,end}}{m_{reac\,0}\,WIS_0 X_{Glu\,0}} \quad (1)$$

In order to further, calculate the ethanol yield based on measured HPLC concentrations, liquid densities were measured for both start and final samples (in addition to the WIS and glucan content). The fermentation yield, Y_{EtOH}, (based on the amount of consumed glucose and xylose) could then be calculated on a mass basis according to Equation 2, and the technical ethanol yield, $Y_{EtOH,Tech}$, (based on the total amount of added glucose and xylose) could be calculated according to Equation 3. Both equations are based on an equation previously presented by Kristensen *et al.* (Kristensen et al. 2009a).

$$Y_{EtOH} = \frac{[EtOH]_{End} V_{Liq_{End}}}{m_{reac_0} WIS_0 \left(\varphi_{Glu} X_{Glu0} + \varphi_{Xyl} X_{Xyl0}\right) - m_{reac_{end}} WIS_{end} \left(\varphi_{Glu} X_{Glu_{end}} + \varphi_{Xyl} X_{Xyl_{end}}\right) + \left([Glu_{tot}]_0 + [Xyl_{tot}]_0\right) V_{Liq_0} - \left([Glu]_{End} + [Xyl]_{End}\right) V_{Liq_{End}}}$$

(2)

$$Y_{EtOH_{tech}} = \frac{[EtOH]_{End} V_{Liq_{End}}}{m_{reac_0} WIS_0 \left(\varphi_{Glu} X_{Glu0} + \varphi_{Xyl} X_{xyl0}\right) + \left([Glu_{tot}]_0 + [Xyl_{tot}]_0\right) V_{Liq_0}}$$

(3)

The liquid volume was calculated according to Equation 4.

$$V_{Liq} = \frac{m_{reac}(1 - WIS)}{\rho_{liq}}$$

(4)

A calculation of the carbon recovery (CR), on c-mole basis, was done for both the hybrid SSCF and shake flask experiments in order to see if all major products had been accounted for. Since carbon dioxide was not measured the amount was instead estimated based on stoichiometry from the ethanol production, hence the factor 1.5 (c-mole c-mole^{-1}) in Equation 5. Furthermore, due to (well established) difficulties in estimating cell dry weight in fiber slurries, *biomass was not included* in the recovery calculations for the SSCF/HYBRID experiments, i.e. part of the missing carbon will be biomass. Biomass was, however, included in carbon recovery for the shake flask experiments.

$$CR = \frac{\left(\frac{1.5[EtOH]_{End}}{M_{EtOH}} + \frac{[Gly]_{End}}{M_{Gly}} + \frac{[Xyl]_{End}}{M_{Xyl}} + \frac{[Xyli]_{End}}{M_{Xyli}} + \frac{[Cel]_{End}}{M_{Cel}} + \frac{[Glu_{End}]}{M_{Glu}} + \frac{[X]_{End}}{M_X}\right) V_{Liq_{End}} + m_{reac_{end}} WIS_{end} \left(\frac{\varphi_{Glu} X_{Glu_{End}}}{M_{Glu}} + \frac{\varphi_{xyl} X_{Xyl_{End}}}{M_{Xyl}}\right)}{m_{reac_0} WIS_0 \left(\frac{\varphi_{Glu} X_{Glu0}}{M_{Glu}} + \frac{\varphi_{xyl} X_{Xyl0}}{M_{Xyl}}\right) + \left(\frac{[Glu_{tot}]_0}{M_{Glu}} + \frac{[Xyl_{tot}]_0}{M_{Xyl}} + \frac{[X]_0}{M_X}\right) V_{Liq_0}}$$

(5)

Results

Experiments were made to compare the standard SSCF strategy to a HYBRID scenario in which a 48 hour high temperature hydrolysis step was conducted prior to continued hydrolysis and fermentation, at lower temperature, keeping a total process time of 96 hours. The strategies were investigated at two different pH levels, in order to elucidate how the inhibitory effect of the present acetic acid affected especially xylose consumption during the different process designs (Table 2). Post-fermentation material analyses of the residual solids were furthermore conducted for all the experiments and with these analysis it was possible to calculate the amount of released sugars and hence also the amount consumed by the yeast during fermentation. Based on this, the hydrolysis performance could be separated from the fermentation performance and be assessed individually for the different process designs. Furthermore the analysis allowed for the calculation of carbon recovery for each experiment by taking volume and density changes during the process into account (Eq 5). The carbon balance could be closed around 95% for the enzymatic hydrolysis experiment, whereas for the HYBRID/SSCF experiments where biomass could not be measured (and CO2 only estimated), the carbon recovery was slightly reduced, to around 90% (Table 2). This indicates that all major compounds have been considered.

The effects of changing pH at different process designs

Increasing the pH from 5.0 to 5.5 during the SSCF design resulted in approximately a 10% increase in final ethanol concentration – i.e. from 35 to 39 g L^{-1} (Figure 1). Likewise, a significant increase in xylose consumption was achieved at the higher pH level. It should be noted that the xylose concentration in Figure 1 (and Figure 2) is the pseudo-steady state concentration generated by simultaneous xylan hydrolysis and xylose fermentation. Post material analysis however showed no significant difference in hydrolysis degree between the different pH levels (Table 2), hence the consumption can be a assumed to have increased. The post-fermentation material analysis did not show any significant difference in the degree of hydrolysis (Table 2), suggesting that the increased ethanol yield is a result of increased xylose fermentation. In contrast, when the pH in the fermentation phase of the HYBRID process was increased, no corresponding increase in ethanol could be observed (Figure 2), despite enhanced xylose consumption.

Table 2 Hydrolysis and fermentation performance for the different process designs

	SSCF pH 5.0	HYBRID pH 5.0	SSCF pH 5.5	HYBRID pH 5.5	Hydrolysis 34°C
Hydrolysis performance					
Degree of glucan hydrolysis (%)	49.4 ± 0.8	54.2 ± 0.5	50.9 ± 3.8	54.0 ± 1.1	45.2 ± 0.3
Degree of xylan hydrolysis (%)	35.9 ± 2.7	43.0 ± 1.8	38.1 ± 5.3	43.3 ± 0.5	31.1 ± 0.6
Fermentation performance					
Ethanol yield (g/g consumed sugars)	0.42 ± 0.01[A]	0.43 ± 0.00 [A]	0.40 ± 0.02 [A]	0.39 ± 0.02 [A]	-
Ethanol yield (% of theoretical)	81.7 ± 1.9	85.0 ± 0.1	79.3 ± 4.7	76.4 ± 3.5	-
Glycerol yield (g/g consumed sugars)	0.015 ± 0.002	0.031 ± 0.004	0.028 ± 0.004	0.039 ± 0.003	
Consumed xylose [%]	40.2 ± 3.8	33.7 ± 9.2	78.3 ± 2.0	55.8 ± 5.5	-
Xylitol production (% of consumed xylose)	32.0 ± 0.1	15.9 ± 0.2	25.5 ± 0.2	19.2 ± 0.4	-
Technical ethanol yield [%]	40.3 ± 0.7 [A]	45.2 ± 1.3 [B]	44.4 ± 0.4 [B]	42.7 ± 0.8 [A B]	-
Calculated carbon recovery[*] (excluding cell growth)	0.93 ± 0.01	0.94 ± 0.00	0.92 ± 0.03	0.90 ± 0.02	0.96 ± 0.00

[*]Carbon recovery calculated according to Eq. 5 in material and methods, with CO_2 estimated based on ethanol production.
Yields are based on the final (96 hour) values, and standard deviations are based on duplicate experiments. All statistically compared mean values are denoted with one or several letters (A, B). Values labeled with the same letter are not significantly different at a confidence level of 95%.

Since no significant differences in hydrolysis performance was observed (Figure 2A and Table 2), this points towards a decrease in fermentation yield (although no statistically significant difference was found) at the higher pH. A potential yield reduction could be related to the faster initial fermentation rate observed at the less inhibiting conditions at the higher pH (Figure 2B). At less inhibiting conditions, cell growth and associated glycerol production is favored, which typically reduces the ethanol yield. Cell growth could not be measured in the slurries, but an increase was indicated by the higher glycerol production at the higher pH level (Table 2).

Furthermore, the degree of hydrolysis was higher in the HYBRID process design in comparison to the SSCF

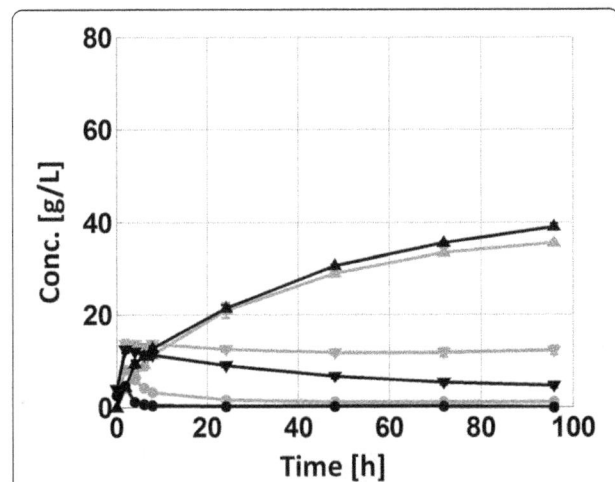

Figure 1 Concentration profiles throughout the SSCF experiments at pH 5.0 (grey) and 5.5 (black). Glucose (●), xylose (▼) and ethanol (▲). The error bars represent standard deviation of duplicate experiments.

case. As a reference to the SSCF experiment a 96 hour enzymatic hydrolysis experiment was run at 34°C. As expected, a very significant temperature effect on the enzymatic hydrolysis was found and a similar sugar release was obtained after a 48 hour hydrolysis at 45°C as after 96 h at 34°C (data not shown). It was also found that the overall glucan conversion was increased in SSCF compared to an enzymatic hydrolysis at the same temperature, indicating end product inhibition (Table 2).

Fermentation at different pH and acetic acid concentrations

Shake flask fermentations were set up in order to further evaluate the effects of pH on acetic acid inhibition at two different acetic acid concentrations (4 and 8 g L^{-1}) during fermentation of hydrolyzate liquid. Three different pH levels (5.0, 5.5 and 6.0) were used, resulting in a set of 6 different conditions. The xylose consumption was clearly enhanced by increased pH, regardless of acetic level (Table 3). However, it was also seen that the ethanol yield (based on consumed sugars) was reduced at the less inhibiting conditions, i.e. high pH and low acetic acid concentration. By measuring OD it could be confirmed that cell growth was indeed promoted together with enhanced glucose consumption when increasing pH at high acetic acid concentrations (Figure 3). However, under milder acetic acid conditions, i.e. 4 g L^{-1}, the effect of increasing pH above 5.0 was not as evident on cell growth (data not shown). One need also keep in mind that these experiments were made without fibers.

Discussion

The present work showed that the HYBRID design, i.e. a high temperature enzymatic hydrolysis step prior to

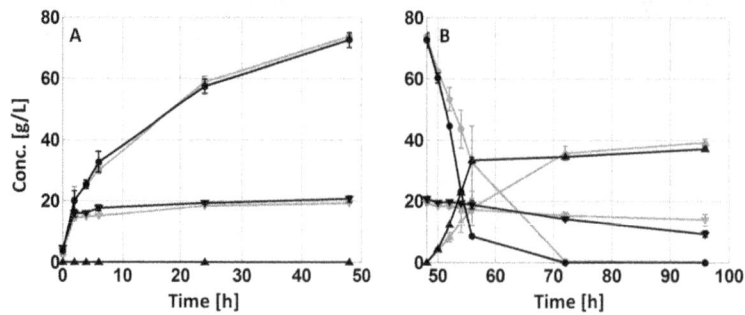

Figure 2 Concentration profiles throughout the HYBRID process design at pH 5.0 (grey) and 5.5 (black). **A)** The 48 hour high temperature hydrolysis. **B)** The SSCF phase during the final 48 hours. Glucose (●), xylose (▼) and ethanol (▲). The error bars represent standard deviation of duplicate experiments.

fermentation, gave a higher overall glucan conversion compared to a 96 hours SSCF process (Table 2). This is well in-line with results reported by Cannella and Jorgensen for pretreated wheat straw (Cannella and Jørgensen 2013), and supports their claim that commercial enzyme mixtures available today are less end-product inhibited and exhibit improved long time temperature stability. It deserves to be pointed out, that the advantage comes from the high temperature for enzymatic hydrolysis. If SSCF and pure hydrolysis is compared at the same (permissible) temperature, e.g. 34°C as done in this study, a better glucan conversion is indeed obtained for the SSCF case (Table 2–first and last columns). This confirms that SSCF

most likely still gives a reduced end-product inhibition of the enzymes. However, this decreased end-product inhibition does not outweigh the enhanced enzymatic activity at the higher temperatures, which resulted in a better overall glucan conversion in the HYBRID process design (Table 2).

With respect to fermentation, the fundamental difference between the two process designs is the availability of glucose throughout the fermentation. During the HYBRID design, a large fraction of the total amount of both glucose and xylose that are to be fermented to ethanol are present from the start of the fermentation, whereas during SSCF the glucose is slowly being released from the

Table 3 Summary of the fermentation performance during the shake flask fermentations

	4 g L^{-1} acetic acid		
pH	5.0	5.5	6.0
Ethanol yield (g/g consumed sugars)	0.44 ± 0.01 [A]	0.43 ± 0.00 [A B]	0.40 ± 0.02[C]
Ethanol yield (% of theoretical)	85.5 ± 1.2	83.8 ± 0.6	78.5 ± 3.6
Glycerol yield (g/g consumed sugars)	0.043 ± 0.007	0.042 ± 0.002	0.045 ± 0.004
Consumed xylose (%)	27.4 ± 1.9[C]	38.7 ± 2.5 [A]	50.2 ± 1.2 [B]
Xylitol production (% of consumed xylose)	23.9 ± 0.2	25.2 ± 4.0	28.5 ± 2.6
Calculated carbon recovery[a] (including cell growth)	0.99 ± 0.00	0.99 ± 0.01	0.97 ± 0.04
	8 g L^{-1} acetic acid		
pH	5.0	5.5	6.0
Ethanol yield (g/g consumed sugars)	0.43 ± 0.03 [A D]	0.43 ± 0.01 [A D]	0.42 ± 0.01 [B C D]
Ethanol yield (% of theoretical)	84.9 ± 5.4	84.5 ± 1.9	82.3 ± 0.9
Glycerol yield (g/g consumed sugars)	0.039 ± 0.009	0.042 ± 0.003	0.046 ± 0.000
Consumed xylose (%)	10.7 ± 1.2 [D]	37.0 ± 0.6 [A]	48.0 ± 2.0 [B]
Xylitol production (% of consumed xylose)	32.1 ± 7.8	23.8 ± 0.7	27.2 ± 2.0
Calculated carbon recovery[*] (including measured cell growth)	0.95 ± 0.04	0.99 ± 0.01	1.01 ± 0.02

[*]Carbon recovery calculated according to Eq. 5 in material and methods, with CO_2 estimated based on ethanol production.
Yields are based on the final (48 hour) values, and standard deviations are based on duplicate experiments. All statistically compared mean values are denoted with one or several letters (A, B, C, D). Values labeled with the same letter are not significantly different at a confidence level of 95%. Note that in this table, yields for all six set-ups are compared.

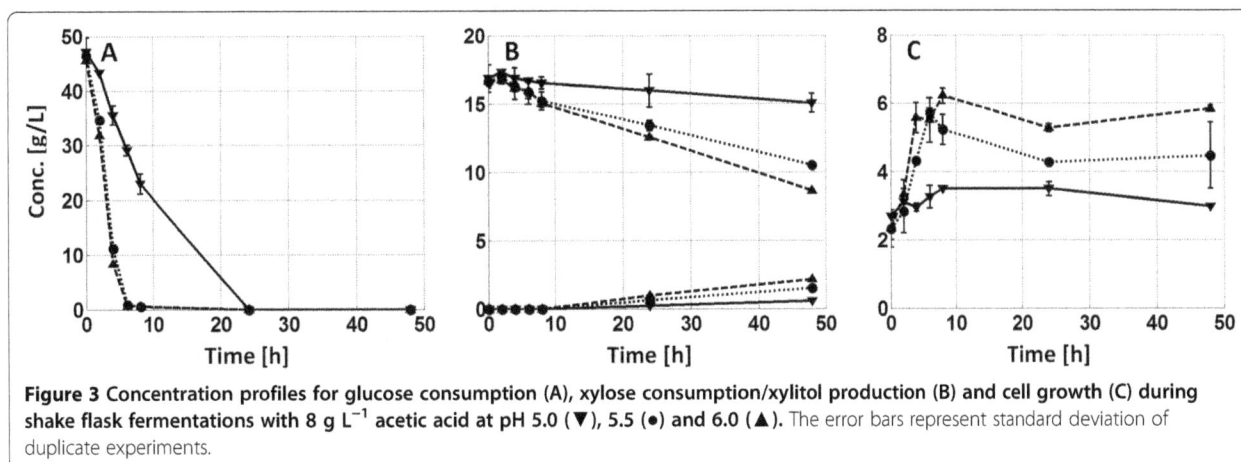

Figure 3 Concentration profiles for glucose consumption (A), xylose consumption/xylitol production (B) and cell growth (C) during shake flask fermentations with 8 g L^{-1} acetic acid at pH 5.0 (▼), 5.5 (●) and 6.0 (▲). The error bars represent standard deviation of duplicate experiments.

fibers throughout the process, hence limiting the fermentation rate. One observed effect of this difference in available glucose was that xylose consumption was better with the SSCF design in comparison to the HYBRID process, most likely since high initial glucose levels are known to inhibit xylose uptake (Lee et al. 2002; Saloheimo et al. 2007). The effect on xylose consumption from process design was, however, not as large as when increasing the fermentation pH. A higher pH was found to clearly reduce the inhibitory effect of the acetic acid, as previously shown by for example Casey et al. (2010). An increased pH significantly enhanced xylose consumption for both process designs, although the SSCF design was more favored than the HYBRID design (Table 2). The strong correlation between xylose utilization and pH in the presence of acetic acid was furthermore confirmed in shake flask fermentations (Figure 3B and Table 3).

The increase in pH for the HYBRID design resulted not only in improved xylose consumption, but also the glucose uptake rates were significantly improved at the less inhibiting conditions (Figure 2). Coupled to the increased sugar uptake rate, a reduced fermentation yield was observed together with increased glycerol production, potentially indicating increased cell growth. Since cell growth is very difficult to quantify in fiber suspensions, shake flask fermentations where carried out on fiber free hydrolyzate where it could be concluded that an overall faster glucose consumption at less inhibiting conditions was correlated to enhanced cell growth (Figure 3). This agrees well with previous studies on both pure glucose and mixed sugar fermentations with other strains of S. cerevisiae, where ethanol yields were reduced at increased pH in the presence of acetic acid as a consequence of improved anaerobic cell growth (and associated glycerol production) (Casey et al. 2010; Taherzadeh et al. 1997). However, for the SSCF experiment the fermentation yield seemed less affected compared to

the HYBRID design, possibly since the glucose uptake rate was limited by the rate of hydrolysis.

Conclusion

In conclusion, this study shows that the impact of an increased pH on the overall ethanol yield is not necessarily in agreement with an a priori expectation, but will depend on the process design. The overall ethanol yield increased significantly for the SSCF strategy at increased pH, whereas no significant difference (or even a slight reduction in yield) was observed with the HYBRID strategy. This can be explained by the difference in available sugars during the fermentation. The high sugar concentrations in the HYBRID design resulted in both a reduced fermentation yield at the higher pH as well as a hampered xylose consumption in comparison to the SSCF strategy.

Competing interests
The authors declare that they have no competing interests.

Acknowledgements
This study was made within the EU Project "Second generation BIOethanol process: demonstration scale for the step of Lignocellulosic hYdrolysis and FErmentation" (BIOLYFE, FP7, EU contract No. 239204) funded by the European Union. Biochemtex S.p.A. Italia, Novozymes A/S and Taurus Energy, are gratefully acknowledged for providing pretreated *Arundo donax*, enzymes and the yeast strain used in the work.

References
Alkasrawi M, Rudolf A, Lidén G, Zacchi G (2006) Influence of strain and cultivation procedure on the performance of simultaneous saccharification and fermentation of steam pretreated spruce. Enzyme Microb Technol 38(1–2):279–286
Almeida JRM, Runquist D, Sànchez Nogué V, Lidén G, Gorwa-Grauslund MF (2011) Stress-related challenges in pentose fermentation to ethanol by the yeast Saccharomyces cerevisiae. Biotechnol J 6(3):286–299
Balan V, Chiaramonti D, Kumar S (2013) Review of US and EU initiatives toward development, demonstration, and commercialization of lignocellulosic biofuels. Biofuels, Bioproducts and Biorefining. doi:10.1002/bbb.1436
Bellissimi E, Van Dijken JP, Pronk JT, Van Maris AJA (2009) Effects of acetic acid on the kinetics of xylose fermentation by an engineered, xylose-isomerase-based Saccharomyces cerevisiae strain. FEMS Yeast Res 9(3):358–364

Bertilsson M, Andersson J, Lidén G (2008) Modeling simultaneous glucose and xylose uptake in Saccharomyces cerevisiae from kinetics and gene expression of sugar transporters. Bioprocess Biosyst Eng 31(4):369–377

Cannella D, Jørgensen H (2013) Do new cellulolytic enzyme preparations affect the industrial strategies for high solids lignocellulosic ethanol production? Biotechnol Bioeng. doi:10.1002/bit.25098

Casey E, Sedlak M, Ho NWY, Mosier NS (2010) Effect of acetic acid and pH on the cofermentation of glucose and xylose to ethanol by a genetically engineered strain of Saccharomyces cerevisiae. FEMS Yeast Res 10(4):385–393

Galbe M, Sassner P, Wingren A, Zacchi G (2007) Process Engineering Economics of Bioethanol Production. In: Olsson L (ed) Advances in Biochemical Engineering/Biotechnology, vol 108. Springer, Berlin Heidelberg, pp 303–327. doi:10.1007/10_2007_063

Hahn-Hägerdal B, Karhumaa K, Fonseca C, Spencer-Martins I, Gorwa-Grauslund M (2007) Towards industrial pentose-fermenting yeast strains. Appl Microbiol Biotechnol 74(5):937–953

Humbird D, Mohagheghi A, Dowe N, Schell DJ (2010) Economic impact of total solids loading on enzymatic hydrolysis of dilute acid pretreated corn stover. Biotechnol Prog 26(5):1245–1251

Janssen R, Turhollow AF, Rutz D, Mergner R (2013) Production facilities for second-generation biofuels in the USA and the EU – current status and future perspectives. Biofuels Bioprod Bioref 7(6):647–665

Jorgensen H, Vibe-Pedersen J, Larsen J, Felby C (2007) Liquefaction of lignocellulose at high-solids concentrations. Biotechnol Bioeng 96(5):862–870

Knutsen JS, Liberatore MW (2009) Rheology of high-solids biomass slurries for biorefinery applications. J Rheol 53(4):877–892

Kristensen J, Felby C, Jørgensen H (2009a) Determining Yields in High Solids Enzymatic Hydrolysis of Biomass. Appl Biochem Biotechnol 156(1):127–132

Kristensen J, Felby C, Jorgensen H (2009b) Yield-determining factors in high-solids enzymatic hydrolysis of lignocellulose. Biotechnol Biofuels 2(1):11

Lee WJ, Kim MD, Ryu YW, Bisson L, Seo JH (2002) Kinetic studies on glucose and xylose transport in Saccharomyces cerevisiae. Appl Microbiol Biotechnol 60(1–2):186–191

Macrelli S, Mogensen J, Zacchi G (2012) Techno-economic evaluation of 2nd generation bioethanol production from sugar cane bagasse and leaves integrated with the sugar-based ethanol process. Biotechnol Biofuels 5 (1):1–18

McMillan JD, Jennings EW, Mohagheghi A, Zuccarello M (2011) Comparative performance of precommercial cellulases hydrolyzing pretreated corn stover. Biotechnol Biofuels 4:29

Meinander NQ, Boels I, Hahn-Hägerdal B (1999) Fermentation of xylose/glucose mixtures by metabolically engineered Saccharomyces cerevisiae strains expressing XYL1 and XYL2 from Pichia stipitis with and without overexpression of TAL1. Bioresour Technol 68(1):79–87

Olofsson K, Rudolf A, Lidén G (2008) Designing simultaneous saccharification and fermentation for improved xylose conversion by a recombinant strain of Saccharomyces cerevisiae. J Biotechnol 134(1–2):112–120

Olofsson K, Palmqvist B, Liden G (2010a) Improving simultaneous saccharification and co-fermentation of pretreated wheat straw using both enzyme and substrate feeding. Biotechnol Biofuels 3(1):17

Olofsson K, Wiman M, Lidén G (2010b) Controlled feeding of cellulases improves conversion of xylose in simultaneous saccharification and co-fermentation for bioethanol production. J Biotechnol 145(2):168–175

Roche CM, Dibble CJ, Knutsen JS, Stickel JJ, Liberatore MW (2009) Particle Concentration and Yield Stress of Biomass Slurries During Enzymatic Hydrolysis at High-Solids Loadings. Biotechnol Bioeng 104(2):290–300

Saloheimo A, Rauta J, Stasyk O, Sibirny A, Penttilä M, Ruohonen L (2007) Xylose transport studies with xylose-utilizing Saccharomyces cerevisiae strains expressing heterologous and homologous permeases. Appl Microbiol Biotechnol 74(5):1041–1052

Sassner P, Galbe M, Zacchi G (2008) Techno-economic evaluation of bioethanol production from three different lignocellulosic materials. Biomass Bioenergy 32(5):422–430

Sluiter A, Hames B, Ruiz R, Scarlata C, Sluiter J, Templeton D (2008a) Determination of sugars, byproducts and degradation products in liquid fraction process samples (LAP). NREL, Golden, CO.

Sluiter A, Hames B, Ruiz R, Scarlata C, Sluiter J, Templeton D, Crocker D (2008b) Determination of structural carbohydrates and lignin in biomass (LAP). NREL, Golden, CO.

Taherzadeh MJ, Lidén G, Gustafsson L, Niklasson C (1996) The effects of pantothenate deficiency and acetate addition on anaerobic batch fermentation of glucose by Saccharomyces cerevisiae. Appl Microbiol Biotechnol 46(2):176–182

Taherzadeh MJ, Niklasson C, Lidén G (1997) Acetic acid—friend or foe in anaerobic batch conversion of glucose to ethanol by Saccharomyces cerevisiae? Chem Eng Sci 52(15):2653–2659

Van Vleet JH, Jeffries TW (2009) Yeast metabolic engineering for hemicellulosic ethanol production. Curr Opin Biotechnol 20(3):300–306

Viamajala S, McMillan JD, Schell DJ, Elander RT (2009) Rheology of corn stover slurries at high solids concentrations - Effects of saccharification and particle size. Bioresour Technol 100(2):925–934

Viamajala S, Donohoe BS, Decker SR, Vinzant TB, Selig MJ, Himmel ME, Tucker MP (2010) Heat and Mass Transport in Processing of Lignocellulosic Biomass for Fuels and Chemicals. Sustainable Biotechnology: Sources of Renewable Energy. Springer. doi:10.1007/978-90-481-3295-9_1

Wahlbom CF, van Zyl WH, Jönsson LJ, Hahn-Hägerdal B, Otero RRC (2003) Generation of the improved recombinant xylose-utilizing Saccharomyces cerevisiae TMB 3400 by random mutagenesis and physiological comparison with Pichia stipitis CBS 6054. FEMS Yeast Res 3(3):319–326

Wiman M, Palmqvist B, Tornberg E, Lidén G (2011) Rheological characterization of dilute acid pretreated softwood. Biotechnol Bioeng 108(5):1031–1041

Wingren A, Galbe M, Zacchi G (2003) Techno-economic evaluation of producing ethanol from SSF and SHF and identification of bottlenecks. Biotechnol Prog 19(4):1109–1117

Wiselogel A, Tyson S, Johnson D (1996) Biomass feedstock resources and composition. In: CE W (ed) Handbook on bioethanol: production and utilization. Taylor & Francis, Washington, DC, pp 105–118

Zhu Y, Malten M, Torry-Smith M, McMillan JD, Stickel JJ (2011) Calculating sugar yields in high solids hydrolysis of biomass. Bioresour Technol 102(3):2897–2903

17

Induction, expression and characterisation of laccase genes from the marine-derived fungal strains *Nigrospora* sp. CBMAI 1328 and *Arthopyrenia* sp. CBMAI 1330

Michel Rodrigo Zambrano Passarini[1], Cristiane Angelica Ottoni[2], Cledir Santos[2,3], Nelson Lima[2,3] and Lara Durães Sette[1,4*]

Abstract

The capability of the fungi *Nigrospora* sp. CBMAI 1328 and *Arthopyrenia* sp. CBMAI 1330 isolated from marine sponge to synthesise laccases (Lcc) in the presence of the inducer copper (1–10 μM) was assessed. In a liquid culture medium supplemented with 5 μM of copper sulphate after 5 days of incubation, *Nigrospora* sp. presented the highest Lcc activity (25.2 $U·L^{-1}$). The effect of copper on Lcc gene expression was evaluated by reverse transcriptase polymerase chain reaction. *Nigrospora* sp. showed the highest gene expression of Lcc under the same conditions of Lcc synthesis. The highest Lcc expression by the *Arthopyrenia* sp. was detected at 96 h of incubation in absence of copper. Molecular approaches allowed the detection of Lcc isozymes and suggest the presence of at least two undescribed putative genes. Additionally, Lcc sequences from the both fungal strains clustered with other Lcc sequences from other fungi that inhabit marine environments.

Keywords: Copper sulphate; Gene expression; Laccase; Marine fungi

Introduction

Marine environments host a huge diversity of microorganisms. Fungi constitute a large part of this microbiota and are also diverse and are important from ecological and biotechnological points of view (Panno et al. 2013). Sponge-derived fungi have repeatedly been shown as interesting sources of novel bioactive metabolites previously not found from terrestrial strains of the same species (Subramani et al. 2013), because they are adapted, amongst others factors, to the harsh marine environment (e.g., high pressure, low temperature, oligotrophic nutrients, high salinity) (Chen et al. 2011). Among the metabolites produced by marine fungi stand out enzymes, that represent enormous potential for the production of pharmaceutical compounds, food, beverages, aromas, fragrances, agrochemicals, and fine chemicals (Rocha et al. 2012).

Laccases (Lcc, EC 1.10.3.2) are found across kingdoms of life (e.g., plants, insects, bacteria and fungi) and, in the Nature act preferably on phenolic compounds. They belong to the class of oxidoreductases and to the family multicopper oxidases as well as ferroxidases, bilirubin oxidases and ascorbate oxidases (Ramos et al. 2011). Coding genes from this family appear to be redundant in fungal genomes, probably due to the different physiological roles played by their coding products and their regulation depending on environmental conditions (Ramos et al. 2011).

The widespread occurrence of Lcc in fungi and their versatility, especially in white-rot fungi, contribute to further research to obtain new sources of improved enzymes (Haibo et al. 2009). Although the copper catalytic centres are similar for all fungal Lcc, significant differences are observed on thermodynamic and kinetic properties depending on the microorganism (de Oliveira et al. 2009). The fungal Lcc often occur as multiple isoenzymes expressed under different growth conditions. Various strategies have been employed to improve the production of Lcc, such as medium optimization,

* Correspondence: larasette@rc.unesp.br
[1]Divisão de Recursos Microbianos, CPQBA/UNICAMP, CP 6171, 13083-970 Campinas, SP, Brazil
[4]Departamento de Bioquímica e Microbiologia, Instituto de Biociências - UNESP, Campus Rio Claro, Av. 24A, n°1515, 13506-900, Rio Claro, SP, Brazil
Full list of author information is available at the end of the article

isolation and breeding of high-producing strains, the utilization of inducers, and the heterologous expression of Lcc genes (Wang et al. 2013). Furthermore, most of Lcc are extracellular inducible enzymes, their rates of synthesis and activity being strongly dependent on the presence of suitable inductor which plays an important role in increasing their production (Kocyigit et al. 2012).

Reports investigating enzymes produced by marine-derived fungi, and their potential applications are very few. Recently, Chen et al. (2011) reported the production of laccases by a *Pestalotiopsis* strain. These authors obtained highest level of Lcc activity (2.0 $U \cdot l^{-1}$) under submerged growth using the untreated sugarcane bagasse. Bonugli-Santos et al. (2010) identified Lcc, manganese peroxidase (MnP) and lignin peroxidase (LiP) activities from marine-derived fungi such as *Aspergillus esclerotiorum*, *Cladosporium cladosporioides* and *Mucor racemosus*. Lcc are of particular interest with regard to potential industrial applications due to their ability to oxidize a wide range of toxic and polluting substrates, including polycyclic aromatic hydrocarbons (PAH) derived from petroleum, textile dyes and pesticides.

Fungal Lcc production is highly regulated by the media composition. Thus, medium optimization has become one of the main methods to enhance Lcc production. Carbon sources and copper ions are the two most critical factors in improving or stimulating Lcc (Wang et al. 2014). Different studies have shown that Lcc production are regulated by metal ions such as Cu^{2+} and Fe^{3+} by gene expression induction or through translational or post-translational regulation (Fonseca et al. 2010). Recently, Manavalan et al. (2013) described significantly increased (1.5 U ml^{-1}) for Lcc production by *Ganoderma lucidum* when the culture medium was amended with 0.4 mM $CuSO_4$. Nakade et al. (2013) using *Pycnoporus cinnabarinus* assessed the effect of different inducers (anicidine, catechol, guaiyacol, 2,5-xylidine, ferulic acid, ethanol ,H_2O_2, Cu^{+2}, Mn^{+2}, Fe^{+2}) in different concentrations for Lcc activity. These authors concluded that the most effective inducer was $CuSO_4$ (0.25 mM) where the Lcc activity detected (34.6 $U \cdot ml^{-1}$) was 20 times higher than in the absence of this inducer. In work by Lorenzo et al. (2006) the addition of copper to the growth medium stimulated Lcc production by *Trametes versicolor*. The cultures treated with 3.5 mM copper sulphate showed the highest Lcc activities of approximately 8 $U \cdot ml^{-1}$. This represented an increase of more than 12-fold in relation to the control culture and was nearly 25% higher than those obtained in the cultures with copper sulphate at 2 mM.

The fungal strains investigated in the present paper (*Nigrospora* sp. CBMAI 1328 and *Arthopyrenia* sp. CBMAI 1330) were isolated from the marine sponge *Dragmacidon reticulatum* and previously selected because of their capacity to synthesise great levels of Lcc (7.7 and 6.5 $U \cdot l^{-1}$), respectively (Passarini 2012). In the present work, the effect of copper sulphate on laccase synthesis by the strains was investigated to further enhance activity. In addition, the laccase gene diversity was determined.

Materials and methods
Microorganisms
The marine-derived fungi *Nigrospora* sp. and *Arthopyrenia* sp. were isolated from the sponge *Dragmacidon reticulatum* and identified according to Passarini et al. (2013). Both strains were deposited at the Brazilian culture collection "CBMAI/Coleção Brasileira de Microorganismos de Ambiente e Indústria" (CPQBA/UNICAMP) under the accession number CBMAI 1328 and CBMAI 1330, respectively.

For the assays related to the present work the strains were maintained on Malt Extract Agar (MEA, Oxoid malt extract 20 $g \cdot l^{-1}$, Sigma glucose monohydrate 20 $g \cdot l^{-1}$, Oxoid bacto peptone1 $g \cdot l^{-1}$, Oxoid agar-3 15 $g \cdot l^{-1}$) at 4°C and subcultured every month.

Culture conditions
Influence of copper sulphate on laccase gene expression
Lcc synthesis assays were performed in 250 ml Erlenmeyer flasks containing 100 ml of liquid culture medium (LCM: Oxoid malt extract 20 g l^{-1}, Panreac NaCl 30 g l^{-1}, Sigma copper sulphate 1–10 μM). Five plug disks of 8-mm diameter collected from the 7 days old colony of the fungi *Nigrospora* sp. CBMAI 1328 and *Arthopyrenia* sp. CBMAI 1330, previously growth on pre-adaption medium (PAM: LCM added by Oxoid agar-3 15 $g \cdot l^{-1}$), were used as inocula. The submerged cultures were incubated in a Certomat rotary shaker (150 rpm) for 7 days at 28°C. Controls were carried out under identical conditions without copper sulphate.

Molecular analyses
Genomic DNA isolation and amplification
Five 8-mm diameter plugs were cut as described above and used to inoculate 250 ml Erlenmeyer flasks with 100 ml GYP (Oxoid malt extract 3 $g \cdot l^{-1}$, Panreac glucose 10 $g \cdot l^{-1}$, Oxoid yeast extract 3 $g \cdot l^{-1}$ and Oxoid peptone 5 $g \cdot l^{-1}$). The cultures were incubated on a Certomat rotary shaker for 7 days at 28°C and 150 rpm. Mycelia from the *Nigrospora* sp. CBMAI 1328 and *Arthopyrenia* sp. CBMAI 1330 were frozen at −80°C and later used for genomic DNA extraction as previously described by Raeder and Broda (1985).

The degenerate primer pair LAC2FOR (5′-GGIACI-WIITGGTAYCAYWSICA-3′) and LAC3REV (5′-CCRT GIWKRTGIAWIGGRTGIGG-3′) (Invitrogen) were used for amplifying two of the four copper binding regions (designated II and III) (Lyons et al. 2003). The housekeeping gene β-tubulin from the white rot fungus *Trametes versicolor* was used as reference (GenBank accession no. AY944858.1: 5′-CGGTGAGAGGCGTCGGACAC-3′). DNA amplification was performed using a thermocycler (Bio-Rad, MyCycler) with an initial denaturation of 3 min at 94°C; followed by 35 cycles of 0.5 min at 94°C, 0.5 min at x°C (x = Lcc: 48; β-tubulin = 47), 2 min at 72°C; final extension of 10 min at 72°C and cooled to 4°C. PCR reactions were done in a 50 µl final volume containing: 0.5 µl of DNA (97 ng µl^{-1}), 1 µl of 200 mM deoxynucleotide triphosphates (Promega), 3 µl of MgCl$_2$ solution (25 mM), 10 µl of GoTaq Flexi buffer (10×), 1 µl of each pair of primers (10 mM), 0.25 µl of enzyme GoTAQ Hot Start Polymerase (Promega) and MilliQ water.

Amplified products were visualized in 1.2% agarose gels stained with ethidium bromide. The bands of the expected sizes were removed and purified with GFX PCR DNA and Gel Band Purification Kit (GE Healthcare).

Cloning, sequencing and sequence analysis

PCR products were cloned into pGEM-T Easy Vector (Promega) according to the manufacturer's instructions and transformed into *E. coli* JM109 competent cells (Promega). About 15 clones per insert were sequenced with primers M13f (5′-CGCCAGGGTTTTCCCAGTCAC GAC-3′) and M13r (5′-TTTCACACAGGAAACAGC TATGAC-3′). Amplified products were purified using GFX PCR DNA and Gel Band Purification Kit (GE Healthcare) for subsequent sequencing with DYEnamic ET Dye Terminator Cycler Sequencing Kit in an automated MegaBace DNA Analysis System 1000 (GE Healthcare) according to the manufacturer's instructions.

The Phred/Phrap/CONSED software was used to assemble the sequences into a contig. Sequences were identified using BALSTn and BLASTx search (with Lcc gene references in the GenBank database) and aligned using ClustalX (Thompson et al. 1997). The determination of the intron was done according to Bonugli-Santos et al. (2010). For that end, the target sequences were aligned with known Lcc cDNA (e.g., *Neurospora crassa* AAA33591 or *Trametes versicolor* U44431) and were manually corrected with BIOEDIT 7 (Hall, 1999) and ClustalW (Thompson et al. 1997). The introns were discarded and the deduced protein sequences were determined before uploading the alignment into phylogenetic programmes. A distance approach using the Kimura 2-parameter model (Kimura 1980) as implemented in MEGA software version 5.0 (Tamura et al. 2011) was used as a substitution model.

Sequences were also used to create a picture of Lcc gene structures using FancyGene v1.4 (Rambaldi and Ciccarelli 2009).

RNA extraction and RT-PCR

As described above five 8-mm diameter plugs of the *Nigrospora* sp. CBMAI 1328 and *Arthopyrenia* sp. CBMAI 1330 grown in PAM were inoculated in a 250 ml Erlenmeyer flask with 100 ml GYP and incubated on a Certomat rotary shaker for 5 days at 28°C and 150 rpm. For each fungus,

biomass was then retrieved and carefully washed by vacuum filtration under sterile conditions for three times with 150 ml sterile water. From that biomass, approximately 1 g was transferred into 3 different sets of four Erlenmeyer flasks, each containing 100 ml of the LCM with the following conditions: copper sulphate 1 µM, 5 µM and control without the inducer. The twelve Erlenmeyer flasks were then incubated for a period of 48, 72, 96 and 120 h, at 150 rpm and 28°C. For each condition, total RNA was extracted according Chomczynski and Sacchi (1987).

For the cDNA syntheses were used the SuperScript™ III Reverse Transcriptase kit (Invitrogen). PCR amplifications were performed as previously described in Section 2.3.1 (x = Lcc: 48; β-tubulin = 47). PCR reactions were done in a final volume of 50 µl, containing: 2 µl of cDNA of those samples, 1 µl of 200 mM deoxynucleotide triphosphates (Promega), 3 µl of MgCl$_2$ solution (25 mM), 10 µl of GoTaq Flexi buffer (10×), 1 µl of each pair of primers (10 mM), 0.25 µl of enzyme GoTAQ Hot Start Polymerase (Promega) and MilliQ water. PCR products were separated by electrophoresis. Amplified products were visualized by using Gel Doc XR System (Bio-Rad). The quantification of expression levels of the gene was performed by densitometry. Documented images of the amplicons were treated with the support of the program ImageJ 1.44f (http://imagej.net/). Results were normalized by densitometry according to the constitutive gene expression of β-tubulin. Thus, it was possible to establish the ratio between the Lcc gene expressions relative to β-tubulin under different conditions.

Analytical methods

Biomass

For the dry weight biomass measurement the samples were filtered on filter paper n° 41 (45-mm of diameter) under vacuum and kept at 105°C for 8 hours. Biomass was calculated by subtracting the initial weight from the final weight.

Enzymatic assays

The enzymatic activities of Lcc ($\varepsilon_{525nm} = 65000$ M^{-1} cm^{-1}) and proteases ($\varepsilon_{440nm} = 4600$ M^{-1} cm^{-1}) were determined as described by Martins et al. (2003). For each enzymatic activity assay, the same reaction mixtures containing boiled supernatant samples were used as control. One unit (U) of enzyme activity was defined as the amount of the enzyme for changing the absorbance by 0.01 per minute. Enzyme activities of all the samples were expressed as U·l^{-1}.

Determination of total proteins

Total proteins were determined by Bradford method using Coomassie Protein Assay Kit (Thermo Scientific Pierce)

Table 1 Data related to the biomass, total protein and enzymatic activity by the selected marine fungi after incubation in LCM supplemented with different [CuSO₄] during 7 days at 28°C and 150 rpm

Fungal strain	Time (d)	Control*			[CuSO₄] 1 µM			[CuSO₄] 5 µM			[CuSO₄] 10 µM		
		Biomass** (g·l⁻¹)	Total protein (µg·ml⁻¹)	Lcc (U·l⁻¹)	Biomass (g·l⁻¹)	Total protein (µg·ml⁻¹)	Lcc (U·l⁻¹)	Biomass (g·l⁻¹)	Total protein (µg·ml⁻¹)	Lcc (U·l⁻¹)	Biomass (g·l⁻¹)	Total protein (µg·ml⁻¹)	Lcc (U·l⁻¹)
Nigrospora sp. CBMAI 1328	1	0.3	6.0	1.2	0.1	5.6	2.7	1.4	6.4	5.5	0.1	5.1	1.4
	3	1.6	5.2	1.8	2.7	5.8	2.8	3.8	6.5	5.6	3.6	5.2	4.3
	5	1.8	5.7	6.5	3.9	5.4	9.5	4.0	7.0	25.2	4.0	5.6	8.5
	7	2.6	8.5	5.3	4.2	6.6	6.8	4.5	7.0	22.5	4.5	6.4	7.0
Arthopyrenia sp. CBMAI 1330	1	0.6	2.6	3.9	0.7	3.0	4.4	0.5	4.8	1.8	0.7	4.8	1.7
	3	1.6	2.6	7.5	0.9	4.6	5.2	0.7	5.4	3.0	0.9	5.0	2.1
	5	2.1	3.3	7.7	2.9	5.5	5.1	2.9	5.5	3.3	2.5	5.1	2.8
	7	2.0	3.4	6.0	2.8	5.4	4.4	2.3	6.5	3.5	2.2	5.2	2.7

*Control = the same culture conditions without CuSO₄.

**Biomass = dry matter.

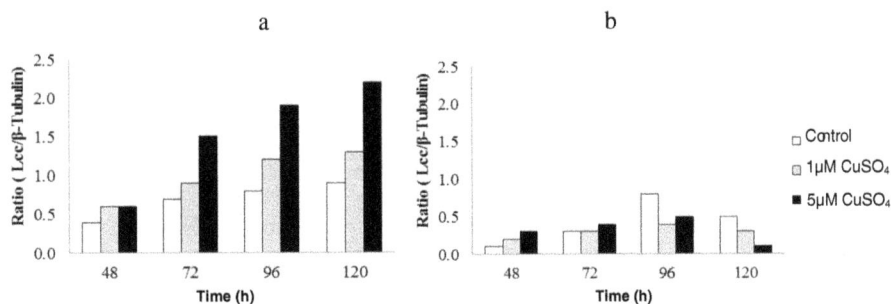

Figure 1 Detection of Lcc gene expression by *Nigrospora* sp. CBMAI 1328 (a) and *Arthopyrenia* sp. CBMAI 1330 (b) in different conditions by using RT-PCR. The Lcc gene expression was determined by the relative intensity based on the ratio between the Lcc gene expressions relative to β-tubulin.

according to the manufacturer. A standard calibration curve was constructed with three replicates. Total proteins content were presented as $\mu g\cdot ml^{-1}$.

Results

Effect of copper sulphate on laccase synthesis

Results showed the efficiency of these two fungal strains to synthesize Lcc in a LCM supplemented with different copper sulphate concentrations (Table 1). *Nigrospora* sp. CBMAI 1328 displayed the greatest amount of Lcc activity (25.2 $U\cdot l^{-1}$) when the medium was supplemented with copper sulphate at 5 μM. In the presence of the highest copper sulphate concentrations (10 μM) there was a decrease in the Lcc activity, indicating a possible enzymatic inhibition. In contrast, *Arthopyrenia* sp. CBMAI 1330 presented highest (3.9 to 7.7 $U\cdot l^{-1}$) Lcc activities without copper sulphate (the control). The Lcc production by this fungus in the presence of copper sulphate (1–10 μM) ranged from 1.7 to 5.5 $U\cdot l^{-1}$. In fact, Lcc activity decreased with increased copper sulphate (Table 1).

The relationship between copper induction, biomass and total protein was evaluated in the present study (Table 1). *Arthopyrenia* sp. CBMAI 1330 presented no significant variation in the biomass production for the assays with and without $CuSO_4$. However, an increasing in total protein was observed for all concentrations of $CuSO_4$ in comparison to the control. In contrast, the *Nigrospora* sp. CBMAI 1328 produced more biomass in the presence of $CuSO_4$. In the culture conditions where the best results of laccase activities were achieved by this fungus (5 μM of $CuSO_4$), the total protein and biomass obtained were highest than that one from the other assays, with the exception of the total protein produced in the control after 7 days and the biomass produced in the presence of $CuSO_4$ 10 μM after 5 and 7 days.

Laccase genes expression

By using the RT-PCR approach, the levels of Lcc gene expression by the *Nigrospora* sp. CBMAI 1328 and

Arthopyrenia sp. CBMAI 1330 were determined. Laccase gene expression was detected in the presence and absence of $CuSO_4$ (Figure 1). These results allow the inference that these fungi the enzyme is constitutively and inductively expressed. The highest rate of Lcc gene expression by *Nigrospora* sp. CBMAI 1328 was achieved in the presence of 5 μM $CuSO_4$ after 120 h of incubation in LCM. In these same conditions this fungus produced the higher level of laccase (25.2 $U\cdot L^{-1}$), as showed in Table 1. For *Arthopyrenia* sp. CBMAI 1330 higher Lcc gene expression was detected in LCM without supplementation of $CuSO_4$ (control) after 96 h of incubation (Figure 1b). Coincidently, the higher rates of Lcc activity by this fungus were also observed in submerged LMC without $CuSO_4$ after 96 and 120 h of incubation (Table 1).

Laccase genes characterization

Lcc genes from *Nigrospora* sp. CBMAI 1328 and *Arthopyrenia* sp. CBMAI 1330 were characterized based on cloning and sequencing analysis. An amount of 17 and 20 clone sequences were recovered from the fungi *Nigrospora* and *Arthopyrenia*, respectively. Amongst all of them, only tree clones presented similarity with sequences from fungal Lcc genes (Table 2): (i) two (C9 and F9) from *Arthopyrenia* sp. CBMAI 1330 which showed high similarity (between 70 to 100%) with a Lcc gene from *Clavariopsis aquatica*, a marine ascomycete fungus

Table 2 Similarity of clones recovered by PCR of laccase gene from strains *Nigrospora* sp. CBMAI 1328 and *Arthopyrenia* sp. CBMAI 1330

CBMAI	Clone	Size intron (pb)	ID	Similarity (%)
1328	A1	0	*Stagonospora* sp. (AAN17288)	45
1330	C9	47	*Clavariopsis aquatica* (ACR20672)	71
	F9	47	*Clavariopsis aquatica* (ACR20672)	100

representative of order *Microascales* and, (ii) one (A1) from *Nigrospora* sp. CBMAI 1328 that showed 45% similarity with the fungus *Stagonospora* sp., a representative of order *Pleosporales* that can be found in marine environments such as swamps (Lyons et al. 2003).

Data derived from phylogenetic analyses showed that only the sequence of clone F9 formed a cluster with the Lcc gene from fungus *Clavariopsis aquatica*. The other two sequences (from clone C9 and A1) clustered together and separated from the sequences recovered from GenBank database (Figure 2).

Molecular characterisation of Lcc genes from the *Nigrospora* sp. CBMAI 1328 and *Arthopyrenia* sp. CBMAI 1330 are showed in Figure 3. Comparative schematics of Lcc genes from both marine-derived fungi and *Clavariopsis aquatica* are presented in Figure 4.

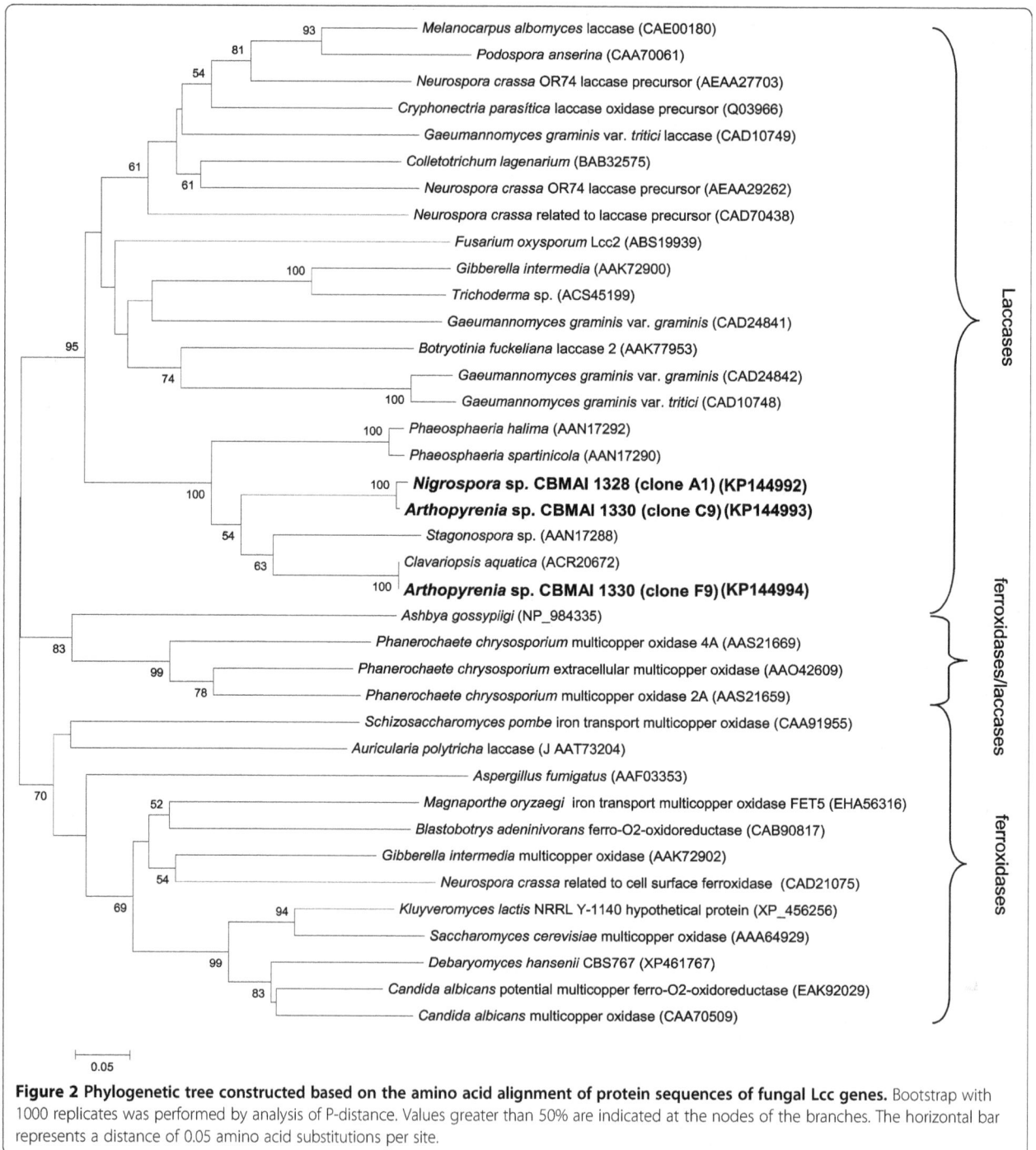

Figure 2 **Phylogenetic tree constructed based on the amino acid alignment of protein sequences of fungal Lcc genes.** Bootstrap with 1000 replicates was performed by analysis of P-distance. Values greater than 50% are indicated at the nodes of the branches. The horizontal bar represents a distance of 0.05 amino acid substitutions per site.

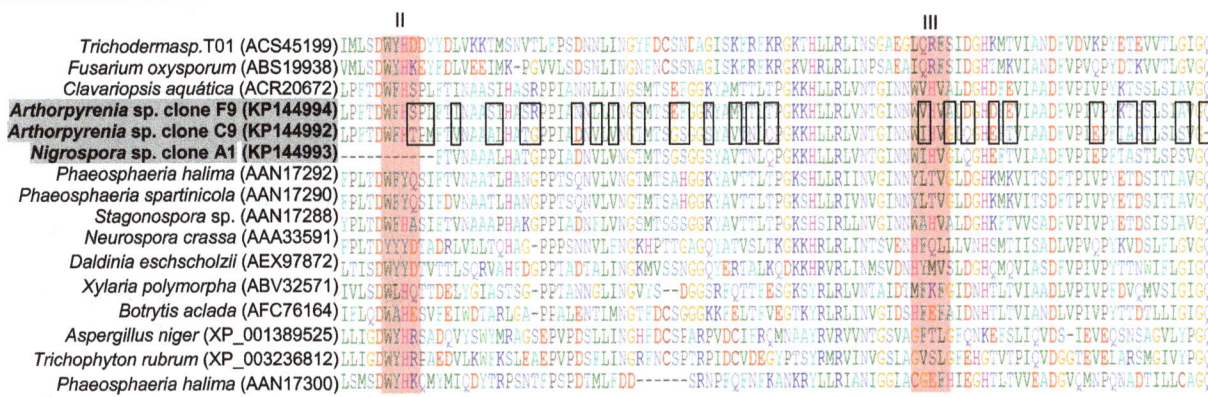

Figure 3 Lcc gene sequences from *Nigrospora* sp. CBMAI 1328 (clone A1) and *Arthopyrenia* sp. CBMAI 1330 (clones F9 and C9) aligned together with sequences recovered from GenBank. The red boxes show the possible II and III regions of binding of copper ions and the black rectangles show the differences between the amino acids sequences from clones F9 and C9.

Discussion

Copper is an essential micronutrient for most living organisms, and requirements by fungi are usually satisfied by very low concentrations of the metal, in the order of 1–10 µM. Copper present in higher concentrations of its free, cupric form is extremely toxic to fungal cells (Galhaup and Haltrich 2001). The mechanism of the Lcc induction by copper is associated with its role in the Lcc active centre, and its participation in the regulation of Lcc genes transcription and post-transcription modifications; whereas, copper toxicity is attributed to the interaction of copper ions with proteins, enzymes, nucleic acids, and metabolites associated with cell functions and viability, and due to oxidative (Kannaiyan et al. 2012).

Copper is also often a strong inducer of Lcc gene transcription, and this may be related to a defence mechanism against oxidative stress caused by free copper ions (Viswanath et al., 2008). In a recent study, Pezzella et al. (2013) performed a transcriptional analysis of nine *Pleurotus ostreatus* Lcc genes by RT-PCR in different growth conditions. The authors concluded that the addition of copper to the culture medium resulted in strong induction of *lcc9/lcc10* and *lcc2* genes, and a lesser induction of other *lcc* genes. In some cases these responses were dependent on the time of growth. Santo et al. (2012) evaluated the degradation of polyethylene by polyethylene-degrading *Rhodococcus ruber* and concluded that copper

markedly affected the induction and activity of Lcc, resulting in polyethylene degradation. The mRNA quantification by RT-PCR revealed a 13-fold increasing in Lcc mRNA levels from copper-treated cultures in comparison with the untreated control. Additionally, the authors emphasise that the addition of copper to *R. ruber* cultures containing polyethylene, enhanced by 75% the biodegradation of this compound.

According to Kannaiyan et al. (2012), the induction of *Dichomitus squalens* Lcc activity required copper sulphate addition as low as 0.06 mM. Palanisami and Lakshmanan (2011) using the marine filamentous non-heterocystous cyanobacterium *Phormidium valderianum* observed a negative effect of Lcc activity when the concentration of copper was greater than 10 µM and, consequently, a gradual decrease in the rate of Poly-R478 dye decolourisation. Singhal et al. (2009) reported a seven fold enhancement of Lcc production by *Cryptococcus albidus* after the optimizing of the growth media, which contained 2 mmol·l^{-1} of copper sulphate. However, for *C. albidus*, Lcc production was inhibited with increasing copper sulphate concentration. Cordi et al. (2007) showed that the addition of 0.07-0.1 mM copper sulphate during the cultivation of *Trametes versicolor* promoted higher Lcc activity, reaching a maximum value of 40.44 U·l^{-1} on day 12 of incubation. In contrast, in the absence of this inductor the values detected for Lcc were considered

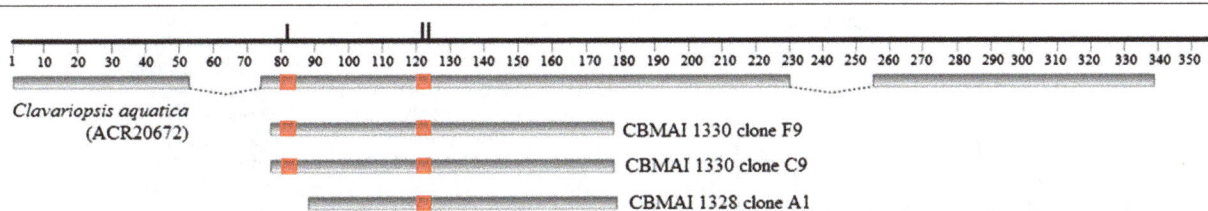

Figure 4 Comparative schematics of Lcc genes from *Nigrospora* sp. CBMAI 1328 (clone A1), *Arthopyrenia* sp. CBMAI 1330 (clones F9 and C9) and *Clavariopsis aquatic* (ACR20672). The red boxes show the possible copper binding regions (I and II).

insignificant. In Sun et al. (2009) the relationship between $CuSO_4$ and fungal biomass was discussed. The authors observed that at high levels of copper sulphate (0.8 mM) the activity of crude Lcc was significantly inhibited possibly due to the weak fungal growth.

Data derived from phylogenetic analyses may be explained by the absence of sequences close to the *Nigrospora* sp. CBMAI 1328 and *Arthopyrenia* sp. CBMAI 1330 Lcc genes or may illustrate the recovery of potential new Lcc genes from fungi associated with marine sponges. In a previous work carried out by our research group, three putative new Lcc genes were detected in a marine-derived basidiomycete (Bonugli-Santos et al. 2010). In addition, Lyons et al. (2003) identified 15 distinct sequences of Lcc from ascomycetes isolated from salt marshes in the southeastern U.S.

A study by Hoegger et al. (2006) reported that the composition of multicopper oxidases in fungal species can be very different and some species appear to encode only one type of enzyme such as, ferroxidases, as well as other species can produce other types of enzymes belonging to this family. Based on this report, representative sequences of Lcc, ferroxidase and "ferroxidases/Lcc" (enzymes with strong ferroxidase activity and weak Lcc activity) were recovered and aligned with *Nigrospora* sp. CBMAI 1328 and *Arthopyrenia* sp. CBMAI 1330 Lcc protein sequences (Figure 2). Results from phylogenetic analyses showed that Lcc genes from the studied marine fungi formed a cluster with protein sequences that encoding strictly Lcc enzymes. In addition, all Lcc sequences from *Nigrospora* sp. CBMAI 1328 and *Arthorpyrenia* sp. CBMAI 1330 grouped with Lcc protein sequences from ascomycetes derived from marine environments.

Since the primer set used was designed to be specific for fungal Lcc and targeted conserved sequences around two pairs of histidine involved in two of the four copper binding regions designated II and III, the alignment of protein sequences from *Arthopyrenia* clones F9 and C9 and from *Nigrospora* clone A1 may be related to regions II and III of binding copper ions (Figures 3 and 4).

The comparison between *Arthopyrenia* clones F9 and C9 protein sequences (Figure 3) and gene sequences alignment (data not shown) revealed that, five different nucleotides and 29 different amino acids were found despite these two Lcc genes being very similar. These results confirm the hypothesis of divergent copies of Lcc genes within the genome of this fungus, where more than one isoform can be expressed under specific experimental conditions.

The presence of multiple Lcc genes in the fungal genome has been discussed previously. According to Ramos et al. (2011), this is probably due to the different physiological roles played by their products and their regulation by environmental conditions. Susla et al. (2007)

reported that the diversity of genes can be attributed to post-transcriptional modifications of enzymes within the cell. In conclusion, the use of Lcc gene libraries from filamentous fungi has been a powerful technology for the characterization of this gene (D'Souza et al. 1996; Dedeyan et al. 2000; Dubé et al. 2008).

The detection of two different Lcc genes in the genome of *Arthopyrenia* sp. CBMAI 1330, suggests the presence of two isoforms. For this fungus, the expression of Lcc was higher when the copper ion was absent in the culture medium. However, in the screening experiments (data not shown) Lcc activity in the range of 15 $U{\cdot}l^{-1}$ was achieved when 1 mM of $CuSO_4$ was added to de medium. It is reasonable to speculate that a different isoform may have been expressed by the use of copper ion.

The present work revealed how Lcc activity from two marine-derived fungal strains, identified as belonging to the genera *Nigrospora* and *Arthopyrenia* could be increased by the addition of copper sulphate. Molecular and phylogenetic analyses allow the classification of these enzymes and suggest the presence of putative new enzymes. In addition, Lcc from the *Nigrospora* sp. CBMAI 1328 and *Arthopyrenia* sp. CBMAI 1330 clustered with other Lcc from fungi that inhabit marine environments.

Fungi from marine environments can produce different enzymes from those produced by the same terrestrial species since they are adapted to oceans where in some cases includes extremes of pH and salinity. The two marine-derived fungal strains studied could be considered strategic for the biotechnology, since they showed to be potential genetic resources, which could be applied in saline environments and/or technological processes.

Competing interests
The authors declare that they have no competing interests.

Acknowledgments
M. Passarini was supported by Ph.D. grant from FAPESP (2008/06720-7), São Paulo, Brazil. The authors thank FAPESP for financial support (BIOTA-FAPESP grant 2010/50190-2 and FAPESP grant 2013/19486-0) and Roberto G.S. Berlinck and CEBIMAR for the support related to samples collecting. L.D. Sette thanks CNPq for Productivity Fellowships 304103/2013-6.

Author details
[1]Divisão de Recursos Microbianos, CPQBA/UNICAMP, CP 6171, 13083-970 Campinas, SP, Brazil. [2]CEB-Centre of Biological Engineering, University of Minho, Campus of Gualtar, 4710-057 Braga, Portugal. [3]Post-Graduate Programme in Agricultural Microbiology, Federal University of Lavras, Lavras, MG, Brazil. [4]Departamento de Bioquímica e Microbiologia, Instituto de Biociências - UNESP, Campus Rio Claro, Av. 24A, n°1515, 13506-900, Rio Claro, SP, Brazil.

References

Bonugli-Santos RC, Durrant LR, Da Silva M, Sette LD (2010) Production of laccase, manganese peroxidase and lignin peroxidase by Brazilian marine-derived fungi. Enzym Microbiol Technol 46:32–37

Chen H-Y, Xue D-S, Feng X-Y, Yao S-J (2011) Screening and production of ligninolytic enzyme by a marine-derived fungal *Pestalotiopsis* sp. J63. Appl Biochem Biotechnol 165:1754–1769

Chomczynski P, Sacchi N (1987) Single-step method of RNA isolation by acid guanidinium thiocyanate–phenol–chloroform extraction. Anal Biochem 162:156–159

Cordi L, Minussi RC, Freire RS, Durán N (2007) Fungal laccase: copper induction, semi-purification, immobilization, phenolic effluent treatment and electrochemical measurement. Afr J Biotechnol 6:1255–1259

D'Souza TM, Boominathan K, Reddy CA (1996) Isolation of laccase gene-specific sequences from white rot and brown rot fungi by PCR. Appl Environ Microbiol 62:3739–3744

de Oliveira IRWZ, Fatibello-Filho O, Fernandes SC, Vieira IC (2009) Imobilização da lacase em micropartículas de quitosana obtidas por spray drying e usadas na construção de biossensores. Quim Nova 32:1195–1201

Dedeyan B, Klonowska A, Tagger S, Tron T, Iacazio G, Gil G, Le Petit J (2000) Biochemical and molecular characterization of a laccase from *Marasmius quercophilus*. Appl Environ Microbiol 66:925–929

Dubé E, Shareck F, Hurtubise Y, Daneault C, Beauregard M (2008) Homologous cloning, expression, and characterization of a laccase from *Streptomyces coelicolor* and enzymatic decolourisation of an indigo dye. Appl Microbiol Biotechnol 79:597–603

Fonseca MI, Shimizu E, Zapata PD, Villalba LL (2010) Copper inducing effect on laccase production of white rot fungi native from Misiones (Argentina). Enzyme Microb Tech 46:534–539

Galhaup C, Haltrich D (2001) Enhanced formation of laccase activity by the white-rot fungus *Trametes pubescens* in the presence of copper. Appl Microbiol Biotechnol 56:225–232

Haibo Z, Yinglong Z, Feng H, Peiji G, Jiachuan C (2009) Purification and characterization of a thermostable laccase with unique oxidative characteristics from *Trametes hirsute*. Biotechnol Lett 31:837–843

Hall TA (1999) BioEdit: a user-friendly biological sequence alignment editor and analysis program for Windows 95/98/NT. Nucleic Acids Symp Ser 41:95–98.

Hoegger PJ, Kilaru S, James TY, Thacker JR, Rües U (2006) Phylogenetic comparison and classification of laccase and related multicopper oxidase protein sequences. EBS J 273:2308–2326

Kannaiyan R, Mahinpey N, Mani T, Martinuzzi RJ, Kostenko V (2012) Enhancement of *Dichomitus squalens* tolerance to copper and copper-associated laccase activity by carbon and nitrogen sources. Biochem Eng J 67:140–147

Kimura M (1980) A simple method for estimating evolutionary rates of base substitutions through comparative studies of nucleotide sequence. J Mol Evol 16:111–120

Kocyigit A, Pazarbasi MB, Yasa Y, Ozdemir G, Karaboz I (2012) Production of laccase from Trametes trogii TEM H2: a newly isolated white-rot fungus by air sampling. J Basic Microbiol 52:1–9

Lorenzo M, Moldes D, Sanromán MÁ (2006) Effect of heavy metals on the production of several laccase isoenzymes by *Trametes versicolor* and on their ability to decolourise dyes. Chemosphere 63:912–917

Lyons JI, Newell SY, Buchan A, Moran MA (2003) Diversity of ascomycete laccase gene sequences in a Southeastern US salt. Microb Ecol 45:270–281

Manavalan T, Manavalan A, Thangavelu KP, Heese K (2013) Characterization of optimized production, purification and application of laccase from *Ganoderma lucidum*. Biochem Eng J 70:106–114

Martins MAM, Lima N, Silvestre AJD, Queiroz MJ (2003) Comparative studies of fungal degradation of single or mixed bioaccessible reactive azo dyes. Chemosphere 52:967–973

Nakade K, Nakagawa Y, Yano A, Konno N, Sato T, Sakamoto Y (2013) Effective induction of pblac1 laccase by copper ion in *Polyporus brumalis* ibrc05015. Fungal Biol 117:52–61

Palanisami S, Lakshmanan U (2011) Role of copper in poly R-478 decolorization by the marine cyanobacterium *Phormidium valderianum* BDU140441. World J Microbiol Biotechnol 27:669–677

Panno L, Bruno M, Voyron S, Anastasi A, Gnavi G, Miserere L, Varese GC (2013) Diversity, ecological role and potential biotechnological applications of marine fungi associated to the seagrass *Posidonia oceanica*. N Biotechnol 30:686–694

Passarini MRZ (2012) Caracterização da diversidade de fungos filamentosos associados a esponjas marinhas e avaliação da produção de lacase. Universidade Estadual de Campinas, Thesis

Passarini MRZ, Santos C, Lima N, Berlinck RGS, Sette LD (2013) Filamentous fungi from the Atlantic marine sponge *Dragmacidon reticulatum*. Arch Microbiol 195:99–111

Pezzella C, Lettera V, Piscitelli A, Giardina P, Sannia G (2013) Transcriptional analysis of *Pleurotus ostreatus* laccase genes. Appl Microbiol Biotechnol 97:705–717

Raeder U, Broda P (1985) Rapid preparation of DNA from filamentous fungi. Lett Appl Microbiol 1:17–20

Rambaldi D, Ciccarelli FD (2009) FancyGene: dynamic visualization of gene structures and protein domain architectures on genomic loci. Bioinformatics 17:2281–2282

Ramos JAT, Barends S, Verhaert RMD, de Graaff LHT (2011) The Aspergillus niger multicopper oxidase family: analysis and overexpression of laccase-like encoding genes. Microb Cell Fact 10:1–11

Rocha LC, Luiz RF, Rosset IG, Raminelli C, Seleghim MHR, Sette LD, Porto ALM (2012) Bioconversion of Iodoacetophenones by Marine Fungi. Mar Biotechnol 14:396–401

Santo M, Weitsman R, Sivan A (2012) The role of the copper-binding enzyme e laccase e in the biodegradation of polyethylene by the actinomycete *Rhodococcus ruber*. Int Biodeter Biodegr 84:204–210

Singhal A, Choudhary G, Thakur IS (2009) Optimization of growth media for enhanced production of laccase by *Cryptococcus albidus* and its application for bioremediation of chemicals. Can J Civ Eng 36:1253–1264

Subramani R, Kumar R, Prasad P, Aalbersberg W (2013) Cytotoxic and antibacterial substances against multi-drug resistant pathogens from marine sponge symbiont: Citrinin, a secondary metabolite of *Penicillium* sp. Asian Pac J Trop Biomed 3:291–296

Sun QY, Hong YZ, Xiao YZ, Fang W, Fang J (2009) Decolorization of textile reactive dyes by the crude laccase produced from solid-state fermentation of agro-byproducts. World J Microbiol Biotechnol 25:1153–1160

Susla M, Novotn C, Svobodová K (2007) The implication of Dichomitus squalens laccase isoenzymes in dye decolorization by immobilized fungal cultures. Bioresour Technol 98:2109–2115

Tamura K, Peterson D, Peterson N, Stecher G, Nei M, Kumar S (2011) MEGA5-molecular evolutionary genetics analysis using maximum likelihood, evolutionary distance, and maximum parsimony methods. Mol Biol Evol 28:2731–2739

Thompson JD, Gibson TJ, Plewniak F, Jeanmou-gin F, Higgins DG (1997) The ClustalX windows interface: flexible strategies for multiple sequence alignment aided by quality analysis tools. Nucleic Acids Res 24:4876–4882

Viswanath B, Chandra MS, Kumar KP, Pallavi H, Reddy BR (2008) Fungal laccases and their biotechnological applications with special reference to bioremediation. Dyn Biochem Process Biotech Mol Biol 2:1–13

Wang F, Guo C, Liu C-Z (2013) Immobilization of *Trametes versicolor* cultures for improving laccase production in bubble column reactor intensified by sonication. J Ind Microbiol Biotechnol 40:141–150

Wang J, Zheng X, Lin S, Lin J, Guo L, Chen X, Chen Q (2014) Identification of differentially expressed genes involved in laccase production in tropical white-rot fungus *Polyporus* sp. PG15. J Basic Microbiol 54:142–151

18

Contribution of soil esterase to biodegradation of aliphatic polyester agricultural mulch film in cultivated soils

Kimiko Yamamoto-Tamura, Syuntaro Hiradate, Takashi Watanabe, Motoo Koitabashi, Yuka Sameshima-Yamashita, Tohru Yarimizu and Hiroko Kitamoto[*]

Abstract

The relationship between degradation speed of soil-buried biodegradable polyester film in a farmland and the characteristics of the predominant polyester-degrading soil microorganisms and enzymes were investigated to determine the BP-degrading ability of cultivated soils through characterization of the basal microbial activities and their transition in soils during BP film degradation. Degradation of poly(butylene succinate-co-adipate) (PBSA) film was evaluated in soil samples from different cultivated fields in Japan for 4 weeks. Both the degradation speed of the PBSA film and the esterase activity were found to be correlated with the ratio of colonies that produced clear zone on fungal minimum medium-agarose plate with emulsified PBSA to the total number colonies counted. Time-dependent change in viable counts of the PBSA-degrading fungi and esterase activities were monitored in soils where buried films showed the most and the least degree of degradation. During the degradation of PBSA film, the viable counts of the PBSA-degrading fungi and the esterase activities in soils, which adhered to the PBSA film, increased with time. The soil, where the film was degraded the fastest, recorded large PBSA-degrading fungal population and showed high esterase activity compared with the other soil samples throughout the incubation period. Meanwhile, esterase activity and viable counts of PBSA-degrading fungi were found to be stable in soils without PBSA film. These results suggest that the higher the distribution ratio of native PBSA-degrading fungi in the soil, the faster the film degradation is. This could be due to the rapid accumulation of secreted esterases in these soils.

Keywords: Aliphatic polyester; Biodegradable plastics; Esterase; PBSA

Introduction

Plastics have spread and persisted around the world today because they have been widely used as basic materials in various industries. Used non-degradable plastic products cause waste management problems. Agricultural mulch films made have greatly contributed to the increase in the production of high-quality vegetables as they have been used to cover cultivated fields to maintain stable soil temperature and humidity, as well as to prevent weed growth. However, after harvesting, the recovery and recycling of used non-degradable mulch films would require a lot of energy and are labor intensive (Kyrikou and Briassoulis 2007).

* Correspondence: kitamoto@affrc.go.jp
National Institute for Agro-Environmental Sciences, 3-1-3 Kannondai, Tsukuba, Ibaraki 305-8604, Japan

Biodegradable plastics (BPs) have been developed as a possible solution to such environmental problems caused by the persistent plastic wastes. To date, various aliphatic polyesters that can be degraded by microorganisms in the natural environment have been commercialized as BP materials. Mulch films made from BPs are already in the market and are used to help farmers save time and labor as well as to reduce plastic wastes (Ngouajio et al. 2008). Chemical, physical, and biological degradability of BP mulch films are largely affected by the composition of the BP materials (Kyrikou and Briassoulis 2007). Although degradation speed of BP mulch films depends largely on environmental conditions (Kariyazono et al. 2000; Hoshino et al. 2001), the key factor controlling the degradation speed of BP mulch film in soil has not yet been elucidated. Knowledge of the mechanisms involved would enable the development of an indicator that would help farmers predict the degradability of mulch films under a given farmland condition before planting, and thus, select the suitable mulch film for their farm use.

Degradation of poly(butylene succinate-*co*-adipate) (PBSA) films (Bionolle® #3001) in three types of uncultivated soils (soil from extinct volcano crater, waste coal, and forest) had already been investigated (Nowak et al. 2011). To date, however, there are only a few papers reporting on the quantitative information about the degradation speeds of BP films and population size of BP-degrading microorganisms in different soils from cultivated fields. In this study, chemical properties, degradation behavior of the PBSA film, microbial populations, and esterase activities of 11 soil samples from different cultivated fields in Japan were analyzed under laboratory condition.

Materials and methods

Substrate and chemicals

To select the PBSA-degrading microorganisms from soils, emulsified PBSA (Bionolle® EM-301; average molecular weight, 12 to 15×10^4; Showa Denko K. K., Tokyo, Japan) was used as substrate. To evaluate their solid polymer-degrading activity, black PBSA film (Bionolle® 3001 G, Showa Denko K. K.) was used. It contains carbon black as additive, and has an average molecular weight of 20 to 25×10^4 and a thickness of 20 μm. *p*-Nitrophenol valerate (*p*NP-valerate) was purchased from Sigma-Aldrich (St. Louis, MO, USA) for the assay of esterase activity.

Soil analysis

Soils collected from the plowed layers of 11 cultivated fields in Japan (Table 1) were stocked at 4°C before use. For chemical analysis, these soil samples were air-dried and sieved through a 2-mm mesh. Water content was determined by drying 10 g of soil sample in an oven at 105°C overnight. To measure soil pH(H_2O), 2 g of soil sample were mixed with 5 ml of deionized water and allowed to stand for 1 d, after which the pH(H_2O) of the suspension was measured using a standard pH meter (F-23II; Horiba, Ltd., Kyoto, Japan). For the determination of total carbon (C) and nitrogen (N) contents in the soils, visible plant residues in the soil samples were carefully removed by using tweezers. The soil samples were thoroughly ground using a mortar and subjected to NC analysis (Sumigraph NC-22 F, Sumika Chemical Analysis Service, Osaka, Japan).

Degradation assay of PBSA films in soils

The degree of degradation of the PBSA film in the soil was evaluated using the procedure used in our previous study (Kitamoto et al. 2011), with modifications as follows. Fresh sieved soil samples were used for the analysis after being brought to a water content of 50% (w/w) of maximum water holding capacity. Pieces of the PBSA film (2 × 2 cm) were packed in between two layers of moistened soil (20-g lower layer and 20-g upper layer) in a sterilized plastic petri dish (φ90 × D20 mm) and incubated at 25°C. The dishes were wrapped with parafilm, and packaged in polyethylene bags in order to keep the moisture during the entire investigation periods. Three dishes with four pieces of film in each were prepared. For the control, sterilized soil was prepared by autoclaving (121°C, 15 min) or gamma ray irradiation (30 kGy).

One piece of film was collected from each dish at 1-week interval for 4 weeks. The mean degradation ratio of three pieces of film collected each time was calculated from mean gray values of digital images containing each collected piece. An image of the residual black film was scanned with a film scanner and saved in TIFF format

Table 1 Properties of soils used in this study

Soil sample	Soil type	Soil texture	pH(H_2O)	Total carbon (%)	Total nitrogen (%)	Area of soil sampling
CHI	alluvial	sandy loam	7.17	1.30	0.13	Chiba
TKB	volcanic ash	loam	6.30	6.22	0.59	Ibaraki
HIO	alluvial	loam	7.05	2.08	0.17	Kagoshima
MJO	volcanic ash	sandy loam	5.36	6.95	0.48	Miyazaki
TAK	alluvial	sandy loam	7.30	NA*	NA	Miyazaki
KIB	alluvial	clay loam	6.15	3.37	0.25	Okayama
OKA	alluvial	loam	7.14	4.52	0.39	Okayama
AKA	alluvial	loam	5.78	1.39	0.10	Shimane
MIY	alluvial	sandy loam	5.91	0.97	0.09	Shimane
YM1	volcanic ash	clay loam	7.26	2.61	0.27	Yamanashi
YM2	alluvial	clay loam	7.17	2.07	0.22	Yamanashi

*NA: not analyzed.

(300 dpi). A mean gray value (from completely black = 0 to completely white = 255) of 300 × 300 pixels containing an image of residual film was compared with that of a fresh film by the Image J (Schneider et al. 2012). We then obtained threshold values from the mean gray value of the background image without the film from each image files. Degradation ratio (%) was calculated using the following equation:

$$Degradation\ ratio\ (\%) =$$
$$\frac{(gray\ value\ of\ residual\ film)-(gray\ value\ of\ fresh\ film)}{(gray\ value\ of\ background)-(gray\ value\ of\ fresh\ film)} \times 100$$

To represent the degradation speed of a film in each soil, the degradation rate (ratio/week) was calculated as average degradation ratio during the first 3 weeks when our results showed linear increase in the degradation ratio.

Media and cultural conditions and microbial viable counts

Soils adhering to the PBSA film during the degradation tests were collected with sterilized spoon and used as peripheral soil for various analyses. Viable counts of microorganisms in the field soil and peripheral soil of the PBSA film were carried out on solid agarose media, and determined as colony forming unit (CFU) of soil suspension. The soil suspension was prepared from one gram of wet soil sample by shaking it in 10 ml of distilled water at 25°C for 10 minutes at 160 rpm. The suspension was diluted with distilled water, and spread on two kinds of selective agarose media: RFMM (fungal minimal medium with rose bengal) and DNB (diluted nutrient broth), designed for determination of total viable counts of fungi and bacteria, respectively. The RFMM agar (liter^{-1}) was composed of 2 g NaNO$_3$, 0.2 g MgSO$_4$·7H$_2$O, 0.2 g KH$_2$PO$_4$, 1 g yeast extract, and 15 g agar dissolved in tap water before autoclaving. After autoclaving, 40 µg ml^{-1} chloramphenicol and 33 µg ml^{-1} rose bengal were added aseptically to the medium to inhibit growth of bacteria and fungi, respectively. The DNB agar was composed of commercial nutrient broth (Difco, Becton, Dickinson and Company, Franklin Lakes, New Jersey) diluted 100-fold with distilled water, and then added with 15 g liter^{-1} agar. It was added with cycloheximide (50 µg ml^{-1}) to inhibit growth of eukaryotic microorganisms.

Microorganisms, which degrade emulsified PBSA, from soil suspensions were evaluated on two kinds of double layered-selective media (RFMM and DNB). To form the bottom layer, 15 ml of each medium with appropriate antibiotics were poured onto separate plates. Upon solidification, each plate was poured with 10 ml of solution (upper layer) containing 1% (w/v) emulsified PBSA, and 1.5% (w/v) agarose. The RFMM and DNB plates were incubated at 25°C for 4 days and 7 days, respectively. The microbial counts were calculated as CFU per gram of dry soil.

Identification of microorganisms

The microorganisms that grew on plates were observed under the microscope to distinguish their morphological characteristics; and they were identified based on 5.8S rDNA-ITS (for fungi) sequences as described previously (Marchesi et al. 1998; White et al. 1990).

Soil esterase assay

Esterase activity for p-nitrophenyl acetate (pNP-acetate) in the soil can be used as an indicator of poly(butylene succinate) biodegradation (Sakai et al. 2002). However, in our additional experiment, the esterase activity in the OKA and TKB soils with PBSA with pNP-acetate as substrate showed lower increase than those that with pNP-valerate (data not shown). Furthermore, the previously reported biodegradable polyester-degrading enzymes from bacteria (Akutsu-Shigeno et al. 2003) and fungi (Kodama et al. 2009; Maeda et al. 2005; Shinozaki et al. 2013; Suzuki et al. 2012) preferred to hydrolyze longer-chain esters of pNP rather than pNP-acetate. These observations suggest that esterases production by microorganisms present in the soils tested increased in the samples with PBSA, and that they preferred to degrade pNP-valerate, like the other microbial esterases identified previously. Considering these results, we therefore estimated soil esterase activities with pNP-valerate instead of pNP-acetate as substrate in the present study. Esterase activity was assayed as described in a previous study (Sakai et al. 2002) with modifications as follows: moist soil samples (100 mg) were incubated with ester substrate (2 mmol liter^{-1} pNP-valerate) in 0.6 ml tris(hydroxymethyl) aminomethane (Tris)-maleic buffer (0.5 mol liter^{-1}, pH 6.0) in 2-ml plastic tubes. The tubes were shaken continuously during incubation at 15 rpm with a shaker (RT-30mini, TAITEC Co. Ltd, Saitama, Japan) at 30°C for 30 min. After incubation, they were centrifuged at 20,000 × g for 5 min, and from each tube, 75 µl of supernatant was collected and was mixed with 200 µl of 100% ethanol. Then 55 µl of 2 mol liter^{-1} Tris was added to the mixture and vortexed for a few seconds. The absorbance of the mixture was measured at 405 nm by multi-spectrophotometer (Benchmark Plus, Bio-Rad Laboratories, Hercules, CA, USA). The absorbance of pNP-valerate in the buffer without soil and that of each soil extract mixture without the substrate were subtracted as a blank and as a background, respectively, from the measured absorbance of the mixture. The data shown represent geometric means of at least three independent assays.

Statistical analysis

All statistical analyses were performed by R program (version 2.15.0) (R Development Core Team 2012). For statistical testing of normality, the Kolmogorov-Smirnov test was used. The Spearman rank correlations were calculated to assess the correlation between the isolation rate of PBSA-degrading fungi and degradation speed of PBSA film in the soil; the difference with $P < 0.05$ was considered significant. Data for viable counts of the microorganisms were normalized by logarithmic transformation, and nontransformed values were presented in the Results. The changes in the population of the PBSA-degrading fungi present in the peripheral soil of the PBSA film were examined by Williams' test (Williams 1971, 1972) for multiple comparisons with a source code (Aoki 2004). The increase in the esterase activities during the degradation of the PBSA film in soils were analyzed by the Shirley-Williams' test (Shirley 1977) for multiple comparisons with a source code (Aoki 2012). Williams' test and the Shirley-Williams' test were conducted at the one-tailed significance level of 2.5% ($\alpha = 0.025$). Statistical difference in the esterase activity between the soils with and without the PBSA film at 3 weeks after incubation was determined by two-sided Mann-Whitney's U-test.

Results

Soil characterization

Soil characteristics, such as type, texture, soil pH(H_2O), total C and N contents, as well as the sampling area of the tested soils are presented in Table 1. All soil samples were found to be either alluvial soil or volcanic ash soil, with textures ranging from sandy loam to clay loam. The soil pH(H_2O) values ranged from 5.78 to 7.26. Soil sample TKB had the highest total C and N contents, and sample MIY, the lowest.

Visual observation of the PBSA film degradation in different soils

Degradation speed of soil-buried PBSA films varied with the soil samples. However, by at least 2 weeks of incubation, tiny holes, tears, or thinned areas were observed in all the incubated films (Figure 1). The weekly degradation ratio of film in each soil sample, covering a 4-week incubation period is presented in Table 2. Among all tested soils, the soil sample OKA showed the highest degradation ratio of films as well as the highest degradation speed after 4 weeks of incubation. The degradation speed of the PBSA film incubated in soil sample TKB was the slowest of all. Films incubated in autoclaved or gamma ray-irradiated soils were not usually degraded, but after a few weeks, some of them showed signs of degradation which could be attributed to incomplete sterilization of soil (data not shown).

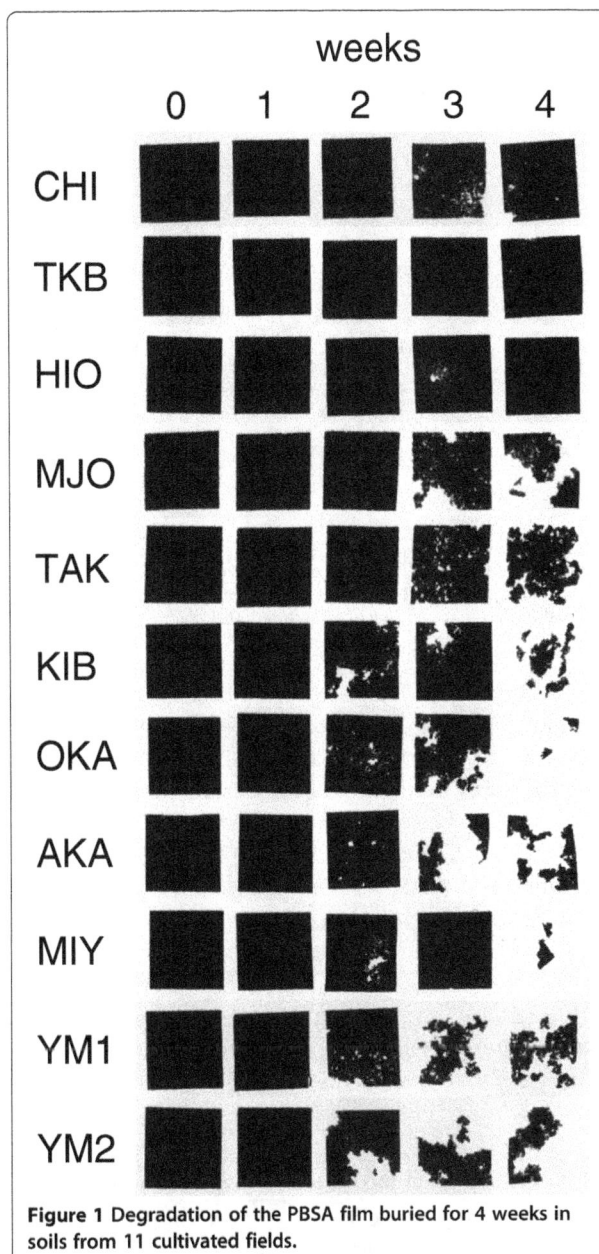

Figure 1 Degradation of the PBSA film buried for 4 weeks in soils from 11 cultivated fields.

Microbial viability in different soil samples

Viable counts of the total culturable microorganisms and the PBSA degraders in each soil sample are listed in Table 3. Based on the results on RFMM agarose plate culture, the fungal population in 11 soil samples ranged from 2.88×10^4 to 2.10×10^5 CFU g^{-1}, while the population of PBSA emulsion-degrading fungi ranged from 2.74×10^3 to 8.12×10^4 CFU g^{-1}. These results demonstrate that PBSA-degraders constituted 4.1 to 42.3% of the total fungal populations. In many cases, lower counts of PBSA degraders were isolated from DNB plate containing cycloheximide compared to those from RFMM with chloramphenicol.

Table 2 Degradation ratios (%) of soil-buried PBSA films

Soil sample	Degradation ratio (%)*			
	Incubation period (week)			
	1	2	3	4
CHI	−2.4 ± 2.5	−0.9 ± 0.9	7.4 ± 7.3	33.1 ± 9.0
TKB	1.0 ± 3.5	−0.7 ± 1.3	0.8 ± 1.5	1.4 ± 3.4
HIO	−1.7 ± 1.1	1.2 ± 1.4	−0.2 ± 0.2	7.3 ± 8.1
MJO	0.6 ± 1.7	0.6 ± 3.0	4.0 ± 2.0	7.9 ± 1.8
TAK	1.8 ± 2.5	3.3 ± 1.7	5.4 ± 3.2	30.1 ± 6.6
KIB	−4.0 ± 2.1	7.8 ± 5.7	34.5 ± 8.3	58.9 ± 36.3
OKA	0.4 ± 2.6	3.9 ± 5.8	60.3 ± 23.2	95.9 ± 6.6
AKA	−4.7 ± 2.5	0.5 ± 2.3	54.3 ± 22.9	66.0 ± 17.3
MIY	1.2 ± 0.2	5.2 ± 3.2	28.5 ± 27.7	69.7 ± 33.2
YM1	2.7 ± 0.9	16.9 ± 3.4	62.0 ± 14.1	48.5 ± 11.6
YM2	0.9 ± 1.0	21.4 ± 5.0	44.1 ± 9.2	61.3 ± 8.4

*The ratios represent the means ± standard error of triplicate assays.
Degradation ratios were calculated by image scanning method as indicated in
the equation given in the Materials and methods section.

We expected to isolate PBSA-degrading bacteria on DNB plates containing cycloheximide. However, most of the microorganisms that formed clear zone around their colonies on the plates were confirmed under the microscope that they were not bacteria, and their 5.8S rDNA-ITS sequences [DDBJ: LC009007, and LC009008] showed the highest similarity (99%) to *Purpureocillium lilacinum* (formerly *Paecilomyces lilacinus*). This species has been reported as a cycloheximide-resistant fungus (Ali-Shtayeh et al. 1998). Under the conditions in this study, it was found that fungi were predominantly responsible for emulsified PBSA degradation. *Purpureocillium* strains were isolated from different soil samples; TKB, HIO, MJO, and OKA.

Correlation between the isolation rate of PBSA-degrading fungi and degradation speed of PBSA film in the soil

The isolation rates of the PBSA-degrading fungi were shown to be correlated with the degradation speed of the PBSA film (Figure 2). Soil sample with higher isolation rate for PBSA-degrading fungi showed a tendency toward higher degradation speed (Spearman's $\rho = 0.63$, $P = 0.044$).

Correlation between the isolation rate of PBSA-degrading fungi and esterase activity in the soil

The isolation rates of the PBSA-degrading fungi were also found to be correlated with the esterase activities in the different soils (Spearman's $\rho = 0.67$, $P = 0.028$) (Figure 3). The highest esterase activity was recorded in soil sample KIB (159.3 nmol g^{-1} min^{-1}), and the lowest was found in MIY (22.9 nmol g^{-1} min^{-1}).

Effect of soil-buried film on the population of the PBSA-degrading fungi and esterase activity in peripheral soils

Viable counts of the PBSA-degrading fungi in the peripheral soils of the PBSA film buried in two soil samples (OKA and TKB) for 4 weeks are presented in Figure 4. Significant increase in PBSA degrader populations was detected after 3 and 4 weeks of incubation in OKA. The population of the PBSA-degrading fungi isolated from OKA was higher than that from TKB in all the sampling periods.

The esterase activities of these two soils monitored at each sampling period are shown in Figure 5. A significant increase of the esterase activity was detected in both soil samples with PBSA during the incubation periods. The basal esterase activity was higher in OKA than in TKB. In OKA, significant differences in esterase activities were detected between treatments with and without PBSA at 3 weeks of incubation (Mann-Whitney's U-test, $P = 0.037$).

Table 3 Viable counts of the total microorganisms and the PBSA degraders, and the isolation rates

Soil	Medium for screening fungi			Medium for screening bacteria		
	Total (CFU g^{-1})	PBSA degrading (CFU g^{-1})	Colonies clearing PBSA (%)	Total (CFU g^{-1})	PBSA degrading (CFU g^{-1})	Colonies clearing PBSA (%)
CHI	6.28×10^4	1.55×10^4	24.7	4.37×10^7	2.83×10^3	0.0
TKB	6.06×10^4	8.47×10^3	14.0	2.59×10^7	2.49×10^4	0.1
HIO	6.41×10^4	2.63×10^3	4.1	NA*	NA	NA
MJO	2.10×10^5	8.12×10^4	38.6	3.86×10^7	0.0	0.0
TAK	9.48×10^4	1.45×10^4	15.3	1.82×10^8	6.13×10^2	0.0
KIB	1.01×10^5	2.71×10^4	26.8	NA	NA	NA
OKA	4.08×10^4	1.72×10^4	42.3	5.78×10^7	9.66×10^3	0.0
AKA	5.88×10^4	8.99×10^3	15.3	NA	NA	NA
MIY	2.88×10^4	2.74×10^3	9.5	NA	NA	NA
YM1	6.09×10^4	2.45×10^4	40.3	2.82×10^7	2.95×10^2	0.0
YM2	5.62×10^4	2.88×10^3	51.3	2.53×10^7	4.64×10^3	0.0

*NA: not analyzed.

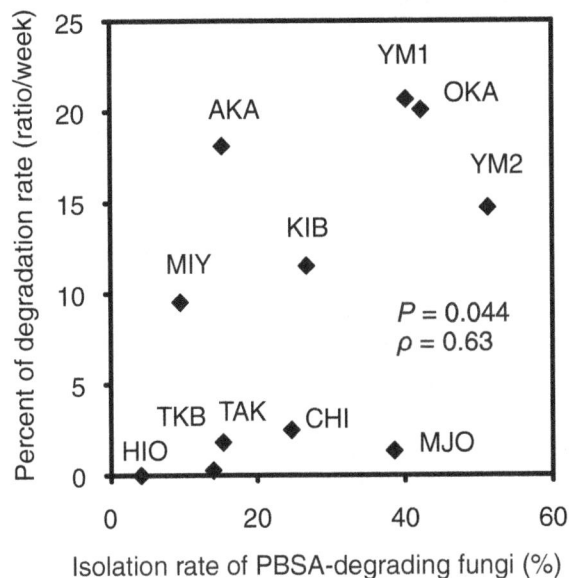

Figure 2 The isolation rate of the PBSA-degrading fungi and the degradation rate of the PBSA film. The scatter diagram shows the correlation between the isolation rate of the PBSA-degrading fungi and the degradation rate of the PBSA film in different soils. P represents the significance, and ρ represents Spearman's correlation coefficient. Degradation rate (%) is the average degradation during the first 3 weeks (ratio/week).

Without buried PBSA film, the esterase activity of each soil sample was found to be stably low.

Discussion

In agricultural fields in Japan, the degradation speed of BP mulch films is sometimes substantially early or

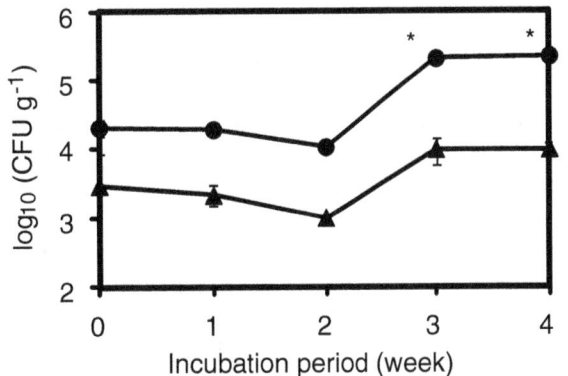

Figure 3 The isolation rate of the PBSA-degrading fungi and the esterase activity. The scatter diagram shows the correlation between the isolation rate of the PBSA-degrading fungi and the esterase activity in different soils. P represents the significance, and ρ represents Spearman's correlation coefficient.

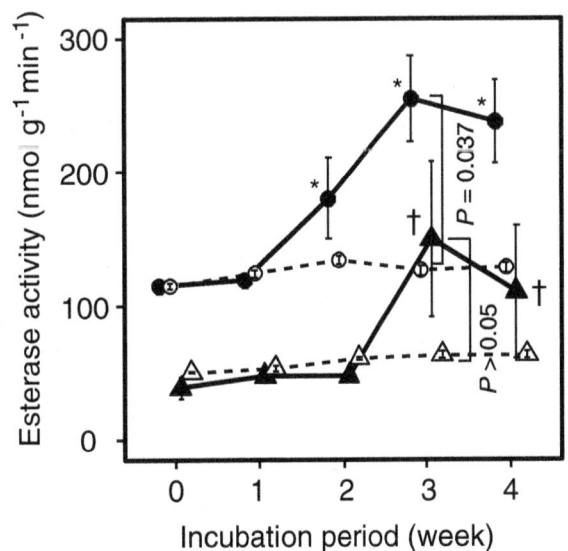

Figure 4 Changes in the viable counts of the PBSA-degrading fungi in peripheral soils for 4 weeks. The PBSA film buried in OKA (circles) and TKB (triangles) soil samples. The data shown are geometric means with error bar (1 standard deviation) of triplicate assays. Asterisk represents a significant increase ($\alpha = 0.025$) in the population of the PBSA-degrading fungi from the 0-day control of each soil sample.

slower than what is desired for practical use, and that hold back farmers from using BP mulch films in place of non-degradable ones. In this study, we showed the correlations between BP film degradation rate in cultivated soil and esterase activity and ratio of the PBSA-degrading soil fungi in the total soil fungal population. No PBSA-degrading bacteria have been isolated from

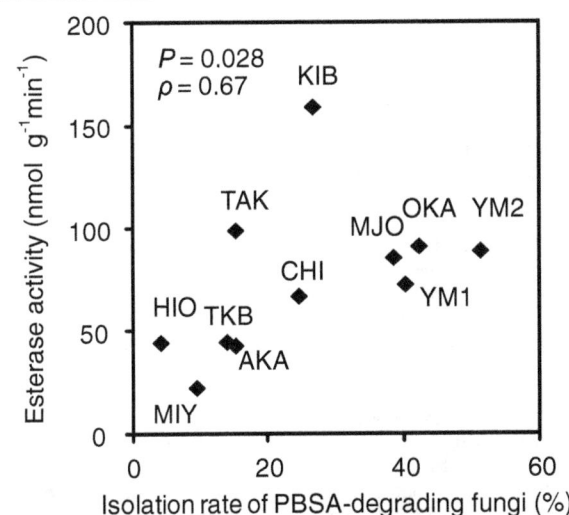

Figure 5 Changes in the esterase activities in soil for 4 weeks with or without PBSA. The peripheral soils of the PBSA film: OKA (black circles) and TKB (black triangles); the no treatment soils (without PBSA film) for control: OKA (white circles) and TKB (white triangles). The data represent geometric means with error bar indicating 1 standard error of 6 independent assays. Asterisk and dagger represent a significant increase ($\alpha = 0.025$) in the esterase activity from the 0-day controls of each soil sample. P values represent significance levels between the esterase activity in the soils with and without PBSA film.

the same soil samples. We were not able to find any significant correlation between the analyzed soil characteristics (pH[H$_2$O], total carbon and nitrogen content) and the degradation rate of the PBSA films, esterase activities, and the isolation rates of the PBSA-degrading fungi. Aliphatic polyesters can be degraded non-enzymatically via simple chemical hydrolysis in the environment (Vert 2005). However, the lower degradation speed of the PBSA film in sterilized soil compared to that in unsterilized one after four weeks incubation also support our contention that PBSA film degradation is mainly caused by the polyester-degrading activity of enzymes produced by soil microorganisms rather than by non-enzymatic chemical hydrolysis. Further studies are expected to provide information about the chemical and physical characteristics of soils that influence BP film degradation speed. The possibility remains that there are some undetected characteristics of soils that promote BP degradation by soil microorganisms.

We have previously reported that 2 ~ 100% of yeast populations isolated from rice husks (Kitamoto et al. 2011) and 4.5% of fungal strains isolated from gramineous plants (Koitabashi et al. 2012) degrade PBSA emulsion. Similarly, previous investigators reported that fungi are the major degraders of BPs, including poly(3-hydroxybutyrate-co-3-hydroxyvalerate) (Sang et al. 2002), polyester polyurethane (Barratt et al. 2003; Cosgrove et al. 2007), and poly(butylene adipate-co-terephthalate) (Kasuya et al. 2009) in soil environments. Nowak *et al.* (2011) reported a higher increment and diversity of fungal population in the soil containing PBSA film compared to that of soil bacteria. Their results also support our observation that soil fungi greatly contribute to BP film degradation.

We found a significant correlation between the isolation rate of the PBSA-degrading fungi and the degradation rate of the PBSA film in the field soil samples (Figure 2). Likewise, the isolation rate of the PBSA-degrading fungi was shown to be significantly correlated with the esterase activity (Figure 3). These results indicate that the degradation of biodegradable mulch films in the soils is strongly influenced by its native distribution ratio of the PBSA-degrading fungi in the soils. The soil sample containing higher population ratio of PBSA-degrading fungi (OKA) showed relatively higher basal esterase activities (Figure 5). These extracellular esterases are expected to break the ester bonds of plant residues and other natural materials in the soils, as well as of BPs, thus, providing the necessary nutrients for the growth of soil microorganisms during cultivation.

Degradation ratios of soil-buried PBSA films were found to be not highly correlated with soil esterase activity. In this study, we measured soil esterase activity by using *p*NP-valerate as substrate. Some soil microorganisms produce a variety of enzymes having esterase activity with specific substrate preferences. For example, a cutinase of *Fusarium solani* prefers *p*NP-butyrate to *p*NP-acetate as substrate, and cutinase-like enzyme of *Cryptococcus* sp. S-2 prefers *p*NP-caproate to *p*NP-butyrate and *p*NP-acetate (Kodama et al. 2009). The enzymes substrate spectrum, optimum temperature, pH and other conditions are expected to be varied as well. The abundance ratio of PBSA-degrading esterases in each soil is still unknown. Currently, we are trying to evaluate PBSA-degradation activities in different soils.

Burying PBSA films in the soils stimulated the esterase production through enhanced proliferation of the PBSA degraders during the incubation period (Figures 4 and 5). Quicker and more drastic increase of esterase activity in the OKA soil sample compared to that in TKB is attributed to the larger distribution ratio of the basal PBSA degraders in the former than in the latter, resulting in the increase in the total esterase activity in the OKA soil.

This study has confirmed our knowledge that fungi contribute to mulch film-degradation in cultivated soils under laboratory conditions. A high isolation rate of PBSA-degrading fungi in cultivated soil could potentially serve as an indicator of the soil's ability to promote BP film degradation. In the light of our findings, there is a need to conduct further studies in order to identify other physical and chemical properties of soil that greatly affect the speed of enzymatic degradation of BP film in soil environments.

Abbreviations

BP: Biodegradable plastic; C: Carbon; CFU: Colony forming unit; DNB: Diluted nutrient broth; N: Nitrogen; PBSA: Poly(butylene succinate-co-adipate); *p*NP: *p*-nitrophenol; *p*NP-acetate: *p*-nitrophenyl acetate; *p*NP-valerate: *p*-nitrophenyl valerate; rDNA: ribosomal deoxyribonucleic acid; rDNA-ITS: rDNA internal transcribed spacers; RFMM: Fungal minimal medium with rose bengal; Tris: Tris(hydroxymethyl) aminomethane.

Competing interests

The authors declare that they have no competing interests.

Authors' contributions

KY designed the study, carried out most of the biological studies and statistical analyses, and drafted the manuscript. SH designed the study, carried out the soil analysis, and drafted the manuscript. TW designed the degradation assay of BP films in soils. MK participated in the sampling of soils, and helped identify microorganisms isolated from soils. YS participated in the characterization of microorganisms from soils. TY participated in the soil esterase assay. HKK conceived and designed the study, and helped draft the manuscript. All authors read and approved the final manuscript.

Acknowledgments

The authors thank Dr. Takeshi Fujii and Dr. Noriko Yamaguchi in the National Institute for Agro-Environmental Sciences, and Dr. Elvira G. Suto for their helpful discussion; Dr. Tetsuhisa Miwa for his advice on statistical analyses; Ms. Xiao-hong Cao and Ms. Shoko Yamazaki for their technical assistance on the research; Showa Denko K. K. for providing the polymer materials; and UNYCK Co. for the soil samples. This research was financially supported by the National Institute for Agro-Environmental Sciences; A-STEP (Adaptive and Seamless Technology Transfer Program through Target-driven R&D) provided by the Japan Science and Technology Agency; and a grant from Science and Technology Research Promotion Program for agriculture, forestry, fisheries, and food industry.

References

Akutsu-Shigeno Y, Teeraphatpornchai T, Teamtisong K, Nomura N, Uchiyama H, Nakahara T, Nakajima-Kambe T (2003) Cloning and sequencing of a poly(DL-lactic acid) depolymerase gene from *Paenibacillus amylolyticus* strain TB-13 and its functional expression in *Escherichia coli*. Appl Environ Microbiol 69:2498–2504, doi:10.1128/AEM.69.5.2498–2504.2003

Ali-Shtayeh MS, Jamous RM, Abu-Ghdeib SI (1998) Ecology of cycloheximide-resistant fungi in field soils receiving raw city wastewater or normal irrigation water. Mycopathologia 144:39–54, doi:10.1023/A:1006952926293

Aoki S (2004) Williams no houhou ni yoru taju hikaku. (Williams' multiple comparison test). http://aoki2.si.gunma-u.ac.jp/R/Williams.html Accessed 13 January 2015

Aoki S (2012) Shirley-Williams no houhou ni yoru taju hikaku. (Shirley-Williams' multiple comparison test). http://aoki2.si.gunma-u.ac.jp/R/Shirley-Williams.html Accessed 13 January 2015

Barratt SR, Ennos AR, Greenhalgh M, Robson GD, Handley PS (2003) Fungi are the predominant micro-organisms responsible for degradation of soil-buried polyester polyurethane over a range of soil water holding capacities. J Appl Microbiol 95:78–85, doi:10.1046/j.1365-2672.2003.01961.x

Cosgrove L, McGeechan PL, Robson GD, Handley PS (2007) Fungal communities associated with degradation of polyester polyurethane in soil. Appl Environ Microbiol 73:5817–5824, doi:10.1128/aem. 01083-07

Hoshino A, Sawada H, Yokota M, Tsuji M, Fukuda K, Kimura M (2001) Influence of weather conditions and soil properties on degradation of biodegradable plastics in soil. Soil Sci Plant Nutr 47:35–43, doi:10.1080/00380768.2001.10408366

Kariyazono H, Nishimoto K, Hamaishi K (2000) Study of decomposition behavior of biodegradable plastic films in Kagoshima area soil. Rep Kagoshima Pref Inst Ind Technol 14:39–43

Kasuya K, Ishii N, Inoue Y, Yazawa K, Tagaya T, Yotsumoto T, Kazahaya J, Nagai D (2009) Characterization of a mesophilic aliphatic-aromatic copolyester-degrading fungus. Polym Degrad Stab 94:1190–1196, doi:10.1016/j.polymdegradstab.2009.04.013

Kitamoto H, Shinozaki Y, X-h C, Morita T, Konishi M, Tago K, Kajiwara H, Koitabashi M, Yoshida S, Watanabe T, Sameshima-Yamashita Y, Nakajima-Kambe T, Tsushima S (2011) Phyllosphere yeasts rapidly break down biodegradable plastics. AMB Express 1:44, doi:10.1186/2191-0855-1-44

Kodama Y, Masaki K, Kondo H, Suzuki M, Tsuda S, Nagura T, Shimba N, Suzuki E, Iefuji H (2009) Crystal structure and enhanced activity of a cutinase-like enzyme from *Cryptococcus* sp. strain S-2. Proteins 77:710–717, doi:10.1002/prot.22484

Koitabashi M, Noguchi MT, Sameshima-Yamashita Y, Hiradate S, Suzuki K, Yoshida S, Watanabe T, Shinozaki Y, Tsushima S, Kitamoto HK (2012) Degradation of biodegradable plastic mulch films in soil environment by phylloplane fungi isolated from gramincous plants. AMB Express 2:40, doi:10.1186/2191-0855-2-40

Kyrikou I, Briassoulis D (2007) Biodegradation of agricultural plastic films: a critical review. J Polym Environ 15:125–150, doi:10.1007/s10924-007-0053-8

Maeda H, Yamagata Y, Abe K, Hasegawa F, Machida M, Ishioka R, Gomi K, Nakajima T (2005) Purification and characterization of a biodegradable plastic-degrading enzyme from *Aspergillus oryzae*. Appl Microbiol Biotechnol 67:778–788, doi:10.1007/s00253-004-1853-6

Marchesi JR, Sato T, Weightman AJ, Martin TA, Fry JC, Hiom SJ, Wade WG (1998) Design and evaluation of useful bacterium-specific PCR primers that amplify genes coding for bacterial 16S rRNA. Appl Environ Microbiol 64:795–799

Ngouajio M, Auras R, Fernandez RT, Rubino M, Counts JW, Kijchavengkul T (2008) Field performance of aliphatic-aromatic copolyester biodegradable mulch films in a fresh market tomato production system. Horttechnology 18:605–610

Nowak B, Pajak J, Drozd-Bratkowicz M, Rymarz G (2011) Microorganisms participating in the biodegradation of modified polyethylene films in different soils under laboratory conditions. Int Biodeterior Biodegrad 65:757–767, doi:10.1016/j.ibiod.2011.04.007

R Development Core Team (2012) R: a language and environment for statistical computing. R Foundation for Statistical Computing, Vienna, Austria. http://www.R-project.org/ Accessed 13 January 2015

Sakai Y, Isokawa M, Masuda T, Yoshioka H, Hayatsu M, Hayano K (2002) Usefulness of soil *p*-nitrophenyl acetate esterase activity as a tool to monitor biodegradation of polybutylene succinate (PBS) in cultivated soil. Polym J 34:767–774

Sang BI, Hori K, Tanji Y, Unno H (2002) Fungal contribution to in situ biodegradation of poly(3-hydroxybutyrate-co-3-hydroxyvalerate) film in soil. Appl Microbiol Biotechnol 58:241–247, doi:10.1007/s00253-001-0884-5

Schneider CA, Rasband WS, Eliceiri KW (2012) NIH Image to ImageJ: 25 years of image analysis. Nat Methods 9:671–675, doi:10.1038/nmeth.2089

Shinozaki Y, Morita T, Cao XH, Yoshida S, Koitabashi M, Watanabe T, Suzuki K, Sameshima-Yamashita Y, Nakajima-Kambe T, Fujii T, Kitamoto HK (2013) Biodegradable plastic-degrading enzyme from *Pseudozyma antarctica*: cloning, sequencing, and characterization. Appl Microbiol Biotechnol 97:2951–2959, doi:10.1007/s00253-012-4188-8

Shirley E (1977) A non-parametric equivalent of Williams' test for contrasting increasing dose levels of a treatment. Biometrics 33:386–389

Suzuki K, Sakamoto H, Shinozaki Y, Tabata J, Watanabe T, Mochizuki A, Koitabashi M, Fujii T, Tsushima S, Kitamoto H (2012) Affinity purification and characterization of a biodegradable plastic-degrading enzyme from a yeast isolated from the larval midgut of a stag beetle, *Aegus laevicollis*. Appl Microbiol Biotechnol 97:7679–7688, doi:10.1007/s00253-012-4595-x

Vert M (2005) Aliphatic polyesters: great degradable polymers that cannot do everything. Biomacromolecules 6:538–546, doi:10.1021/bm0494702

White TJ, Bruns ST, Lee SF, Taylor J (1990) Amplification and direct sequencing of fungal ribosomal RNA genes for phylogenetics. In: Innis MA, Gelfand DH, Sninsky JJ, White TJ (eds) PCR protocols : a guide to methods and applications. Academic Press, San Diego, pp 315–322

Williams DA (1971) A test for differences between treatment means when several dose levels are compared with a zero dose control. Biometrics 27:103–117

Williams DA (1972) Comparison of several dose levels with a zero dose control. Biometrics 28:519–531

19

Strain and process development for poly(3HB-co-3HP) fermentation by engineered *Shimwellia blattae* from glycerol

Shunsuke Sato[1], Björn Andreeßen[1] and Alexander Steinbüchel[1,2]*

Abstract

Poly(3-hydroxybytyrate-*co*-3-hydroxypropionate), poly(3HB-*co*-3HP), is a possible alternative to synthetic polymers such as polypropylene, polystyrene and polyethylene due to its low crystallinity and fragility. We already reported that recombinant strains of *Shimwellia blattae* expressing 1,3-propanediol dehydrogenase DhaT as well as aldehyde dehydrogenase AldD of *Pseudomonas putida* KT2442, propionate-CoA transferase Pct of *Clostridium propionicum* X2 and PHA synthase PhaC1 of *Ralstonia eutropha* H16 are able to accumulate up to 14.5% (wt_{PHA}/wt_{CDW}) of poly(3-hydroxypropionate), poly(3HP), homopolymer from glycerol as a sole carbon source (Appl Microbiol Biotechnol 98:7409-7422, 2014a). However, the cell density was rather low. In this study, we optimized the medium aiming at a more efficient PHA synthesis, and we engineered a *S. blattae* strain accumulating poly(3HB-*co*-3HP) with varying contents of the constituent 3-hydroxypropionate (3HP) depending on the cultivation conditions. Consequently, 7.12, 0.77 and 0.32 g_{PHA}/L of poly(3HB-*co*-3HP) containing 2.1, 8.3 and 18.1 mol% 3HP under anaerobic/aerobic (the first 24 hours under anaerobic condition, thereafter, aerobic condition), low aeration/agitation (the minimum stirring rate required in medium mixing and small amount of aeration) and anaerobic conditions (the minimum stirring rate required in medium mixing without aeration), respectively, were synthesized from glycerol by the genetically modified *S. blattae* ATCC33430 strains in optimized culture medium.

Keywords: Copolymerization ratio; Fermentation condition; Glycerol; Poly(3HB-*co*-3HP); *Shimwellia blattae*

Introduction

Polyhydroxyalkanoates (PHA) are polyesters synthesized by a wide range of microorganisms (Anderson et al., 1990). Most of PHA are produced from renewable resources like sugars, plant oils, glycerol, and carbon dioxide (CO_2). As these polyesters are biodegradable, they have been expected to play an important role in environmental protection and in reduction of CO_2 emissions, a cause of global warming (Steinbüchel and Füchtenbusch, 1998). There have been many attempts to investigate industrial production of such polymers to ascertain if they are environmently friendly or biocompatible materials (Lee, 1996; Steinbüchel, 2001). In nature, poly(3-hydroxybutyrate), poly(3HB), a homopolymer of (*R*)-3-hydroxybutyric

acid (3HB), is the most abundant PHA. However, because poly(3HB) is highly crystalline, hard and brittle, its practical applications are limited. Many studies have been undertaken to improve these properties. For example, among other PHA, poly(3-hydroxybutyrate-*co*-3-hydroxyvalerate), poly(3HB-*co*-3HV), poly(3-hydroxybutyrate-*co*-3-hydroxyhexanoate), poly(3HB-*co*-3HH), and poly(3-hydroxybutyrate-*co*-3-hydroxypropionate), poly(3HB-*co*-3HP) are much more flexible and less crystalline than poly(3HB) (Andreeßen et al., 2014b; Chen et al., 2000; Doi et al., 1995; Shimamura et al., 1993; Shimamura et al., 1994). The flexibility depends on the ratio of the constituents in the copolymer. Therefore, these copolymers are accordingly expected to have a broader range of applications in packaging, agriculture and medical materials (Chen et al., 2000). Among these copolymers, poly(3HB-*co*-3HP) is considered to be very promising due to its benefiting material properties (Andreeßen and Steinbüchel, 2010).

* Correspondence: steinbu@uni-muenster.de
[1]Institut für Molekular Mikrobiologie und Biotechnologie, Westfälische Wilhelms-Universität Münster, Corrensstraße 3, D-48149 Münster, Germany
[2]Environmental Sciences Department, Faculty of Meteorgy, Environment and Arid Land Agriculture, King Abdulaziz University, Jeddah, Saudi-Arabia

The global glycerol production has increased rapidly during the last decade due to the increase of biodiesel production. Concomitant with the conversion of about 10 million tons of vegetable oil into biofuel, about 1 million tons of glycerol were produced as a by-product in 2011 (Quispe et al. 2013). Therefore, the aim of this study was the development of strains for poly(3HB-co-3HP) synthesis from glycerol as sole carbon source.

Some processes for synthesis of poly(3HB-co-3HP) have already been reported by Shimamura et al., 1994, Fukui et al., 2009, Wang and Inoue, 2001 and Wang et al., 2013. However, in these studies the use of expensive 3HP as precursor of 3HP-CoA (Shimamura et al., 1994; Wang and Inoue, 2001), insufficient 3HP contents to reduce the crystallinity (Fukui et al., 2009) and the requirement of high cost vitamin B_{12} are major drawbacks (Wang and Inoue, 2001; Wang et al., 2013).

Vitamin B_{12} is a cofactor of the glycerol dehydratase (Martens et al., 2002), which converts glycerol to 3-hyrdoxypropionaldehyde (3HPA), a precursor of 3HP-CoA (Wang et al., 2013), to produce 1,3-propanediol (1,3PD). However, only few bacteria are capable of synthesizing vitamin B_{12} (Sun et al., 2003). To solve this problem, we used the enteric bacterium *Shimwellia blattae* ATCC33430 (Burgess et al., 1973; Priest and Barker, 2010) which cannot naturally produce PHA but synthesizes vitamin B_{12} (Andres et al., 2004) and converts glycerol to 1,3PD.

Recently, we reported that *S. blattae* expressing 1,3-propanediol dehydrogenase (*dhaT*) and aldehyde dehydrogenase (*aldD*) of *Pseudomonas putida* KT2442, propionate-coenzyme A (propionate-CoA) transferase (*pct*) of *Clostridium propionicum* X2, and PHA synthase (*phaC1*) of *Ralstonia eutropha* H16 accumulates poly(3HP) from glycerol as a sole carbon source up to 14.5% (wt_{PHA}/wt_{CDW}) (Andreeßen et al., 2014a; Heinrich et al., 2013). Here, 1,3PD produced by *S. blattae* is oxidized first to 3HPA by DhaT and subsequently to 3HP by AldD. 3HP is then activated by addition of coenzyme A by Pct. In order to synthesize poly(3HB-co-3HP) from glycerol in *S. blattae*, we co-expressed *phaA* and *phaB1* from *R. eutropha* H16 (Budde et al., 2010) together with the enzymes for the already mentioned artificial poly(3HP) pathway (Heinrich et al., 2013). Two molecules of acetyl-CoA are condensed to acetoacetyl-CoA by a β-ketothiolase (PhaA) and acetoacetyl-CoA is then reduced by an (*R*)-specific acetoacetyl-CoA reductase (PhaB1) to generate (*R*)-3HB-CoA. As a result, the recombinant *S. blattae* (*Sb*6BP) is capable of synthesizing poly(3HB-co-3HP) (Figure 1).

However, the residual cell density was less than 6 g/L and therefore too low to produce much poly(3HB-co-3HP). Since PHA are accumulated inside the cells, low residual cell density result in only low PHA productivity.

For example, even if 90 % (wt_{PHA}/wt_{CDW}) of poly(3HB-co-3HP) is accumulated in a cell, less than 54 g/L of polymer is produced under such conditions.

Therefore, an optimized culture medium was needed to overcome this problem. In this study, we report on a new strategy for synthesis of poly(3HB-co-3HP) using optimized cultivation medium and glycerol in genetically modified *S. blattae* without the addition of vitaminB_{12} and 3HP into the culture.

Materials and methods
Strain and plasmid
Table 1 lists all strains and plasmids used in this study. For cloning experiments, plasmids were transformed into *Escherichia coli* TOP10. For synthesis of poly(3HB-co-3HP) or poly(3HB), *S. blattae* ATCC33430 was transformed with plasmid pBBR1MCS-2::p_{lac}::aldD:dhaT::pct::p_{lac}::phaC1AB.

Growth of cells
250-mL Erlenmeyer flasks containing 50 mL MMB medium [3.56 g/L $Na_2PO_4 \cdot 2 H_2O$, 0.68 g/L KH_2PO_4, 0.63 g/L $(NH4)_2SO_4$, 2.47 g/L $MgSO_4 \cdot 7 H_2O$, 1.0% (vol/vol) trace element solution (0.1 N HCl in 4.2 g/L $FeSO_4 \cdot 7 H_2O$, 5.0 g/L $CaCl_2 \cdot 2H_2O$, 2.4 g/L $CoCl_2 \cdot 6 H_2O$, 0.58 g/L $CuCl_2 \cdot 2 H_2O$, 2 mg/L $NiCl_2 \cdot 6 H_2O$, 3 mg/L $MnCl_2 \cdot 4 H_2O$, 0.03 g/L H_3BO_3, 4.3 g/L $ZnSO_4 \cdot 7 H_2O$, 3 mg/L $NaMoO_4 \cdot 2 H_2O$)] with 300 mM of glycerol was used for optimization of culture medium. Cells were cultivated in 250-mL Erlenmeyer flasks at an agitation of 125 rpm and at 30°C. High cell density fed-batch cultivation of *S. blattae* were conducted in a 2 L jar fermenter (Biostat B plus, Sartorius AG, Göttingen, Germany) containing 1.5 L of basal medium (BM) (Andreeßen et al. 2014a,b) or MMB medium with 300 mM of glycerol as carbon source.

Glycerol was intermittently added to the culture medium to maintain a concentration between 50 and 300 mM. 500-mL flasks containing 100 mL BM or MMB medium, 300 mM of glycerol and 50 µg/L of kanamycin were used for seed cultivations. Dissolved oxygen was monitored and pH was controlled in the range of 6.8 – 6.9 by using a 7.5% aqueous solution of ammonium hydroxide.

Plasmid construction and transfer into *E. coli* and *S. blattae*
All processing and manipulation of DNA was carried out as described by Sambrook et al., 1989. Plasmid pBHR68 (Spiekermann et al., 1999) was digested with *Bsp*119I and *Eco*RI to generate a 4.2-kbp fragment comprising the coding regions of *phaC1*, *phaA* and *phaB1*. This fragment was ligated to the *Cla*I and *Eco*RI restriction fragment of pBBR1MCS-2 (Kovach et al., 1995) to generate pBBR1MCS-2::p_{lac}::phaC1AB.

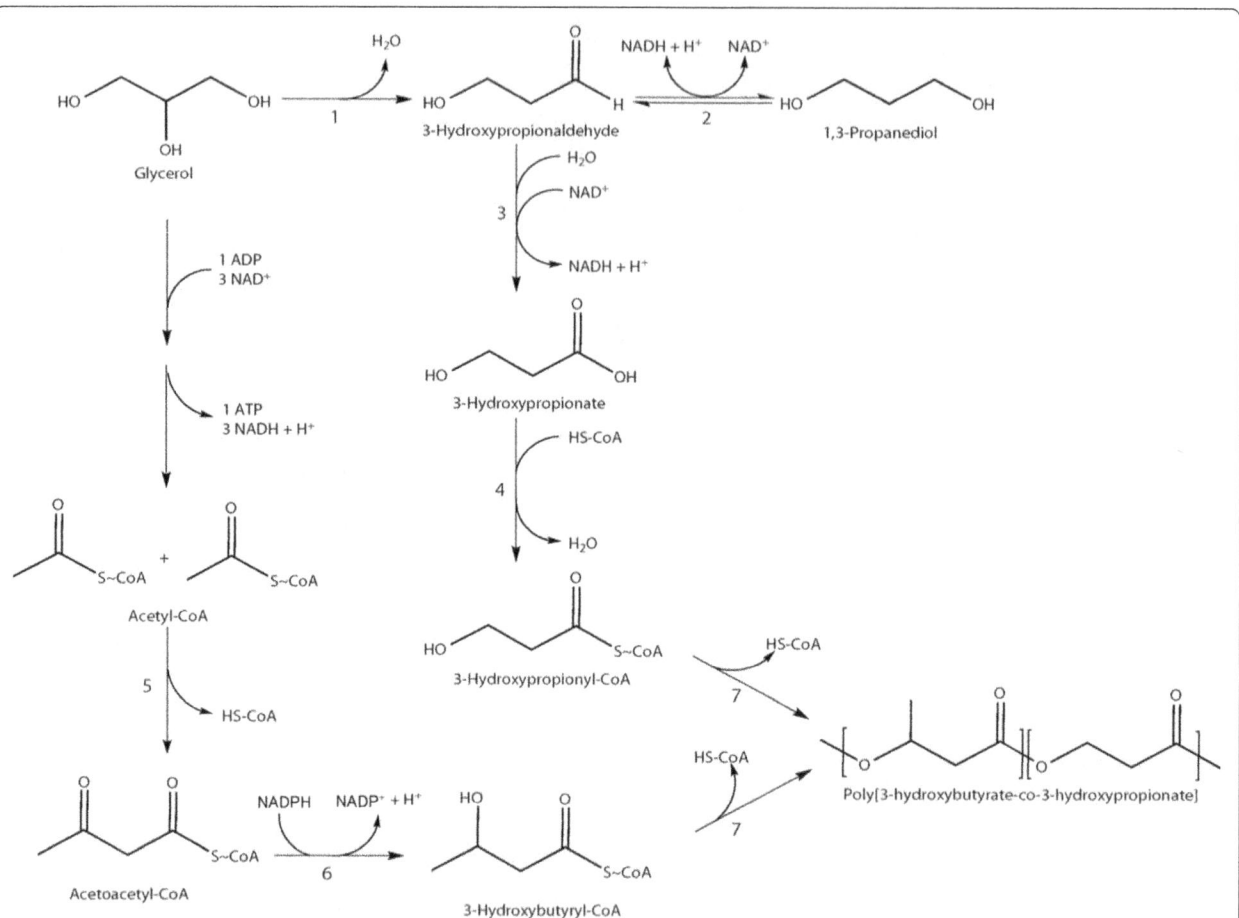

Figure 1 Pathway for conversion of glycerol to poly(3-hydroxybutyrate-*co*-3-hydroxypropionate) in a recombinant strain of *S. blattae*.
1: DhaBCE$_{Sb}$, 2: DhaT$_{Sb}$/DhaT$_{Pp}$, 3: AldD$_{Pp}$, 4: Pct$_{Cp}$, 5: PhaA$_{Re}$, 6: PhaB1$_{Re}$, 7: PhaC1$_{Re}$. Acetyl-CoA is synthesized from glycerol though glycolytic pathway.

Table 1 Bacterial strains and plasmids used in this study

Strains and plasmids	Relevant characteristics	Origin or reference
Strains		
E. coli		
TOP10	F⁻ *mcrA Δ(mrr-hsdRMS-mcrBC) φ80lacZΔM15 ΔlacX74 mupG recA1 araD139 Δ(ara-leu)7697 galE15 galK16 rpsL(StrR) endA1λ*	Life technologies (Darmstadt, D)
S. blattae		
ATCC33430	Wild type strain	ATCC33430
Sb6BP	pBBR1MCS-2 ::*plac::aldD::dhaT:: pct::plac::phaC1AB* in *S. blattae* ATCC33430	This study
Plasmids		
pBBR1MCS-2	Cloning vector, Km^r	Kovach et al., 1995
pBBR1MCS-2 ::*aldD::dhaT::pct*	Km^r; *aldD$_{Pp}$; dhaT$_{Pp}$; pct$_{Cp}$*	Heinrich et al., 2013
pBBR1MCS-2 ::*plac::phaC1AB*	Km^r; *phaC1$_{Re}$; phaA$_{Re}$; phaB1$_{Re}$*	This study
pBBR1MCS-2 ::*plac::aldD::dhaT::pct ::plac::phaC1AB*	Km^r; *aldD$_{Pp}$; dhaT$_{Pp}$; pct$_{Cp}$; phaC1$_{Re}$; phaA$_{Re}$; phaB1$_{Re}$*	This study

Then, pBBR1MCS-2::p_{lac}::phaC1AB was digested with SspI to generate a 4.7-kbp expression cassette of the phaC1, phaA and phaB1 under control of the lac promoter and ligated with the EcoICRI linearized fragment of pBBR1MCS-2::aldD::dhaT::pct (Heinrich et al., 2013) to generate the expression vector pBBR1MCS-2:: p_{lac}::aldD::dhaT::pct::p_{lac}::phaC1AB. In addition, S. blattae ATCC33430 was transformed with pBBR1MCS-2::p_{lac}:: aldD::dhaT::pct::p_{lac}::phaC1AB to generate Sb6BP by electroporation as previously described (Heinrich et al., 2013).

Optimization of cultivation medium

When optimizing the medium, we thought yeast extract is not necessary, because cultivations in complete synthetic medium have been made for bacteria such as *Klebsiella pneumonia* (Brandl et al., 1998), *E. coli* (Enayati et al., 1999) or *R. eutropha* (Sato et al., 2013). Therefore, several different concentrations of yeast extract were tested as described below.

250-mL Erlenmeyer flasks containing 50 mL MMB medium with 0, 0.2 or 2.0 (g/L) of yeast extract, respectively were used to cultivate S. blattae ATCC33430. The optical density at 600 nm and the pH were measured in samples withdrawn from the culture. In order to decide which yeast extract concentration is favorable for high cell density cultivation, 2 L bioreactors containing 1.5 L of MMB medium with 300 mM of glycerol and 0.2, 0.67 or 6.7 g/L of yeast extract and Sb6BP were used. Cell densities (g_{CDW}/L) and polymer contents (% wt_{PHA}/ wt_{CDW}) were measured.

Synthesis and purification of poly(3HB-co-3HP)

PHA was synthesized in a 2 L bioreactors containing 1.5 L of BM or MMB medium for 72 or 48 h. Glycerol was used as sole carbon source. Generally, enteric bacteria also S. blattae synthesize 1,3PD only under anaerobic condition. Thus, 4 different cultivation conditions (aerobic, anaerobic, low aeration/agitation and two-step) were conducted to optimize poly (3HB-co-3HP) synthesize condition in recombinant S. blattae (SB6P).

The operating conditions were as follows: Aerobic condition means an agitation at 800 rpm and an aeration rate of 2.0 L/min. Anaerobic conditions were maintained at an agitation of 150 rpm without any aeration whereas low aeration/agitation conditions were provided at a stirring rate of 150 rpm and an aeration rate of 0.4 L/min. 150 rpm was the minimum stirring rate required in medium mixing. The two-step fermentation (the first 24 hours under anaerobic condition, thereafter, aerobic condition) was performed according to Heinrich et al., 2013.

Cell harvest and extraction of poly(3HB-co-3HP)

After separating the cells from the culture broth, cells were frozen at −30°C and freeze dried. Poly(3HB-co-3HP) or poly(3HB) was isolated from the pulverized dry cell matter by digestion of non-PHA biomass employing a 13% (vol/vol) sodium hypochlorite solution (Heinrich et al., 2013, Heinrich et al., 2012).

Determination of poly(3HB-co-3HP)

Analysis of polymer content and purity of the extracted polymer was done by gas chromatography (GC). For this dried cell mass or samples of isolated poly(3HB-co-3HP) and poly(3HB) were exposed to acidic methanolysis as described before (Brandl et al., 1998; Timm et al., 1990). For microscopic analysis of PHA-granules cells were stained with Nile red (Spiekermann et al., 1999) (Figure 2).

Determination of glycerol and 1,3PD

Concentrations of glycerol and 1,3PD in the media were monitored by HPLC analysis. For this, supernatants were assayed using a Lachrom Elite HPLC-System (VWR-Hitachi, Darmstadt, D) chromatograph with a RI-detector (Type 2490 VWR, Darmstadt, D) and a Metacarb 67H-column (300 × 6.5 mM, VWR-Varian, Darmstadt, D) at 75°C and at a flow rate of 0.8 ml/ min for 20 min. The mobile phase was 4.5 mM sulfuric acid.

Results

Synthesis of poly(3HB-co-3HP) influenced by cultivation conditions

In order to develop a process for poly(3HB-co-3HP) production by a newly engineered Sb6BP strain, cultivation was conducted under four different conditions (aerobic, two-step, low aeration/agitation and aerobic conditions). (i) Aerobic condition occurred at high aeration and agitation, (ii) the two-step condition occurred during the first 24 hours under anaerobic condition and thereafter under aerobic condition, (iii) for low aeration/agitation condition a minimum stirring rate was applied together with low rate of aeration and (iv) for anaerobic condition was a minimum stirring rate was applied without aeration.

Glycerol supplementation started from 8 to 16 hours of cultivation in every culture when glycerol concentration became lower than 100 mM. In most cases, glycerol consumption rate became lower than supplementation rate after 18 h cultivation. If this was the case, the supplementation rate was regulated to maintain a glycerol concentration between 50 and 300 mM during the cultivations. In case of 1,3PD, it might not be effectively transported into the cell after it was transiently formed and excreted into the medium before it could be

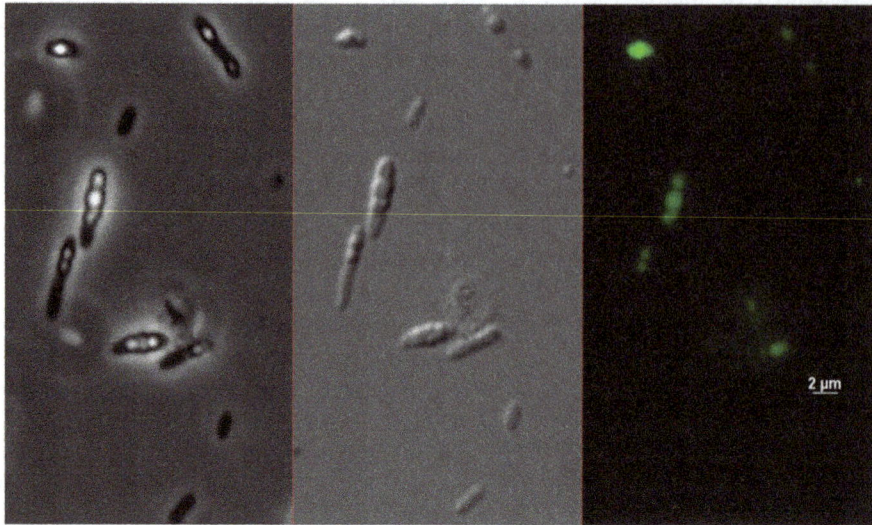

Figure 2 Poly(3HB-co-3HP) accumulation of Shimwellia blattae Sb6BP. After 72 h of fermentation in BM under two-step condition. Hydrophobic inclusions were stained with Nile red and observed with a fluorescence microscope employing phase contrast (left), differential interference contrast (middle), and fluorescence microscopy at 312 nm (right), respectively.

incorporated into the polymer via 3-hydroxypropionyl-CoA; thus the 1,3PD concentration was not so affected during cultivations. Moreover, the reduction of 1,3PD concentration in the later phase of the cultivation period could also occur due to a dilution effect (Figures 3 and 4).

The amount of 1,3PD that was converted into 3-hydroxypropionyl-CoA and polymerized into poly(3HB-co-3HP) was too small to affect the 1,3PD concentration of the culture medium in these studies. We already confirmed that 1,3PD supplemented to the culture medium was converted to 3-hydroxypropiony-CoA which is

polymerized into poly(3HB-co-3HP) in the recombinant strain in which *dhaT* was heterologously expressed. However, efficiency was very low (data not shown). Therefore, most of the 3HP monomer in the accumulated polymer was provided directly from 3-hydroxypropionaldehyde via 3-hydroxypropionate and 3-hydroxyproionyl-CoA by AldD and Pct, respectively.

Strain *Sb*6BP was confirmed as a poly(3HB-co-3HP) producer (Table 2). Moreover, strain *Sb*6BP synthesized the poly(3HB-co-3HP) copolymer only under anaerobic, two-step and low aeration/agitation conditions. Under

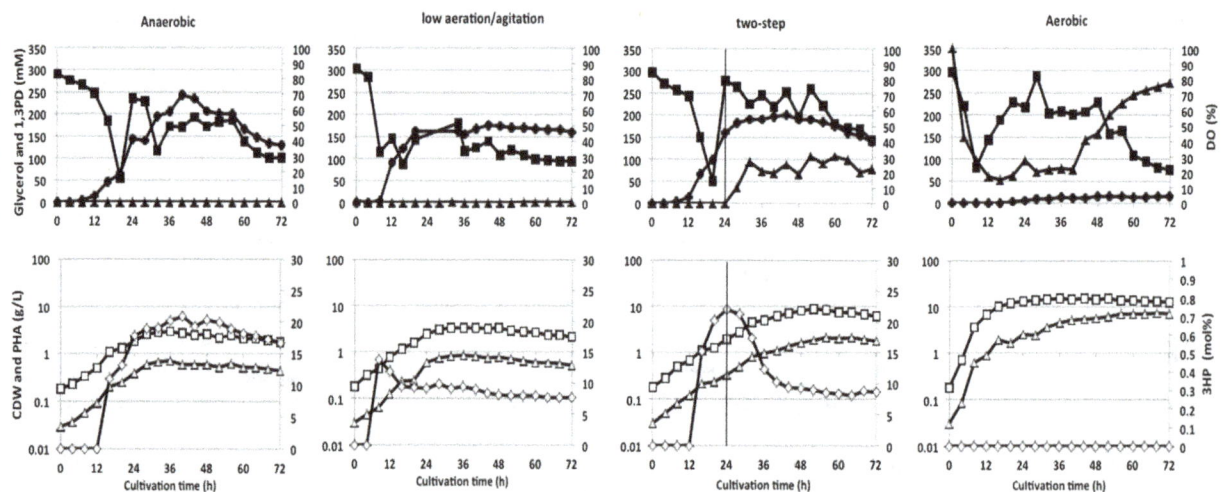

Figure 3 Fed-batch profile of Shimwellia blattae Sb6BP under different cultivation conditions in BM. The cultures were grown in 2 L jar-fermenter at 30°C. 1,3PD (●), glycerol (■), dissolved oxygen (▲), CDW (□), poly(3HB-co-3HP) or poly(3HB) (△), 3HP (◇).

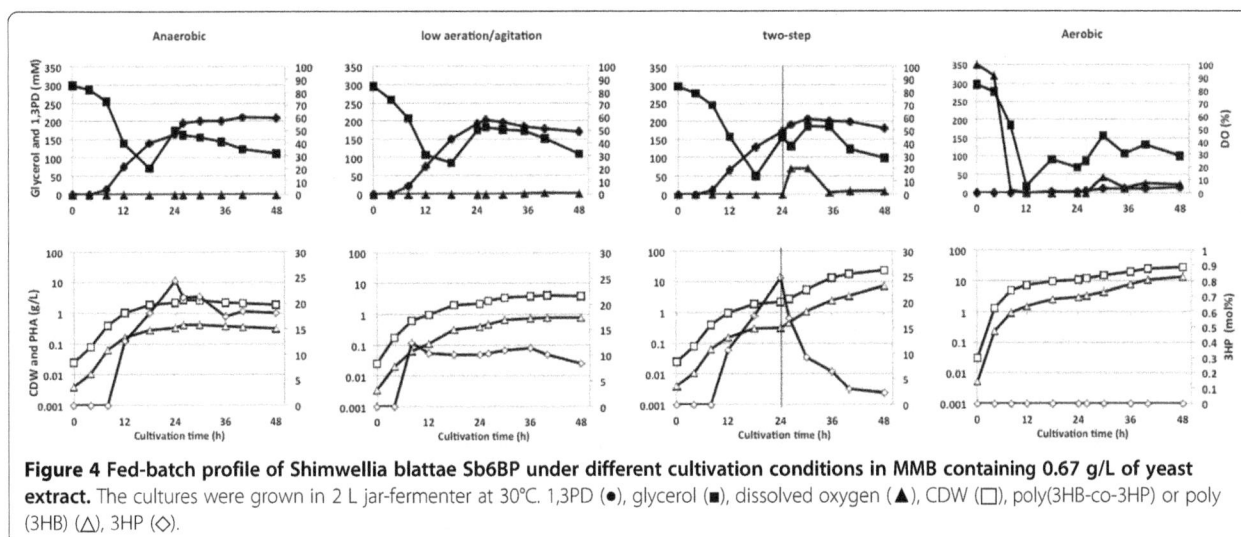

Figure 4 Fed-batch profile of Shimwellia blattae Sb6BP under different cultivation conditions in MMB containing 0.67 g/L of yeast extract. The cultures were grown in 2 L jar-fermenter at 30°C. 1,3PD (●), glycerol (■), dissolved oxygen (▲), CDW (□), poly(3HB-co-3HP) or poly (3HB) (△), 3HP (◇).

aerobic condition no 3HP were incorporated as constituent. The productivity for poly(3HB-co-3HP) in BM medium ranged from 4 mg$_{PHA}$/L/h to 25 mg$_{PHA}$/L/h and was highest under the two-step condition. No copolymer but poly(3HB) homopolymer was formed under aerobic conditions, and the productivity for poly(3HB) was higher in comparison to poly(3HB-co-3HP) in both BM and MMB medium. The polymer content was also highest in the two-step fermentations (29.9% in BM and 30.7% wt$_{PHA}$/wt$_{CDW}$ in MMB, respectively), and lowest (14.7% in BM and 17.2% wt$_{PHA}$/wt$_{CDW}$ in MMB, respectively) in the anaerobic fermentation. On the other hand, poly(3HB-co-3HP) with the highest 3HP content was obtained (16.6 mol% in BM and 18.1 mol% in MMB, respectively) in anaerobic fermentations.

The time courses of polymer synthesis for each cultivation condition in BM medium were as follows (Figure 3): (1) Anaerobic: cell growth and addition of ammonium hydroxide almost stopped after 48 h cultivation time. The highest 3HP composition (20.2 mol%),

1,3PD concentration (242 mM) and polymer production was recorded after about 40 h cultivation time. These data indicate that 3HP-CoA and 1,3PD synthesis occurred only during exponential cell growth. Moreover, it was indicated that both, (R)-3HB-CoA and 3HP-CoA are not supplied after the stop of cell growth, because the 3HP content was very stable after 40 h. (2) Low aeration/agitation: the time courses of every parameter except 3HP monomer fraction were very similar when compared to anaerobic conditions as explained above. The highest molar 3HP monomer fraction was recorded earlier than 12 hour of culture time. The residual cell mass was only marginally influenced by aeration, but the 3HP monomer content of the polyester was decreased to less than 50% in comparison to anaerobic condition. (3) Two-step: Interestingly, the highest molar 3HP fraction in the copolymer was recorded at the time when the cultivation conditions were just switched. After that, the fraction of 3HP moieties rapidly dropped but the cell dry weight (CDW) increased on the other hand. Formation of 1,3PD was maintained for some hours

Table 2 Results of fed-batch cultivation using Sb6BP and glycerol as sole carbon source

Strain	Medium	Condition	Cell density (g$_{CDW}$/L)	PHA content (wt%)	PHA (g$_{PHA}$/L)	Monomer composition (mol%)		Cultivation time (h)	PHB(P) productivity (mg$_{PHA}$/L/h)	Glycerol yield
						3HB	3HP			
Sb6BP	BM	Anaerobic	1.72	14.7	0.25	83.4	16.6	72	3.51	0.0041
		Low aeration/agitation	2.11	24.6	0.52	92.4	7.3		7.21	0.0075
		Two-step	6.12	29.9	1.83	91.4	8.6		25.4	0.032
		Aerobic	12.6	55.2	6.96	100	0		96.6	0.097
	MMB	Anaerobic	1.84	17.2	0.32	82.9	18.1	48	6.59	0.0033
		Low aeration/agitation	3.78	20.3	0.77	91.7	8.3		16.0	0.011
		Two-step	23.2	30.7	7.12	97.9	2.1		148.4	0.093
		Aerobic	27.3	48.1	13.1	100	0		273.6	0.15

Cells were cultivated in MB or MMB medium with glycerol that was intermittently added. MMB contained 0.67 g/L yeast extract. Kanamycin was added as 50 μg/ml final concentration at the beginning of the cultivation. All cultivations were conducted in a 2 L jar-fermenter.

after the cultivation condition was changed from anaerobic to aerobic and then stopped. (4) Aerobic: only the poly(3HB) homopolymer was synthesized under aerobic conditions. Furthermore, the cell density (12.6 g_{CDW}/L) and the PHA productivity (96.6 mg_{PHA}/L/h) were the highest. Only a small amount of 1,3PD (10 to 15 mM) was synthesized after 24 h of cultivation, and the 3HP monomer was not detected by GC analysis at any period.

GPC analysis of the isolated poly(3HB-co-3HP) obtained in BM after two-step cultivation for 72 h revealed an average molecular weight of 765,293 Da with a polydispersity index (M_w/M_n) of 2.49.

Optimization of the medium for cultivation

Although cells of strain *Sb*6BP could be successfully cultivated, poly(3HB-co-3HP) productivity was still low due to a low residual cell mass. The cell density was only 12.6 g_{CDW}/L under aerobic condition in BM medium. From these results it was suspected that some substances required for cell growth were missing or that at least a severe shortage had occurred in BM medium. Therefore, the medium was optimized.

It was confirmed that the growth rate of the cells was influenced by the concentration of yeast extract in the medium (Figure 5). The growth rates (μ) of the cells in MMB between 0 to 12 h cultivation time containing 0, 0.2 or 2.0 g/L of yeast extract were 0.32, 0.36 and 0.41 h^{-1}, respectively. It was also confirmed that the trace element solution used in this study is necessary for cell growth (Figure 5A). These results indicate in contrast to our expectations that yeast extract is very important for cell growth of *S. blattae*. Yeast extract contains many vitamins like biotin, 4-aminobenzoic acid, pantothenic acid, pyridoxine, riboflavin and thiamine (data from DIFCO) in addition to amino acids and some of these compounds are likely to be growth limiting.

Although yeast extract was confirmed as a key factor for growth of *S. blattae*, the optimum concentration could not be identified in flask culture experiments since cell growth stopped after 24 h of cultivation time when the pH had dropped below 5. Therefore, higher cell density cultivations were conducted in 2 L fermenter in order to determine the optimum concentration of yeast extract. The cell densities obtained in aerobic cultivations in MMB containing 0.2, 0.67 and 6.7 g/L of yeast extract were 14.6, 27.3 and 44.6 g_{CDW}/L, respectively (Figure 5B, Table 3).

With regard to the cost, we chose of 0.67 g/L to obtain 27.3 g_{CDW}/L.

Synthesis of poly(3HB-co-3HP) in MMB medium

Improved productivities for PHA synthesis were between 1.9 and 5.8 times higher in MMB than in BM; the highest poly(3HB-co-3HP) productivity was obtained in two-step

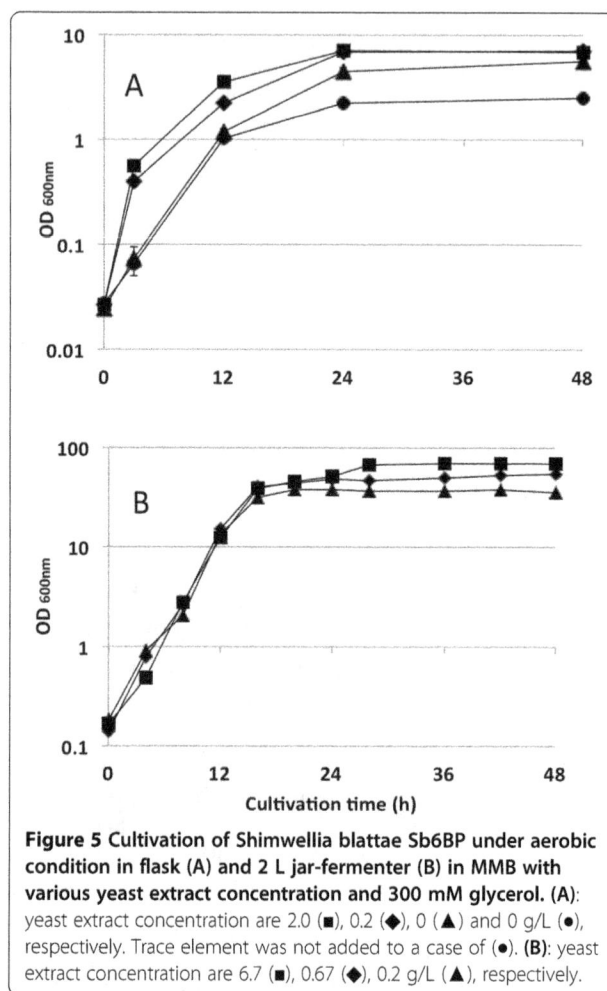

Figure 5 Cultivation of Shimwellia blattae Sb6BP under aerobic condition in flask (A) and 2 L jar-fermenter (B) in MMB with various yeast extract concentration and 300 mM glycerol. **(A)**: yeast extract concentration are 2.0 (■), 0.2 (◆), 0 (▲) and 0 g/L (●), respectively. Trace element was not added to a case of (●). **(B)**: yeast extract are 6.7 (■), 0.67 (◆), 0.2 g/L (▲), respectively.

cultivation (25 mg_{PHA}/L/h in BM medium in comparison to 148 mg_{PHA}/L/h in MMB) (Table 2). The positive effect of MMB medium was highest in two-step cultivation experiments. This is explained by the fact that cell growth was very high after the culture conditions were changed (Figure 4).

Conversely, these effects were less under anaerobic and low aeration/agitation conditions. An explanation is most probably because the rate-limiting factor was oxygen and not one of the compounds included in yeast extract. These observations could not be confirmed in

Table 3 Cultivation of *Sb*6BP in MMB containing various concentrations of yeast extract

Yeast extract (g/L)	Cell density (g_{CDW}/L)	PHA (g_{PHA}/L)	PHA content (wt%)	Cultivation time (h)
0.2	14.6	5.58	38.3	48
0.67	27.3	14.2	48.1	48
6.7	44.6	21.1	47.2	48

During cultivation, glycerol was intermittently added. Kanamycin was added as 50 μg/ml final concentration at the beginning of the cultivation. All cultivations were conducted in a 2 L jar-fermenter.

MMB under aerobic condition as there was no poly (3HB-co-3HP) accumulated. The polymer content and the 3HP fraction in the copolyester were not so much influenced under anaerobic and low aeration/agitation conditions whereas the 3HP fraction in two-step condition dropped rapidly from 8.6 to 2.1 mol% after the cultivation conditions were changed (Figure 4). These results indicate that the decrease of 3HP fraction might be due to a relative decrease of the monomer composition by additional accumulation of 3HB and/or poly(3HB), because, after switching the conditions, cell growth started again and 3HB-CoA provision via acetoacetyl-CoA was started as well. In addition, it was confirmed that the net 3HP amount increased until 30 h; thereafter no increase occurred in case of two-step cultivation when using the MMB medium. Therefore, there is a possibility that the resulting polymer was likely a blend of poly(3HB) and poly(3HB-co-3HP).

Poly(3HB-co-3HP) with the highest 3HP fraction (18.1 mol%) was obtained during anaerobic condition; however, the polymer productivity and cell density were only 6.6 mg_{PHA}/L/h and 1.84 g_{CDW}/L, respectively.

Discussion

In this study we engineered a recombinant strain of *S. blattae* ATCC33430 (*Sb*6BP), which is able to synthesize poly(3HB) and poly(3HB-co-3HP). However, *Sb*6BP synthesized the copolymer only under oxygen limiting conditions because the cells produced only small amounts of 1,3-propanediol (1,3PD) as a precursor of 3-hydroxypropionyl-CoA (3HP-CoA) under aerobic condition. In any case, in comparison with the 1,3PD productivity under oxygen limiting conditions, a lower productivity of 1,3PD under aerobic condition causes non-accumulation of poly(3HB-co-3HP).

However, successful production of poly(3HB-co-3HP) in recombinant *S. blattae* was achieved, 3HP composition was rather low. The glass transition temperature of poly(3HB-co-3HP) decreases rapidly from −3°C to −15°C, and the melting temperature (Tm) decreases from 163°C to 73°C as the 3HP fraction in the copolymer increased from 25.6 to 36.3 mol% (Wang et al., 2013). In particular, the poly(3HB-co-3HP) with a 3HP content exceeding 30 mol% is therefore much more flexible and less crystalline and is expected to approach that of conventional plastics such as polypropylene, polystyrene, and polyethylene. The low 3HP fraction indicates that the strain cannot actively import 1,3PD from culture medium or that 3HP-CoA supply via 3-hydroxypropionaldehyde and 3-hydroxypropionate is insufficient.

This is the first report for the production of poly(3HB-co-3HP) without using vitamin B_{12} and expensive compounds such as 3-hydroxypropionate by recombinant *S. blattae*. We achieved a poly(3HB-co-3HP) productivity

of 148 mg_{PHA}/L/h in this study, which is the highest so far reported (Shimamura et al., 1994; Fukui et al., 2009; Wang and Inoue, 2001; Wang et al., 2013). *S. blattae* is therefore one of the promising bacterial strains for poly (3HB-co-3HP) production from glycerol.

However, the reported processes still require oxygen limitation. Therefore, the polymer productivity is still very low. In addition, the polymer produced by two-step cultivation method was likely a blend of poly(3HB) and poly (3HB-co-3HP) owing to the conditional change from anaerobic to aerobic condition. The remaining challenges are to achieve efficient utilization of 3HPA or 1,3PD and the expression of the genes regulated by *dha* regulon in aerobic conditions for aerobic production of poly(3HB-co-3HP). Moreover, an enhancement of the metabolite flow from 3-hydroxypropionaldehyde to 3HP-CoA will be necessary to increase the 3HP fraction in the copolymer. By using the *pduP* gene from *Salmonella enterica* (Andreeßen et al., 2010) or *Salmonella typhimurium* (Gao et al., 2014), the provision of 3HP-CoA provision might be improved.

Competing interests

The authors declare that they have no competing interests.

Authors' contributions

SS performed the experiments. SS and BA participated in the design of the study. SS and BA drafted the manuscript. AS edited the manuscript. BA and AS conceived the study. All authors read and approved the final manuscript.

Acknowledgement

We thank Rolf Daniel and his laboratory at the Department of Genomic and Applied Microbiology (Georg-August University Göttingen) for providing *S. blattae* ATCC 33430 and Kaneka Corporation, Japan, for GPC analysis. We acknowledge support by Deutsche Forschungsgemeinschaft and Open Access Publication Fund of University of Münster.

References

Anderson AJ, Haywood GW, Dawes EA (1990) Biosynthesis and composition of bacterial poly(hydroxyalkanoates). Int J Biol Macromol 12:102–105

Andres S, Wiezer A, Bendfeldt H, Waschkowitz T, Toeche-Mittler C, Daniel R (2004) Insights into the genome of the enteric bacterium *Escherichia blattae*; cobalamin (B_{12}) biosynthesis, B_{12}-dependent reactions, and inactivation of the gene region encoding B_{12}-dependent glycerol dehydratase by a new Mu-like prophage. J Mol Microbial Biotechnol 8:150–168

Andreeßen B, Johanningmeier B, Burbank J, Steinbüchel A (2014a) Influence of the operon structure on poly(3-hydroxypropionate) synthesis in *Shimwellia blattae*. Appl Microbiol Biotechnol 98(17):7409–7422

Andreeßen B, Lange AB, Robenek H, Steinbüchel A (2010) Conversion of glycerol to Poly(3-hydroxypropionate) in Recombinant *Escherichia coli*. Appl Environ Microbiol 76:622–626

Andreeßen B, Steinbüchel A (2010) Biosynthesis and biodegradation of 3-hydroxypropionate-containing polyesters). Appl Environ Microbiol 76 (15):4919–4925

Andreeßen B, Taylor N, Steinbüchel A (2014b) Poly(3-hydroxypropionate): A promising alternative to fossil fuel-based materials. Appl, Environ. Microbiol. August, 22 doi:10.1128/AEM. 02361-14

Brandl H, Gross RA, Lenz RW, Fuller RC (1998) *Pseudomonas oleovorans* as a source of poly(β-hydroxyalkanoates) for potential applications as biodegradable polyesters. Appl Environ Microbiol 54:1977–1982

Budde CF, Mahan AE, Lu J, Rha C, Sinskey AJ (2010) Role of multiple acetoacetyl coenzyme A reductase in polyhydroxybutyrate biosynthesis in *Ralstonia eutropha* H16. J Bacteriol 192:5319–5328

Burgess NR, McDermott SN, Whiting J (1973) Aerobic bacteria occurring in the hindgut of the cockroach, *Blatta orientalis*. J Hyg 71:1–7

Chen GQ, Wu Q, Zhao K, Yu PH (2000) Functional polyhydroxyalkanoates synthesized by microorganisms. Chinese J Polym Sci 18–5:389–396

Doi Y, Kitamura S, Abe H (1995) Microbial synthesis and characterization of poly (rhydroxybutyrate-*co*-hydroxyhexanoate). Macromolecules 8:4822–4828

Enayati N, Trai C, Parulekar SJ, Stark BC, Webster DA (1999) Production of r-amylase in fed-batch cultures of *vgb*+ and *vgb*-recombinant *Escherichia coli*: some observations. Biotechnol Prog 15:640–645

Fukui T, Suzuki M, Tsuge T, Nakamura S (2009) Microbial synthesis of poly((*R*)-3-hydroxybutyrate-*co*-3-hydroxypropionate) from unrelated carbon sources by engineered *Cupriavidus necator*. Biomacromolecules 10:700–706

Gao Y, Liu C, Ding Y, Sun C, Zhang R, Xian M, Zhao G (2014) Development of genetically stable *Escherichia coli* strains for poly(3-Hydroxypropionate) production. PLoS One 9(5):e97845

Heinrich D, Andreeßen B, Steinbüchel A (2013) From waste to plastic: Synthesis of poly(3-hydroxypropionate). Appl Environ Microbiol 79:3582–3589

Heinrich D, Madkour MH, Al-Ghamdi MA, Shabbaj II, Steinbüchel A (2012) Large scale extraction of poly(3-hydroxybutyrate) from *Ralstonia eutropha* H16 using sodium hypochlorite. AMB Express 2(1):59

Kovach ME, Elzer PH, Hill DS, Robertson GT, Farris MA, Roop RM 2nd, Peterson KM (1995) Four new derivatives of the broad-host-range cloning vector pBBR1MCS, carrying different antibiotic-resistance cassettes. Gene 166:175–176

Lee SY (1996) Bacterial polyhydroxyalkanoates. Biotechnol Bioeng 49:1–14

Martens JH, Barg H, Warren MJ, Jahn D (2002) Microbial production of VitaminB$_{12}$. Appl Microbial Biotechnol 58:275–285

Priest FG, Barker M (2010) Gram-negative bacteria associated with brewery yeasts: reclassification of *Obesumbacterium proteus* biogroup 2 as *Shimwellia pseudoproteus* gen. nov., sp. nov., and transfer of *Escherichia blattae* to *Shimwellia blattae* comb. nov. Int J Envol Microbiol 60:828–833

Quispe C, Coronado CJR, Carcalho JA (2013) Glycerol: Production, consumption, price, characterization and new trends in combustion. Renew Sust Energ Rev 27:475–493

Sambrook J, Fritsch EF, Maniatis T (1989) Molecular cloning: A laboratory manual. Cold spring Harbor Laboratory Press. Cold Spring Harbor, NY

Sato S, Tetsuya F, Matsumoto K (2013) Construction of a stable plasmid vector for industrial production of poly(3-hydroxybutyrate-*co*-3-hydroxyhexanoate) by a recombinant *Cupriavidus necator* H16 strain. J Biosci Bioeng 116:677–681

Shimamura E, Kasuya K, Kobayashi G, Shiotani T, Shima Y, Doi Y (1993) Physical properties and biodegradability of microbial poly(3-hydroxybutyrate-*co*-hydroxyhexanoate). Macromolecules 27:878–880

Shimamura E, Scandola M, Doi Y (1994) Microbial synthesis and characterization of poly(3-hydroxybutyrate-*co*-3-hydroxypropionate). Macromolecules 27:4429–4435

Spiekermann P, Rehm BHA, Kalscheuer R, Baumeister D, Steinbüchel A (1999) A sensitive, viable-colony staining method using Nile red for direct screening of bacteria that accumulate polyhydroxyalkanoic acids and other lipid storage compounds. Arch Microbiol 171:73–80

Steinbüchel A (2001) Perspectives for biotechnological production and utilization of biopolymers: metabolic engineering of polyhydroxyalkanoate biosynthesis pathway as a successful example. Macromol Biosci 1:1–24

Steinbüchel A, Füchtenbusch B (1998) Bacterial and other biological systems for polyester production. Trends Biotechnol 16:419–427

Sun J, Heuvel JVD, Soucaille P, Qu Y, Zeng AP (2003) Comparative genomic analysis of *dha* regulon and related genes for anaerobic glycerol metabolism in bacteria. Biotechnol Prog 19:263–272

Timm A, Byrom D, Steinbüchel A (1990) Formation of blends of various poly(3-hydroxyalkanoic acids) by a recombinant strain of *Pseudomonas oleovorans*. Appl Microbiol Biotechnol 33:296–301

Wang Y, Inoue Y (2001) Effect of dissolved oxygen concentration in the fermentation medium on transformation of the carbon sources during the biosynthesis of poly(3-hydroxybutyrate-*co*-3-hydroxypropionate) by *Alcaligenes latus*. Int J Biol Macromol 28:235–243

Wang Q, Yang P, Xian M, Yang Y, Liu C, Xue Y, Zhao G (2013) Biosynthesis of poly(3-hydroxybutyrate-*co*-3-hydroxypropionate) with fully controllable structure from glycerol. Bioresour Technol 142:741–744

20

Isolation of cellulolytic bacteria from the intestine of *Diatraea saccharalis* larvae and evaluation of their capacity to degrade sugarcane biomass

Karina I Dantur, Ramón Enrique, Björn Welin[*] and Atilio P Castagnaro

Abstract

As a strategy to find efficient lignocellulose degrading enzymes/microorganisms for sugarcane biomass pretreatment purposes, 118 culturable bacterial strains were isolated from intestines of sugarcane-fed larvae of the moth *Diatraea saccharalis*. All strains were tested for cellulolytic activity using soluble carboxymethyl cellulose (CMC) degrading assays or by growing bacteria on sugarcane biomass as sole carbon sources. Out of the 118 strains isolated thirty eight were found to possess cellulose degrading activity and phylogenetic studies of the *16S rDNA* sequence revealed that all cellulolytic strains belonged to the phyla γ-Proteobacteria, Actinobacteria and Firmicutes. Within the three phyla, species belonging to five different genera were identified (*Klebsiella*, *Stenotrophomonas*, *Microbacterium*, *Bacillus* and *Enterococcus*). Bacterial growth on sugarcane biomass as well as extracellular endo-glucanase activity induced on soluble cellulose was found to be highest in species belonging to genera *Bacillus* and *Klebsiella*. Good cellulolytic activity correlated with high extracellular protein concentrations. In addition, scanning microscopy studies revealed attachment of cellulolytic strains to different sugarcane substrates. The results of this study indicate the possibility to find efficient cellulose degrading enzymes and microorganisms from intestines of insect larvae feeding on sugarcane and their possible application in industrial processing of sugarcane biomass such as second generation biofuel production.

Keywords: Bacterial symbionts; Cellulase; Endo-glucanase activity; Gut microbiota; Lignocellulose; Plant-insect-bacterial interaction; Sugarcane

Introduction

The imminent need to replace fossil-based transport fuels with more environment-friendly renewable alternatives, has sparked an increasing interest in finding abundant and cheap resources for biofuel production. One of the most interesting and promising alternatives, in the short- and medium-term perspective, is the second generation bioethanol (B2G) produced from lignocellulosic byproducts of agricultural, forestry and industrial activities or from urban waste residues. Natural occurring lignocellulose material, especially in the form of plant cell wall material, is a renewable, abundant and relatively cheap mixture of organic materials, principally containing polysaccharides (~75% dry weight) and lignin (~25% dry weight).

The carbohydrates consist mainly of fibers of cellulose (glucose units) and hemicellulose (composed of various 5- and 6-carbon sugars), which give strength to plant structures. Lignin on the other hand, is formed by a non-carbohydrate complex structure built from phenylpropanoid units. This phenolic polymer sticks to the polysaccharide components, strengthening the whole structure which renders it extremely resistant to biological degradation (Cheng and Wang 2013).

Although much effort and resources have been directed to develop industrial scale bioethanol production from lignocellulose, there is still no economically viable industrial production system available for any type of biomass (Limayem and Ricke 2012; Morone and Pandey 2014). One of the main reasons for the high production cost of B2G is the recalcitrance of the lignocellulose to enzymatic hydrolysis, which requires a pre-treatment step before an efficient enzyme-based degradation of the complex polysaccharides (cellulose or hemicellulose) into its

* Correspondence: bwelin@gmail.com
Estación Experimental Agroindustrial Obispo Colombres (EEAOC) - Consejo Nacional de Investigaciones Científicas y Técnicas (CONICET), Instituto de Tecnología Agroindustrial del Noroeste Argentino (ITANOA), 3150 William Cross Av., Las Talitas PC T4101XAC, Tucumán, Argentina

fermentable monosaccharide components can be carried out (Canilha et al. 2012; Cardoso et al. 2012). Currently available pretreatment methods are biological, chemical (cellulose solvents, acids, or bases), or physical (mechanical size reduction, comminution, steam explosion, vibratory ball milling, compression milling, and hydrothermolysis) (Taherzadeh and Karimi 2007; Sierra et al. 2011). Even though applying a pretreatment of the plant biomass large amounts of enzymes are still needed to obtain a rapid and efficient cellulose and hemicelluloses degradation, and therefore new and more efficient enzyme cocktails are needed in order to generate a more economic degradation process rendering a cheaper total bioethanol production.

Natural lignocellulose degradation, an essential part of the carbon cycle, is carried out by highly specialized wood-degrading microorganisms (fungi and bacteria) and symbiotic microbes found in the intestines of many plant feeding animals (Watanabe and Tokuda, 2010; Cardoso et al. 2012; Gupta et al. 2012; He et al. 2013). Hydrolysis of cellulose is a multi-enzymatic process involving at least three different types of enzymatic activities in order to liberate the smallest basic unit, a glucose molecule. First, endo-β-glucanases (E.C. 3.2.1.4) cleave the cellulose backbone at internal amorphous sites, reducing the chain length while creating numerous ends where exo-β-glucanases (E.C.3.2.1.91) attack to release short-chain glucose oligomers (cellodextrins and cellobiose). There are two forms of exo-glucanases, the first attacking the reducing end of a cellulose chain while the second attacks the non-reducing ends. Finally, the short glucose chains released after exo-glucanase attacks are hydrolyzed to single glucose units by a β-glucosidase (E.C. 3.2.1.21) (Singh and Hayashi 1995; Teeri 1997).

It is thought that the highly effective plant biomass degrading capacity found in intestines of many insect herbivores constitutes one of the most efficient naturally occurring bioreactors and could provide an important and interesting biotechnology source of microorganisms and enzymes for cellulose degradation (Sun and Scharf 2010). The sugarcane stalks borer Diatraea saccharalis is the major sugarcane pest in Argentina causing considerable damage to infested plants, which facilitates secondary fungal and/or bacterial infections, resulting in important economic losses to producers. Due to the high and rapid sugarcane feeding capacity of larvae of this species, we hypothesized that larvae only fed on sugarcane, could have developed a consortium community of symbiotic-bacteria possessing an enzyme arsenal which efficiently degrade sugarcane plant tissue. Supporting this theory are recently published studies where a comparison among gut microbiota from Coleoptera, Lepidoptera and Orthoptera species, show correlation between the cellulolytic enzyme activities in the insect gut with the lignocellulosic biomass composition in the food consumed by the insect (Shi et al.

2011; Cardoso et al. 2012). These results suggest that larvae fed on a specific plant species develops or carries symbiotic bacteria possessing a highly efficient enzyme arsenal that allows the insect to rapidly process a specific plant biomass. Thus, we wanted to isolate cellulolytic bacteria from larvae feeding on sugarcane in order to explore their potential as a source of enzymes for efficient sugarcane biomass degradation that could be employed in second generation bioethanol production or other biomass treatments.

Materials and methods

Isolation of cellulolytic bacteria from the gut of Diatraea saccharalis

Fifth instar larvae of D. saccharalis were collected from sugarcane fields at the "Estación Experimental Agroindustrial Obispo Colombres (EEAOC)", Tucumán, Argentina and surface sterilized with 70% ethanol before guts were aseptically dissected from replicates of ten larvae. Isolated intestines were cut into small pieces, homogenized in an isotonic saline solution before plated on a minimal saline agar medium (Na$_2$HPO$_4$ 1 g; KH$_2$PO$_4$ 1 g; MgSO$_4$ 0.05 g; NaCl 3 g; CaCl$_2$ 0.05 g, (NH$_4$)$_2$SO$_4$ 2 g and agar 20 g in 1 L of H$_2$O) containing 0.5% of glucose, finely milled sugarcane bagasse or harvest trash (HT) were added as sole carbon sources. Cultures were incubated for 3 days at 37°C and bacterial colonies able of grow on cellulose as sole carbon source were isolated and cellulose-degrading ability was confirmed by streaking the bacteria on the same minimal saline agar supplemented with 0.5% carboxymethyl cellulose (CMC). After 4 days of incubation at 30°C, the medium was treated with 0.1% (w/v) aqueous Congo red, a specific indicator for cellulose hydrolysis leaving a distained halo where CMC has been degraded (Teather and Wood 1982).

Identification of bacterial isolates and sequence homology analysis of the 16SrDNA gene

DNA extraction was performed using the PureLink Genomic DNA Kit (Invitrogen, USA) according to the manufacturer's instructions. Bacterial universal primers fD1 (5'-CCGAATTCGTCGACAACAGAGTTTGATCCT GGCTCAG-3') and rP1 (5'-CCCGGGATCCAAGCTTAC GGTTACCTTGTTACGACTT-3') were used to amplify the 16S rDNA gene from genomic DNA (Weisburg et al. 1991). Polymerase chain reaction (PCR) was performed in a Bio-Rad Mycycler Thermalcycler (Hercules, USA). Each reaction mixture (50 µl) contained 1X reaction buffer, 2 mM MgCl$_2$, 0.2uM of each primer, 0.2 mM of dNTPs, 2U of Taq DNA polymerase (Invitrogen, USA) and 40 ng of genomic DNA. PCR cycling parameters were 1 cycle at 95°C (2 min); 35 cycles at 95°C (30 sec), 54°C (30 sec), and 72°C (3 min); and a final cycle at 72°C for 10 min. All amplified DNA products were checked by gel electrophoresis in 1% (w/v) agarose gels stained with GelRed (Biotium, USA). The 1.5Kb amplification products were directly purified from the PCR reaction using the PureLink Quick PCR purification kit (Invitrogen, Germany) according to the manufacturer's instructions. Purified reactions were sequenced by the Biotechnology Institute Sequencing Service, INTA Castelar using an ABI 3130 Capillary DNA analyzer (Applied Biosystems, USA). All sequences were aligned against the NCBI database and in the RDP II database (Cole et al. 2009) using the BLAST (Basic Local Alignment and Search Tool) algorithm (Altschul et al. 1997). The aligned 16S rDNA sequences were analyzed based on their homologous characteristics and a phylogenetic tree was constructed using the software MAFFT 7 online version (Katoh 2013) using the neighbor-joining option with a bootstrap analysis of 1,000 random replications. A sequence from a *Chlamydia pneumoniae* strain CHT16SR (L06108) was used to root the tree. The resulting *16S rDNA* gene sequences for the isolates *Klebsiella oxytoca* Kd70 TUC-EEAOC, *K. pneumoniae* KdB5 TUC-EEAOC, *K. variicola* KdB1 TUC-EEAOC, *Stenotrophomonas maltophilia* Kd3 TUC-EEAOC, *S. rhizophila* Kd46 TUC-EEAOC, *Bacillus pumilus* Kd109 TUC-EEAOC, *Enterococcus casseliflavus* Kd7 TUC-EEAOC, *Microbacterium hominis* KdL49 TUC-EEAOC, *M. schleiferi* KdL45 TUC-EEAOC were submitted to GenBank under the accession numbers of KM096608, KM096599, KM096598, KM096600, KM096602, KM096605, KM096606, KM096603, KM096601 respectively.

Klebsiella oxytoca Kd70 TUC-EEAOC, *Bacillus pumilus* Kd109 TUC-EEAOC, *Stenotrophomonas maltophilia* Kd3 TUC-EEAOC, *Enterococcus casseliflavus* Kd7 TUC-EEAOC have been deposited at the German Collection of Microorganisms and Cell Cultures (DSMZ, http://www.dsmz.de) as accession DSM 27019, DSM 27021, DSM 27017 and DSM 27018 respectively.

Determination of total cellulase activity

Bacteria were grown overnight in Luria-Bertani Broth (LB) and 20 µl of a bacterial suspension adjusted to an O.D. of 0.2 was thereafter applied as a drop on solid agar plates containing the previously described saline minimal medium supplemented with 0.5% (w/v) finely milled bagasse or sugarcane harvest trash residues. Bacteria were grown for a fortnight at 30°C and plates were thereafter colored with Congo red to determine possible cellulose degradation. The diameter of the clear zone around the bacterial drop is indicative of the magnitude of cellulolytic activity.

Extracellular CMCase activity assay

Twenty (20) micro liters of a bacterial growth suspension were adjusted to an O.D. of 0.2 and applied, as a drop, on solid minimal medium containing 0.1% glucose and 0.5%

(w/v) CMC. Plates were thereafter incubated at 30°C during 10 days before plates were photographed to monitor bacterial growth and subsequently colored with Congo red to visualize extracellular endo-glucanase activity. The enzymatic activity index (EIA) was calculated as the diameter of the clear halo plus the colony diameter/diameter of the colony (Hankin and Anagnostakis 1977; Huang et al. 2012). A bacterium with an EIA above of 2.5 is considered as a producer of cellulolytic enzymes (Anagnostakis and Hankin 1975).

For testing extracellular endo-glucanase activity in liquid medium bacterial cells grown in a rich medium were inoculated in a saline minimal media (described previously) supplemented with 0.2% glucose, 0.2% tryptone and 0.5% (w/v) CMC (adaptation medium) and incubated for 24 h at 37°C. Adapted bacterial cells were inoculated in minimal media with different carbon sources as mentioned above and incubated during 14 days in a shaker incubator at 150 rpm at 30°C. Samples were taken at different time (5, 10, 14 days after inoculation) and centrifuged, first at 5,000 rpm for 15 min and thereafter 20 min at 13,000 rpm. Total protein concentration was quantified by the method of Bradford with the Bio-Rad protein assay kit (Bio-Rad Laboratories, Richmond, CA) using bovine serum albumin as protein standard (Bradford 1976). Two hundred microliter (200 µl) containing 10 µg of total protein were applied in small wells in agar plates containing minimal medium supplemented with 0.5% (w/v) CMC. Plates were thereafter incubated for 10 days at 30°C before overlaid with Congo red. A clear halo around a well indicates extracellular cellulolytic activity.

Endo-cellulolytic enzyme activity was measured using the 3,5-dinitrosalicylic acid (DNS) reagent (Miller 1959). The reaction mixture consisted of 100 µl of 2% (w/v) CMC in 100 mM sodium acetate buffer (pH 5.0), to which 100 µl of the obtained supernatants were added before incubation at 40°C for different time points. To stop the reaction, 200 µl of DNS reagent was added and the reaction mixture was heated at 100°C for 10 min. Next, the mixture was left to cool to room temperature, centrifuged and then the absorbance at 540 nm was measured using a spectrophotometer. One unit of enzymatic activity is defined as the amount of enzyme needed to release 1 µmol of reducing sugars (measured as glucose) per ml/min during the reaction.

Scanning electron microscopy

The scanning electron microscopy (SEM) was used following the protocol described by Karnovsky (1965). Briefly, 7 days bacterial cultures were harvested from minimal media with 0.5% bagasse or HT and fixed overnight at 4°C in 8% paraformaldehyde, 0.1 M sodium phosphate buffer at pH 7.4 and 25% glutaraldehyde. The suspension was washed in sodium phosphate buffer 3 times for 10 min each and post-fixed in a solution of a 2% osmium tetroxide in sodium phosphate buffer. After fixation samples were stepwise dehydrated with 30%, 50%, 70%, 80%, 90% and 100% concentrations of ethanol, followed by a final treatment using 100% acetone. To preserve the surface structure of the specimen and to avoid damaging due to surface tension during microscopic observation, critical point procedure was used. The sample was then mounted in an aluminum sample holder and covered with a layer of gold and observed under high vacuum on a Zeiss Supra 55 VP (Carl Zeiss, Oberkochen, Germany) scanning electron microscope at the CIME (Centro Integral de Microscopía Electrónica, INSIBIO, Tucumán, Argentina).

Results

Isolation of cellulolytic bacteria from the intestine of *Diatraea saccharalis*

In order to isolate bacterial symbionts of the intestine of sugarcane-fed *D. saccharalis*, fifth instar larvae were collected in sugarcane production fields in the Province of Tucumán in the Northwest of Argentina. Intestines of larvae were immediately dissected and plated on agar plates containing a bacterial minimal growth medium supplemented either with glucose or sugarcane biomass in the form of bagasse or harvest trash (HT), as sole carbon source. From these initial plates 118 bacterial colonies were isolated showing preliminary cellulose degrading activity as deduced from halo formation around the bacterial colony when treated with Congo red (Figure 1A), eighty from glucose plates (1.2×10^8 colony forming units.ml^{-1}) and thirty eight from the two lignocellulose substrate plates (2.0×10^5 colony forming units.ml^{-1}).

After a round of colony purification all bacterial colonies indicated to possess cellulase activity were re-tested for cellulose degrading capacity using minimal growth medium supplemented with CMC. Thirty eight of the 118

originally isolated bacterial colonies, nineteen from glucose plates and nineteen from sugarcane biomass plates, showed variable zones of clearance surrounding the bacterial colony after incubation with Congo red (Figure 1B) and were selected for further studies.

Bacterial taxonomy and sequencing analysis of the *16S rDNA* gene

In vitro DNA amplification by the polymerase chain reaction (PCR) of the entire *16S rDNA* gene sequence was performed using genomic DNA isolated from the 38 bacterial strains showing cellulolytic activity as a template. Amplified DNA fragments of correct size were sequenced and compared to known 16S rDNA sequences in the GenBank DNA database and the Ribosomal Database Project. Sequence alignment allowed us to identify isolates at the genus level and, in some cases at the species level (Table 1).

The corresponding phylogenetic analysis based on the *16S rDNA* gene sequences obtained is presented in Figure 2 where the homology tree shows two main groups, Gram-negative and Gram-positive bacteria. The former group contained isolates belonging to the phylum Proteobacteria, which included species from genera *Stenotrophomonas* (5.2%) and *Klebsiella* (47.4%). The Gram-positive group included bacteria belonging to the phyla Actinobacteria, represented by species of the genus *Microbacterium* (13.2%), and Firmicutes represented by species of the two genera, *Enterococcus* (23.7%) and *Bacillus* (10.5%).

The *16S rDNA* gene sequences of the isolated strains showed nucleotide homology ranging from 97.0 to 99.8% with bacterial strains published in the data base. The Gram-negative bacteria *Klebsiella oxytoca* Kd70 TUC-EEAOC isolate was phylogenetically most closely related to the *K. oxytoca* KCTC1686 strain (99.0% sequence homology). The *Klebsiella pneumoniae* isolate

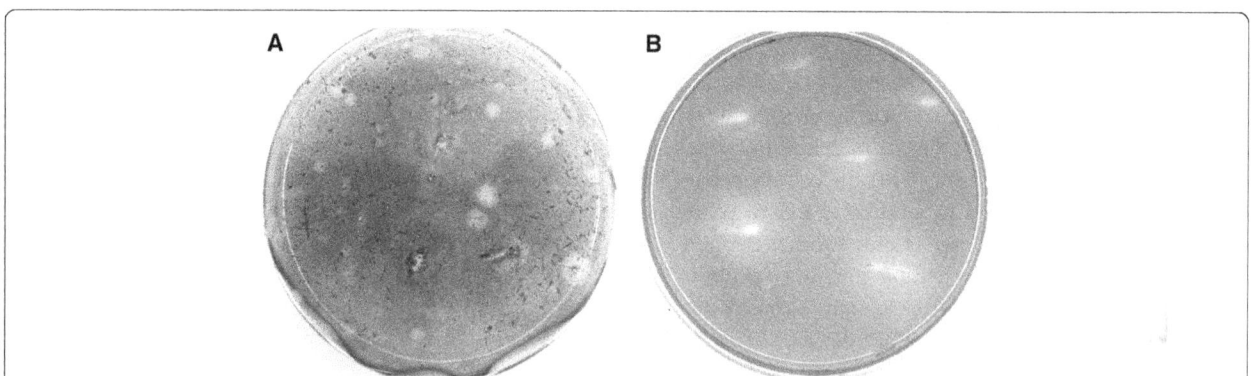

Figure 1 Screening for sugarcane biomass degrading bacteria. A) Bacteria from the intestine of *D. saccharalis* were grown for 3 days on Petri dishes containing minimal medium supplemented with 0.5% (w/v) sugarcane bagasse as sole carbon source. After colonies had developed, plates were stained using Congo red dye to indicate cellulolytic activity of the colony. **B)** Bacterial strains showing cellulolytic activity on plates in **A** were purified and grown on Petri dishes containing CMC, which were later coloured with Congo red. The halo around bacterial colonies indicates total cellulolytic activity of the bacterial isolate.

Table 1 Identification and characterization of bacterial isolates based on 16S rDNA homology analysis and cellulolytic activity

Strain	GenBank/DSMZ accession no.*	Bacteria	Closest relative	16S rDNA identity (%)	Cellulolytic activity		
					Bagasse	HT	CMC (EIA)
Kd70 TUC-EEAOC	KM096608/DSM 27019	Klebsiella oxytoca	K. oxytoca KCTC1686	99.0	+	+	+ (5.0)
KdB5 TUC-EEAOC	KM096599	Klebsiella. pneumoniae	K. pneumoniae MGH78578	99.0	+	+	+ (5.5)
KdB1 TUC-EEAOC	KM096598/DSM 27017	Klebsiella variicola	K. variicola At-22	99.7	+	+	+ (5.0)
Kd3 TUC-EEAOC	KM096600	Stenotrophomonas maltophilia	S. maltophilia R551-3	98.3	-	-	+ (2.5)
Kd46 TUC-EEAOC	KM096602	Stenotrophomonas rhizophila	S. rhizophila ep-10	99.0	-	-	+ (2.5)
Kd109 TUC-EEAOC	KM096605/DSM 27021	Bacillus pumilus	B. pumilus SAFR032	99.5	+	+	+ (7.0)
Kd7 TUC-EEAOC	KM096606/DSM 27018	Enterococcus casseliflavus	E. casseliflavus EC-20	99.0	+ weak	+ weak	+ (2.5)
KdL49 TUC-EEAOC	KM096603	Microbacterium hominis	M. hominis DSM12509	99.8	-	-	+ (1.5) weak
KdL45 TUC-EEAOC	KM096601	Microbacterium schleiferi	M. schleiferi DSM20489	97.0	-	-	+ (1.5) weak

Numbers in parenthesis in the column of cellulolytic activity on CMC indicate diameter of halo in centimeters.

KdB1 TUC-EEAOC showed a 99.0% sequence similarity with *K. pneumoniae* strain MGH78578 whereas *Klebsiella variicola* KdB5 TUC-EEAOC was very closely related to *K. variicola* strain At-22 (99.7% sequence similarity). The *Stenotrophomonas maltophilia* Kd3 TUC-EEAOC isolate shared a 98.3% sequence identity with *S. maltophilia* strain R551-3 and *S. rhizophila* Kd46 TUC-EEAOC showed 99,0% homology with the *S. rhizophila* ep-10 strain.

In the case of the Gram-positive isolates, the Kd109 TUC-EEAOC isolate identified as *Bacillus pumilus* showed a 99.5% of 16S rDNA sequence homology with the *B. pumilus* strain SAFR-032. Isolate KdL49 TUC-EEAOC (*Microbacterium hominis*) is very closely related to strain *M. hominis* DSM-12509 with a 99.8% sequence homology

while KdL45 TUC-EEAOC (*M. schleiferi*) is phylogenetically most closely related with the *M. schleiferi* strain DSM-20489 (97.0% of identity). Finally the Kd7 TUC-EEAOC isolate (*Enterococcus casseliflavus*) showed a 99.0% homology with the *E. casseliflavus* strain EC-20.

Total cellulase activity assay

Endo-glucanases cleave amorphous regions of the cellulose microfibril producing termini of cellulose chains whereas exo-glucanases act on these termini to loosen the crystalline structure of the microfibril. Endo- and exo-glucanase activities are measured using soluble substrates (CMC) and crystalline forms of cellulose (alpha-cellulose or Avicel), respectively. Exo- and endo-glucanase activity

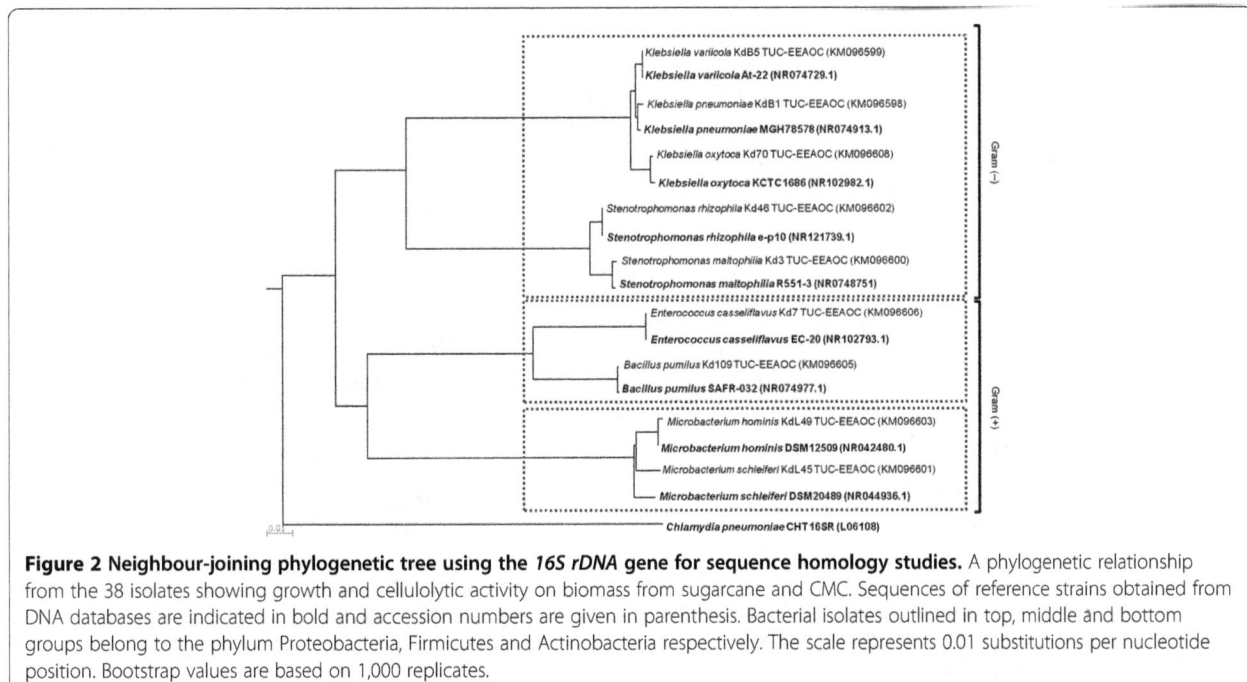

Figure 2 Neighbour-joining phylogenetic tree using the *16S rDNA* gene for sequence homology studies. A phylogenetic relationship from the 38 isolates showing growth and cellulolytic activity on biomass from sugarcane and CMC. Sequences of reference strains obtained from DNA databases are indicated in bold and accession numbers are given in parenthesis. Bacterial isolates outlined in top, middle and bottom groups belong to the phylum Proteobacteria, Firmicutes and Actinobacteria respectively. The scale represents 0.01 substitutions per nucleotide position. Bootstrap values are based on 1,000 replicates.

was analyzed for the thirty-eight (38) bacterial isolates selected using agar plates containing either bagasse, HT or CMC as carbon sources. After 10 days of growth at 30°C, bacterial plates were stained with Congo red and scored for presence of a clearing zone (halo) around each colony, indicating cellulose degradation. Among the isolates tested, only strains belonging to species of the genera *Klebsiella* and *Bacillus* demonstrated clear total cellulase and endoglucanase activities (Figure 3A, 3B, 3C and Table 1). Bacteria belonging to species of the genera *Enterococcus* and *Stenotrophomonas* showed marked endo-glucanase activity on CMC plates (Figures 3C and 4A) but despite that they were able to grow on lignocellulose substrates did not show any noticeable clearing zones of hydrolysis on sugarcane biomass substrates (Figures 3A, 3B and 4A). Bacterial strains of species belonging to the genus *Microbacterium* displayed a very low cellulolytic activity when grown on CMC plates and were unable to grow on any of the two lignocellulose substrates from sugarcane tested (Figure 3A and Figure 3B).

As shown in Figure 3C and Table 1 bacterial isolates of *Klebsiella*, *Bacillus*, *Enterococcus*, *Stenotrophomonas* and *Microbacterium* produced cleared zones of variable size on CMC plates. After 4 days of incubation, the strain of *B. pumilus* generated the largest halo indicating the highest cellulase activity of all the isolated strains tested. The *B. pumilus* strain produced a halo with a 7 cm diameter,

while strains belonging to the genera *Klebsiella* produced halos of 5.5, and 5.0 cm in diameter. Finally, strains belonging to species of the genera *Enterococcus*, *Stenotrophomonas* and *Microbacterium* showed much smaller areas of discoloration with diameters ranging from 2.5 to 1.5 cm (Table 1).

To investigate whether isolates of *Klebsiella oxytoca*, *K. pneumoniae*, *K. variicola* and *Bacillus pumilus*, which showed the highest cellulolytic activity, were constituted by different genotypes, the repetitive extragenic palindromic elements-PCR fingerprinting technique (rep-PCR) was performed based on cluster analysis combining the ERIC, BOX and REP markers. Genetic diversity studies of the eighteen (18) *Klebsiella* isolates revealed 3 clusters separating the three aforementioned species, but in only one of the species, *Klebsiella oxytoca*, genetic difference among strains were detected where strain Kd-70 TUC-EEAOC was found to be genetically distinct to the other isolates (data not shown). The four isolates of *Bacillus pumilus* were found to consist of two different genotypes which grouped into two clusters each formed by two members (data not shown).

Extracellular CMCase activity assay

Total endo-glucanase activity produced by the isolated strains was tested in both solid and liquid medium, while extracellular endo-glucanase activity was evaluated using

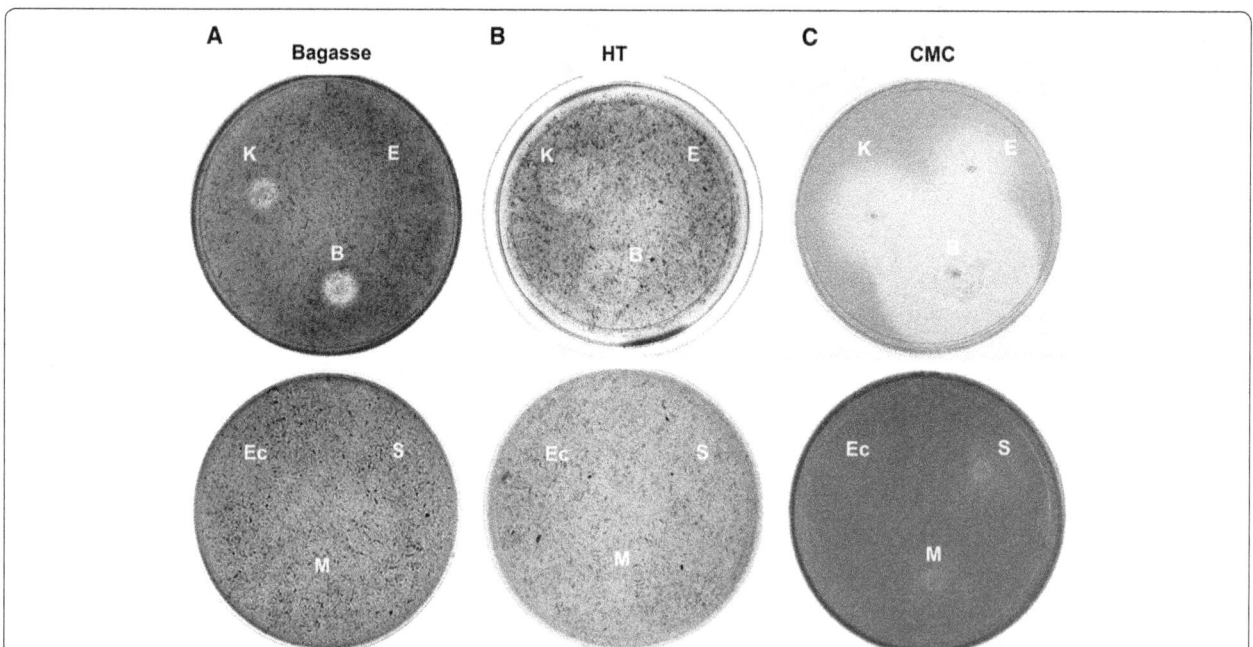

Figure 3 Qualitative cellulolytic assay of bacterial isolates. Solid agar plates containing a saline minimal medium supplemented with **(A)** bagasse, **(B)** harvest trash (HT) and **(C)** carboxymethyl cellulose (CMC) as sole carbon sources. Bacteria were let to grow for 15 days at 30°C before treatment with Congo red. Halos around bacterial colonies are indicative of cellulose degradation. Bacteria abbreviations: E = *Enterococcus casseliflavus* strain Kd7 TUC-EEAOC, K = *Klebsiella oxytoca* strain Kd70 TUC-EEAOC and B = *Bacillus pumilus* strain Kd109 TUC-EEAOC, S = *Stenotrophomonas maltophilia* Kd3 TUC-EEAOC, M = *Microbacterium hominis* KdL49 TUC-EEAOC and Ec = *Escherichia coli* DH5α as a negative control.

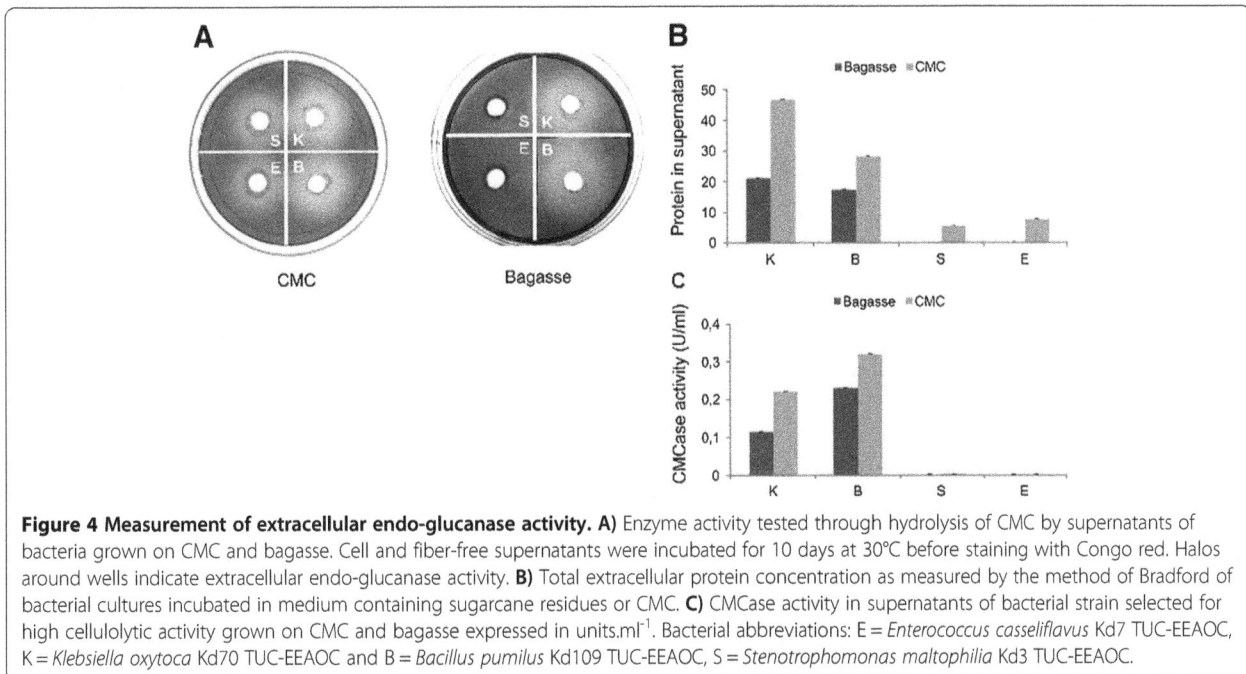

Figure 4 Measurement of extracellular endo-glucanase activity. A) Enzyme activity tested through hydrolysis of CMC by supernatants of bacteria grown on CMC and bagasse. Cell and fiber-free supernatants were incubated for 10 days at 30°C before staining with Congo red. Halos around wells indicate extracellular endo-glucanase activity. **B)** Total extracellular protein concentration as measured by the method of Bradford of bacterial cultures incubated in medium containing sugarcane residues or CMC. **C)** CMCase activity in supernatants of bacterial strain selected for high cellulolytic activity grown on CMC and bagasse expressed in units.ml^{-1}. Bacterial abbreviations: E = *Enterococcus casseliflavus* Kd7 TUC-EEAOC, K = *Klebsiella oxytoca* Kd70 TUC-EEAOC and B = *Bacillus pumilus* Kd109 TUC-EEAOC, S = *Stenotrophomonas maltophilia* Kd3 TUC-EEAOC.

agar plates containing CMC, where 200 μL aliquots of cell and fiber free supernatants obtained from bacterial cultures supplemented with CMC or sugarcane biomass were applied. After incubation at 30°C for 10 days and subsequent staining with Congo red, halo formation around inoculated wells showed extracellular cellulase activity. It is interesting to note that the largest halo diameter, indicating a higher activity (ratio > 5.0), was seen for species of *Klebsiella* and *Bacillus* when cultures were incubated in the presence of either insoluble or soluble cellulosic substrates (Figure 4A), corroborating our results from total cellulase activity measurements (Figure 3A, 3B and 3C).

In contrast, species belonging to the genera *Stenotrophomonas* and *Enterococcus* only showed a moderate extracellular activity when incubated in medium supplemented with CMC and did not show any measureable activity when cultured in insoluble cellulosic substrates (Figure 4A). These results are consistent with the growth and halo profiles obtained on CMC and sugarcane biomass plates in Figure 3A, 3B and 3C. No detectable extracellular cellulase activity was found for species belonging to *Microbacterium* grown in medium containing CMC or sugarcane biomass (data not shown).

When analyzing total protein concentration in the extracellular medium during growth on CMC, highest protein accumulation were found for *Klebsiella oxytoca* Kd70 TUC-EEAOC (46.6 μg/ml) and *Bacillus pumilus* Kd109 TUC-EEAOC (28.3 μg/ml) followed by *Stenotrophomonas maltophilia* Kd3 TUC-EEAOC and *Enterococcus casseliflavus* Kd7 TUC-EEAOC with 5.5 μg/ml and 7.8 μg/ml, respectively (Table 2). When growth was performed in media supplemented with sugarcane bagasse,

the same tendency as with CMC was observed. In this case, similar amounts of extracellular protein was secreted by *K. oxytoca* (21.08 μg/ml) and *B. pumilus* (17.47 μg/ml) while in the case of *E. casseliflavus* and *S. maltophilia* very little or no protein was detected in the supernatant (Table 2).

When quantifying CMCase activity, highest enzymatic activities were observed for *B. pumilus* (0.32 U/ml) and *K. oxytoca* (0.22 U/ml), while much lower activities were obtained for strains belonging to *E. casseliflavus* and *S. maltophilia*. The same pattern of enzymatic activity was observed for strains grown on sugarcane biomass with highest values scored for *B. pumilus* (0.23 U/ml) and *K. oxytoca* (0.13 U/ml) whereas *E. casseliflavus* and *S. maltophilia* showed much lower or no detectable CMCase activity (Table 2). Similar values as determined for *K. oxytoca* and *S. maltophilia* were obtained when the other species of *Klebsiella* and *S. rhizophila* were assayed (data not shown).

Bacterial cells adherence to insoluble cellulosic material

An important step for efficient bacterial hydrolysis of cellulose and hemicellulose is the adhesion of the bacteria to the substrate fiber. To verify that bacterial isolates from the intestine of *D. saccharalis* can attach to sugarcane biomass material, scanning electron microscopy studies were performed on bagasse and HT incubated with one Grampositive (*Enterococcus*) and one Gram-negative bacterium (*Klebsiella*), respectively. As can be seen in Figure 5 both strains attach firmly to both sugarcane bagasse (left) and HT fibers (right). The *Klebsiella oxytoca* strain Kd70 TUC-EEAOC attaches as single rod-shaped bacterial cells

Table 2 Total protein concentrations and CMCase activity determined in the extracellular medium of bacteria isolates

Bacteria	Total secreted protein (µg.ml-1) ± SD		CMCase activity (U.ml⁻¹) ± SD	
	CMC supernatant	Bagasse supernatant	CMC supernatant	Bagasse supernatant
Klebsiella oxytoca Kd70 TUC-EEAOC	46.65 ± 0.04	21.08 ± 0.04	0.22 ± 0.001	0.13 ± 0.001
Bacillus pumilus Kd101 TUC-EEAOC	28.31 ± 0.05	17.47 ± 0.02	0.32 ± 0.002	0.23 ± 0.001
Stenotrophomonas maltophilia Kd3 TUC-EEAOC	5.48 ± 0.01	ND	0.0010 ± 0.0002	ND
Enterococcus casseliflavus Kd7 TUC-EEAOC	7.79 ± 0.01	ND	ND	ND

ND: No detected under the assay conditions used.

while *Enterococcus casseliflavus* strain Kd7 TUC-EEAOC is characteristically found attached in cell-pairs.

Discussion

Out of a total number of 118 cultivable bacterial isolates obtained from intestines of sugarcane-fed *D. saccharalis*, one third (38) were found to possess cellulolytic activity as determined by degradation of CMC. Homology sequence studies of *16S rDNA* from the 38 samples revealed that all cellulolytic strains belonged to the 3 phyla γ-Proteobacteria, Actinobacteria and Firmicutes, which were represented by species belonging to five different genera: *Klebsiella, Stenotrophomonas, Microbacterium, Bacillus* and *Enterococcus*. Members of these genera are wide-spread in nature and all of them have previously been isolated from insect guts (Cardoso et al. 2012; Huang et al. 2012; Shao et al. 2014), shown to possess cellulolytic activity (Teather and Wood 1982; Ohkuma et al. 1995; Charrier et al. 1998; Dillon and Dillon 2004; Prem Anand and Sripathi 2004; Ariffin et al. 2006; Charrier et al. 2006; Mohamed and Huang 2007; Anand et al. 2010; Suen et al. 2010; Okeke and Lu 2011; Ko et al. 2011; Huang et al. 2012) and all except species belonging to the genus *Enterococcus* have been reported as plant endophytes (Zinniel et al. 2002; Asis and Adachi 2004; Dillon and Dillon 2004; Velazquez et al. 2008; Malfanova et al. 2011; Huang et al. 2012; Mingyue et al. 2013; Murugappan et al. 2013; Ren et al. 2013; Chun-Yan et al. 2014; Soni et al. 2014). Taken into account the extensive literature information it seems plausible to conclude that the majority of the cellulolytic bacteria isolated in this study exist in the plant and is colonizing the intestine

upon larval feeding of plant tissue. The only genus of bacteria isolated from the *D. saccharalis* gut that has not been reported in plants is *Enterococcus*, which would suggest a colonization of the larval gut from the adult insect via the egg, as has been shown for *Enterococci* in other Lepidoptera larvae (Brinkmann et al. 2008).

It is interesting to notice that many studies of gut microbiota of herbivorous insects are dominated by members of the phyla Actinobacteria, γ-Proteobacteria and Firmicutes, in which the main symbiotic function suggested involve nitrogen fixation, denitrification, carbohydrate degradation, detoxification, and diverse defensive roles against pathogens (Schloss et al. 2006; Pittman et al. 2008; Hernández et al. 2013; Shao et al. 2014). A similar, predominantly nutritive and defense function for the bacteria isolated from the intestine of *D. saccharalis* is highly probable as the immature sugarcane stalks, from where larvae were collected, have a very low nutritional value containing low amounts of nitrogen, proteins, vitamins, sugars and minerals.

To further study and characterize the cellulolytic activity of the CMC-degrading strains isolated from the intestine of *D. saccharalis*, experiments were performed using non-pretreated sugarcane agricultural residues as a substrate in order to reproduce the natural feeding of larvae. These studies revealed that of all the species of the five bacterial genera isolated; only species from *Bacillus* and *Klebsiella* were found to possess a relatively high cellulolytic ability when grown on sugarcane plant biomass indicating that these genera are able to efficiently metabolize both insoluble and soluble cellulose, a prerequisite for complete digestion of lignocellulose biomass. Another interesting

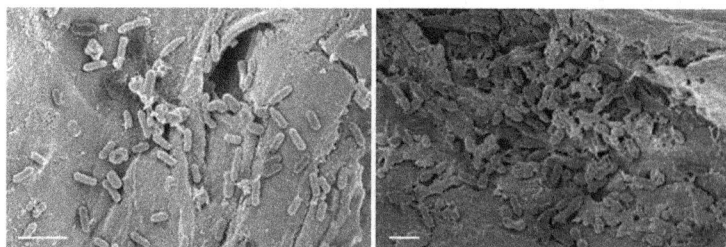

Figure 5 Bacterial adhesion to sugarcane harvest residues. Scanning electron microscopy (SEM) images showing bacterial adherence to plant fibers originating from sugarcane HT. Left image, rod-shaped *Klebsiella oxytoca* strain Kd70 TUC-EEAOC. Right image, diplococci of the *Enterococcus casseliflavus* strain Kd7 TUC-EEAOC. White bars shown in the left bottom corners of both figures indicate 2 µm.

observation from the cellulolytic studies of strains belonging to species of these two genera was that sugarcane biomass seemed to induce a higher extracellular enzymatic activity, as deduced by larger halos seen on bagasse and sugarcane HT plates stained with Congo red (cph), compared to CMC and other carbon sources. This observation indicates that there could be a higher specificity and/or efficiency in degrading sugarcane biomass for these bacteria which would support our initial hypothesis. Supporting this observation are studies showing a clear correlation between changes in bacterial gut symbionts with different biomass feed. As an example, larvae of the Lepidoptera *Spodoptera littoralis* fed on artificial feed, lima bean and barley showed a very different composition of bacterial intestinal species in a metagenomic study (Tang et al. 2012). Interestingly, numerous bacteria belonging to the genera *Klebsiella* were found when larvae were fed on barley but were absent in larvae from the artificial feed or only found in very low representation in lima bean-fed insects, which could indicate an important role for species of this genera in larva feeding on grass species, as was seen in our study on sugarcane. However, more studies are needed to elucidate if these bacteria are indeed more efficient in degrading cellulose from a specific biomass by cloning individual cellulase genes and perform *in vitro* studies on enzymatic degradation capacity of individual proteins using different carbohydrate substrates, before any direct conclusion should be drawn. Nevertheless, these results are encouraging and merits further investigation of the efficiency and cellulolytic activity of cellulases and hemicellulases from strains belonging to *Klebsiella* and *Bacillus* species isolated in this study in order to evaluate their efficiency and possible application in sugarcane biomass degradation for 2nd generation bioetanol production or other applications.

The low cellulolytic activity found for bacterial species belonging to species of *Microbacterium*, *Stenotrophomonas* and *Enterococcus* genera suggest primary roles other than carbohydrate degradation in the larval intestine. As mentioned earlier, the importance of *Enterococcus* as an important colonizer of the insect gut is indicated by the early colonization and that *Enterococcus* is categorized as a lactic acid bacterium (LAB), which are known beneficial organisms of the gut microbiota of animals, including insects (Vasquez et al. 2012; Shao et al. 2014). If *Enterococcus* is already existing in the egg of *D. saccharalis* it is probable that this genera plays a very important symbiotic role similar to the one proposed in the metagenomic study of the intestine of the cotton leaf worm, where *Enterococcus* were found to be the most predominant and metabolic active bacteria throughout the entire larval life-cycle indicating the importance of this genus as gut colonizers of larva from members of Lepidoptera (Shao et al. 2014). As one of the likely founder species of

the gut microbiota it is interesting to speculate of a possible control role for members of this genus in the establishment of other bacteria in the gut. It is well-known that several members of *Enterococci* attach to the mucus layer of gut epithelium to form a biofilm-like structure (Mohamed and Huang 2007), which could protect and prevent the larval gut from being colonized by pathogenic microbes. As members of *Enterococcus* are also known to produce bacteriocins it is possible that a combined action of biofilm formation and production of antimicrobial compounds helps preventing the entrance and establishing of harmful microorganisms in the larval gut (Ennahar et al. 1998; Ruiz-Rodriguez et al. 2012).

Members of the genus *Microbacterium* have been isolated from air, soil, water, humans and plants (Zinniel et al. 2002; Velazquez et al. 2008; Soni et al. 2014). Strains of *M. testaceum*, isolated from the leaf surface of potato plants were shown to produce N-acylhomoserine lactone (AHL)-degrading enzymes (Morohoshi et al. 2011), indicating a protective role against plant pathogens, which frequently uses quorum sensing strategies when colonizing plant tissue. Another interesting feature reported for *Microbacterium* spp. is the chromium detoxification of plants by reducing the bioavailability of toxic chrome IV from soil irrigated with tannery effluent (Soni et al. 2014). Another member, *M. arborescens*, found in the guts of herbivorous caterpillars produces a powerful iron-binding dps (DNA protection during starvation) protein with peroxidase activity, that has been suggested to prevent the formation of cell-damaging oxygen radicals (Pesek et al. 2011). In addition, the same Dps protein can synthesize and hydrolyze amino acid conjugates (N-acyl-glutamines), which have been shown to trigger the plant defense against insect herbivores (Alborn 1997; Baldwin et al. 2001). This latter characteristic could help the larva to escape plant defense actions by evading the detection signaling.

The genus *Stenotrophomonas*, occur ubiquitously in nature and species like S. *maltophilia* and S. *rhizophila* are often found in the rhizosphere and inside many different plant species where they have been associated with beneficial plant interactions. In contrast to the phylogenetically very closely related genera *Xanthomonas* and *Xylella*, no *Stenotrophomonas* species are known to be phytopathogenic, which make *Stenotrophomonas* spp. excellent candidates for biotechnological applications in agriculture. Furthermore, species of the genus *Stenotrophomonas* play an important ecological role in the nitrogen and sulphur cycles and several species of this genus have a high capacity to metabolize a large range of organic compounds which make them ideal candidates for bio- and phytoremediation. Another important feature of which could be beneficial for the larva is the anti fungal activity described for S. *rhizophila* (Wolf et al. 2002).

In accordance with many recent studies (Watanabe and Tokuda, 2010; Cardoso et al., 2012; Gupta et al., 2012; He et al., 2013; Brune, 2014) our results indicate a very complex interaction between bacteria-insect where the bacterial colonizing of the insect gut seem to help the larva in carbohydrate degradation, nutritional access and uptake, protection against pathogens, evasion of plant defense and possible detoxification of chemical compounds. Our understanding of these interactions are only beginning to unfold and many more studies are needed on individual species together with genetic and chemical manipulations of the insect and bacteria in order to advance our knowledge on the role of the larval gut microbiota in herbivorous insects. It is important not to forget the impact of the plant as most of the bacteria isolated from the larva in this study seem to originate from plant endophytes. If this is the case these bacteria show an interesting and highly adoptive life ecology changing a mutual beneficial interaction with its host to a direct antagonistic role protecting and helping an invasive organism. Further understanding of this tri-trophic plant-insect-bacterial interaction should provide valuable information and discoveries that could be employed in novel and more efficient biotechnological solutions for carbohydrate degradation, pest and disease management and phyto- and bio-remediation for example.

Competing interests

The authors declare that they have no competing interests.

Authors' information

Karina I. Dantur; Ramón Enrique; Atilio P. Castagnaro are co-authors.

Acknowledgments

We thank EEAOC and the Consejo Nacional de Investigaciones Científicas y Técnicas (CONICET, Argentina) for their financial support. Dr. K. Dantur, Dr. B. Welin and Dr. A. Castagnaro are career members of CONICET, Argentina. The authors gratefully acknowledge Dr. Raúl Pedraza, Dr. Francisca Perera (career member of CONICET), Dr. Josefina Racedo (post-doctoral fellow of CONICET), Dr. Nadia Chalfoun (career member of CONICET) for useful technical advices and Dr. Conrado Adler for critical reading of the manuscript.

References

Alborn HT (1997) An elicitor of plant volatiles from beet armyworm oral secretion. Science 276:945–949, doi:10.1126/science.276.5314.945

Altschul SF, Madden TL, Schaffer AA, Zhang J, Zhang Z, Miller W, Lipman DJ (1997) Gapped BLAST and PSI-BLAST: a new generation of protein database search programs. Nucleic Acids Res 25:3389–3402, doi:10.1093/nar/25.17.3389

Anagnostakis L, Hankin SL (1975) Use of selective media to detect enzyme production by microorganisms in food products. J Milk Food Technol 38:570–572

Anand AA, Vennison SJ, Sankar SG, Prabhu DI, Vasan PT, Raghuraman T, Geoffrey CJ, Vendan SE (2010) Isolation and characterization of bacteria from the gut of Bombyx mori that degrade cellulose, xylan, pectin and starch and their impact on digestion. J Insect Sci 10:1536–2442

Ariffin H, Abdullah N, Kalsom MSU, Shirai Y, Hassan MA (2006) Production and characterization of cellulase by Bacillus pumilus EB3. Int J Eng Technol 3:47–53

Asis CA, Adachi K (2004) Isolation of endophytic diazotroph Pantoea agglomerans and nondiazotroph Enterobacter asburiae from sweetpotato stem in Japan. Lett Appl Microbiol 38:19–23, doi:10.1046/j.1472-765X.2003.01434.x

Baldwin IT, Halitschke R, Kessler A, Schittko U (2001) Merging molecular and ecological approaches in plant–insect interactions. Curr Opin Plant Biol 4:351–358

Bradford MM (1976) A rapid and sensitive method for the quantitation of microgram quantities of protein utilizing the principle of protein-dye binding. Anal Biochem 72:248–254

Brinkmann N, Martens R, Tebbe CC (2008) Origin and diversity of metabolically active gut bacteria from laboratory-bred larvae of Manduca sexta (Sphingidae, Lepidoptera, Insecta). Appl Environ Microbiol 74:7189–7196, doi:10.1128/AEM. 01464-08

Brune A (2014) Symbiotic digestion of lignocellulose in termite guts. Nat Rev Microbiol 12:168–180, doi:10.1038/nrmicro3182

Canilha L, Chandel AK, Dos Santos Milessi S, Antunes T, Fernandes Antunes AF, Da Costa Freitas LW, Almeida Felipe MG, Silva D, Silverio S (2012) Bioconversion of sugarcane biomass into ethanol: An overview about composition, pretreatment methods, detoxification of hydrolysates, enzymatic saccharification, and ethanol fermentation. J Biomed Biotechnol. doi:10.1155/2012/989572.

Cardoso AM, Cavalcante JJV, Cantao ME, Thompson CE, Flatschart RB, Glogauer AS, Scapin SMN, Sade YB, Beltrao PJMSI, Gerber AL, Martins OB, Garcia ES, de Souza W, Vasconcelos ATR (2012) Metagenomic analysis of the microbiota from the crop of an invasive snail reveals a rich reservoir of novel genes. PLoS One. doi:10.1371/journal.pone.0048505.

Charrier M, Combet-Blanc Y, Ollivier B (1998) Bacterial flora in the gut of Helix aspersa (Gastropoda Pulmonata): evidence for a permanent population with a dominant homolactic intestinal bacterium, Enterococcus casseliflavus. Can J Microbiol 44:20–27

Charrier M, Fonty G, Gaillard-Martinie B, Ainouche K, Andant G (2006) Isolation and characterization of cultivable fermentative bacteria from the intestine of two edible snails, Helix pomatia and Cornu aspersum (Gastropoda: Pulmonata). Biol Res 39:669–681, doi:10.4067/S0716-97602006000500010

Cheng H, Wang L (2013). Lignocelluloses Feedstock Biorefinery as Petrorefinery Substitutes, Biomass Now - Sustainable Growth and Use, Dr. Miodrag Darko Matovic (Ed.), ISBN: 978-953-51-1105-4, InTech, doi: 10.5772/51491 (www.intechopen.com/books/biomass-now-sustainable-growth-and-use/lignocelluloses-feedstock-biorefinery-as-petrorefinery-substitutes)

Chun-Yan W, Li L, Li-Jing L, Yong-Xiu X, Hu C-J, Yang L-T, Yang-Rui L, An Q (2014) Endophytic nitrogen-fixing Klebsiella variicola strain DX120E promotes sugarcane growth. Biol Fertil Soils 50:657–666

Cole JR, Wang Q, Cardenas E, Fish J, Chai B, Farris RJ, Mcgarrell DM, Marsh T, Garrity GM (2009) The ribosomal database project: improved alignments and new tools for rRNA analysis. Nucleic Acids Res 37:141–145, doi:10.1093/nar/gkn879

Dillon RJ, Dillon VM (2004) The gut bacteria of insects: nonpathogenic interactions. Annu Rev Entomol 49:71–92, doi:10.1146/annurev.ento.49.061802.123416

Ennahar S, Aoude-Werner D, Assobhei O, Hasselmann C (1998) Antilisterial activity of enterocin 81, a bacteriocin produced by Enterococcus faecium WHE 81 isolated from cheese. J Appl Microbiol 85:521–526, doi:10.1046/j.1365-2672.1998.853528.x

Gupta P, Samant K, Sahu A (2012) Isolation of cellulose-degrading bacteria and determination of their cellulolytic potential. Int J Microbiol. doi:10.1155/2012/578925.

Hankin L, Anagnostakis SL (1977) Solid media containing carboxymethylcellulose to detect CX cellulose activity of micro-organisms. J Gen Microbiol 98:109–115, doi:10.1099/00221287-98-1-109

He S, Ivanova N, Kirton E, Allgaier M, Bergin C, Scheffrahn RH, Kyrpides NC, Warnecke F, S.G.Tringe, Hugenholtz P (2013) Comparative metagenomic and metatranscriptomic analysis of hindgut paunch microbiota in wood- and dung-feeding higher termites. PLoS One. doi:10.1371/journal.pone.0061126.

Hernández N, Escudero JA, Millán ÁS, González-Zorn B, Lobo JM, Verdú JR, Suárez M (2013) Culturable aerobic and facultative bacteria from the gut of the polyphagic dung beetle Thorectes lusitanicus Jeckel. Insect Sci n/a-n/a. doi:10.1111/1744-7917.12094.

Huang S, Sheng P, Zhang H (2012) Isolation and identification of cellulolytic bacteria from the gut of Holotrichia parallela larvae (Coleoptera: Scarabaeidae). Int J Mol Sci 13:2563–2577, doi:10.3390/ijms13032563

Karnovsky MJ (1965) A formaldehyde-glutaraldehyde fixative of high osmolarity for use in electron microscopy. J Cell Biol 27:137A

Katoh S (2013) MAFFT multiple sequence alignment software version 7: improvements in performance and usability. Mol Biol Evol 30:772–780

Isolation of cellulolytic bacteria from the intestine of Diatraea saccharalis larvae and evaluation of their capacity...

181

Ko K-C, Han YC, Jong Hyun K, Geun-Joong L, Seung-Goo SJJ (2011) A novel bifunctional endo-/exo-type cellulase from an anaerobic ruminal bacterium. Appl Microbiol Biotechnol 89:1453–1462, doi:10.1007/s00253–010–2949–9

Limayem A, Ricke SC (2012) Lignocellulosic biomass for bioethanol production: current perspectives, potential issues and future prospects. Prog Energy Combust Sci 38:449–467, doi:10.1016/j.pecs.2012.03.002

Malfanova N, Kamilova F, Validov S, Shcherbakov A, Chebotar V, Tikhonovich I, Lugtenberg B (2011) Characterization of Bacillus subtilis HC8, a novel plant-beneficial endophytic strain from giant hogweed. Microb Biotechnol 4:523–532, doi:10.1111/j.1751–7915.2011.00253.x

Miller GL (1959) Use of dinitrosalicyclic reagent for determination of reducing sugar. Anal Chem 31:426–428

Mingyue C, Li L, Yanming Z, Li S, Qianli A (2013) Genome sequence of Klebsiella oxytoca SA2, an endophytic nitrogen- fixing bacterium isolated from the pioneer grass Psammochloa villosa. Genome Announc 1:1–2

Mohamed JA, Huang DB (2007) Biofilm formation by enterococci. J Med Microbiol 56:1581–1588, doi:10.1099/jmm. 0.47331–0

Morohoshi T, Wang W-Z, Someya N, Ikeda T (2011) Genome sequence of Microbacterium testaceum StLB037, an N-acylhomoserine lactone-degrading bacterium isolated from potato leaves. J Bacteriol 193:2072–2073, doi:10.1128/JB.00180–11

Morone A, Pandey RA (2014) Lignocellulosic biobutanol production: gridlocks and potential remedies. Renew Sustain Energy Rev 37:21–35

Murugappan RM, Benazir Begum S, Raja RR (2013) Symbiotic influence of endophytic Bacillus pumilus on growth promotion and probiotic potential of the medicinal plant Ocimum sanctum. Symbiosis 60:91–99

Ohkuma M, Noda S, Horikoshi K, Kudo T (1995) Phylogeney of symbiotic methanogens in the gut of the termite Reticulitermes speratus. FEMS Microbiol Lett 134:45–50

Okeke BC, Lu J (2011) Characterization of a defined cellulolytic and xylanolytic bacterial consortium for bioprocessing of cellulose and hemicelluloses. Appl Biochem Biotechnol 163:869–881

Pesek J, Buchler R, Albrecht R, Boland W, Zeth K (2011) Structure and mechanism of iron translocation by a Dps protein from Microbacterium arborescens. J Biol Chem 286:34872–34882, doi:10.1074/jbc.M111.246108

Pittman GW, Brumbley SM, Allsopp PG, O'Neill SL (2008) Assessment of gut bacteria for a paratransgenic approach to control Dermolepida albohirtum larvae. Appl Environ Microbiol 74:4036–4043, doi:10.1128/AEM. 02609–07

Prem Anand A, Sripathi K (2004) Digestion of cellulose and xylan by symbiotic bacteria in the intestine of the Indian flying fox (Pteropus giganteus). Comp Biochem Physiol A Mol Integr Physiol 139:65–69, doi:10.1016/j.cbpb.2004.07.006

Ren IH, Li H, Wang YF, Ye JR, Yan AQ, Wu XQ (2013) Biocontrol potential of an endophytic Bacillus pumilus JK-SX001 against poplar canker. Biol Control 67:421–430, doi:10.1016/j.biocontrol.2013.09.012

Ruiz-Rodriguez M, Valdivia E, Martin-Vivaldi M, Martin-Platero AM, Martinez-Bueno M, Mendez M, Peralta-Sanchez JM, Soler JJ (2012) Antimicrobial activity and genetic profile of enteroccoci isolated from hoopoes uropygial gland. PLoS One. doi:10.1371/journal.pone.0041843.

Schloss PD, Delalibera I, Handelsman J, Raffa KF (2006) Bacteria associated with the guts of two wood-boring beetles: Anoplophora glabripennis and Saperda vestita (Cerambycidae). Environ Entomol 35:625–629, doi:10.1603/0046–225X–35.3.625

Shao Y, Arias-Cordero E, Guo H, Bartram S, Boland W (2014) In vivo Pyro-SIP assessing active gut microbiota of the cotton leafworm, Spodoptera littoralis. PLoS One 9:e85948, doi:10.1371/journal.pone.0085948

Shi W, Ding SY, Yuan JS (2011) Comparison of insect gut cellulase and xylanase activity across different insect species with distinct food sources. Bioenergy Res 4:1–10, doi:10.1007/s12155–010–9096–0

Sierra R, Holtzapple MT, Granda CB (2011) Long-term lime pretreatment of poplar wood. AIChE J 57:1320–1328, doi:10.1002/aic.12350

Singh A, Hayashi K (1995) Microbial cellulases: protein architecture, molecular properties, and biosynthesis. Adv Appl Microbiol 40:1–44, doi:10.1016/S0065–2164(08)70362–9

Soni SK, Singh R, Awasthi A, Kalra A (2014) A Cr(VI)-reducing Microbacterium sp. strain SUCR140 enhances growth and yield of Zea mays in Cr(VI) amended soil through reduced chromium toxicity and improves colonization of arbuscular mycorrhizal fungi. Environ Sci Pollut Res Int 21:1971–1979, doi:10.1007/s11356–013–2098–7

Suen G, Scott JJ, Aylward FO, Adams SM, Tringe SG, Pinto-Tomas AA, Foster CE, Pauly M, Weimer PJ, Barry KW, Goodwin LA, Bouffard P, Li L, Osterberger J,

Harkins TT, Slater SC, Donohue TJ, Currie CR (2010) An insect herbivore microbiome with high plant biomass-degrading capacity. PLoS Genet. doi:10.1371/journal.pgen.1001129.

Sun J-Z, Scharf ME (2010) Exploring and integrating cellulolytic systems of insects to advance biofuel technology. Insect Sci 17:163–165

Taherzadeh MJ, Karimi K (2007) Enzyme-based hydrolysis processes for ethanol from lignocellulosic materials: a review. BioResources 2:707–738, doi:10.15376/biores.2.4.707–738

Tang X, Freitak D, Vogel H, Ping L, Shao Y, Cordero EA, Andersen G, Westermann M, Heckel DG, Boland W (2012) Complexity and variability of gut commensal microbiota in polyphagous lepidopteran larvae. PLoS One. doi:10.1371/journal. pone.0036978.

Teather RM, Wood PJ (1982) Use of Congo red-polysaccharide interactions in enumeration and characterization of cellulolytic bacteria from the bovine rumen. Appl Environ Microbiol 43:777–780

Teeri TT (1997) Crystalline cellulose degradation: new insight into the function of cellobiohydrolases. Trends Biotechnol 15:160–167, doi:10.1016/S0167–7799 (97)01032–9

Vasquez A, Forsgren E, Fries I, Paxton RJ, Flaberg E, Szekely L, Olofsson TC (2012) Symbionts as major modulators of insect health: Lactic acid bacteria and honeybees. PLoS One. doi:10.1371/journal.pone.0033188.

Velazquez E, Rojas M, Lorite MJ, Rivas R, Zurdo-Pineiro JL, Heydrich M, Bedmar EJ (2008) Genetic diversity of endophytic bacteria which could be find in the apoplastic sap of the medullary parenchym of the stem of healthy sugarcane plants. J Basic Microbiol 48:118–124, doi:10.1002/jobm.200700161

Watanabe H, Tokuda G (2010) Cellulolytic systems in insects. Annu Rev Entomol 55:609–632, doi:10.1146/annurev-ento-112408–085319

Weisburg WG, Barns SM, Pelletier D, Lane DJ (1991) 16S ribosomal DNA amplification for phylogenetic study. J Bacteriol 173:697–703

Wolf A, Fritze A, Hagemann M, Berg G (2002) Stenotrophomonas rhizophila sp. nov., a novel plant-associated bacterium with antifungal properties. Int J Syst Evol Microbiol 52:1937–1944, doi:10.1099/ijs. 0.02135–0

Zinniel DK, Lambrecht P, Harris NB, Feng Z, Kuczmarski D, Higley P, Ishimaru CA, Arunakumari A, Barletta RG, Vidaver AK (2002) Isolation and characterization of endophytic colonizing bacteria from agronomic crops and prairie plants. Appl Environ Microbiol 68:2198–2208, doi:10.1128/AEM. 68.5.2198–2208.2002

21

Comparison of cake compositions, pepsin digestibility and amino acids concentration of proteins isolated from black mustard and yellow mustard cakes

Ashish Kumar Sarker[1*], Dipti Saha[2], Hasina Begum[2], Asaduz Zaman[2] and Md Mashiar Rahman[3]

Abstract

As a byproduct of oil production, black and yellow mustard cakes protein are considered as potential source of plant protein for feed applications to poultry, fish and swine industries. The protein contents in black and yellow mustard cakes were 38.17% and 28.80% and their pepsin digestibility was 80.33% and 77.43%, respectively. The proteins were extracted at different pH and maximum proteins (89.13% of 38.17% and 87.76% of 28.80% respectively) isolated from black and yellow mustard cakes at pH 12. The purity of isolated proteins of black and yellow mustard cakes was 89.83% and 91.12% respectively and their pepsin digestibility was 89.67% and 90.17% respectively which assigned the absence of antinutritional compounds. It was found that essential amino acids isoleucine, lysine, methionine, threonine and tryptophan and non essential amino acids arginine and tyrosine were present in greater concentration in black mustard cake protein whereas other amino acids were higher in yellow mustard cake protein.

Keywords: Mustard cake; Protein isolation; Pepsin digestibility; Amino acids analysis

Introduction

Oil cakes/oil meals are by-products obtained after oil extraction from the seeds. Oil cakes are of two types, edible and non-edible. Edible oil cakes have a high nutritional value; especially have protein contents ranging from 15% to 50% (Ramachandran et al. 2007). Black Mustard (*Brassica nigra*) and yellow mustard (*Sinapis alba*) are the third important oilseed crops in the world after soybean (*Glycine max*) and palm (*Elaeis guineensis* Jacq.) oil seed. These seeds are produced in Bangladesh in a large extent. They contain relatively high amount of protein with small amounts of anti nutritional compounds (Clandinin and Heard, 1968). Deoiled groundnut cake is commonly used as poultry feed ingredient. Though high in protein, groundnut cake is a poor source of essential amino acids like lysine and methionine; but commonly

infested with *Aspergillus* sp., which will produce aflatoxins under favourable conditions (Adebesin et al. 2004). Soybean is a major protein source for humans and other animals. The protein composition of soybean seed is not ideal for human and animal nutrition because of the poor content of sulfur containing amino acids (Fukushima, 1991; Sievwright and Shipe, 1986) and presence of large amount of phytic acid. In this circumstance, deoiled mustard cakes appear to be a potential source of protein replacing ground nut and soybean cakes in fish and poultry rations.

Different methods have been examined for the production of protein isolate (>90% protein) and concentrates (>65% protein) from these cakes. Prapakornwiriya and Diosady used yellow mustard flour to successfully develop a microfiltration based process for production of protein concentrate (Prapakornwiriya and Diosady, 2004). Marnoch and Diosady used oriental mustard seed (*Brassica juncea* L.) to develop a membrane-based process that produced three products: a precipitated protein isolate (PPI), a soluble protein isolate (SPI) and a meal

* Correspondence: ashish10608@hotmail.com
[1]Plant Protein Research Section, Institute of Food Science and Technology (IFST), Bangladesh Council of Scientific and Industrial Research(BCSIR), Dr. Kudrat-i-Khuda Road, Dhanmondhi, Dhaka 1205, Bangladesh
Full list of author information is available at the end of the article

residue (Marnoch and Diosady, 2006). The protein isolates were high in protein and free of anti-nutritional compounds. High levels of protein, suitable amount of essential amino acids, minerals and the behavior ability of these nutrients have given mustard seed cakes prime importance as a quality protein source. Unfortunately, cakes or seeds contain some compounds like glucosinolates and their breakdown products, phenolics and phytates which hinder bioavailability of amino acids and minerals (Dijkstra et al. 2003; Naczk et al. 1992). These compounds are responsible for the dark color and astringent flavor and they must be removed. Before incorporating deoiled mustard cakes in poultry ration, they should be analyzed for their proximate compositions, isolation of proteins, and pepsin digestibility of proteins and amino acid profiles of isolated proteins to know their nutritive value.

The objectives of the present work were to determine nutritional composition of cakes, isolation of proteins from cakes, pepsin digestibility of isolated proteins and cakes and analysis of amino acids composition of isolated proteins.

Materials and methods

Materials

Sodium hydroxide (NaOH), sulfuric acid (H_2SO_4), copper sulfate ($CuSO_4.5H_2O$), potassium sulfate (K_2SO_4), hydrochloric acid (HCl), pepsin, petroleum ether, and ethanol were from BDH and used without further purification. Phosphate buffers (pH 8, 9, 10, 11, 12, 13 and 14) were prepared by mixing proper amount of 0.1 M disodium hydrogen phosphate and 0.1 M sodium hydroxide.

Methods

Testing samples preparation

Black and yellow mustard cakes were used for this study and obtained from a local oil mills. Prior to use, the cakes were ground and defatted with petroleum ether (40-60°C), using a Soxhlet apparatus, for 16 h, and then dried overnight in an oven at 80°C. The moisture, minerals, crude fibre and fat of fat were determined by the standard AOAC method 950.46, AOAC method 920.153, AOAC method 985.29 and AOAC method 960.39 respectively (AOAC, 2005). Crude protein was determined by the micro-Kjeldahl method and reported as%N × 6.25 (AACC, 2000).

Removal of allylisothiocyanate

The allylisothiocyanates were removed by the modified method of Singh (Singh, 1988). 10 g defatted sample was grinded and passed through a No 20 sieve. 150 ml 5% ethanol was added in 6 g powdered sample in 300 ml Erlenmeyer flask, stopper tightly and magnetic stirred for 90 m at 37°C. After extraction, the cakes were filtered with mild vacuum and dried at 80°C for 8 h. The raw mustard cakes were also dried same temperature to compare the pepsin digestibility with allylisothiocyanate free cakes.

Determination of allylisothiocyanate

The quantity of allylisothiocyanate was determined by titrimetric method. Exactly 5 g raw mustard cakes were mixed with 12.5 ml absolute ethanol and 237.5 ml distilled water into a 500 ml distillation flask. The mixture was distilled with steam and 150 ml distillate was collected in the 25 ml 0.1 N silver nitrate and 10 ml 10% ammonium hydroxide solution. The distillate mixture was boiled for 1 h under air reflux in water bath, cooled, volume made upto 250 ml and then filtered. 100 ml filtrate was titrated with standard ammonium thiocyanate solution in acidic condition using few drops of ferric ammonium sulfate indicator. A blank titration was also done and calculated the amount of allylisothiocyanate.

Protein extraction

The extraction of protein was carried out according to the method of Marnoch and Diosady with small modification (Marnoch and Diosady, 2006). The protein extractability was determined by contacting 20 g of ground defatted mustard cakes with aqueous NaOH solution at a solvent to cakes ratio of 18 in a preset pH, ranging from 8 to 14. The pH was adjusted using phosphate buffer. The extract and solids were separated by centrifugation at 12000 rpm. The liquid was decanted and vacuum filtered through Whatman 41 paper to a receiving flask. The solids were washed twice with distilled water and each time decanted through the filter paper into the same receiving flask. The extractability was measured as the mass ratio of the recovered protein in the collected extract solution compared with that in the 20 g of starting material.

Protein was precipitated from extract solution by adding 1 M HCl solution. The pH of extracted protein solution was kept constant at 5 and allowed to stand for overnight at 5°C for precipitation. The precipitated protein isolate (PPI) was separated by centrifugation (10,000 rpm), for 15 min in a centrifuge machine (Kokusan, H2000 series). The PPI was then washed with water, dried by freeze drying and stored at 5°C for further analysis.

Protein digestibility

The in vitro protein digestibility of raw defatted cakes, allylisothiocyanate free cakes and PPI was carried out according to the method of Mertz et al. with a minor modification (Mertz et al., 1984). 2.0 g sample was mixed with 490 ml distilled water and 1.5 g pepsin. Then 10 ml 25% HCl was added and the final solution was incubated for 24 h at 37°C in incubator. After this treatment, further 6 h incubation at 37°C was done with additional 10 ml 25% HCl. After incubation the reaction was stopped by addition of 15 ml of 10% trichloroacetic acid (TCA). The mixture was filtered and washed with

Table 1 Proximate composition of black mustard cake and yellow mustard cake

Materials	Moisture, %	Minerals, %	Acid insoluble ash, %	Oil, %	Crude fiber, %	Crude protein, %	Allylisothiocyanate %
Black mustard cake	9.20 ± 0.5	7.10 ± 0.3	1.93 ± 0.4	8.70 ± 0.8	12.17 ± 1.3	38.17 ± 1.0	0.086 ± 0.009
Yellow mustard cake	9.73 ± 0.6	5.90 ± 0.3	1.23 ± 0.3	15.67 ± 0.6	14.80 ± 0.2	28.80 ± 0.7	0.077 ± 0.003

All data are the average of three replicate independent experiments and the standard deviation was calculated using one-way ANOVA.

distilled water. The residue was collected and estimated the nondigest nitrogen by micro-Kjeldahl method.

Amino acids analysis

Amino acid composition of protein isolates was determined by using an amino acid analyzer (Shimadzu, Japan) and only fourteen amino acids were determined due to the limitation of the instrument. 0.5 g isolated protein was pasted with 50 ml 6 N HCl by mortar pestle, filter and then the filtrate was hydrolyzed 22–24 h in a hydrolysis tube. After hydrolyzing, HCl was removed from the filtrate by evaporating and three times re-evaporating with water in water bath. After evaporation, the solution was volume to 25 ml in volumetric flask by 0.1 N HCl. The stock solution was used for amino acids analysis using Shimadzu Amino Acid Analyzer.

Statistical analysis

Three replicates were carried out in each experiment. All data were analyzed by SPSS software, version 15 using one-way ANOVA analysis. The level of statistical significance was set at 5% ($p < 0.05$).

Results
Proximate chemical analysis

Proximate compositions of black mustard cake and yellow mustard cake are presented in Table 1. The black mustard cake contains higher crude protein of 38.17% than that of yellow mustard cake protein of 28.80% hence appeared to be a moderately good source of protein. On the other hand, ether extract content was high as 15.67% for yellow mustard cake than 8.70% for black mustard cake. Crude fibers content were also so high for both types of cakes. Fiber and oil content can be used to approximate energy values of the cakes, since utilizable energy content decreases as fiber content increases, but increases as the oil content increases. Thus, the cakes are a good source of energy, since they have high oil content however; the energy content is limited by the high fiber content. Total minerals content of black mustard cake and yellow mustard cake were found to be 7.10% and 5.90% among which contribution of acid insoluble ash were only 1.93% and 1.23% indicating that it is a good source of minerals. Allylisothiocyanate content in cakes is 0.086% and 0.077% respectively. It can bind with protein and decreased the digestibility of protein.

After removal of this antinutritional compound the digestibility of protein increased.

Protein extraction

Protein extraction is normally governed by the pH values which influence the ratio of free to neutralized charges. The data of the study indicates that protein solubility was gradually enhanced with the increase in pH values from 8. However, maximum solubility of mustard cakes protein was increased rapidly from 38.17% to 89.13% at pH 12 for black mustard cake and from 28.80% to 87.76% at pH 12 for yellow mustard cake. The effect of pH on protein extractability is presented in Figure 1.

Protein digestibility

The pepsin digestibility of three forms of protein: protein in raw mustard cakes, protein free from allylisothiocyanate and precipitated protein isolate (PPI) of black and yellow mustard cakes are shown in Table 2. In both mustard cakes and anti nutritional compounds free mustard cakes there were significant differences in protein digestibility. Protein concentrations in the raw black and yellow cakes were 38.17% and 28.80% respectively. In vitro protein digestibility values were 80.33% for black mustard cake and 77.43% for yellow mustard cake; whereas the antinutritional compounds free type mustard cakes had digestibility values of 89.67% for black mustard cake and

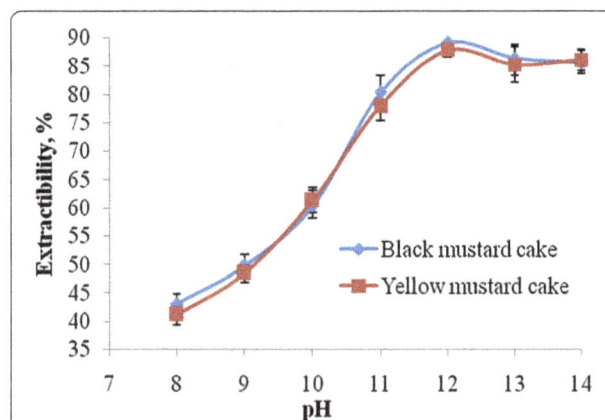

Figure 1 Protein extraction from defatted ground mustard cakes by aqueous sodium hydroxide solutions as a function of pH (All data are the average of three replicate independent experiments and the standard deviation was calculated using one-way ANOVA).

Table 2 Pepsin digestibility of protein in black mustard and yellow mustard cakes

Parameters	Black mustard cake	Yellow mustard cake
Pepsin Digestibility of protein before removal of antinutritional compounds,%	80.33 ± 2.08	77.43 ± 1.53
Pepsin digestibility of protein after removal of antinutritional compounds,%	89.67 ± 1.15	90.17 ± 2.51
Pepsin digestibility of isolated protein,%	96.67 ± 1.53	95.27 ± 2.08

All data are the average of three replicate independent experiments and the standard deviation was calculated using one-way ANOVA.

90.17% for yellow mustard cake. Except for two entries, the isolated protein had higher digestibility values 96.67% for black mustard cake and 95.27% for yellow mustard cake. The protein digestibility was slightly changed after preparation of protein isolate than cakes without antinutritional compounds. From the Table 2 it is shown that the protein digestibility both yellow and black mustard cakes increased about 16% in case isolated protein than the crude protein in mustard cakes.

Amino acids composition

Amino acid composition of protein isolates is an indicator of their nutritive value. The concentrations of essential amino acids in black mustard cake protein isolate and yellow mustard cake protein isolate differ from each other considerably (Table 3). The isoleucine (5.57%), lysine (4.55%), methionine (2.52%), threonine (19.17%) and tryptophan (1.96%) content in black mustard cake protein isolate have greater concentration than in yellow mustard cake protein isolate while leucine (1.12%), valine (1.74%) and histidine (0.90%) content in yellow mustard protein isolate have higher concentration than black mustard cake protein isolate. The non essential amino acids such as arginine (2.74%) and tyrosine (1.96%) in black mustard cake

protein isolate and alanine (4.26%), aspartate (7.11%), glycine (5.55%) and serine (5.03%) in yellow mustard cake isolate are found in higher concentration.

Discussion

Mustard cakes is a potential source of protein for animals and it is the first study to compare the nutritional, isolation of protein, pepsin digestibility of protein and amino acids pattern of black and yellow mustard cakes. As shown in Table 1, moisture content of black and yellow mustard cakes were 9.20 ± 0.5% and 9.73 ± 0.6%. The results of the moisture content were slightly higher than the moisture content 8.3 ± 0.2% reported in literature. (Al Mahmud et al. 2012; Marnoch and Diosady, 2006). Excess water used in the moisturizing of mustard seeds during oil extraction is responsible for this reason. The Crude protein content of mustard cakes obtained were 38.17% and 28.80% which were lower (45.0% and 34.0%) than those reported by many other authors (Marnoch and Diosady, 2006; Prapakornwiriya and Diosady, 2004). However, Chowdhury et al. and Kumar ent al. reported comparatively equal amount of crude protein in mustard cakes (Anil Kumar et al., 2002; Chowdhury et al. 2010). The variation observed could be because of the variety of mustard raised and the differences in sampling adopted, influence of season of harvest etc. Mean ether extract content of black and yellow mustard cakes were also low compared to many other authors (Latif et al., 2008; Ramachandran et al., 2007). Crude fiber content of mustard cakes were 12.17 ± 2.25% which was also found to vary among report of different authors (Latif et al., 2008; Sharma et al., 2012). Variations due to hulling procedure and time, variety etc., would have contributed to these differences. Ash content of mustard cakes were 7.10 ± 0.3% and 5.90 ± 0.3% respectively. These data are comparable (7.12 ± 0.12%) to the results reported in literature (Datta et al., 2013). Allyl isothiocyanates content of black and yellow mustard cakes were

Table 3 The composition of amino acids in black and yellow mustard cake protein isolates

Essential amino acids			Other non-essential amino acids		
Amino acids	Precipitated protein isolate (PPI) from black mustard cake (%)	Precipitated protein isolate (PPI) from yellow mustard cake (%)	Amino acids	Precipitated protein isolate (PPI) from black mustard cake (%)	Precipitated protein isolate (PPI) from yellow mustard cake (%)
Isoleucine	5.57	2.95	Alanine	3.56	4.26
Leucine	0.83	1.12	Arginine	2.74	2.28
Lysine	4.55	2.70	Aspartate	4.49	7.11
Methionine	2.52	1.50	Glycine	2.54	5.55
Threonine	19.17	14.31	Serine	2.95	5.03
Tryptophan	1.96	1.39	Tyrosine	1.96	1.39
Valine	1.20	1.74			
Histidine	0.43	0.90			

$0.086 \pm 0.009\%$ and $0.077 \pm 0.003\%$ which were comparable to the report of Sharma et al. (Sharma et al., 2012).

Various techniques for extraction and precipitation of protein isolates have been published. The alkaline extraction and acidic precipitation were chosen to produce PPI because this technique ensured maximum protein recovery and minimum protein denaturation. When pH value is above 5, the solubility of protein will be increased and hence the extraction of protein from mustard cakes gradually increased with increasing pH values from 9. The effect of pH demonstrated in this experiment is in agreement with the pH values reported in the literature (Lindeboom and Wanasundara, 2007; Marnoch and Diosady, 2006). The maximum extractability of proteins 89.13% and 87.76% was observed at pH 12.0. These extractions of proteins were also comparable to the reports of Marnoch and Diosady (Marnoch and Diosady, 2006). Buffer composition and filtration method had no effect for separation of protein from mustard cakes whereas pH is the only factor for separation of protein from cakes.

Pepsin digestibility of protein is an important chemical property of proteins, as it determines their nutritional value. The digestibility increased after separation of allylisothiocyanates from cakes. The pepsin digestibility of PPI was highest compared with the protein of allylisothiocyanates free cakes and protein of raw cakes. Therefore, the antinutritional compounds have negative effect on protein pepsin digestibility (Aparicio-Saguilán et al., 2015; Sun et al., 2012). The high fibre content in raw cakes and allylisothiocyanate free cakes may also affect the pepsin digestibility of protein. From the Table 2 it is shown that the protein digestibility both yellow and black mustard cakes increased about 15% in case of isolated protein than the crude protein in mustard cakes. This results assigned that PPI was free of anti nutritional compounds and there was no scope for lysinoalanine formation during extraction Some examples of naturally occurring antinutritional factors include glucosinolates in mustard and rapeseed protein products which would adversely affect nutrient utilization and may contribute to growth depression in animals (Fenwick et al., 1982). All of these tend to make the amino acids less available and, in general, the protein less digestible (Talati et al., 2004). Amino acid composition of PPI from black and yellow mustard cakes indicated that it is a good source of lysine and methionine. However, essential amino acids such as isoleucine, lysine, methionine, threonine and tryptophan contents were found to be higher in black mustard protein than in yellow mustard protein. The decrease of the lysine amounts in the protein isolates is probably due to the interaction of the respective amino acid with other plant components during oil processing.

Our experimental results indicate that black mustard cake is more beneficial than yellow mustard cake for preparation of feeds. The use of this byproduct, as a part of a protein extraction process would increase the viability of the linked industrial processes.

Competing interests

The authors declare that they have no competing interest.

Authors' contributions

AKS coordinate the study, designed experiments, analyzed results and wrote the manuscript. DS coordinate the study and designed experiments. HB performed experiments and analyzed results. AZ analyzed results and edited the manuscript. MMR performed experiments and analyzed the amino acid pattern data. All authors read and approved the final manuscript.

Acknowledgement

The authors deeply acknowledge the financial support provided by the Institute of Food Science and Technology, Bangladesh Council of Scientific and Industrial Research, Dhaka, Bangladesh.

Author details

[1]Plant Protein Research Section, Institute of Food Science and Technology (IFST), Bangladesh Council of Scientific and Industrial Research(BCSIR), Dr. Kudrat-i-Khuda Road, Dhanmondhi, Dhaka 1205, Bangladesh. [2]Department of Applied Chemistry & Chemical Engineering, Dhaka University, Dhaka 1000, Bangladesh. [3]Enzymlogy Research Section, IFST, BCSIR, Dhaka, 1205, Bangladesh.

References

AACC (2000) Approved Methods of the American Association of Cereal Chemists, 10th edn. American Association of Cereal Chemists, St. Paul, MN

Adebesin A, Saromi O, Amusa N, Fagade S (2004) Microbiological quality of some groundnut products hawked in Bauchi, a Nigerian City. J Food Technol Africa 6(2):53–55

Al Mahmud N, Hasan M, Hossain M, Minar M (2012) Proximate composition of fish feed ingredients available in Lakshmipur region, Bangladesh. American-Eurasian J Agric & Environ Sci 12(5):556–560

Anil Kumar G, Panwar V, Yadav K, Sihag S (2002) Mustard cake as a source of dietary protein for growing lambs. Small Rumin Res 44(1):47–51

AOAC (2005) Official Methods of Analysis of Association of Official Analytical Chemists (AOAC) International Method 950.46, Method 920.153, Method 985.29, Method 960.39, 18th edn. AOAC, Gaithersburg, MD, USA

Aparicio-Saguilán, A., Valera-Zaragoza, M., Perucini-Avendaño, M., Páramo-Calderón, D. E., Aguirre-Cruz, A., Ramírez-Hernández, A., & Bello-Pérez, L. A. (2015). Lintnerization of banana starch isolated from underutilized variety: morphological, thermal, functional properties, and digestibility. CyTA-Journal of Food 13(1):3-9.

Clandinin D, Heard J (1968) Tannins in prepress-solvent and solvent processed rapeseed meal. Poult Sci 47(2):688–688

Datta SN, Kaur VI, Dhawan A, Jassal G (2013) Estimation of length-weight relationship and condition factor of spotted snakehead Channa punctata (Bloch) under different feeding regimes. SpringerPlus 2(1):436

Dijkstra DS, Linnemann AR, van Boekel TA (2003) Towards sustainable production of protein-rich foods: appraisal of eight crops for Western Europe. PART II: Analysis of the technological aspects of the production chain. Crit Rev Food Sci Nutr 43(5):481–506

Fenwick GR, Heaney RK, Mullin WJ, VanEtten CH (1982) Glucosinolates and their breakdown products in food and food plants. Crit Rev Food Sci Nutr 18 (2):123–201

Fukushima D (1991) Recent progress of soybean protein foods: chemistry, technology, and nutrition. Food Reviews International 7(3):323–351

Latif K, Alam M, Sayeed M, Hussain MA, Sultana S, Hossain M (2008) Comparative study on the effects of low cost oil seed cakes and fish meal as dietary protein sources for Labeo rohita (Hamilton) fingerling. University Journal of Zoology, Rajshahi University 27:25–30

Lindeboom N, Wanasundara P (2007) Interference of phenolic compounds in < i > Brassica napus</i > < i > Brassica rapa</i > and < i > Sinapis alba</i > seed extracts with the Lowry protein assay. Food Chem 104(1):30–38

Chowdhury MFN, Ahmed MSKU, Nuruddin MM, Hosen M (2010) Study on fatty acid composition, oil and protein of different varieties and advanced lines of ustard and rapeseeds. Bangladesh Research Publication Journal 4(1):7

Marnoch R, Diosady LL (2006) Production of mustard protein isolates from oriental mustard seed (Brassica juncea L.). J Am Oil Chem Soc 83(1):65–69

Mertz ET, Hassen MM, Cairns-Whittern C, Kirleis AW, Tu L, Axtell JD (1984) Pepsin digestibility of proteins in sorghum and other major cereals. Proc Natl Acad Sci 81(1):1–2

Naczk M, Wanasundara P, Shahidi F (1992) Facile spectrophotometric quantification method of sinapic acid in hexane-extracted and methanol-ammonia-water-treated mustard and rapeseed meals. J Agric Food Chem 40(3):444–448

Prapakornwiriya N, Diosady LL (2004) Isolation of yellow mustard proteins by a microfil tration-based process. Int J Appl Sci Eng 2(2):127–135

Ramachandran S, Singh SK, Larroche C, Soccol CR, Pandey A (2007) Oil cakes and their biotechnological applications–A review. Bioresour Technol 98(10):2000–2009

Sharma HK, Ingle S, Singh C, Sarkar BC, Upadhyay A (2012) Effect of various process treatment conditions on the allyl isothiocyanate extraction rate from mustard meal. J Food Sci Technol 49(3):368–372

Sievwright CA, Shipe W (1986) Effect of storage conditions and chemical treatments on firmness, in vitro protein digestibility, condensed tannins, phytic acid and divalent cations of cooked black beans (Phaseolus vulgaris). J Food Sci 51(4):982–987

Singh U (1988) Antinutritional factors of chickpea and pigeonpea and their removal by processing. Plant Foods Hum Nutr 38(3):251–261

Sun M, Mu T, Zhang M, Arogundade LA (2012) Nutritional assessment and effects of heat processing on digestibility of Chinese sweet potato protein. J Food Compos Anal 26(1):104–110

Talati J, Patel K, Patel B (2004) Biochemical composition, in vitro protein digestibility, antinutritional factors and functional properties of mustard seed, meal and protein isolate. Journal of Food Science and Technology-Mysore 41(6):608–612

22

Enhanced ethanol production and reduced glycerol formation in *fps1Δ* mutants of *Saccharomyces cerevisiae* engineered for improved redox balancing

Clara Navarrete, Jens Nielsen and Verena Siewers[*]

Abstract

Ethanol is by volume the largest fermentation product. During ethanol production by *Saccharomyces cerevisiae* about 4-5% of the carbon source is lost to glycerol production. Different approaches have been proposed for improving the ethanol yield while reducing glycerol production. Here we studied the effect of reducing glycerol export/formation through deletion of the aquaglyceroporin gene *FPS1* together with expressing *gapN* encoding $NADP^+$-dependent non-phosphorylating glyceraldehyde-3-phosphate dehydrogenase from *Streptococcus mutans* and overexpressing the ATP-NADH kinase gene *UTR1* from *S. cerevisiae*. This strategy will allow reducing the redox balance problem observed when the glycerol pathway is blocked, and hereby improve ethanol production. We found that our strategy enabled increasing the ethanol yield by 4.6% in the case of the best producing strain, compared to the reference strain, without any major effect on the specific growth rate.

Keywords: *Saccharomyces cerevisiae*; Ethanol production; Glycerol; Redox balancing

Introduction

Ethanol is both in terms of market value and volume one of the most important products from the biotechnology industry. Even though this process is highly optimized there is still interest to improve the productivity, the robustness of the strains and the product yield (van Maris et al. 2006; Hahn-Hagerdal et al. 2007). There are different parameters that determine the economy of this industrial bioprocess; one of the most important ones is the price of the feedstock (Wyman and Hinman 1990). Therefore, it is of utmost importance to increase the ethanol yield as well as the carbon source utilization. During ethanol production by *Saccharomyces cerevisiae*, glycerol is a major by-product, representing 4-5% of the carbon source consumption, in addition to biomass, carbon dioxide and a number of other by-products such as acetic acid, pyruvic acid or succinic acid (Nissen et al. 2000b; Wyman and Hinman 1990; Zhang and Chen 2008; Oura 1977).

During anaerobic fermentation, the respiratory chain is not functional and the NADH generated in connection with cell growth must be re-oxidized to NAD^+ by formation of glycerol, in order to avoid an imbalance in the NAD^+/NADH ratio (Nissen et al. 2000a). Furthermore, under osmotic stress conditions, glycerol is produced and accumulated in the cell as an osmolyte, to protect cells against cell lysis (Andre et al. 1991; Larsson et al. 1993; Ansell et al. 1997). Glycerol is synthethized from dihydroxyacetone phosphate in two steps catalysed by Gpd1/Gpd2 (glycerol-3-phosphate dehydrogenases) and Gpp1/Gpp2 (glycerol-3-phosphate phosphatases), respectively (Figure 1). Expression of *GPD1* and *GPP2* is induced by high osmolarity, whereas expression of *GPD2* and *GPP1* is stimulated under anaerobic conditions (Larsson et al. 1993; Eriksson et al. 1995; Nissen et al. 2000a). It has been reported that the formation of glycerol could be decreased by the consumption of NADH by alternative metabolic pathways (Vemuri et al. 2007; Bro et al. 2006). It has also been shown that deletion of either the *GPD1* or *GPD2* gene led to a decrease in the glycerol yield (Guo et al. 2009; Michnick et al. 1997; Nissen et al.

---

[*] Correspondence: siewers@chalmers.se
Department of Chemical and Biological Engineering, Chalmers University of Technology, Kemivägen 10, SE-41296 Göteborg, Sweden

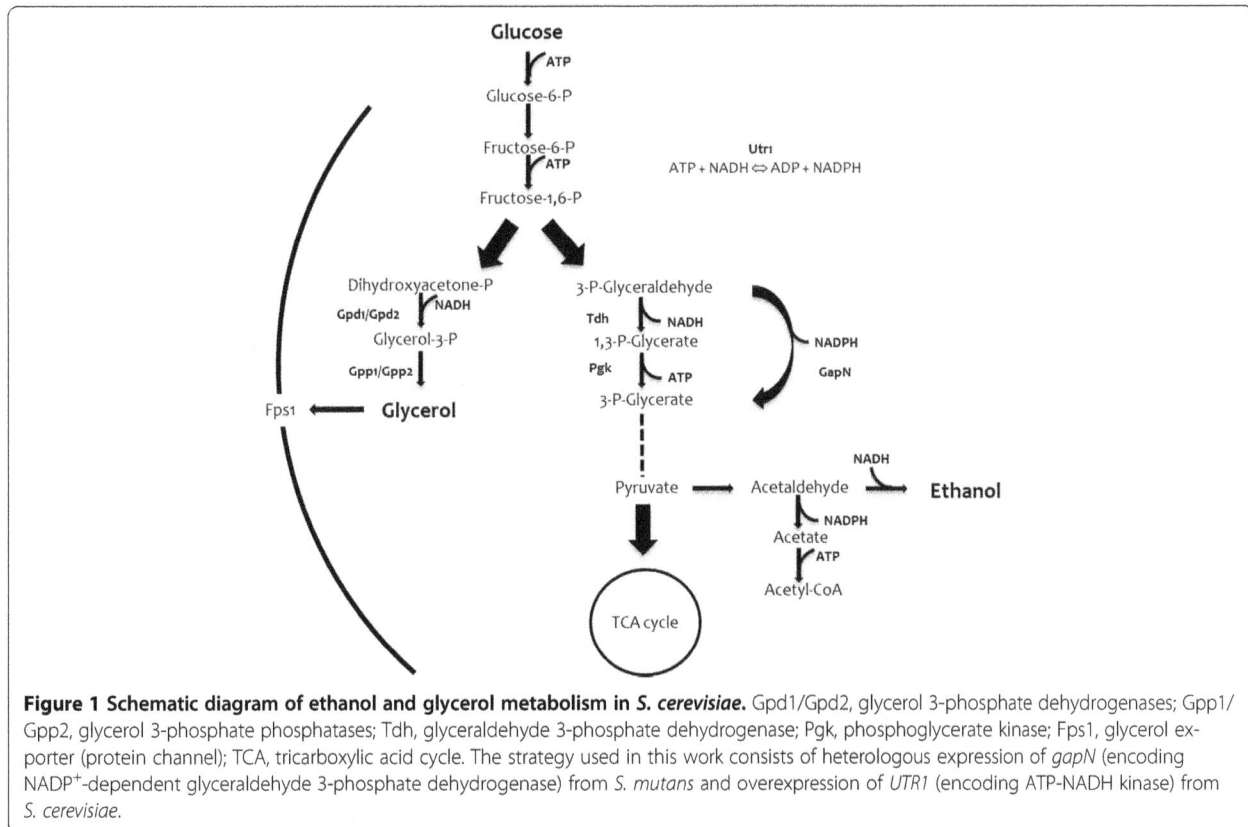

Figure 1 Schematic diagram of ethanol and glycerol metabolism in *S. cerevisiae.* Gpd1/Gpd2, glycerol 3-phosphate dehydrogenases; Gpp1/Gpp2, glycerol 3-phosphate phosphatases; Tdh, glyceraldehyde 3-phosphate dehydrogenase; Pgk, phosphoglycerate kinase; Fps1, glycerol exporter (protein channel); TCA, tricarboxylic acid cycle. The strategy used in this work consists of heterologous expression of *gapN* (encoding NADP⁺-dependent glyceraldehyde 3-phosphate dehydrogenase) from *S. mutans* and overexpression of *UTR1* (encoding ATP-NADH kinase) from *S. cerevisiae.*

2000a), but the double *gpd1Δgpd2Δ* mutant had a dramatically reduced specific growth rate under aerobic conditions with growth being completely abolished at anaerobic conditions (Bjorkqvist et al. 1997). To improve the ethanol yield while reducing glycerol formation, different approaches have been reported. To show whether a reduced formation of surplus NADH and an increased consumption of ATP in biosynthesis would result in a decreased glycerol yield and an increased ethanol yield in anaerobic cultivations, a yeast strain was constructed in which *GLN1* (glutamine synthetase) and *GLT1* (glutamate synthase) were overexpressed, and *GDH1* (NADP⁺-dependent glutamate dehydrogenase) was deleted (Nissen et al. 2000b), which resulted in a 38% reduced glycerol yield. A genome-scale reconstructed metabolic network of *S. cerevisiae* was used to score the best strategies for metabolic engineering of the redox metabolism that would lead to decreased glycerol and increased ethanol yields, and this showed that expressing a non-phosphorylating, NADP⁺-dependent glyceraldehyde-3-phosphate dehydrogenase (GapN) was one of the best strategies tested (Bro et al. 2006). This has been confirmed in several studies, and it has also been shown that expression of GapN can rescue the negative effects from deletion of the glycerol export system Fps1 (Bro et al. 2006; Guo et al. 2009; Wang et al. 2011; Zhang et al. 2011). GapN catalyses the irreversible conversion of glyceraldehyde-3-phosphate and NADP⁺

into 3-phosphoglycerate and NADPH in glycolysis (Figure 1). With this strategy, production of glycerol is substituted with production of ethanol involving a net oxidation of NADH (Bro et al. 2006; Arnon et al. 1954). In another strategy to reduce glycerol production the *Escherichia coli mhpF* gene, encoding an acetylating NAD-dependent acetaldehyde dehydrogenase, was expressed in a *gpd1Δ gpd2Δ* strain, and it was shown that anaerobic growth could be restored by supplementation with 2 g/l acetic acid accompanied by reduced glycerol production (Guadalupe Medina et al. 2010).

Although it has been also shown that glycerol can cross the plasma membrane through a H⁺ symport, detected in cells grown on non-fermentable carbon sources, and by passive diffusion (Gancedo et al. 1968; Lages and Lucas 1997; Lages et al. 1999; Oliveira et al. 2003), glycerol is mainly exported across the plasma membrane through the protein channel Fps1 (Remize et al. 2001) regulated by extracellular osmolarity (Tamas et al. 1999). Fps1 is a member of the major intrinsic protein (MIP) family of channel proteins. MIP channels have been reported to contain six putative transmembrane domains (Walz et al. 1997; Li et al. 1997; Stamer et al. 1996). Although Fps1 can transport glycerol in both directions, the main role of this glycerol facilitator is to regulate glycerol export rather than its uptake (Luyten et al. 1995; Tamas et al. 1999). The stress-activated protein kinase Hog1 can phosphorylate

Fps1 triggering its endocytosis and further degradation under different stress conditions such as high acetic acid levels (Mollapour and Piper 2007). Fps1 is required for glycerol export under anaerobic conditions, and mutants lacking this protein grow poorly under anaerobiosis (Tamas et al. 1999). In an *fps1Δ* mutant, intracellular glycerol accumulates since it cannot be exported. As a result, the accumulation of glycerol inside the yeast cells may induce other regulatory systems in order to reduce glycerol biosynthesis, thus resulting in an increase of the ethanol yield (Zhang et al. 2007).

NAD(H) kinases catalyse NAD(H) phosphorylation by using ATP or inorganic polyphosphate, constituting the last step of the NADP(H) biosynthetic pathway (Kawai et al. 2001; McGuinness and Butler 1985). *S. cerevisiae* is endowed with three NAD(H) kinase homologues, namely Utr1, Pos5 and Yel041w (Kawai et al. 2001; Outten and Culotta 2003; Strand et al. 2003). The ATP-NAD kinase Utr1 was proposed to participate in the ferrireductase system by supplying NADP (Kawai et al. 2001; Lesuisse et al. 1996). Pos5 was described as an ATP-NADH kinase, located in the mitochondrial matrix and reported to play an important role in several mitochondrial processes requiring NADPH. Cells lacking Pos5 accumulate mutations in the mitochondrial DNA and show poor growth in the presence of glycerol, oxidative damage and when growing in medium without arginine (Strand et al. 2003; Outten and Culotta 2003). Later on, Yel041w (re-named as Yef1) and Utr1 were identified as ATP-NADH kinases (Shi et al. 2005). Analysis of the single, double and triple mutants, which were found viable, showed the important

contribution of Pos5 to the mitochondrial function and survival at high temperature (37°C). The contribution of Utr1 to growth in low iron medium was also reported to be critical (Shi et al. 2005).

Here, we explored the combinatorial effect of reducing glycerol export and formation through deletion of *FPS1*, while expressing *gapN* from *S. mutans* and overexpressing the ATP-NADH kinase gene *UTR1* from *S. cerevisiae*. With this strategy we aimed to solve the resulting redox balance problem and, at the same time, to increase the ethanol yield. We also analysed the effect on glycerol reduction/ethanol production, by using a set of different plasmids for varying expression levels of *gapN* and *UTR1*.

Materials and methods

Strain construction and media

The *S. cerevisiae* strains and plasmids used in this study are described in Table 1. CN1, which was used as the reference strain and CN6, where *FPS1* was knocked-out, were transformed with empty plasmids. The *FPS1* gene was deleted in CN2 to CN6 strains. CN2 to CN5 were transformed with different combinations of plasmids carrying *gapN* and *UTR1* under control of the *TEF1* promoter, and expressing the mentioned genes in high or low copy number (Table 1). Agar plates of synthetic medium (6.7 g/l yeast nitrogen base w/o aminoacids, 0.75 g/l complete supplement mixture w/o uracil, 2% glucose, 2% agar, pH 6) or YPD (2% glucose, 2% peptone, 1% yeast extract, 2% agar) supplemented with 200 mg/l of G418, were used for selection of the strains. For the fermentations, yeast cells were grown in minimal synthetic

Table 1 *S. cerevisiae* strains and plasmids used in this work

Strains	Genotype description	Reference
CEN.PK 113-11C	*MATa his3Δ1 ura3-52 MAL2-8^c SUC2*	(van Dijken et al. 2000)
CEN.PK 113-11C; p426; p423 (CN1)	*MATa his3Δ1 ura3-52 MAL2-8^c SUC2* p426TEF1 p423TEF1	This work
fps1Δ; p426-GapN; p423-UTR1 (CN2)	*MATa his3Δ1 ura3-52 MAL2-8^c SUC2 fps1Δ* p426TEF1-GapN p423TEF1-UTR1	This work
fps1Δ; p426-GapN; p413-UTR1 (CN3)	*MATa his3Δ1 ura3-52 MAL2-8^c SUC2 fps1Δ* p426TEF1-GapN p413TEF1-UTR1	This work
fps1Δ; p416-GapN; p423-UTR1 (CN4)	*MATa his3Δ1 ura3-52 MAL2-8^c SUC2 fps1Δ* p416TEF1-GapN p423TEF1-UTR1	This work
fps1Δ; p416-GapN; p413-UTR1 (CN5)	*MATa his3Δ1 ura3-52 MAL2-8^c SUC2 fps1Δ* p416TEF1-GapN p413TEF1-UTR1	This work
fps1Δ; p426; p423 (CN6)	*MATa his3Δ1 ura3-52 MAL2-8^c SUC2 fps1Δ* p426TEF1 p423TEF1	This work
Plasmids		
pCIChE-KK004-Gap	*neo^r*, integrative plasmid	(Kocharin 2013)
p426TEF1	*URA3*, 2 μ plasmid, *TEF1* promoter, *CYC1* terminator	(Mumberg et al. 1995)
p416TEF1	*URA3*, centromeric plasmid, *TEF1* promoter, *CYC1* terminator	(Mumberg et al. 1995)
p423TEF1	*HIS3*, 2 μ plasmid, *TEF1* promoter, *CYC1* terminator	(Mumberg et al. 1995)
p413TEF1	*HIS3*, centromeric plasmid, *TEF1* promoter, *CYC1* terminator	(Mumberg et al. 1995)
p426TEF1-GapN	*URA3*, 2 μ plasmid, *gapN* from *S. mutans*	This work
p416TEF1-GapN	*URA3*, centromeric plasmid, *gapN* from *S. mutans*	This work
p423TEF1-UTR1	*HIS3*, 2 μ plasmid, *UTR1*	This work
p413TEF1-UTR1	*HIS3*, centromeric plasmid, *UTR1*	This work

medium (Verduyn et al. 1992) containing 20 g/l glucose. The pH was adjusted to 5.2 with 2 M KOH before sterilization. In bioreactors, 10 mg/l of ergosterol and 420 mg/l of Tween80 were added to the medium for the anaerobic growth of *S. cerevisiae*.

Deletion of *FPS1*

The *FPS1* gene was replaced, by using the *loxP-kanMX-loxP* cassette obtained by PCR from the plasmid pUG6 (Wach et al. 1994) and a bipartite strategy (Erdeniz et al. 1997). Primers containing upstream and downstream homologous regions of *FPS1* were used for that purpose (Table 2). The PCR products were used for transformation of CEN.PK 113-11C, and the transformed cells were selected in G418-YPD medium. The selectable marker was removed afterwards by expression of the Cre recombinase from plasmid pSH47 under the control of a galactose inducible promoter (Guldener et al. 1996). The correct integration of the fragment disrupting the coding sequence was further tested by PCR and sequencing.

Heterologous expression of *gapN* and overexpression of *UTR1*

The sequence of the NADP$^+$-dependent glyceraldehyde 3-phosphate dehydrogenase gene *gapN* from *S. mutans* (Gene ID: 1028095) was obtained from plasmid pCIChE-KK004-Gap and cloned into p426TEF1 and p416TEF1, respectively, by using PCR and restriction enzymes (REs). *Bam*HI and *Cla*I sites were additionally included in the designed PCR-primers for that purpose (Table 2). The *UTR1* sequence (Gene ID: 853508) was obtained by PCR from CEN.PK 113-11C genomic DNA and cloned into p423TEF1 and p413TEF1 plasmids by REs. In this case, the *Bam*HI/*Eco*RI combination was used (Table 2). The constructs were checked by PCR for *gapN* or *UTR1* genes, digestion with the proper RE combination and sequencing.

Batch cultivations in flasks and analytical methods

Seed cultures of yeast cells were grown during 20–24 hours in shake flasks containing 25 ml of minimal medium

at 30°C and constant shaking (120 rpm). Micro-aerobic conditions were obtained by using 150 ml flasks covered with fermentation bungs and containing 100 ml of medium (Arnon et al. 1954). The shake flasks were inoculated with an initial OD_{600} of 0.05 and grown at 30°C (120–140 rpm). Samples of 2 ml of culture were taken at the selected time-points for optical density and HPLC measurements. Supernatants centrifuged for 5 min at high speed (1 ml) were loaded to a HPX-87G column (Biorad, Hercules, CA) on a Dionex Ultimate 3000 HPLC (Dionex Softron GmbH, Jülich, Germany) to measure the concentrations of glucose, ethanol, glycerol and acetate in the cultures at the different time points. The samples were run at a flow rate of 0.6 ml/min at 65°C, using 5 mM H_2SO_4 as mobile phase. Fermentation experiments were repeated three independent times.

Fermentation analysis in bioreactors

Yeast cell seed cultures were grown in 25 ml of minimal medium supplemented with 20 g/l of glucose, and incubated at 30°C for 22–24 hours (130 rpm). The experiment was performed in 1 l stirrer-pro vessels (DasGip GmbH, Jülich, Germany) with a working volume of 0.5 l, and bioreactors were inoculated at initial OD_{600} of 0.02. The cells were centrifuged and washed with 5 ml of water before inoculation.

The temperature was controlled at 30°C using a DasGip Bioblock integrated heating and cooling thermo well. Agitation was maintained at 600 rpm using an overhead drive stirrer with one Rushton impeller. The bioreactors were flushed with N_2 at 1 vvm when anaerobic growth conditions were required. The pH was maintained constant at 5.0 by the automatic addition of 2 M KOH.

Results

Yeast cells expressing the *gapN/UTR1* genes show an increased ethanol production capacity and reduced glycerol production.

Strains deleted in *FPS1* and expressing *gapN* and *UTR1* from either high or low copy number plasmids were cultivated in shake flasks under micro-aerobic conditions.

Table 2 Oligonucleotide primers used in this study

Primer name	Primer sequence (5'-3')	Restriction sites[*]
kanMX_FPS1_F	**TATTTTCGATCAGATCTCATAGTGAGAAGGCGCAATT**CAGTAGTTGGCATCAGAGCAGATTGTACTGAGAG	
kanMX_R	AAACTCACCGAGGCAGTTCCATAG	
kanMX_F	ATGGTCAGACTAAACTGGCTGACG	
kanMX_FPS1_R	**CATCCATGCGATACATCATGTATAGTAGGTGACCAGGCTGAGTTC**TACCGCCTTTGAGTGAGCTGATAC	
UTR1_F	CG*GGATCCC*GAAAACAATGAAGGAGAATGACATGAATA	*Bam*HI
UTR1_R	CG*GAATTC*CGGTAACATTATACTGAAAACCTTGCTTGA	*Eco*RI
GapN_F	CG*GGATCCC*GAAAACAATGACAAAACAATACAAAAACT	*Bam*HI
GapN_R	CC*ATCGAT*GGAGATCTTCACTTTATGTCAAAGACAACA	*Cla*I

Primer overhangs containing part of *FPS1* up/down sequences are depicted in bold letters.
[*]Restriction sites with corresponding restriction enzyme are represented in italics.

Glucose consumption was completed in CN1 (reference strain) after around 30 h of fermentation. For the rest of strains, this level was reached first after about 37 h. The CN6 strain (*FPS1* deletion strain carrying both empty plasmids) showed 0.5-1 g/l of remaining glucose still after 45 hours of fermentation and the final ethanol production, for this strain, was much lower in comparison to the rest of the transformants tested (Figure 2). This correlates with the specific growth rate of the analysed strains (Table 3).

In the case of the reference strain, the highest level of ethanol was reached already after 25 h of fermentation, whereas for the rest of the strains we could observe a more gradual production over time until reaching the highest ethanol production (Figure 2). Ethanol production after 35–40 h of fermentation was improved in strains CN2 to CN5 expressing the alternative enzymes to reduce NADH levels (Figure 2). These results were confirmed when the specific ethanol productivity was calculated for these strains (Table 3). Strain CN6 grew very poorly hence affecting the calculation of ethanol productivity, and data for this strain were therefore not included in the further analysis.

Glycerol production was reduced in strains CN2 to CN6, confirming our initial hypothesis, showing only a minor effect on cell growth. An exception is strain CN6,

which suffers from the lack of an NADH sink and hence has a much reduced growth rate (Figure 2 and Table 3).

Fermentation analysis in bioreactors showed that, except for the reference strain (CN1), cells were not able to grow under strict anaerobic conditions. Adaptation was necessary in order to complete the fermentation profile of the strains. This problem was partially solved by sparging air containing 12% oxygen at the beginning of fermentation. Under these conditions cells were able to grow, consume the carbon source and produce ethanol.

The ATP consumption by Utr1 and the inhibition of ATP formation by GapN, could be one of the reasons why the engineered strains showed a reduction in their ability to grow anaerobically.

Analytical methods revealed that results from individual experiments were similar to those obtained in shake flasks, with a higher ethanol production and glycerol reduction in the engineered strains (CN2-CN5) compared to the reference strain. But as a consequence of the previous adaptation needed, results were not very consistent between experiments and cells did not always grow at the same level, making it very difficult to represent the data as average yields of the different fermentation experiments. We concluded that more investigations are needed when growing cells in bioreactors, in order to use this strategy for industrial purposes, and therefore,

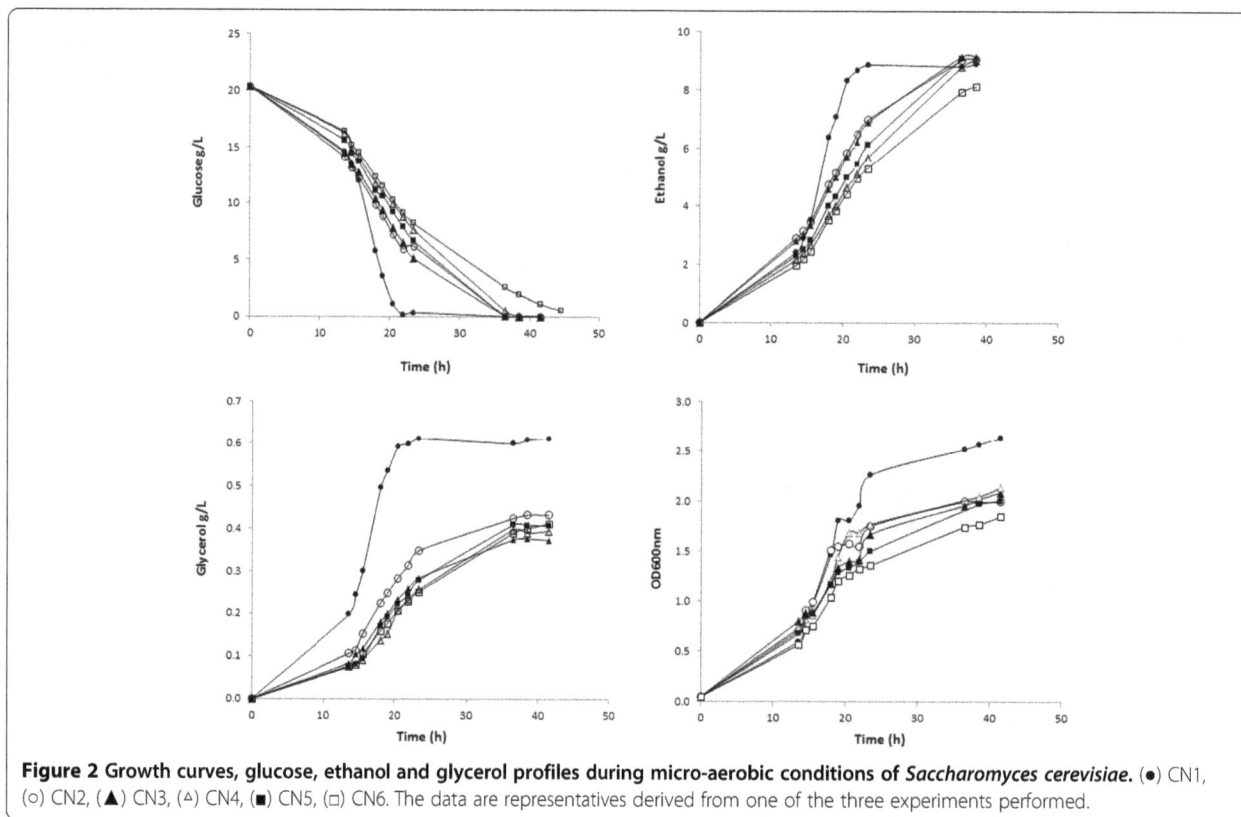

Figure 2 Growth curves, glucose, ethanol and glycerol profiles during micro-aerobic conditions of *Saccharomyces cerevisiae*. (●) CN1, (○) CN2, (▲) CN3, (△) CN4, (■) CN5, (□) CN6. The data are representatives derived from one of the three experiments performed.

Table 3 Compound yields, growth and specific ethanol productivity in engineered yeast strains

Strain	Ethanol yield (g/g)*	Ethanol increase (%)	Glycerol yield (g/g)*	Acetate yield (g/g)*	μ (h^{-1})	q_{sp} (g g^{-1} h^{-1})
CEN.PK 113-11C; p426; p423 (CN1)	0.449 ± 0.011	-	0.042 ± 0.018	0.0033 ± 0.00028	0.18 ± 0.048	26.55 ± 0.293
fps1Δ; p426-GapN; p423-UTR1 (CN2)	0.466 ± 0.013	3.78	0.030 ± 0.012	0.0038 ± 0.00014	0.11 ± 0.059	51.71 ± 0.440
fps1Δ; p426-GapN; p413-UTR1 (CN3)	0.470 ± 0.018	4.67	0.026 ± 0.010	0.0039 ± 0.00085	0.11 ± 0.052	52.02 ± 0.648
fps1Δ; p416-GapN; p423-UTR1 (CN4)	0.457 ± 0.014	1.78	0.027 ± 0.012	0.0046 ± 0.00134	0.13 ± 0.046	40.47 ± 0.609
fps1Δ; p416-GapN; p413-UTR1 (CN5)	0.458 ± 0.012	2	0.029 ± 0.013	0.0032 ± 0.00021	0.11 ± 0.038	52.37 ± 1.208
fps1Δ; p426; p423 (CN6)	0.443 ± 0.0040	−1.33	0.023 ± 0.00049	0.0096 ± 0.00021	0.070 ± 0.0011	-

Data represent the means of results from triplicates ± SD.
*g/g is equivalent to gram of produced compound per gram of glucose.
μ represents the specific growth rate.
q_{sp} represents the specific ethanol productivity.

we focused on the results we obtained in micro-aerobic conditions to analyse our strategy.

Physiological characterization of the engineered yeast strains. Effect of the gene copy number of *gapN* and *UTR1* genes on ethanol yield

At the end of fermentation process, the ethanol yield was improved in the transformant strains by 1.7-4.6% (depending on the strain analysed) compared to the reference strain (CN1). CN2 and CN3 were the best ethanol producers (p-value < 0.05 after Student's t-test), showing ethanol yields up to 0.47 grams per gram of glucose (Table 3). After the analysis of the possible effect of high/low expression of *gapN* and *UTR1*, by using different expression plasmids, we could observe the improvement in CN2 and CN3 strains both carrying *gapN* on a high-copy plasmid. In the case of *UTR1*, the copy number of the kinase gene did not seem to be crucial for a better ethanol yield, since the best producer (CN3) expressed *UTR1* from a low-copy plasmid. These results suggest a more important role of the NADP$^+$-dependent glyceraldehyde 3-phosphate dehydrogenase in the redox balancing for improved ethanol production.

To further check this hypothesis, we performed an analysis of the *fps1Δ* strain expressing *gapN* or *UTR1* individually (both from a high-copy plasmid). The *fps1Δ* strain expressing *gapN* showed similar growth and final ethanol production compared to CN2 and CN3 strains, (7.98 g/l versus 8.14 and 8.43 g/l respectively) after ~40 h of fermentation in 20 g/l of glucose. On the other hand, the *fps1Δ* strain expressing *UTR1* grew similar to the CN6 control strain. Ethanol production at the end of fermentation was also at the same level as measured in the CN6 strain (5.6 g/l versus 6.7 g/l).

Furthermore, for the transformant strains the glycerol yield was successfully reduced by 25-40% compared to the reference strain. Although the transformant strains

CN2-CN5 showed a specific growth rate reduced by 27-38%, they could still grow at a good level and entirely consume the carbon source. As explained before, the CN6 strain with highest glycerol reduction was the one most affected in terms of specific growth rate and also regarding ethanol production (Table 3 and Figure 2).

Discussion

In *S. cerevisiae*, it is well known from previously published studies that, under aerobic conditions, *fps1Δ* mutants show a growth profile similar to a wild type, due to NADH re-oxidation by the respiratory chain (Zhang et al. 2007). Under anaerobic conditions, glycerol production is the main redox-sink for the excess of NADH (Valadi et al. 2004). When glycerol export is blocked, the *fps1Δ* mutants produce less glycerol and biomass, and show a lower glucose uptake rate compared to the reference strain. However, in case of the mutant strain, there is residual glucose in the growth medium, resulting in a decrease of total ethanol produced (Wang et al. 2011). As already described in the results, the final ethanol production for the CN6 strain was much lower in comparison to the rest of the transformants tested (Figure 2). Moreover, it has been previously described that acetic acid enters glucose-repressed yeast cells primarily by facilitated diffusion through Fps1. The total loss of the channel also creates an acetate resistant phenotype by elimination of the major source of the acetic acid flux into the cell (Mollapour and Piper 2007). The increased acetate yield observed in the CN6 strain might be connected to an inhibited acetate import in these cells (Table 3).

GapN catalyses the irreversible conversion of glyceraldehyde-3-phosphate and NADP$^+$ into 3-phosphoglycerate and NADPH in glycolysis, and in contrast to the native yeast pathway, no NADH and ATP are released in this reaction (Figure 1). Expression of *gapN* has a high potential to reduce NADH generation, and may therefore lead to a complete redirection of flux from formation of glycerol to ethanol (Bro et al. 2006). When alternative routes for NADH re-oxidation were expressed, ethanol production was improved in strains CN2 to CN5 (Figure 2).

The intracellular concentration of the $NADP^+/NADPH$ pool is much lower than that of the $NAD^+/NADH$ pool, and shifting part of the glycolytic flux from using NAD^+ to using $NADP^+$ as co-factor may therefore result in limitation of supply of this co-factor. We therefore evaluated whether increasing the $NADP^+/NADPH$ pool by over-expressing the UTR1 gene encoding an NADH-kinase may have a positive effect on top of the use of GapN as a glycolytic enzyme. As already mentioned and represented in Figure 1, NAD(H) kinases catalyse NAD(H) phosphorylation by using ATP or inorganic polyphosphate in yeast, constituting the last step of the NADP biosynthetic pathway (McGuinness and Butler 1985; Kawai et al. 2001). UTR1 overexpression does, however, not seem to lead to an effect on ethanol and glycerol production.

Regarding glycerol production, it is believed that the complete elimination of its production is not practical due to its importance in osmoregulation and maintenance of the intracellular redox balance (Blomberg and Adler 1992; van Dijken and Scheffers 1986). Moreover, it is a precursor used to synthesize the cellular membrane. On the other hand, it can be reduced to a minimal level in order to improve the ethanol yield (Bjorkqvist et al. 1997; Nissen et al. 2000b). Our hypothesis was confirmed and glycerol production was reduced in the strains CN2-CN6, Except for strain CN6, which suffers from the lack of an NADH sink, the cell growth was only mildly affected (Figure 2 and Table 3).

The present work evidences that the function of glycerol as a redox sink for anaerobic growth can be successfully replaced by introducing a heterologous GapN pathway from S. mutans in addition to deletion of the FPS1 gene (responsible for glycerol export) and overexpression of the NADH-kinase Utr1, resulting in a reduction of glycerol (by-product) formation together with an increase in the ethanol yield.

By expressing gapN (in combination with the overexpression of UTR1) we can solve the redox problem in the fps1Δ mutant thereby increasing the specific growth rate and reducing the formation of NADH. Thus, the carbon source was redirected towards ethanol production in the engineered strains, which increased the ethanol yield by up to 4.6% compared to the reference strain.

Competing interests
The authors declare that they have no competing interest.

Authors' contributions
CN, JN and VS participated in the design of the experiment. CN performed all the experiments, analyzed the data and wrote the manuscript. JN and VS edited the manuscript. All authors read and approved the final manuscript.

Acknowledgements
The authors would like to thank The Swedish Research Council (Vetenskapsrådet) and Novozymes A/S for financial support. Francesco Gatto is acknowledged for his help with the statistical analysis of the data. We also thank Michael L. Nielsen, Pia Francke Johannesen and Carsten Hjort for helpful suggestions and comments about this work.

References
Andre L, Hemming A, Adler L (1991) Osmoregulation in Saccharomyces cerevisiae. Studies on the osmotic induction of glycerol production and glycerol-3-phosphate dehydrogenase (NAD+). FEBS Lett 286(1–2):13–17

Ansell R, Granath K, Hohmann S, Thevelein JM, Adler L (1997) The two isoenzymes for yeast NAD + –dependent glycerol 3-phosphate dehydrogenase encoded by GPD1 and GPD2 have distinct roles in osmoadaptation and redox regulation. EMBO J 16(9):2179–2187, doi:10.1093/emboj/16.9.2179

Arnon DI, Rosenberg LL, Whatley FR (1954) A New Glyceraldehyde Phosphate Dehydrogenase from Photosynthetic Tissues. Nature 173(4415):1132–1134

Bjorkqvist S, Ansell R, Adler L, Liden G (1997) Physiological response to anaerobicity of glycerol-3-phosphate dehydrogenase mutants of Saccharomyces cerevisiae. Appl Environ Microbiol 63(1):128–132

Blomberg A, Adler L (1992) Physiology of osmotolerance in fungi. Adv Microb Physiol 33:145–212

Bro C, Regenberg B, Forster J, Nielsen J (2006) In silico aided metabolic engineering of Saccharomyces cerevisiae for improved bioethanol production. Metab Eng 8(2):102–111, doi:10.1016/j.ymben.2005.09.007

Erdeniz N, Mortensen UH, Rothstein R (1997) Cloning-free PCR-based allele replacement methods. Genome Res 7(12):1174–1183

Eriksson P, Andre L, Ansell R, Blomberg A, Adler L (1995) Cloning and characterization of GPD2, a second gene encoding sn-glycerol 3-phosphate dehydrogenase (NAD+) in Saccharomyces cerevisiae, and its comparison with GPD1. Mol Microbiol 17(1):95–107

Gancedo C, Gancedo JM, Sols A (1968) Glycerol Metabolism in Yeasts - Pathways of Utilization and Production. Eur J Biochem 5(2):165, doi:10.1111/j.1432-1033.1968.tb00353.x

Guadalupe Medina V, Almering MJ, van Maris AJ, Pronk JT (2010) Elimination of glycerol production in anaerobic cultures of a Saccharomyces cerevisiae strain engineered to use acetic acid as an electron acceptor. Appl Environ Microbiol 76(1):190–195, doi:10.1128/AEM.01772-09

Guldener U, Heck S, Fielder T, Beinhauer J, Hegemann JH (1996) A new efficient gene disruption cassette for repeated use in budding yeast. Nucleic Acids Res 24(13):2519–2524

Guo ZP, Zhang L, Ding ZY, Wang ZX, Shi GY (2009) Interruption of glycerol pathway in industrial alcoholic yeasts to improve the ethanol production. Appl Microbiol Biotechnol 82(2):287–292, doi:10.1007/s00253-008-1777-7

Hahn-Hagerdal B, Karhumaa K, Fonseca C, Spencer-Martins I, Gorwa-Grauslund MF (2007) Towards industrial pentose-fermenting yeast strains. Appl Microbiol Biotechnol 74(5):937–953, doi:10.1007/s00253-006-0827-2

Kawai S, Suzuki S, Mori S, Murata K (2001) Molecular cloning and identification of UTR1 of a yeast Saccharomyces cerevisiae as a gene encoding an NAD kinase. FEMS Microbiol Lett 200(2):181–184

Kocharin K (2013) Metabolic engineering of Saccharomyces cerevisiae for polyhydroxybutyrate production. In: Thesis. Chalmers University of Technology, Sweden

Lages F, Lucas C (1997) Contribution to the physiological characterization of glycerol active uptake in Saccharomyces cerevisiae. Bba-Bioenergetics 1322(1):8–18, doi:10.1016/S0005-2728(97)00062-5

Lages F, Silva-Graca M, Lucas C (1999) Active glycerol uptake is a mechanism underlying halotolerance in yeasts: a study of 42 species. Microbiol-Uk 145:2577–2585

Larsson K, Ansell R, Eriksson P, Adler L (1993) A gene encoding sn-glycerol 3-phosphate dehydrogenase (NAD+) complements an osmosensitive mutant of Saccharomyces cerevisiae. Mol Microbiol 10(5):1101–1111

Lesuisse E, Casteras-Simon M, Labbe P (1996) Evidence for the Saccharomyces cerevisiae ferrireductase system being a multicomponent electron transport chain. J Biological Chem 271(23):13578–13583

Li H, Lee S, Jap BK (1997) Molecular design of aquaporin-1 water channel as revealed by electron crystallography. Nat Struct Mol Biol 4(4):263–265

Luyten K, Albertyn J, Skibbe WF, Prior BA, Ramos J, Thevelein JM, Hohmann S (1995) Fps1, a yeast member of the MIP family of channel proteins, is a facilitator for glycerol uptake and efflux and is inactive under osmotic stress. EMBO J 14(7):1360–1371

McGuinness ET, Butler JR (1985) NAD+ kinase–a review. Int J Biochem 17(1):1–11

Michnick S, Roustan JL, Remize F, Barre P, Dequin S (1997) Modulation of glycerol and ethanol yields during alcoholic fermentation in Saccharomyces cerevisiae strains overexpressed or disrupted for GPD1 encoding glycerol 3-phosphate dehydrogenase. Yeast 13(9):783–793, doi:10.1002/(SICI)1097-0061(199707)13:9 < 783::AID-YEA128 > 3.0.CO;2-W

Mollapour M, Piper PW (2007) Hog1 mitogen-activated protein kinase phosphorylation targets the yeast Fps1 aquaglyceroporin for endocytosis, thereby rendering cells resistant to acetic acid. Mol Cell Biol 27(18):6446–6456, doi:10.1128/MCB.02205-06

Mumberg D, Muller R, Funk M (1995) Yeast vectors for the controlled expression of heterologous proteins in different genetic backgrounds. Gene 156(1):119–122

Nissen TL, Hamann CW, Kielland-Brandt MC, Nielsen J, Villadsen J (2000a) Anaerobic and aerobic batch cultivations of Saccharomyces cerevisiae mutants impaired in glycerol synthesis. Yeast 16(5):463–474, doi:10.1002/(SICI)1097-0061(20000330)16:5 < 463::AID-YEA535 > 3.0.CO;2-3

Nissen TL, Kielland-Brandt MC, Nielsen J, Villadsen J (2000b) Optimization of ethanol production in Saccharomyces cerevisiae by metabolic engineering of the ammonium assimilation. Metab Eng 2(1):69–77, doi:10.1006/mben.1999.0140

Oliveira R, Lages F, Silva-Graca M, Lucas C (2003) Fps1p channel is the mediator of the major part of glycerol passive diffusion in Saccharomyces cerevisiae: artefacts and re-definitions. Bba-Biomembranes 1613(1–2):57–71, doi:10.1016/S0005-2736(03)00138-X

Oura E (1977) Reaction products of yeast fermentations. Process Biochem 12(35):19–21

Outten CE, Culotta VC (2003) A novel NADH kinase is the mitochondrial source of NADPH in Saccharomyces cerevisiae. EMBO J 22(9):2015–2024, doi:10.1093/emboj/cdg211

Remize F, Barnavon L, Dequin S (2001) Glycerol export and glycerol-3-phosphate dehydrogenase, but not glycerol phosphatase, are rate limiting for glycerol production in Saccharomyces cerevisiae. Metab Eng 3(4):301–312, doi:10.1006/mben.2001.0197

Shi F, Kawai S, Mori S, Kono E, Murata K (2005) Identification of ATP-NADH kinase isozymes and their contribution to supply of NADP(H) in Saccharomyces cerevisiae. FEBS J 272(13):3337–3349, doi:10.1111/j.1742-4658.2005.04749.x

Stamer WD, Snyder RW, Regan JW (1996) Characterization of the transmembrane orientation of aquaporin-1 using antibodies to recombinant fusion proteins. Biochemistry 35(50):16313–16318, doi:10.1021/bi9619536

Strand MK, Stuart GR, Longley MJ, Graziewicz MA, Dominick OC, Copeland WC (2003) POS5 gene of Saccharomyces cerevisiae encodes a mitochondrial NADH kinase required for stability of mitochondrial DNA. Eukaryotic Cell 2(4):809–820

Tamas MJ, Luyten K, Sutherland FC, Hernandez A, Albertyn J, Valadi H, Li H, Prior BA, Kilian SG, Ramos J, Gustafsson L, Thevelein JM, Hohmann S (1999) Fps1p controls the accumulation and release of the compatible solute glycerol in yeast osmoregulation. Mol Microbiol 31(4):1087–1104

Valadi H, Valadi A, Ansell R, Gustafsson L, Adler L, Norbeck J, Blomberg A (2004) NADH-reductive stress in Saccharomyces cerevisiae induces the expression of the minor isoform of glyceraldehyde-3-phosphate dehydrogenase (TDH1). Curr Genet 45(2):90–95, doi:10.1007/s00294-003-0469-1

van Dijken JP, Bauer J, Brambilla L, Duboc P, Francois JM, Gancedo C, Giuseppin ML, Heijnen JJ, Hoare M, Lange HC, Madden EA, Niederberger P, Nielsen J, Parrou JL, Petit T, Porro D, Reuss M, van Riel N, Rizzi M, Steensma HY, Verrips CT, Vindelov J, Pronk JT (2000) An interlaboratory comparison of physiological and genetic properties of four Saccharomyces cerevisiae strains. Enzym Microb Technol 26(9–10):706–714

van Dijken JP, Scheffers WA (1986) Redox balances in the metabolism of sugars by yeasts. FEMS Microbiol Lett 32(3–4):199–224, doi:10.1111/j.1574-6968.1986.tb01194.x

van Maris AJ, Abbott DA, Bellissimi E, van den Brink J, Kuyper M, Luttik MA, Wisselink HW, Scheffers WA, van Dijken JP, Pronk JT (2006) Alcoholic fermentation of carbon sources in biomass hydrolysates by Saccharomyces cerevisiae: current status. Antonie Van Leeuwenhoek 90(4):391–418, doi:10.1007/s10482-006-9085-7

Vemuri GN, Eiteman MA, McEwen JE, Olsson L, Nielsen J (2007) Increasing NADH oxidation reduces overflow metabolism in Saccharomyces cerevisiae. Proc Natl Acad Sci U S A 104(7):2402–2407, doi:10.1073/pnas.0607469104

Verduyn C, Postma E, Scheffers WA, Van Dijken JP (1992) Effect of benzoic acid on metabolic fluxes in yeasts: a continuous-culture study on the regulation of respiration and alcoholic fermentation. Yeast 8(7):501–517, doi:10.1002/yea.320080703

Wach A, Brachat A, Pohlmann R, Philippsen P (1994) New heterologous modules for classical or PCR-based gene disruptions in Saccharomyces cerevisiae. Yeast 10(13):1793–1808

Walz T, Hirai T, Murata K, Heymann JB, Mitsuoka K, Fujiyoshi Y, Smith BL, Agre P, Engel A (1997) The three-dimensional structure of aquaporin-1. Nature 387(6633):624–627, doi:10.1002/jctb.2634

Wang P-M, Zheng D-Q, Ding R, Chi X-Q, Tao X-L, Min H, Wu X-C (2011) Improvement of ethanol production in Saccharomyces cerevisiae by hetero-expression of GAPN and FPS1 deletion. J Chem Technol Biotechnol 86(9):1205–1210, doi:10.1002/jctb.2634

Wyman C, Hinman N (1990) Ethanol. Appl Biochem Biotechnol 24–25(1):735–753, doi:10.1007/BF02920291

Zhang A, Chen X (2008) Improve Ethanol Yield Through Minimizing Glycerol Yield in Ethanol Fermentation of Saccharomyces cerevisiae. Chin J Chem Eng 16(4):620–625, doi:http://dx.doi.org/10.1016/S1004-9541(08)60130-5

Zhang A, Kong Q, Cao L, Chen X (2007) Effect of FPS1 deletion on the fermentation properties of Saccharomyces cerevisiae. Lett Appl Microbiol 44(2):212–217, doi:10.1111/j.1472-765X.2006.02041.x

Zhang L, Tang Y, Guo ZP, Ding ZY, Shi GY (2011) Improving the ethanol yield by reducing glycerol formation using cofactor regulation in Saccharomyces cerevisiae. Biotechnol Lett 33(7):1375–1380, doi:10.1007/s10529-011-0588-6

Permissions

List of Contributors

Birgit Jovanović
Department for Biotechnology and Microbiology, Institute of Chemical Engineering, Vienna University of Technology, Gumpendorfer Str. 1a, A-1060 Wien, Austria

Robert L Mach
Department for Biotechnology and Microbiology, Institute of Chemical Engineering, Vienna University of Technology, Gumpendorfer Str. 1a, A-1060 Wien, Austria

Astrid R Mach-Aigner
Department for Biotechnology and Microbiology, Institute of Chemical Engineering, Vienna University of Technology, Gumpendorfer Str. 1a, A-1060 Wien, Austria

Leiv M Mortensen
Department of Plant Science, The University of Life Sciences, Ås NO-1432, Norway

Hans R Gislerød
Department of Plant Science, The University of Life Sciences, Ås NO-1432, Norway

Valeria Wallace-Salinas
Applied Microbiology, Department of Chemistry, Lund University, P.O. Box 124, SE-22100 Lund, Sweden

Lorenzo Signori
University of Milano Bicocca, Piazza della Scienza 2, 20126 Milan, Italy

Ying-Ying Li
Laboratory of Molecular Cell Biology, Institute of Botany and Microbiology, Leuven, KU, Belgium
Department of Molecular Microbiology, VIB, Kasteelpark Arenberg 31, Leuven, B-3001 Heverlee, Flanders, Belgium

Magnus Ask
Department of Chemical and Biological Engineering, Industrial Biotechnology, Chalmers University of Technology, SE-41296 Gothenburg, Sweden

Maurizio Bettiga
Department of Chemical and Biological Engineering, Industrial Biotechnology, Chalmers University of Technology, SE-41296 Gothenburg, Sweden

Danilo Porro
University of Milano Bicocca, Piazza della Scienza 2, 20126 Milan, Italy

Johan M Thevelein
Laboratory of Molecular Cell Biology, Institute of Botany and Microbiology, Leuven, KU, Belgium
Department of Molecular Microbiology, VIB, Kasteelpark Arenberg 31, Leuven, B-3001 Heverlee, Flanders, Belgium

Paola Branduardi
University of Milano Bicocca, Piazza della Scienza 2, 20126 Milan, Italy

María R Foulquié-Moreno
Laboratory of Molecular Cell Biology, Institute of Botany and Microbiology, Leuven, KU, Belgium
Department of Molecular Microbiology, VIB, Kasteelpark Arenberg 31, Leuven, B-3001 Heverlee, Flanders, Belgium

Marie Gorwa-Grauslund
Applied Microbiology, Department of Chemistry, Lund University, P.O. Box 124, SE-22100 Lund, Sweden

Ren Wei
Department of Microbiology and Bioprocess Technology, Institute of Biochemistry, University of Leipzig, Johannisallee 21-23, D-04103 Leipzig, Germany

Thorsten Oeser
Department of Microbiology and Bioprocess Technology, Institute of Biochemistry, University of Leipzig, Johannisallee 21-23, D-04103 Leipzig, Germany

Johannes Then
Department of Microbiology and Bioprocess Technology, Institute of Biochemistry, University of Leipzig, Johannisallee 21-23, D-04103 Leipzig, Germany

Nancy Kühn
Department of Microbiology and Bioprocess Technology, Institute of Biochemistry, University of Leipzig, Johannisallee 21-23, D-04103 Leipzig, Germany

Markus Barth
Department of Microbiology and Bioprocess Technology, Institute of Biochemistry, University of Leipzig, Johannisallee 21-23, D-04103 Leipzig, Germany

Juliane Schmidt
Department of Microbiology and Bioprocess Technology, Institute of Biochemistry, University of Leipzig, Johannisallee 21-23, D-04103 Leipzig, Germany

Wolfgang Zimmermann
Department of Microbiology and Bioprocess Technology, Institute of Biochemistry, University of Leipzig, Johannisallee 21-23, D-04103 Leipzig, Germany

Judit Willenbacher
Institute of Process Engineering in Life Sciences, Section II: Technical Biology, Karlsruhe Institute of Technology (KIT), Engler-Bunte-Ring 1, 76131, Karlsruhe, Germany

Jens-Tilman Rau
Institute of Process Engineering in Life Sciences, Section II: Technical Biology, Karlsruhe Institute of Technology (KIT), Engler-Bunte-Ring 1, 76131, Karlsruhe, Germany

Jonas Rogalla
Institute of Process Engineering in Life Sciences, Section II: Technical Biology, Karlsruhe Institute of Technology (KIT), Engler-Bunte-Ring 1, 76131, Karlsruhe, Germany

Christoph Syldatk
Institute of Process Engineering in Life Sciences, Section II: Technical Biology, Karlsruhe Institute of Technology (KIT), Engler-Bunte-Ring 1, 76131, Karlsruhe, Germany

Rudolf Hausmann
Institute of Food Science and Biotechnology (150), Section Bioprocess Engineering (150 k), University of Hohenheim, Garbenstr. 25, 70599 Stuttgart, Germany

Amir Hussain
Downstream Bioprocessing Laboratory, School of Engineering and Science, Jacobs University, Campus Ring 1, 28759 Bremen, Germany

Martin Kangwa
Downstream Bioprocessing Laboratory, School of Engineering and Science, Jacobs University, Campus Ring 1, 28759 Bremen, Germany

Ahmed Gad Abo-Elwafa
Department of Biotechnology, Faculty of Agriculture, Al-Azhar University, Naser City, Cairo 11884, Egypt

Marcelo Fernandez-Lahore
Downstream Bioprocessing Laboratory, School of Engineering and Science, Jacobs University, Campus Ring 1, 28759 Bremen, Germany

Xiang Li
School of Chemical and Biomedical Engineering, Nanyang Technological University, 62 Nanyang Drive, Singapore 637459, Singapore

Wei Ning Chen
School of Chemical and Biomedical Engineering, Nanyang Technological University, 62 Nanyang Drive, Singapore 637459, Singapore

Ken'ichiro Matsumoto
Division of Biotechnology and Macromolecular Chemistry, Graduate School of Engineering, Hokkaido University, N13-W8, Kita-ku, Sapporo 060-8628, Japan
PRESTO-JST, K's Gobancho, Building 7, Gobancho Chiyoda-ku, Tokyo 102-0076, Japan

Kota Tobitani
Division of Biotechnology and Macromolecular Chemistry, Graduate School of Engineering, Hokkaido University, N13-W8, Kita-ku, Sapporo 060-8628, Japan

Shunsuke Aoki
Division of Biotechnology and Macromolecular Chemistry, Graduate School of Engineering, Hokkaido University, N13-W8, Kita-ku, Sapporo 060-8628, Japan

Yuyang Song
Division of Biotechnology and Macromolecular Chemistry, Graduate School of Engineering, Hokkaido University, N13-W8, Kita-ku, Sapporo 060-8628, Japan
College of Enology, Northwest A&F University, 22 Xinong Road, Yangling, 712100 Shaanxi, China

Toshihiko Ooi
Division of Biotechnology and Macromolecular Chemistry, Graduate School of Engineering, Hokkaido University, N13-W8, Kita-ku, Sapporo 060-8628, Japan
CREST, JST, 4-1-8 Honcho, Kawaguchi, Saitama 332-0012, Japan

Seiichi Taguchi
Division of Biotechnology and Macromolecular Chemistry, Graduate School of Engineering, Hokkaido University, N13-W8, Kita-ku, Sapporo 060-8628, Japan
CREST, JST, 4-1-8 Honcho, Kawaguchi, Saitama 332-0012, Japan

Martina Schedler
Institute of Technical Biocatalysis, Hamburg University of Technology, Hamburg 21073, Germany

Robert Hiessl
Institute of Technical Biocatalysis, Hamburg University of Technology, Hamburg 21073, Germany

Ana Gabriela Valladares Juárez
Institute of Technical Biocatalysis, Hamburg University of Technology, Hamburg 21073, Germany

Giselher Gust
Institute for Product Development and Mechanical Engineering Design, Hamburg University of Technology, Hamburg 21073, Germany

Rudolf Müller
Institute of Technical Biocatalysis, Hamburg University of Technology, Hamburg 21073, Germany

Claire M Hull
Institute of Life Science, College of Medicine, Swansea University, Swansea SA2 8PP, Wales, UK

E. Joel Loveridge
Institute of Life Science, College of Medicine, Swansea University, Swansea SA2 8PP, Wales, UK

Iain S Donnison
Institute of Biological, Environmental & Rural Sciences, Aberystwyth University, Gogerddan, Aberystwyth SY23 3EE, Wales, UK

Diane E Kelly
Institute of Life Science, College of Medicine, Swansea University, Swansea SA2 8PP, Wales, UK

Steven L Kelly
Institute of Life Science, College of Medicine, Swansea University, Swansea SA2 8PP, Wales, UK

Kohei Esaka
Division of Applied Life Sciences, Graduate School of Agriculture, Kyoto University, Sakyo-ku, Kyoto, Japan

Shunsuke Aburaya
Division of Applied Life Sciences, Graduate School of Agriculture, Kyoto University, Sakyo-ku, Kyoto, Japan

Hironobu Morisaka
Division of Applied Life Sciences, Graduate School of Agriculture, Kyoto University, Sakyo-ku, Kyoto, Japan
Kyoto Integrated Science and Technology Bio-Analysis Center, Shimogyo-ku, Kyoto, Japan

Kouichi Kuroda
Division of Applied Life Sciences, Graduate School of Agriculture, Kyoto University, Sakyo-ku, Kyoto, Japan

Mitsuyoshi Ueda
Division of Applied Life Sciences, Graduate School of Agriculture, Kyoto University, Sakyo-ku, Kyoto, Japan
Kyoto Integrated Science and Technology Bio-Analysis Center, Shimogyo-ku, Kyoto, Japan

Yang Jiang
Department of Biotechnology, Delft University of Technology, Julianalaan 67, 2628 BC Delft, The Netherlands

Gizela Mikova
Polymer Technology Group Eindhoven BV, De Lismortel 31, 5612 AR Eindhoven, The Netherlands

Robbert Kleerebezem
Department of Biotechnology, Delft University of Technology, Julianalaan 67, 2628 BC Delft, The Netherlands

Luuk AM van der Wielen
Department of Biotechnology, Delft University of Technology, Julianalaan 67, 2628 BC Delft, The Netherlands

Maria C Cuellar
Department of Biotechnology, Delft University of Technology, Julianalaan 67, 2628 BC Delft, The Netherlands

Yutaro Baba
Graduate School of Life and Environmental Sciences, Osaka Prefecture University, 1-1 Gakuen-cho, Naka-ku, Sakai, Osaka 599-8531, Japan

Jun-ichi Sumitani
Graduate School of Life and Environmental Sciences, Osaka Prefecture University, 1-1 Gakuen-cho, Naka-ku, Sakai, Osaka 599-8531, Japan

Shuji Tani
Graduate School of Life and Environmental Sciences, Osaka Prefecture University, 1-1 Gakuen-cho, Naka-ku, Sakai, Osaka 599-8531, Japan

Takashi Kawaguchi
Graduate School of Life and Environmental Sciences, Osaka Prefecture University, 1-1 Gakuen-cho, Naka-ku, Sakai, Osaka 599-8531, Japan

Hironaga Akita
Biomass Refinery Research Center, National Institute of Advanced Industrial Sciences and Technology (AIST), 3-11-32 Kagamiyama, Higashi-Hiroshima, Hiroshima 739-0046, Japan

Masahiro Watanabe
Biomass Refinery Research Center, National Institute of Advanced Industrial Sciences and Technology (AIST), 3-11-32 Kagamiyama, Higashi-Hiroshima, Hiroshima 739-0046, Japan

Toshihiro Suzuki
Biomass Refinery Research Center, National Institute of Advanced Industrial Sciences and Technology (AIST), 3-11-32 Kagamiyama, Higashi-Hiroshima, Hiroshima 739-0046, Japan

Nobutaka Nakashima
Bioproduction Research Institute, National Institute of Advanced Industrial Sciences and Technology (AIST), 2-17-2-1 Tsukisamu-Higashi, Toyohira-ku, Sapporo 062-8517, Japan
Department of Biological Information, Graduate School of Bioscience and Biotechnology, Tokyo Institute of Technology, 2-12-1-M6-5 Ookayama, Meguro-ku, Tokyo 152-8550, Japan

Tamotsu Hoshino
Biomass Refinery Research Center, National Institute of Advanced Industrial Sciences and Technology (AIST), 3-11-32 Kagamiyama, Higashi-Hiroshima, Hiroshima 739-0046, Japan
Bioproduction Research Institute, National Institute of Advanced Industrial Sciences and Technology (AIST), 2-17-2-1 Tsukisamu-Higashi, Toyohira-ku, Sapporo 062-8517, Japan

Leonardo de Figueiredo Vilela
Department of Biochemistry, Institute of Chemistry, Federal University of Rio de Janeiro, Rio de Janeiro, Brazil

Verônica Parente Gomes de Araujo
Department of Biochemistry, Institute of Chemistry, Federal University of Rio de Janeiro, Rio de Janeiro, Brazil

Raquel de Sousa Paredes
Department of Biochemistry, Institute of Chemistry, Federal University of Rio de Janeiro, Rio de Janeiro, Brazil

Elba Pinto da Silva Bon
Department of Biochemistry, Institute of Chemistry, Federal University of Rio de Janeiro, Rio de Janeiro, Brazil

Fernando Araripe Gonçalves Torres
Department of Cellular Biology, Institute of Biology, University of Brasília, Brasília, DF, Brazil

Bianca Cruz Neves
Department of Biochemistry, Institute of Chemistry, Federal University of Rio de Janeiro, Rio de Janeiro, Brazil

Elis Cristina Araújo Eleutherio
Department of Biochemistry, Institute of Chemistry, Federal University of Rio de Janeiro, Rio de Janeiro, Brazil

Benny Palmqvist
Department of Chemical Engineering, Lund University, Box 124, SE-221 00 Lund, Sweden

Gunnar Lidén
Department of Chemical Engineering, Lund University, Box 124, SE-221 00 Lund, Sweden

Michel Rodrigo Zambrano Passarini
Divisão de Recursos Microbianos, CPQBA/UNICAMP, CP 6171, 13083-970 Campinas, SP, Brazil

Cristiane Angelica Ottoni
CEB-Centre of Biological Engineering, University of Minho, Campus of Gualtar, 4710-057 Braga, Portugal

Cledir Santos
CEB-Centre of Biological Engineering, University of Minho, Campus of Gualtar, 4710-057 Braga, Portugal
Post-Graduate Programme in Agricultural Microbiology, Federal University of Lavras, Lavras, MG, Brazil

Nelson Lima
CEB-Centre of Biological Engineering, University of Minho, Campus of Gualtar, 4710-057 Braga, Portugal
Post-Graduate Programme in Agricultural Microbiology, Federal University of Lavras, Lavras, MG, Brazil

Lara Durães Sette
Divisão de Recursos Microbianos, CPQBA/UNICAMP, CP 6171, 13083-970 Campinas, SP, Brazil
Departamento de Bioquímica e Microbiologia, Instituto de Biociências - UNESP, Campus Rio Claro, Av. 24A, n°1515, 13506-900, Rio Claro, SP, Brazil

Kimiko Yamamoto-Tamura
National Institute for Agro-Environmental Sciences, 3-1-3 Kannondai, Tsukuba, Ibaraki 305-8604, Japan

Syuntaro Hiradate
National Institute for Agro-Environmental Sciences, 3-1-3 Kannondai, Tsukuba, Ibaraki 305-8604, Japan

Takashi Watanabe
National Institute for Agro-Environmental Sciences, 3-1-3 Kannondai, Tsukuba, Ibaraki 305-8604, Japan

Motoo Koitabashi
National Institute for Agro-Environmental Sciences, 3-1-3 Kannondai, Tsukuba, Ibaraki 305-8604, Japan

Yuka Sameshima-Yamashita
National Institute for Agro-Environmental Sciences, 3-1-3 Kannondai, Tsukuba, Ibaraki 305-8604, Japan

Tohru Yarimizu
National Institute for Agro-Environmental Sciences, 3-1-3 Kannondai, Tsukuba, Ibaraki 305-8604, Japan

Hiroko Kitamoto
National Institute for Agro-Environmental Sciences, 3-1-3 Kannondai, Tsukuba, Ibaraki 305-8604, Japan

Shunsuke Sato
Institut für Molekular Mikrobiologie und Biotechnologie, Westfälische Wilhelms-Universität Münster, Corrensstraße 3, D-48149 Münster, Germany

Björn Andreeßen
Institut für Molekular Mikrobiologie und Biotechnologie, Westfälische Wilhelms-Universität Münster, Corrensstraße 3, D-48149 Münster, Germany

Alexander Steinbüchel
Institut für Molekular Mikrobiologie und Biotechnologie, Westfälische Wilhelms-Universität Münster, Corrensstraße 3, D-48149 Münster, Germany
Environmental Sciences Department, Faculty of Meteorogy, Environment and Arid Land Agriculture, King Abdulaziz University, Jeddah, Saudi-Arabia

Karina I Dantur
Estación Experimental Agroindustrial Obispo Colombres (EEAOC) – Consejo Nacional de Investigaciones Científicas y Técnicas (CONICET), Instituto de Tecnología Agroindustrial del Noroeste Argentino (ITANOA), 3150 William Cross Av., Las Talitas PC T4101XAC, Tucumán, Argentina

Ramón Enrique
Estación Experimental Agroindustrial Obispo Colombres (EEAOC) – Consejo Nacional de Investigaciones Científicas y Técnicas (CONICET), Instituto de Tecnología Agroindustrial del Noroeste Argentino (ITANOA), 3150 William Cross Av., Las Talitas PC T4101XAC, Tucumán, Argentina

Björn Welin
Estación Experimental Agroindustrial Obispo Colombres (EEAOC) – Consejo Nacional de Investigaciones Científicas y Técnicas (CONICET), Instituto de Tecnología Agroindustrial del Noroeste Argentino (ITANOA), 3150 William Cross Av., Las Talitas PC T4101XAC, Tucumán, Argentina

Atilio P Castagnaro
Estación Experimental Agroindustrial Obispo Colombres (EEAOC) – Consejo Nacional de Investigaciones Científicas y Técnicas (CONICET), Instituto de Tecnología Agroindustrial del Noroeste Argentino (ITANOA), 3150 William Cross Av., Las Talitas PC T4101XAC, Tucumán, Argentina

Ashish Kumar Sarker
Plant Protein Research Section, Institute of Food Science and Technology (IFST), Bangladesh Council of Scientific and Industrial Research(BCSIR), Dr. Kudrat-i-Khuda Road, Dhanmondhi, Dhaka 1205, Bangladesh

Dipti Saha
Department of Applied Chemistry & Chemical Engineering, Dhaka University, Dhaka 1000, Bangladesh

Hasina Begum
Department of Applied Chemistry & Chemical Engineering, Dhaka University, Dhaka 1000, Bangladesh

Asaduz Zaman
Department of Applied Chemistry & Chemical Engineering, Dhaka University, Dhaka 1000, Bangladesh

Md Mashiar Rahman
Enzymlogy Research Section, IFST, BCSIR, Dhaka, 1205, Bangladesh

Clara Navarrete
Department of Chemical and Biological Engineering, Chalmers University of Technology, Kemivägen 10, SE-41296 Göteborg, Sweden

Jens Nielsen
Department of Chemical and Biological Engineering, Chalmers University of Technology, Kemivägen 10, SE-41296 Göteborg, Sweden

Verena Siewers
Department of Chemical and Biological Engineering, Chalmers University of Technology, Kemivägen 10, SE-41296 Göteborg, Sweden

www.ingramcontent.com/pod-product-compliance
Lightning Source LLC
Chambersburg PA
CBHW050446200326
41458CB00014B/5084